MW00345703

The Handbook of Global Energy Policy

Handbooks of Global Policy Series

Series Editor
David Held
Master of University College and Professor of Politics and International Relations at Durham University

The *Handbook of Global Policy* series presents a comprehensive collection of the most recent scholarship and knowledge about global policy and governance. Each Handbook draws together newly commissioned essays by leading scholars and is presented in a style which is sophisticated but accessible to undergraduate and advanced students, as well as scholars, practitioners, and others interested in global policy. Available in print and online, these volumes expertly assess the issues, concepts, theories, methodologies, and emerging policy proposals in the field.

Published

The Handbook of Global Climate and Environment Policy
Robert Falkner

The Handbook of Global Energy Policy
Andreas Goldthau

The Handbook of Global Companies
John Mikler

The Handbook of
Global Energy Policy

Edited by

Andreas Goldthau

A John Wiley & Sons, Ltd., Publication

Library of Congress Cataloging-in-Publication Data

The handbook of global energy policy / edited by Andreas Goldthau.
        pages cm
    Includes bibliographical references and index.
    ISBN 978-0-470-67264-8
    1. Energy policy.   I. Goldthau, Andreas.
    HD9502.A2H255 2013
    333.79–dc23
                                                                            2012045558

A catalogue record for this book is available from the British Library.

Cover image: Night time satellite view of Asia. Image © NASA/Corbis.
Cover design by Design Deluxe.

Set in 9.5/12pt Sabon by Aptara Inc., New Delhi, India
Printed in Malaysia by Ho Printing (M) Sdn Bhd

1   2013

Contents

Figures and Tables

Figures

Tables

Notes on Contributors

Christopher Allsopp, Director of the Oxford Institute for Energy Studies, is a Fellow of New College and Senior Research Fellow at the Department of Economics at the University of Oxford. He was leader of an independent Review of Statistics for Economic Policymaking, 2003 (the "Allsopp Review"). He is a former member of the Court of Directors of the Bank of England (1997–2000) and of the Bank's Monetary Policy Committee (2000–2003). He is the (founding) Editor of the *Oxford Review of Economic Policy* and a Director of Oxford Economics Ltd. Previous activities include working at HM Treasury, the OECD, and the Bank of England (where he was Advisor from 1980–1983) as well as extensive involvement with domestic and international policy issues as consultant to international institutions and private sector organizations. He has published widely on monetary, fiscal, and exchange rate issues as well as the problems of economic reform and transition.

Govinda Avasarala is a Senior Research Assistant in the Energy Security Initiative at the Brookings Institution. His research focuses on the geopolitics of energy in emerging markets, domestic and international oil and natural gas markets, and multilateral energy frameworks. He has a BSc in Economics from the University of Mary Washington.

Robert Bailey is Senior Research Fellow, Food and Environmental Security, at Chatham House. Before this, he was head of Economic Justice at Oxfam GB and authored a number of reports on biofuels, climate change, and food security including *Another Inconvenient Truth: How Biofuel Policies Are Deepening Poverty and Accelerating Climate Change*, and *Growing a Better Future: Food Justice in a Resource-Constrained World*. In 2011, he was named one of the Devex 40-under-40 leading thinkers on international development.

Andrew Bauer is an economic analyst at the Revenue Watch Institute where he focuses on economic technical assistance and research, including advising governments and civil society on the economic implications of various extractive sector policy options and helping policy-makers improve their natural resource revenue management. Prior to joining, he served on Canada's G7/8 and G20 teams as an international economist at the Department of Finance, where he provided economic policy advice and participated in the planning and execution of the G8 and G20 Summits as well as the preparatory ministerial meetings during Canada's host year. He has held positions in government, nonprofits, and the private sector, having worked for Debt Relief International, UNICEF-Canada, Transparency International-Kenya, and the Commission on Human Rights and Administrative Justice (Ghana), among others. Andrew holds an MSc in Economics for Development from Oxford University and a BA in Economics and International Development Studies from McGill University.

Subhes Bhattacharyya is Professor of Energy Economics and Policy at the Institute of Energy and Sustainable Development, De Montfort University, Leicester, UK. He is leading a consortium of British and Indian universities to develop business models for off-grid electrification in South Asia. Dr Bhattacharyya has more than 25 years of expertise on energy-related issues with a specific focus on the developing economies. He is the author of *Energy Economics: Concepts, Issues, Markets and Governance* (Springer, 2011) and edited the volume *Rural Electrification Through Decentralised Off-grid Systems in Developing Countries* (Springer, 2013).

Michael Bradshaw is Professor of Human Geography and former Head in the Department of Geography at the University of Leicester, UK. His PhD is from the University of British Columbia, Canada. His research is on resource geography with a particular focus on the economic geography of Russia and global energy security. Most recently his research has focused on global energy issues. From 2008 to 2011 he was engaged in a programme of research on Global Energy Dilemmas, funded by a Leverhulme Trust Major Research Fellowship that examined the relationship between energy security, globalization, and climate change. In January 2011 he started a two-year research project on global gas security funded by the UK Energy Research Centre. He is Editor-in-Chief of Wiley-Blackwell's *Geography Compass* and Contributing Editor of *Eurasian Geography and Economics*. He is an Honorary Senior Research Fellow in the Centre for Russian and East European Studies at the University of Birmingham and a Visiting Senior Research Fellow at the Oxford Institute for Energy Studies.

Albert Bressand is Professor of International Strategic Management at Rijksuniversiteit in Groningen, Netherlands, Special Adviser to the EU Commissioner on Development at the EU Commission in Brussels, and Senior Fellow at the Vale Columbia Center on Sustainable International Investment. Previously he was Professor in the Practice of International and Public Affairs at Columbia University in New York and Executive Director of Columbia's Center for Energy, Marine Transportation and Public Policy. He formerly led the Global Business Environment department in Royal Dutch Shell's global headquarters in London and served as a Special Advisor to the EU Commissioner in charge of energy. Bressand is a member of the faculty of the World Economic Forum, and serves on the Board of the New York Energy Forum and on the Advisory Board of the European Center for Energy and Resources Security (EUCERS) at King's College London. He is a member of the Oxford Energy Policy Club at St Antony's College and of the Scientific Committee, Official Monetary and Financial Institutions Forum (OMFIF), London. He has published in *Foreign Affairs, International Affairs, Futuribles, Politique Internationale, Revue d'Économie Financière, Le Monde*, etc. Bressand earned advanced degrees in both mathematics and engineering at École Polytechnique in Paris, École Nationale des Ponts et Chaussées, and Paris-Sorbonne, and an MPA and a PhD in Political Economy at the Kennedy School of Government at Harvard University.

Fritz Brugger has recently received his PhD from the Graduate Institute of International and Development Studies in Geneva based on research on the CSR performance of Asian state-owned and Western oil companies in Sudan and Chad. His interest focuses on natural resources and development, and development cooperation.

Gilles Carbonnier is Professor of Development Economics at the Graduate Institute of International and Development Studies Geneva, and Editor-in-Chief of *International Development Policy*. He has 20 years of experience in international trade policy, development, and humanitarian action. He is President of the Centre for Education and Research in Humanitarian Action and Vice-President of the European Association of Development Institutes. His research focuses on the energy–development nexus and the governance of extractive resources, as well as on humanitarianism, the political economy of armed conflict, and international development cooperation.

Debalina Chakravarty is pursuing her PhD in Economics at Jadavpur University, Kolkata, India and serves as a Junior Research Fellow. Her area of specialization is Climate Change and Rebound Effect. She completed her MPhil work on rebound estimation for Indian consumers in the context of energy use. She has presented her work in conferences and journals.

Dag Harald Claes is Professor at the Department of Political Science at the University of Oslo and Adjunct Professor at Molde University College. He holds a doctoral degree in Political Science from the University of Oslo. Among his publications are studies of oil-producer cooperation, the energy relations between Norway and the EU, and the role of oil in Middle East conflicts. At present he is Head of the Department of Political Science at the University of Oslo.

Suani T. Coelho graduated in Chemical Engineering from the Armando Alvares Penteado Foundation (FAAP) and holds a Master and Doctorate degree in Energy from the University of São Paulo (USP). She serves as a Professor of the Postgraduate Program in Energy (PPGE) of USP, as a coordinator of the Brazilian Reference Center on Biomass (CENBIO), a center created to implement biomass usage as an energy source in the country, as a researcher of the Institute of Electrotechnics and Energy of USP (in Energy, Energy Conservation, Biomass), and co-coordinator of the Lato-Sensu Graduate Course in Environmental Management in Energy Sector Business, Institute of Electrotechnics and Energy, USP.

Shyamasree Dasgupta is pursuing her PhD at the Department of Economics, Jadavpur University, Kolkata as a SYLFF (Ryoichi Sasakawa Young Leaders' Fellowship Fund) Doctoral Fellow. The primary focus of her research is to understand sustainability transition challenges on the developmental pathway for energy intensive industries in India. Her interests are on domestic and international mitigation policy induced low carbon growth potential assessment. Previously she earned her MPhil in Economics from Jadavpur University, also with the SYLFF Fellowship, in 2010. She has publications in peer reviewed journals, as well as overseas research and conference participation experience.

Charles Ebinger is a Senior Fellow and Director of the Energy Security Initiative at the Brookings Institution. He has more than 35 years of experience specializing in international and domestic energy markets (oil, gas, coal, and nuclear) and the geopolitics of energy, and has served as an energy policy advisor to over 50 governments. He has served as an Adjunct Professor in Energy Economics at the Johns Hopkins School of Advanced International Studies and Georgetown University's Walsh School of Foreign Service.

Christian Egenhofer is an Associate Senior Research Fellow and Head of Energy and Climate at the Brussels-based Centre for European Policy Studies (CEPS). He is also Visiting Professor at the College of Europe in Bruges (Belgium) and Natolin (Poland), SciencesPo (Paris), and LUISS University in Rome, where he teaches on energy and climate change policy and regulation. From 1997 to 2010 he was Senior Research Fellow and Jean Monnet Lecturer at the Centre for Energy, Petroleum and Mineral Law and Policy at the University of Dundee in Scotland (part-time). Christian Egenhofer has more than 20 years of experience working with EU institutions on numerous policy areas. In the last decade he has been specializing in EU energy and climate change policy with a particular focus on EU energy, climate, and transport policies.

Bassam Fattouh is the Director of the Oil and Middle East Programme at the Oxford Institute for Energy Studies and a Professor at the School of Oriental and African Studies, University of London. He has published a variety of articles on the international oil pricing system and the dynamics of oil markets and prices. His papers have appeared in the *Energy Journal, Energy Economics, Energy Policy, Journal of Development Economics, Oxford Review of Economic Policy, Economic Inquiry, Empirical Economics, Journal of Financial Intermediation, Economics Letters, Macroeconomic Dynamics*, and other journals and books.

José Goldemberg earned his PhD in Physical Sciences from the University of São Paulo in 1954, where he held the position of Full Professor in the Engineering School's Physics Department. He was Rector of the University from 1986 to 1991. A member of the Brazilian Academy of Sciences, he has served as the President of the Brazilian Association for the Advancement of Science and President of the Energy Company of the State of São Paulo (CESP). Between 1990 and 1992, he was Brazil's Secretary of State for Science and Technology and Minister of State for Education. He did research and taught at the University of Illinois, Stanford University of Paris (Orsay), and Princeton University. From 1998 to 2000, he served as Chairman of the World Energy Assessment. Between 2002 and 2006, he was Secretary for the Environment of the State of São Paulo. In 2007 *Time* honored him as one of its "Heroes of the Environment." In 2008 he was awarded the Blue Planet Prize 2008 of the Asahi Glass Foundation (Japan). In 2010 he received the Trieste Science Prize of the Third World Academy of Sciences. He is presently Professor Emeritus of the University of São Paulo.

Andreas Goldthau is Head of the Department of Public Policy and Associate Professor at Central European University, an American graduate school based in Budapest. He is also a Fellow with the Global Public Policy Institute (Berlin/Geneva) and an Adjunct Professor with Johns Hopkins University's MSc in Energy Policy and Climate. His academic career includes stints with the RAND Corporation, and SAIS, Johns Hopkins University. Andreas Goldthau's research interests focus on energy security and on global governance issues related to oil and gas. He has authored, co-authored or co-edited *Dynamics of Energy Governance in Europe and Russia* (Palgrave Macmillan, 2012), *Global Energy Governance: The New Rules of the Game* (Brookings Institution Press, 2010), *Imported Oil and U.S. National Security* (RAND, 2009), *OPEC* (Hanser, 2009), and *Domestic Trends in the United States, China and Iran* (RAND, 2008).

Neil Gunningham holds Professorial positions in the Regulatory Institutions Network and the Fenner School of Environment and Society at the Australian National University and is co-director of the Climate and Environmental Governance Network (CEGNet). His books include *Shades of Green: Business, Regulation and Environment* (with Kagan and Thornton, 2003) and *Smart Regulation* (with Grabosky, 1999). He works principally in the areas of environmental and energy law, regulation, and governance.

Mel Horwitch is Dean and University Professor at Central European University Business School, Budapest. His recent work is focused on cleantech and sustainability management, global innovation (especially with regard to emerging economies), global entrepreneurship in both stand-alone and corporate venues, and the future configuration of modern innovation. Previously, he was Professor of Technology Management, Director of the Institute for Technology and Enterprise, and former Chair of the Department of Technology Management at the Polytechnic Institute of New York University.

Matthew Hulbert is a political advisor to Saudi Aramco having previously been Lead Analyst for European Energy Review and consultant to numerous governments, most recently as Senior Research Fellow at Clingendael International Energy Programme (The Hague) where he wrote their flagship book, *Age of Paradox: Exploring the Uncertain World of Energy 2000–2020*. He was previously Senior Fellow at ETH Zurich, leading its work on energy security and political risk, having originally come from the City of London where he was Senior Energy Analyst at Datamonitor and head of Global Risk Analysis at Control Risks Group in London and Washington. Prior to this, he held political consulting positions at Weber Shandwick Worldwide; was policy director of the largest All Party Parliamentary Group in the UK Parliament, and started his career at the Foreign Policy Centre. He publishes widely in leading policy journals and popular outlets and also holds a permanent energy and commodities column with *Forbes Magazine* called "Old-School Energy, New World Order." He went to Cambridge and Durham Universities.

Jaap Jansen (MSc, MIF) is a senior researcher at the Energy Research Centre of the Netherlands (ECN), focusing on energy policy from a socio-economic perspective. He (co-)authored many papers, articles, reports, and book chapters, mainly on policy stimulation, market and system integration of renewables, security of supply, and energy-related environmental markets.

Andrey A. Konoplyanik is Adviser to the Director General of Gazprom export LLC. He received his PhD (1978) and Doctor of Science (1995), both in international energy economics, from the Moscow State Academy of Management, and the title Professor in International Oil and Gas Business in 2012. During the Soviet era he worked in the Moscow Institute of World Economy and International Relations (IMEMO), the USSR Academy of Sciences (1979–1990), and in the USSR State Planning Committee – GOSPLAN (1990–1991). Dr Konoplyanik served as Russia's Deputy Minister of Fuel and Energy with particular responsibility for external economic relations and direct foreign investments (1991–1993), as a non-staff Advisor to the Deputy Prime Minister E. Gaidar, several ministries in the Russian government (Energy, Economy, Finance), and the Russian State Duma (1993–2002). From 1996 to 1999 Dr Konoplyanik worked as an Executive Director of the Russian Bank for Reconstruction and Development. From 1999 to 2002 he was the President of the Moscow-based Energy and Investment Policy & Project Financing Development Foundation (ENIP&PF). From 2002 to 2008 he served as Deputy Secretary General of the Energy Charter Secretariat in Brussels, Belgium. Since then he was an Advisor to the Board, GPB Neftegas Services BV, Moscow branch, a Consultant to the Board, Gazprombank, Russia (2008–2011), and Director for Energy Markets Regulation and Project Leader at the Foundation Institute for Energy and Finance, Moscow (2011–2012). Having served as a visiting Professor at Moscow

Academy of Management since 1997, as of 2008 he holds a visiting Professorship at the Chair "International Oil and Gas Business" at Gubkin Russian State Oil and Gas University, offering a special course on evolution of international oil and gas markets and investment protection instruments in energy.

Michael LaBelle is an assistant professor at Central European University. He holds a joint appointment at CEU Business School and in the CEU Department of Environmental Sciences and Policy. His research is focused on how institutions and organizations foster change in the energy sector to develop a low carbon future. This includes issues of risk governance, investment strategies, and regulations in the energy sector. Previously, he worked in the CEU Center for Climate Change and Sustainable Energy Policy (3CSEP) at Central European University and at the Regional Center for Energy Policy Research (REKK) at Corvinus University. He holds an MSc and PhD in Geographical Sciences from the University of Bristol.

Honoré Le Leuch has over 40 years of professional experience in international oil and gas business and is an acknowledged petroleum expert. He acts as Senior Consultant and Advisor on petroleum legislation, taxation and contracts, institutional and regulatory issues, economics, financing, and petroleum contract negotiations. He has advised more than 60 countries and companies in all parts of the world for their petroleum policy, law, tax, and investment strategy.

Alvin Lin is China Climate and Energy Policy Director at the Natural Resources Defense Council (NRDC), based in NRDC's Beijing office. His work focuses on analysis of China's climate and clean energy policies, including how China can expand its energy efficiency and renewables resources and put in place incentives and systems for low carbon development. His work has included analyzing China's renewable energy law and policies, providing recommendations on amendments to China's air pollution law, and seeking ways for the US and China to cooperate on clean energy bilaterally and in the international climate negotiations. His recent projects have included organizing workshops on power sector GHG emissions monitoring and strengthening nuclear safety post-Fukushima. Mr Lin has a BA in Political Science and East Asian studies from Yale University, an MPhil. in Government and Public Administration from the Chinese University of Hong Kong, and a JD from New York University School of Law. Prior to joining NRDC, he worked in the law firm Morrison & Foerster as a commercial litigator and clerked for a federal judge in Brooklyn, New York.

Sudha Mahalingam, an energy economist and lawyer with over three decades of professional experience, has recently completed her five-year tenure as downstream regulator with the Petroleum and Natural Gas Regulatory Board, India's statutory downstream regulatory body. She served a term as member of the National Security Advisory Board, giving policy inputs on energy to the Prime Minister through the National Security Advisor. She has held positions in academia and business journalism. She is currently Visiting Scholar at George Washington University and member of the Scientific Advisory Board of Delft University.

Charles McPherson consults internationally on petroleum and mineral policies. He was Tax Policy Adviser at the International Monetary Fund (IMF) from 2007 to 2010 with particular responsibilities for fiscal and financial policies in natural resource-rich countries. Prior to taking up his position at the Fund, he was Senior Advisor on Oil and Gas at the World Bank. His work at the Bank focused on petroleum sector reform and sector lending activities in a wide range of countries. He also managed the Bank's participation in the Extractive Industries Transparency Initiative (EITI). Before joining the Bank, Mr McPherson spent 15 years at two international oil companies, holding a variety of senior positions in international negotiations and government agreements. He received his BA in Economics and Political Science from McGill University, MSc in International Economics from the London School of Economics and Political Science, and PhD in Economics from the University of Chicago.

Emanuela Menichetti is Director of the Renewable Energy Division at the Observatoire Méditerranéen de l'Énergie (OME), an industry association gathering some 30 energy companies of the Mediterranean. Prior to joining OME, Mrs Menichetti worked as Associate Programme Officer at UNEP, Division of Technology, Industry and Economics in Paris, and as a researcher and consultant at Ambiente Italia, in Rome. She has over 12 years of experience in the field of renewable energy and sustainable development. An economist by background, Mrs Menichetti received a PhD from the University of St Gallen (Switzerland).

Øystein Noreng is Professor Emeritus at the BI Norwegian School of Management where he held the TOTAL E&P chair in Petroleum Economics and Management. Professor Noreng holds a PhD in Political Science from Sorbonne University, Paris and was a post-doctoral research fellow at the Institute of Energy Studies at Stanford University. He has been a visiting research scholar at JFK School of Government, Harvard University. Professor Noreng is famous for numerous publications on the petroleum industry. He is a researcher and advisor to international organizations, governments, and oil and gas companies.

Ike Okonta is Coordinating Fellow of the New Centre for Social Research, Abuja, Nigeria. He was until recently Leverhulme Early Career Fellow in the Department of Politics and International Relations, University of Oxford, where he also took a doctorate in 2002. Dr Okonta has held research fellowships at the University of Cambridge, University of California, Berkeley, Columbia University, New York, and Open Society Institute, New York. He is the author of *When Citizens Revolt: Nigerian Elites, Big Oil, and the Ogoni Struggle for Self Determination* (Africa World Press, 2009), co-author of *Shell, Human Rights and Oil* (Verso, 2003), and author of *Biafran Ghosts: The MASSOB Ethnic Militia and Nigeria's Democratisation Process* (Nordic Africa Institute, 2012). He advises a wide range of African civil society organizations including the Movement for the Survival of the Ogoni People (MOSOP) and Friends of the Earth Nigeria. He was chief policy advisor to Dr Nuhu Ribadu, the presidential candidate of the Action Congress of Nigeria (ACN), Nigeria's leading opposition party during the 2011 presidential election.

Meghan L. O'Sullivan is the Jeane Kirkpatrick Professor of the Practice of International Affairs and Director of the Geopolitics of Energy Project at Harvard University's Kennedy School. Between 2004 and 2007, she was special assistant to President George W. Bush and Deputy National Security Advisor for Iraq and Afghanistan during the last two years of her tenure. Prior to this, Dr O'Sullivan was senior director for strategic planning and Southwest Asia in the NSC; political advisor to the Coalition Provisional Author-ity administrator and deputy director for governance in Baghdad; chief advisor to the presidential envoy to the Northern Ireland peace process; and a fellow at the Brookings Institution. Her publications include *Shrewd Sanctions: Statecraft and State Sponsors of Terrorism* (2003). Dr O'Sullivan is an adjunct senior fellow at the Council on Foreign

Relations, a consultant to the National Intelligence Council, and a strategic advisor to Hess Corporation, an American independent oil and gas company. She is a member of the Council of Foreign Relations, the Trilateral Commission, and the Aspen Strategy Group. In 2008, *Esquire* voted her one of the most influential people of the century. She holds a doctorate in Politics and a master's in Economics from Oxford University and a BA from Georgetown University.

Philipp Pattberg is an Associate Professor (tenured) of Transnational Environmental Governance with the Department of Environmental Policy Analysis, Institute for Environmental Studies, VU University Amsterdam. He is the author of *Private Institutions and Global Governance: The New Politics of Environmental Sustainability* (Edward Elgar, 2007); co-editor of *Global Climate Governance beyond 2012: Architecture, Agency and Adaptation* (Cambridge University Press, 2010); co-editor of *Global Environmental Governance Reconsidered* (MIT Press, 2012); and co-editor of *Public–Private Partnerships for Sustainable Development: Emergence, Influence and Legitimacy* (Edward Elgar, 2012).

Jason Portner is working with the Natural Resources Defense Council's (NRDC) China Program Climate and Energy team as a Nuclear Safety Research Associate. His work focuses on strengthening nuclear safety post-Fukushima and researching norm building in international relations. Mr Portner graduated from Northeastern University with a BA in International Affairs and Asian Studies.

Juan Carlos Quiroz is Senior Policy Analyst at Revenue Watch Institute, New York. Before joining Revenue Watch in January 2007, he worked analyzing energy policies in Latin America and helping to implement the Extractive Industries Transparency Initiative (EITI) at the World Bank. He has published articles about the oil sector in Mexico and his professional experience includes some time as speechwriter at the Policy Planning Staff of the Ministry of Foreign Affairs of Mexico. Currently, he is leading the research, development, and publication of Revenue Watch's new Revenue Transparency Index. He studied at El Colegio de Mexico and Johns Hopkins University's School of Advanced International Studies.

Joyashree Roy is Professor of Economics at Jadavpur University, Kolkata, India. Dr Roy has been one of the two Coordinating Lead Authors of the Chapter on "Climate Change 2007: Mitigation of Climate Change" of the Nobel Peace Prize-winning Intergovernmental Panel on Climate Change (IPCC). Currently she serves as CLA in WG III and in steering committees of selected special reports of the IPCC 5th assessment report. She has been involved in preparation of the Stern Review, Global Energy Assessment, and many other national and global efforts. Dr Roy initiated and also coordinates the Global Change Programme at Jadavpur University which focuses on selected aspects of Climate Change research. She directs the SYLFF (Ryoichi Sasakawa Young Leaders' Fellowship Fund) Project on "Tradition, Social Change, and Sustainable Development: A Holistic Approach." Dr Roy was also a Ford Foundation Post-Doctoral Fellow in Environmental Economics at the Lawrence Berkeley National Laboratory, California, USA.

Roman Sidortsov holds BA and Master's degrees in Law from Law Faculty of Irkutsk State University, and JD and LLM degrees from Vermont Law School. He currently works as a Senior Global Energy Fellow at the Institute for Energy and the Environment of Vermont Law School where he also teaches energy law and policy courses in the distance learning program.

Benjamin K. Sovacool is a Visiting Associate Professor at Vermont Law School, where he manages the Energy Security and Justice Program at their Institute for Energy and the Environment. He is a Contributing Author to the Intergovernmental Panel on Climate Change's forthcoming Fifth Assessment (AR5). He is the co-editor of *Energy and American Society* (Springer, 2007) and the editor, author, or co-author of *The Dirty Energy Dilemma* (Praeger, 2008), *Powering the Green Economy* (Earthscan, 2009), *The Routledge Handbook of Energy Security* (Routledge, 2010), *Contesting the Future of Nuclear Power* (World Scientific, 2011), *Climate Change and Global Energy Security* (MIT Press, 2011), *The National Politics of Nuclear Power* (Routledge, 2012), *The Governance of Small-Scale Renewable Energy in Developing Asia* (Ashgate, 2012), and *The Governance of Energy Megaprojects: Politics, Hubris, and Energy Security* (Edward Elgar, 2013).

Hannes Stephan is a Lecturer in Environmental Politics and Policy at the University of Stirling, Scotland. He is a co-convenor of the Environmental Politics Standing Group of the European Consortium for Political Research (ECPR). His co-authored article "International Climate Policy after Copenhagen" appeared in 2010 in *Global Policy*. Another article on the transatlantic cultural politics of GM foods and crops was published in 2012 in *Global Environmental Politics*.

Harro van Asselt is a Research Fellow at the Stockholm Environment Institute in Sweden, Visiting Researcher at the Environmental Change Institute of the University of Oxford, and Visiting Researcher at the Institute for Environmental Studies of the VU University Amsterdam. Recent publications include *Climate Change Policy in the European Union: Confronting the Dilemmas of Mitigation and Adaptation?* (Cambridge University Press, 2010).

Adriaan van der Welle (MSc) is a scientific researcher at the Energy Research Centre of the Netherlands (ECN) since 2006. He is (co-)author of papers and book chapters about the integration of renewable energy in electricity systems (INTECH, energy policy) and security of supply (FEEM, IEA).

Rolf Wüstenhagen is a Director of the Institute for Economy and the Environment (IWÖ-HSG) and holds the Good Energies Chair for Management of Renewable Energies at the University of St Gallen. He graduated in Management Science and Engineering (TU Berlin) and holds a PhD in Business. In 2005, 2008, and 2011, respectively, he held visiting faculty positions at University of British Columbia (Vancouver), Copenhagen Business School, and National University Singapore. His research focuses on decision-making under uncertainty by energy investors, consumers, and entrepreneurs. He embarked on an academic career after retiring from one of the leading European energy venture capital funds. From 2008 to 2011 he served as a lead author for the Intergovernmental Panel on Climate Change on the Special Report on renewable energy and climate change mitigation. Since 2011, he is a member of the advisory board for the Swiss government's energy strategy 2050. Professor Wüstenhagen is the academic director of St Gallen University's executive education programme in Renewable Energy Management.

Fuqiang Yang is senior advisor of NRDC on climate change, energy, and environment, and has been involved in energy and environmental issues for more than three decades. He was director of Global Climate Solutions, WWF International, from 2008 to 2010, and vice president of Energy Foundation and chief representative of Energy Foundation Beijing Office from 2000 to 2008. The EF China Sustainable Energy Program (CSEP) was dedicated to public policy development in China aimed at cost-effective carbon emissions reductions through the deployment of energy efficiency and renewable energy technologies. Fuqiang Yang had worked with Lawrence Berkeley National Laboratory on China's energy and environmental issues. In 1984, Fuqiang Yang worked on regional energy planning at Cornell University as a Fellow of the World Bank. Before he moved to the US in 1984, Fuqiang Yang worked with Energy Research Institute, State Planning Commission, on renewable and rural energy policy, energy modeling and forecasting, project evaluation, and long-term planning. Fuqiang Yang has published dozens of papers and reports in those areas. He took his PhD in Industrial Engineering at West Virginia University in 1991 and his BSc in Physics at Jilin University in China in 1977.

Richard Youngs is Director of FRIDE. He is also Professor at the University of Warwick in the UK. Prior to joining FRIDE, he was EU Marie Curie Research Fellow at the Norwegian Institute for International Relations, Oslo (2001-4), and Senior Research Fellow at the UK Foreign and Commonwealth Office (1995–1998). He has a PhD and MA in International Studies from the University of Warwick and a BA in Social and Political Science from the University of Cambridge. His research focuses mainly on democracy promotion and democratization, European foreign policy, energy security, and the MENA region. He has written several books on different elements of European external policy and published over 40 articles and working papers, while writing regularly in national and international media. His latest work is *Europe's Decline and Fall: The Struggle Against Global Irrelevance* (Profile Books, 2010).

Yury Yudin is a Senior Researcher at UNIDIR and manager of the Multilateral Approaches to the Nuclear Fuel Cycle project. Previously, he was Director of a Russian NGO, the Analytical Center for Non-proliferation, and Senior Researcher at the Russian Federal Nuclear Center–All-Russian Research Institute of Experimental Physics. He graduated from the Moscow Engineering Physics Institute as a nuclear physicist and holds a PhD in nuclear engineering. He has special expertise in nuclear engineering, nuclear non-proliferation, nuclear disarmament, and international security.

Fariborz Zelli is an Assistant Professor at the Department of Political Science at Lund University, Sweden. He is also an associate fellow of the German Development Institute in Bonn. His recent publications include the co-edited volumes *Global Climate Governance Beyond 2012* (Cambridge University Press, 2010) and *Climate and Trade Policies in a Post-2012 World* (UNEP, 2009).

Introduction: Key Dimensions of Global Energy Policy

Andreas Goldthau

Energy as a Global Policy Field

Energy policy scholars live in exiting times. At the time of writing the introduction to this book, ExxonMobil is back to top the world's biggest, most powerful, and most valuable public companies again, gathered in Forbes' Global 2000 ranking. At the same time, PetroChina has for the first time surpassed ExxonMobil in terms of total crude production. Lately, Argentina decided to nationalize its petroleum industry, expropriating Spanish company REPSOL of its oil assets in the country. Bolivia followed suit by seizing control of the country's electricity sector and power grids from foreign investor Red Electric. In the Gulf, conflict over Iran's nuclear program is on the brink of turning into open warmongering. As a corollary, Tehran threatens to stop oil transit through the Strait of Hormuz, the world's premier chokepoint and home to a good third of seaborne oil trade. In the US, homeland of modern petroleum industry, yet another technological breakthrough just revolutionized global hydrocarbon extraction: hydraulic fracking. Not only did this turn gas markets upside down, trigger a global "dash for shale," and have serious repercussions on Gazprom's business model in Europe. It is also likely to make the US "oil import independent" in a few years time. Meanwhile, scandal-torn investment house Goldman Sachs is hit by corruption allegations over offshore crude reserves in Angola. Angola stands for a larger group of oil-rich nations that have so far not managed to convert resource wealth into economic well-being. Worse, and again representing a larger sample, the country has failed to manage lifting its own population out of energy poverty. Reacting to the energy poverty challenge, and embracing the issue of some 1.3 billion people living without access to clean energy, the United Nations declared 2012 the "Sustainable Energy for All Year." Neatly, this initiative comes 20 years after the Rio conference on climate change, kick-starting the UNFCC process. Rio+20, the follow-up conference two decades later, has now put energy into the spotlight of the climate challenge. The list goes on.

The Handbook of Global Energy Policy, First Edition. Edited by Andreas Goldthau.
© 2013 John Wiley & Sons, Ltd. Published 2013 by John Wiley & Sons, Ltd.

Yet, as exciting as these developments are, there is very little common ground among scholars and policy analysts over how to make sense of them. The policy community remains focused on selective aspects of global energy challenges. Essentially subscribing to economic assumptions on markets, pricing, and resource allocation, public policy scholars address energy primarily as an issue of market failure and externalities warranting public intervention. Works include studies on energy market imbalances such as international cartels (Alhajji and Huettner 2000; Hallouche 2006), the public goods characteristics of energy infrastructure (Andrews-Speed 2011; Goldthau 2011; Helm 2011; Kuenneke 1999) or the challenges and consequences for energy production, investment, and pricing relating to state ownership of energy companies (Baker Institute 2007; Marcel 2006; Stevens 2008; Victor *et al.* 2011). Some works also adopt a socio-technical approach to public policy (Brown and Sovacool 2011). Prevalent public policy prescriptions therefore tend to focus on making markets work, fixing shortcomings of market-based solutions and fostering technological innovation.

Foreign policy analysts, by contrast, primarily adopt the perspective of states and model energy as subject to international geopolitical scheming. Analyses tend to center on the degree to which energy can be used as a foreign policy tool by energy-exporting nations (Bilgin 2011; Orban 2008; Rutland 2008; Stulberg 2008) or on the consequences of a nation's dependence on foreign sources of energy (Crane *et al.* 2009; Zha 2006; Deutch *et al.* 2006; Youngs 2009). From this perspective, energy as a global security challenge may even invoke the option of "resource wars," with countries engaging in armed conflict to secure access to crucial supplies of energy for their population, economies, and military machinery (Klare 2001).

Adopting yet another but still selective lens, some contributions have come to center on aspects related to the global governance of energy. They tend to focus on international organizations such as the G8, G20, the International Energy Forum (IEF) or the International Energy Agency (IEA), assessing their capacity to address crucial global energy problems (Graaf and Lesage 2009; Jong 2011; Lesage *et al.* 2009). Focusing more on market governance and "smart" rules of the game in energy, other works have examined the degree to which international regimes are susceptible to buffering price shocks and providing for long-term planning security in global energy (Goldthau and Witte 2010; Victor and Yueh 2010). A separate strand of this literature also assesses the global institutional architecture addressing the climate change challenge (Biermann *et al.* 2010; Newell and Bulkeley 2010). Policy implications from these works are primarily geared towards establishing appropriate institutional solutions on a global level.

Each of these strands of literature comes with its own set of assumptions that serve as a prism to reality and allow building of coherent models. This approach has merits. Scholarly prisms, rooted in academic disciplines, reduce complexity. They allow generating consistent narratives on how to make sense of challenges surrounding global energy and how to address them in terms of policy. Yet, they also come at a cost, as global energy challenges are not one-dimensional, but indeed multi-faceted. In this, a single-focused approach may miss the target, fail to comprehensively capture the problem, and consequently sketch inappropriate solutions. Energy is about more than regulating the production and consumption of a commodity or about making it part of foreign policy calculations. It is of crucial importance to the welfare and economic development of nations, commonly considered nothing less than the lifeblood of modern societies. It stands front and center of fighting climate change due to the externalities traditional energy consumption produces – greenhouse gas (GHG) emissions. And tackling energy poverty requires accommodating the rather conflicting goals of facilitating better energy access by at the same time safeguarding climate targets, as the additional consumption is not necessarily "green."

Contemporary energy challenges are characterized by global interconnectedness. Externalities stemming from energy production and use no longer remain local or regional, but have truly global repercussions. Further, modern socio-economic systems are characterized by a sophisticated division of labor on a global scale; national energy systems exhibit similar interconnectedness with regional or global systems. Local changes in energy production, transmission or consumption thus easily impact on global value chains, or on infrastructure or even welfare of third countries, ignoring national jurisdictions.

In short, the production of public goods such as energy security or universal energy access becomes a global policy challenge, and so does the prevention of public bads such as climate change. To be sure, these challenges play out very differently on various levels. On the international level, a spike in oil prices calls for mechanisms to smoothen the crude market; on the national level, the challenge may rather consist in buffering the impacts on local food markets as affected through the food–fuel link. Likewise, on the domestic level, the challenge of lifting 1.3 billion people out of energy poverty consists in mobilizing sufficient resources to improve energy infrastructure and enhance energy access; at the same time, it calls for international regimes to facilitate clean technology transfer in order to mitigate the correlating negative climate effects.

Global energy policy, therefore, addresses energy challenges of transnational nature. In doing so, it not only cuts across issue areas, it also embraces all levels, from the international to the domestic and local. This crucially necessitates going beyond scholarly prisms and organizing academic inquiry in ways that account for the multi-faceted and multi-level nature of global energy challenges.

Dimensions of Global Energy Policy

Global energy policy spans four deeply intertwined key dimensions: markets; security; sustainability; and development. To start with the first dimension, markets have for the past three decades been the dominant mechanism to make the supply side react to demand increments. On an international level, the oil market globalized; regionally, gas markets became more integrated. Electricity markets became liberalized in most of the OECD world, and the energy sectors privatized. Rather than public agencies, private companies were put in charge of pricing electrons and molecules. This proved to be by and large effective. The oil market, for instance, effectively translated demand increases into signals for the supply side and now delivers some 86 million barrels per day (mbd), up from 65 mbd just 30 years ago (IEA 2011). Yet it remains to be seen whether the market will remain a dominant governance mechanism in global energy also in the future. As noted elsewhere, the liberal paradigm that guided energy policy-making for the past decades is clearly on the retreat. The state is back in the game, and with it comes some strong scepticism towards market-based principles regarding the provision of secure and affordable supply of energy, a private good with significant public good characteristics (Goldthau 2012).

In addition, there is evidence that the shift of the global energy gravity center to Asian consumers adds to the trend away from the liberal market model. In 2035, the world is projected to consume some 40% more energy than it does today (IEA 2011). Almost the entire future energy demand increment is projected to come from non-OECD – and notably Asian – countries. OECD energy consumption, by contrast, is set to peak by the end of this decade. China will consume nearly 70% more energy by 2035 than the US, representing 23% of global consumption by then. Meanwhile, India's energy consumption is set to double, which places the country in a firm third place in global energy demand, after China and the US (IEA 2011). In some markets, such as the oil market, these numbers will clearly come with a shift in rule-setting power. Probably more importantly, however, the question emerges what this shift implies for the mode of energy governance. A mercantilist approach to oil supply has of late become a model of choice not only for China, but also India. State-centered models also feature prominently in emerging economies and energy producers such as Brazil. In this context the question arises what characterizes current trends in the oil and gas markets; not only whether shifts can be observed in oil and gas upstream regimes but also what role state or market actors assume in driving innovation in energy technology. In this context, special attention needs to be placed on the role of national energy companies.

At the same time, the projected significant increase in global energy demand, coupled with concerns – justified or not – over the finite nature of fossil fuels, comes with a significant security dimension. In addition to being a traded commodity, energy has always been subject to a complex interplay with "hard" national security strategies. Recurring conflict over, related to or triggered by resources is a historical fact. The rise of new consumers such as India and China, an eastward shift in the global energy gravity center, and a growing assertiveness of some producing but also consuming nations to apply state intervention in the energy sector, therefore trigger questions relating to access to vital (fossil) resources. Persisting debates over a looming global "scramble for resources" very much reflect the strong security dimension entailed in the nature of energy as the lifeblood of modern economies. Questions arising in this context center on conflict and cooperation in global energy and the role institutions can play to facilitate win-win situations. In this context, it is important to examine the ability of international energy organizations and regimes to accommodate newcomers in the global energy system and to give them voice but also responsibility; whether they are capable of effectively addressing threats – perceived or real – arising from national energy security concerns; and whether existing institutions are favorable to generate policies tackling important side effects coming with modern energy technologies, such as nuclear proliferation.

As another corollary of sharply rising global energy demand, climate change has moved to the center of global energy debates. Starting with the Industrial Revolution, GHG emissions related to combustion of fossil fuels have constantly increased CO_2 concentrations in the atmosphere, causing global warming. In order to stabilize the climate at sustainable levels – mostly associated with the 450 ppm scenario limiting global warming to 2 °C – the world has to go "low carbon" (IPCC 2007). The challenge is indeed daunting. To achieve the 2 °C goal, the carbon intensity of global energy use would have to decrease by a factor of 21 by 2050 (Hoffmann 2011). At the same time, the non-OECD world will account for virtually all additional greenhouse gas emissions in the decades to come – countries which have not yet decoupled their GDP growth from energy use. To be sure, decarbonizing the energy sector by increasing energy efficiency or beefing up the share of renewables in the energy mix is not necessarily driven by sustainability concerns alone. In fact, low carbon policies can be informed by various additional motives – improving energy security (by reducing dependence on imported fossil fuels), industrial policy goals (giving domestic industry a competitive edge in a nascent technological sector) or as a response to external price shocks (such as the 1970s oil shocks, triggering Brazil's bioethanol program), to name a few. A new strand of literature has recently also emerged on the "climate security" challenge, reflecting that there are clear links to the security dimension as well (Barnett and Adger 2007; Campbell *et al.* 2007). Whatever the drivers, decarbonizing energy systems comes with tremendous investment challenges. According to some estimates, some \$25 trillion of investment will be needed, a price tag which gets exponentially higher in case of non-action (IEA 2011). Questions arising in this context center on the effectiveness of various policy tools available in decreasing the carbon intensity of energy systems, ranging from regulation to fiscal incentives such as subsidies to economic instruments such as cap and trade systems; on the role of global climate regimes informing energy choices and facilitating the deployment of clean energy technologies; and on the potential of global best practice and policy transfer, both from industrialized nations to emerging economies (such as the EU's efforts in decarbonizing its electricity sector) and from "energy newcomers" (such as Brazil) to developed and developing countries.

Global energy, finally, entails a strong development component. Evolving from rural communities to modern societies, mankind has replaced low-quality energy fuels (such as biofuels) with high-quality ones such as oil (Grübler 2008; Smil 2010). In other words, the development potential of nations is inextricably linked to the availability of modern fuels, and access thereof. Energy poverty seriously hinders the development prospects of nations and societies (IEA 2011; UN 2010). Providing them with access to modern forms of energy is a key global policy challenge, as recently acknowledged by the United Nations. At the same time, developing nations tend to be more energy intensive than developed ones, in that they need comparably more energy to produce a given unit of GDP. Energy efficiency programs may therefore have a positive effect on development. Ironically, the availability of energy resources can, however, also have negative effects on a country's development path if not well managed. The infamous "resource curse" is a case in point and illustrates the need for appropriate governance to successfully turn energy access into economic development. Moreover, given their interconnectedness, swings on the oil market may well impact on other commodity markets, notably corn, wheat or rice. This may have direct implications for households in poorer countries, as rising food costs eat into their available budgets, impacting for instance on their ability to invest in the education of the next generation. The global shift towards more sustainable energy, notably biofuels, has similar effects. Questions surrounding the development nexus in energy center on the "call on policy" of the energy access challenge as well as the "call on governance" of the resource wealth challenge; whether and to what extent sustainability policies may be at odds with development and human security goals; and what role energy technology can play in fostering economic goals.

Organization of This Book

Conceptualizing energy as a global policy issue obviously raises a set of essential questions for organizing scholarly inquiry: How to account for the multi-faceted nature of energy policy challenges by at the same time limiting the assessment to clearly defined research areas; how to bridge "silo thinking" in scholarly and policy research to account for the complexity underpinning global energy challenges; how to grasp the local dimensions of global energy issues; and what implications for global policy follow from them.

This book addresses these issues by, first, organizing Parts II–V along the four dimensions discussed above: markets, security, development, and sustainability. These dimensions serve as a prism to study global energy challenges whilst analyses in individual chapters are informed by multiple disciplines. Second, case studies in Part VI complement these dimensional assessments by offering regional perspectives on global energy

challenges. Cases comprise major incumbent or emerging energy producers (Russia, Nigeria, Brazil), newly emerging energy consumers (China, India), as well as established energy
players (US and European Union). Each of these cases offers its own specific insights
into key aspects of global energy challenges and energy transition. Thus, four sections
adopting a top-down approach and organized along the dimensions of markets, security, development, and sustainability, are complemented by a fifth section embracing a
bottom-up approach, bringing to the debate rich empirical insights from a country level.
Third, this book features an interdisciplinary group of renowned scholars in energy policy research, as well as practitioners deeply embedded in energy policy-making and the
energy business. Contributors comprise lawyers, economists, political analysts, natural
scientists, and engineers, from all five continents. Their insights allow bridging between
theory and policy practice; ensure balanced and representative views beyond the OECD
world; and account for the fact that the subject as such – global energy policy – also
entails a normative component and calls for a highly contextualized problem-solving
approach.

Before these more detailed analyses, Part I starts off by painting the bigger picture.
Albert Bressand sets the stage for inquiry into the market dimension by offering a conceptual approach to "The Role of Markets and Investment in Global Energy" (Chapter 1). He
argues that the energy industry differs from other sectors due to its capital intensity, the
endogenous nature of most energy transportation infrastructure, the importance of rent
and of conflict over rent distribution, and the eminent role of the state in the ownership,
control, and development of energy resources. In order to understand the emergence and
transition of energy markets three dimensions are therefore key: the institutions, rules,
and regimes structuring interaction in energy; the power structure determining which
actors are capable of influencing the rules of the game; and the very transactions of
energy assets, products, and services. Meghan O'Sullivan complements this analysis by
sketching "The Entanglement of Energy, Grand Strategy, and International Security"
(Chapter 2). Focusing on energy as a "hard" security issue, she conceptualizes modes in
which energy may influence national security policy as an end (energy shaping political,
military or economic strategies), way (energy being a tool to achieve non-energy goals),
and means (energy as a resource allowing pushing foreign or domestic agendas). She
argues that a grand strategy framework does not diminish the importance of international cooperation on pressing energy challenges; yet, it offers a different perspective on
them as it points to the important role of national political, military or economic policies
in shaping the global energy landscape.

Next, Michael Bradshaw discusses key aspects of "Sustainability, Climate Change,
and Transition in Global Energy" (Chapter 3). Focusing on key global energy dilemmas,
the chapter highlights the central importance of overhauling the energy system to mitigate
climate change. It argues that policies must be aimed at transiting the system into a low
carbon paradigm. Against the backdrop of an eastward shift in energy demand, and a
vast majority of future emissions originating from the non-OECD world, the chapter
concludes that it will be particularly crucial to decarbonize energy systems of developing
economies such as China and India. Gilles Carbonnier and Fritz Brugger conclude the
first section mapping this field by sketching "The Development Nexus of Global Energy"
(Chapter 4). Their chapter points to three main linkages: the causes and consequences
of energy poverty in the developing world; the link between energy, food, and water;
and the specific development challenges facing hydrocarbon producer countries. They
argue that the twin constraints of dwindling low-cost oil reserves and climate change
fundamentally put in question the traditional development model based on an abundant

availability of cheap coal, oil, and gas. Domestic and international efforts are needed to enable development without tapping exhaustible resource stocks, producing detrimental greenhouse gases or putting in question biological diversity.

Assessing the dimension of "Global Energy and Markets" in more detail, Part II starts with a discussion by Christopher Allsopp and Bassam Fattouh on "The Oil Market: Context, Selected Features, and Implications" (Chapter 5). The authors stress the importance of energy security and climate change policies in determining the future of oil in global energy consumption, as well as the role of regulation and market design. Against this backdrop, the oil market may develop in various ways in the future, triggering uncertainties about market fundamentals. Coupled with a decreasing reliability of feedback loops between the oil price and crude demand and supply, this, they argue, calls for policies able to limit the uncertainty surrounding energy security and climate change policies. Next, Matthew Hulbert and Andreas Goldthau address the question "Natural Gas Going Global? Potential and Pitfalls" (Chapter 6). They argue that soaring unconventional gas production in North America and depressed demand in the West created a perfect storm: an enormous liquidity in gas volumes available across the world. As a result, gas-to-gas competition is in the making, and spot market trade challenges the decade old model of oil-indexed long-term contracts (LTCs) firmly tying producers and consumers into bilateral relationships. In the short run, they argue, gas markets going global should benefit consumers, driving competition and fostering energy security. But they warn that this trend might eventually tilt towards greater supply side collusion, favoring a potential gas cartel in which producers coordinate prices and volumes. Picking up a key aspect of changing gas markets, Mel Horwitch and Mike LaBelle address "The Breakout of Energy Innovation: Accelerating to a New Low Carbon Energy System" (Chapter 7). They argue that innovation may break technological and regulatory lock-in and contribute to achieving a low carbon future. Representing a bottom-up approach, shale gas exemplifies such a technological innovation as it does away with established ways of doing business and government regulation in the natural gas markets. Top-down approaches prove to be equally important as demonstrated by the European smart grids programs. They conclude by assessing how technological innovation in energy occurring at a global scale can become embedded in the local context.

Next, addressing a core question of the extracting industry – how to strike a fair balance between the interests of the reserve holding country and foreign energy corporations – Honoré Le Leuch examines "Recent Trends in Upstream Petroleum Agreements: Policy, Contractual, Fiscal, and Legal Issues" (Chapter 8). The chapter focuses on fiscal regimes, the cornerstone of the relationships between reserve holding countries, their national oil companies (NOCs), and the investing international oil companies (IOCs). It argues that given the uncertainties surrounding the global oil market it is imperative to put in place progressive fiscal schemes, protecting all the involved contractual partners during price fluctuations – a precondition for future upstream investments and further ventures into unconventional oil. Finally, Charles McPherson looks at NOCs in more detail in "National Oil Companies: Ensuring Benefits and Avoiding Systemic Risks" (Chapter 9). Often dubbed "resource nationalism," NOCs have come to control the vast majority of oil reserves, flanked by renewed enthusiasm for strong state control over hydrocarbon extraction. The chapter discusses the various roles NOCs have been assigned beyond pure commercial responsibilities and assesses the systemic governance risks that may come with the continuing trend towards state-run models. The chapter concludes that NOCs are likely to stay, as they offer elites the prospect of easy money, patronage, and political influence. This, it is argued, calls for domestic but also international efforts to improve the institutional capacity of NOCs, crucial for their performance both in the economic realm and in governance.

Part III deals with "Global Energy and Security." Opening up the discussion, Øystein Noreng puts the peak oil theory at the center of his analysis "Global Resource Scramble and New Energy Frontiers" (Chapter 10). Offering a critical review of the peak oil debate, he argues that flawed assumptions on the future availability of crude reserves may misinform policy. He warns that using the scarcity argument as a basis of energy policy would come with economic costs and may even trigger unneeded military intervention. Following up on this latter theme, Dag Harald Claes discusses "Cooperation and Conflict in Oil and Gas Markets" (Chapter 11). He offers a historical assessment by looking at the company level (NOCs versus IOCs), the (inter-)regional level (natural gas trade), and the international organization level (OPEC, IEA, GECF). He argues that future periods will no longer be characterized by clear market power exerted by one party. This might increase instability in oil and gas markets and call for mechanisms to address this challenge. Complementing this discussion, Charles Ebinger and Govinda Avasarala assess the role of G2, G8, and G20 in coping with the deep eastward shift in global energy. Their chapter "The Gs and the Future of Energy Governance in a Multipolar World" (Chapter 12) discusses the potential of these groups to contribute to achieving more effective global energy governance. It argues that while the G8 loses relevance, the G20 is likely to emerge as a global high-level forum for energy issues, provided it recognizes the shift in global energy consumption and acknowledges the priorities of the new major energy consumers. Concluding Part III, Yury Yudin addresses the pressing challenge of "Nuclear Energy and Non-Proliferation" (Chapter 13). An established part of the global energy mix, the dual-use nature of atomic energy technology constitutes a persisting security threat related to spreading nuclear weapon capabilities. Assessing the various elements of the global non-proliferation regime, he argues in favor of strengthening the safeguards system of the International Atomic Energy Agency (IAEA) in order to effectively identify sensitive nuclear technologies and to prevent their further global dissemination.

Investigating the "Global Energy and Development" dimension, Part IV first of all offers insights into the causes and consequences of energy poverty. In Chapter 14, "Energy Access and Development," Subhes Bhattacharyya reviews available measures to grasp the energy access and development link. He argues that electricity access has so far received disproportionate attention from policy-makers, neglecting the cooking and heating aspects. As a result, 2.7 billion people still rely on biomass for basic daily needs. Policy imperatives that follow from this are providing for adequate planning and financing as well as balancing local needs with international level efforts. Next, Juan Carlos Quiroz and Andrew Bauer address policy challenges arising for countries holding abundant energy resources. Looking into "Resource Governance" (Chapter 15), they argue that energy resources are different from other sources of state income, because they are non-renewable, characterized by high upfront costs and production timelines, and generating substantial but volatile economic rents. In order to realize their welfare potential, however, they advocate improving transparency and accountability across the natural resource value chain, and for a transparent and rule-based resource revenue management, promoting an appropriate balance of public savings, consumption, and investment. They conclude by highlighting that global support by development partners and the international financial institutions is key for capacity building and standard setting in resource rich countries.

Shedding light on a perilous side effect of the world's efforts to go low carbon, Robert Bailey investigates the impact of continued growth in biofuel production for food security.

In his contribution addressing "The Food Versus Fuel Nexus" (Chapter 16), he dismantles US and EU biofuel policies as mere financial transfers to domestic agriculture and biofuel sectors. Instead of catering for climate goals, biofuel programs tend to have serious impact on the food security of developing nations, drive up food prices, and increase price volatility. The call on policy is to protect vulnerable populations from food price volatility, increase efficiency in the food system, and reduce direct competition between food and fuel. Finally, Joyashree Roy, Shyamasree Dasgupta, and Debalina Chakravarty assesses the role of energy efficiency in fostering economic goals. Their chapter "Energy Efficiency: Technology, Behavior, and Development" (Chapter 17) argues that energy efficiency entails a broad range of opportunities to foster energy savings by at the same time achieving environmental goals and other macro-level benefits. It concludes stressing that a systemic transition towards more efficient energy regimes requires a strategically designed sequence of actions involving all policy levels, from local to global.

Starting Part V on "Global Energy and Sustainability," Neil Gunningham assesses a broad range of policy tools possibly advancing progress towards a low carbon economy. The straightforward question of "what works, when and why?" guides his analysis of economic instruments such as tradable emission rights, taxes, and subsidies, as well as regulation such as feed-in tariffs for renewable energy production. While solutions need to account for domestic circumstances, his contribution on "Regulation, Economic Instruments, and Sustainable Energy" (Chapter 18) calls for greater policy learning and transfer to facilitate best practice on a transnational level. Next, Jaap Jansen and Adriaan van der Welle offer a case study on "The Role of Regulation in Integrating Renewable Energy: The EU Electricity Sector" (Chapter 19). Aimed at effectively squaring the circle – limiting carbon emissions by at the same time facilitating a competitive energy market and security of supply – the EU has embarked on a transition of the EU electricity sector towards a low carbon paradigm. This, as the authors argue, provides for an interesting opportunity for policy learning and transfer regarding financial aspects but also regulatory design and implementation.

Exploring the link between "Global Climate Governance and Energy Choices" (Chapter 20), Fariborz Zelli, Philipp Pattberg, Hannes Stephan, and Harro van Asselt then assess the capacity of the existing global climate governance architecture to promote clean energy technologies, thus influencing the way energy is used. They argue that a coherent and comprehensive global energy regime is unlikely to emerge, given diverging interests and global power constellations. Yet they suggest that the climate regime could serve as an "orchestrator" to support country-level transitions toward low carbon. Assessing a specific aspect of the global climate regime, and linking the discussion back to the potential of economic incentives to combat climate change, Christian Egenhofer explores "The Growing Importance of Carbon Pricing in Energy Markets" (Chapter 21). According to his analysis, carbon markets are set to become an essential element of governments' energy and climate change policies. Although a global system as envisaged by the Kyoto Protocol is unlikely to emerge in the near future, sub-national, national, regional or sector-specific trading systems are forming, generating increasingly important price signals. He suggests that the European Trading System (ETS) may offer important insights into how regional markets can be created and made to work. Part V concludes with a focus on the financial aspects of a low carbon future. In their chapter "The Influence of Energy Policy on Strategic Choices for Renewable Energy Investment" (Chapter 22), Rolf Wüstenhagen and Emanuela Menichetti ask how policy frameworks can be appropriately designed to facilitate sufficient private capital flowing into clean energy investment. They point to the fact that public funding will not do the trick when it comes to finding a required annual investment volume in renewables of some $235bn by 2020. Rather, it will be private actors that will be key. For this, they argue, it will be essential to properly understand the determinants of private strategic investment choices, which are not only characterized by risk-return considerations, but also by policy and perceptions; and to create effective frameworks for clean energy investment.

Complementing the above chapters structured along the lines of four dimensions of global energy policy, Part VI adopts a bottom-up approach and adds regional perspectives to the debate. Providing seven in-depth case studies from a country-level perspective, the authors highlight how key energy challenges play out on a domestic level, and what implications follow for global policy. Starting off with a contribution on "Global Energy Policy: A View from China" (Chapter 23), Alvin Lin, Fuqiang Yang, and Jason Portner analyze key trends in the development of China's energy sector during the past decade. They argue that China's rapid economic growth was accompanied by both energy over-consumption and heavy carbonization of the economy. On an international level, China succeeded in entering oil and gas markets and became an important player. Yet, with regards to the sustainability challenge, they conclude that China needs to embrace the entire policy toolbox, ranging from domestic carbon markets to global clean technology deployment policies. Assessing the case of another emerging Asian power, Sudha Mahalingam highlights a different problem: persisting energy subsidies. Her contribution "Dismounting the Subsidy Tiger. A Case Study of India's Fuel Pricing Policies" (Chapter 24) poses that fuel subsidies severely restrict India's ability to switch to renewables, hinder the deployment of fuel-efficient technologies, and lock the country in a fossil fuel-based growth trajectory. While difficult to implement for reasons of strong vested interests, she argues that the only way round the impasse is to fully integrate India – an overly energy import dependent country – into the global market place.

Shifting to Europe, a front-runner in low carbon transition, Richard Youngs argues that the EU's climate and energy policies have indeed clearly gained traction during the past years. Yet, as he explains in "The EU's Global Climate and Energy Policies: Gathering or Losing Momentum?" (Chapter 25), the EU risks compromising its leadership in global climate policy. This is due to tensions between domestic and external policies, coupled with diverging goals underpinning traditional energy security and climate change policies. As Youngs concludes, this centrally raises the question what kind of (global) energy actor the EU should seek to be – aimed at exporting its regulatory state approach to an international level, or at complementing its climate diplomacy approach with a hard security one. Next, Benjamin Sovacool and Roman Sidortsov assess "Energy Governance in the United States" (Chapter 26) and to what extent it interacts with the global energy governance architecture. As they highlight, no overarching theme or principle guides US energy policy; rather, the country seems characterized by a portfolio approach featuring a variety of technologies and energy systems at the same time, some contradictory. Besides being a key actor in international energy markets, energy organizations, and energy diplomacy, the authors argue that the US may exert global influence and leadership in mobilizing its potential for policy innovation.

Next, Suani Coelho and José Goldemberg explore which factors contributed to making Brazil a global front-runner in the production and use of biofuels and renewables. Their chapter "Global Energy Policy: A View From Brazil" (Chapter 27) also explores the possibility and potential of replicating the Brazilian experience in other developing nations. Among other important lessons learnt, the authors stress the importance of ensuring food security and contributing to rural development through biofuel projects, providing for adequate local incentive policies, and complementing them with foreign funding and capacity building where needed. Switching from a major producer in renewables to a global heavyweight in oil and gas, Andrei Konoplyanik's contribution "Global Oil Market Developments and Their Consequences for Russia" (Chapter 28) argues that Russia may be an important producer but effectively is a rule-taker in global oil. In a detailed historical assessment of market development, he traces Russian energy policies back to major changes in global oil. He concludes that Russia, far from being an energy superpower, needs to embrace the emergence of the oil global derivatives market, adopt a sustainable fiscal paradigm, and get away from inefficient models in state-run oil production and use of oil revenues. Finally, concluding Part VI, Ike Okonta assesses Nigeria, Africa's most populous country, one of the world's largest oil exporters and (in)famous for its "resource curse." In "Nigeria: Policy Incoherence and the Challenge of Energy Security" (Chapter 29), Okonta argues that the major cause of Nigeria's current energy security crisis is a lack of coherent policies, caused by domestic struggles between ethnic and regional elites competing for a share of the oil bonanza. The results, he argues, are authoritarian rule, damaging consequences for social and economic life, and a limited ability to fully achieve the country's crude production potential in the future.

Following Part VI, Andreas Goldthau presents a conclusion (Chapter 30) which extracts key insights from the book. In "Global Energy Policy: Findings and New Research Agendas," he summarizes what can be learnt from a four-dimensional approach to global energy policy; to what extent bottom-up analyses have complemented this top-down approach; and sketches further routes of research.

References

Alhajji, Anas F., and David Huettner. 2000. OPEC and Other Commodity Cartels: A Comparison. *Energy Policy* 28: 1151–1164.

Andrews-Speed, Philip. 2011. *Energy Market Integration in East Asia: A Regional Public Goods Approach*. ERIA Discussion Paper 2011-06.

Baker Institute. 2007. *The Changing Role of National Oil Companies in International Energy Markets*. Baker Institute Policy Report 35.

Barnett, Jon, and W. Neil Adger. 2007. Climate Change, Human Security and Violent Conflict. *Political Geography* 26: 639–655.

Biermann, Frank, Philipp Pattberg, and Fariborz Zelli. 2010. *Global Climate Governance Beyond 2012. Architecture, Agency and Adaptation*. Cambridge: Cambridge University Press.

Bilgin, Mert. 2011. Energy Security and Russia's Gas Strategy: The Symbiotic Relationship between the State and Firms. *Communist and Post-Communist Studies* 44, 2: 119–127.

Brown, Marilyn A., and Benjamin K. Sovacool. 2011. *Climate Change and Global Energy Security: Technology and Policy Options*. Cambridge: MIT Press.

Campbell, Kurt M., Jay Gulledge, John R. McNeill, *et al.* 2007. *The Age of Consequences: The Foreign Policy and National Security Implications of Global Climate Change*. Washington, DC: Center for Strategic and International Studies.

Crane, Keith, Andreas Goldthau, Michael Toman, *et al.* 2009. *Imported Oil and U.S. National Security*. Washington, DC: RAND Corporation.

Deutch, John M., James R. Schlesinger, and David G. Victor. 2006. *National Security Consequences of U.S. Oil Dependency: Report of an Independent Task Force*. New York: Council on Foreign Relations.

Goldthau, Andreas. 2011. A Public Policy Perspective on Global Energy Security. *International Studies Perspectives* 13 (December): 64–83.

Goldthau, Andreas. 2012. From the State to the Market and Back. Policy Implications of Changing Energy Paradigms. *Global Policy* 3, 2: 198–210.

Goldthau, Andreas, and Jan Martin Witte. 2010. *Global Energy Governance. The New Rules of the Game*. Washington, DC: Brookings Institution.

Graaf, Thijs Van de, and Dries Lesage. 2009. The International Energy Agency After 35 Years: Reform Needs and Institutional Adaptability. *Review of International Organizations* 4: 293–317.

Grübler, Arnulf. 2008. Energy Transitions. In C. J. Cleveland, ed. *Encyclopedia of Earth*.

Hallouche, Hadi. 2006. *The Gas Exporting Countries Forum: Is It Really a Gas OPEC in the Making?* Oxford: Oxford Institute for Energy Studies.

Helm, Dieter. 2011. Infrastructure and Infrastructure Finance: The Role of the Government and the Private Sector. *EIB Papers* 15, 2: 8–27.

Hoffmann, Ulrich. 2011. *Some Reflections on Climate Change, Green Growth Illusions and Development Space*. Geneva: UNCTAD.

IEA. 2011. *World Energy Outlook 2011*. Paris: OECD.

IPCC. 2007. *IPCC Fourth Assessment Report: Climate Change*.

Jong, Sijbren De. 2011. Vers une gouvernance mondiale de l'énergie: comment compléter le puzzle. *Revue internationale de politique de développement* 2: 29–54.

Klare, Michael. 2001. *Resource Wars: The New Landscape of Global Conflict*. New York: Henry Holt.

Kuenneke, Rolf W. 1999. Electricity Networks: How Natural Is the Monopoly? *Utilities Policy* 8, 2: 99–108.

Lesage, Dries, Thijs Van de Graaf, and Kirsten Westphal. 2009. The G8's Role in Global Energy Governance since the 2005 Gleneagles Summit. *Global Governance* 15, 2: 259–277.

Marcel, Valerie. 2006. *Oil Titans: National Oil Companies in the Middle East*. Baltimore, MD: Brookings Institution.

Newell, Peter, and Harriet Bulkeley. 2010. *Governing Climate Change*. London: Routledge.

Orban, Anita. 2008. *Power, Energy, and the New Russian Imperialism*. Westport, CT: Praeger.

Rutland, Peter. 2008. Russia as an Energy Superpower. *New Political Economy* 13, 2: 203–210.

Smil, Vaclav. 2010. *Energy Transitions: History, Requirements, Prospects*. St Barbara: Praeger.

Stevens, Paul. 2008. National Oil Companies and International Oil Companies in the Middle East: Under the Shadow of Government and the Resource Nationalism Cycle. *Journal of World Energy Law & Business* 1, 1: 5–30.

Stulberg, Adam N. 2008. *Well-Oiled Diplomacy: Strategic Manipulation and Russia's Energy Statecraft in Eurasia*. New York: State University of New York Press.

UN. 2010. *Energy for a Sustainable Future: The Secretary-General's Advisory Group on Energy and Climate Change (AGECC): Summary Report and Recommendations*. New York: United Nations.

Victor, David G., David R. Hults, and Mark C. Thurber, eds. 2011. *Oil and Governance. State-Owned Enterprises and the World Energy Supply*. Cambridge: Cambridge University Press.

Victor, David, and Linda Yueh. 2010. The New Energy Order. Managing Insecurities in the Twenty-first Century. *Foreign Affairs* January/February.

Youngs, Richard. 2009. *Energy Security: Europe's New Foreign Policy Challenge*. New York: Routledge.

Zha, Daojiong. 2006. China's Energy Security: Domestic and International Issues. *Survival* 48, 1: 179–190.

Part I Global Energy: Mapping the Policy Field

The Handbook of Global Energy Policy, First Edition. Edited by Andreas Goldthau.
© 2013 John Wiley & Sons, Ltd. Published 2013 by John Wiley & Sons, Ltd.

Chapter 1

The Role of Markets and Investment in Global Energy

Albert Bressand

Introduction: A Political Economic Perspective on Energy Markets

The term "energy markets" is misleadingly simple. True, in a world of 7 billion people, energy is one of the fundamental factors of production, comparable in importance to labor, capital, technology, and commodities in the satisfaction of human needs. Yet, at first look, the energy industry differs from the rest of the economy in four essential ways. The first of these is its capital intensity, the second the endogenous nature of many energy transportation infrastructures and the elements of natural monopoly they often exhibit. The third differentiating feature is the importance of rent and of conflict over rent distribution. Resulting in part from these first three specificities, but also rooted in security concerns, the fourth distinctive feature is the eminent role of the state in the ownership, control, and development of energy resources, with major implications regarding the role of policy, regulation, and geopolitics. In many countries, the role of the state extends to setting prices and conditions for the consumption of, notably, petroleum products and renewable electricity. Let us consider briefly these features before we put forward a political economic framework adapted to the study of energy markets and investment.

The World's Most Capital Intensive Industry

Energy is the world's most capital intensive sector, with major implications for relations over energy resources. A half-century perspective often governs energy investment and it is not infrequent to have to gather in excess of $10 billion for a given project, with major undertakings absorbing over $100 billion as is the case for the development of the Kashagan oilfield in the Kazakh part of the Caspian Sea or of the redesign of national electricity grids in Europe to adapt to intermittent and decentralized power sources. Turning enough in-place deposits (coal, oil, gas, uranium . . .) and natural forces (wind, sun, hydro . . .) into actual energy resources and bringing them to consumers in various

The Handbook of Global Energy Policy, First Edition. Edited by Andreas Goldthau.
© 2013 John Wiley & Sons, Ltd. Published 2013 by John Wiley & Sons, Ltd.

markets can be an amazingly expensive proposition. According to the International Energy Agency (IEA), and subject to the policy assumptions in its New Policies scenario, cumulative investment in energy supply infrastructures will amount to no less than $38 trillion (year 2010 basis) over the 2011–2035 period, the equivalent every year of a tenth of the US GDP, or of the whole GDP of Spain. Of this investment, about $17 trillion will happen in the power sector and $24 trillion in total in non-OECD countries. The term infrastructure as used by the IEA is closer to fixed capital, as it encompasses upstream investments for the replacement and exploitation of reserves ($8.7 trillion for oil, $6.8 trillion for gas, $1.1 trillion for coal mining (IEA 2011a: 96–97)) as well as transport and distribution infrastructures in the more usual usage of the term.

A Largely Industry Specific and Industry Produced Energy Transport Infrastructure

Whereas labor, capital, technology, and even raw materials are easily mobile using general purpose infrastructures, many forms of energy can only reach their market through the construction of dedicated and costly pipelines, power grids or distribution networks that often exhibit natural monopoly features and are also typically subject to stronger forms of political control than is the case for most other industries. Although oil and coal can be carried in trains and trucks, the provision of the infrastructure needed to transport and distribute almost all of natural gas and electricity, most of crude oil, and a part of petroleum products is endogenous to the sector, provided as part of a three-tier value chain of upstream, midstream, and downstream.

No other industry features such a clearly identified midstream, which introduces an additional layer of connection between energy resources and territory, and therefore political control. This gives transit states nuisance power that may be exercised in the non-economic manner, well illustrated in Russia–Ukraine relations among others. The oil and gas midstream abounds indeed in "pipeline wars." While landlocked producer countries have little alternative, others can opt for seaborne transport of energy, for which investment is limited to vessels and port infrastructure, with fewer claims on territory and more limited options to exercise territorial control. In the case of natural gas, however, and leaving aside the still limited use of compressed natural gas, seaborne trade is not the magic bullet for depoliticization it is often made to sound; liquefaction is a prerequisite and comes at a high cost of $4–20 billion dollars depending on specific site circumstances, which makes it uneconomic compared to pipelines over short and medium distances and which also calls for complex licensing in light of national interest considerations, as can be seen in the US presently. Nevertheless, seaborne transportation gives producers access to a diversity of markets, thereby reducing the potential for counterparty opportunism so frequent in the case of dedicated, land-based infrastructures.

In the case of electricity, seaborne transportation is not an option and grids exhibit features similar to oil and gas pipelines. The difference however is that, currently, international trade in electricity is limited to a few regions and is often of a more symmetrical nature (as any country can produce electricity, unlike oil and gas), reflecting differences in daily and seasonal patterns of consumption and production. This may change however, as can be seen from massive projects to carry "green electrons" across the Mediterranean and as large-scale interconnections are needed to offset some at least of the unpredictability of intermittent solar and wind energy. One should be prepared for the possibility that midstream issues will become as important, and potentially as politicized, in electricity as they are in oil and gas.

Rents and Subsidies

A third distinctive feature of energy markets and investment is the importance of economic rent. Indeed, as the point is restated by Bassam Fattouh and Coby van der Linde, "sizable economic rents have been a prize deemed worth fighting for, far beyond the normal competition among market players. They have guaranteed persistent involvement by governments everywhere, either as producers or tax collectors" (Fattouh and van der Linde 2011). Rents exist in all markets but, under normal competition, they tend to reflect comparative and competitive advantages that can be reduced or eliminated through innovation, economies of scale, marketing, and other techniques endogenous to productive activities. Rent is far more pervasive, lasting, and protected in energy markets than in most other markets.

Deliberately limiting a company's or a whole country's production below the production that could meet demand at a given price, as happens in oil and gas, is fairly uncommon in most markets, especially if competition rules are in place to proscribe abuse of market power.

In oil markets, production by the low-cost OPEC producers is subject to production quotas. Even if they tend to be only loosely respected, such quotas make higher-cost production indispensable, placing high-cost producers in the position of marginal suppliers to the market. OPEC members meanwhile denounce the petroleum taxes levied notably by European countries as unfair tools to snatch rent away from them. The high rent embedded in oil and to some extend natural gas prices reinforces calls for subsidized energy. To the public in the Arabian Peninsula countries, paying oil at a price close to extraction cost rather than reflecting rent captured in the international market sounds natural even if the national oil companies are keenly aware of what this means in terms of foregone export revenues. Several hundred million final hydrocarbon customers benefit from subsidies that the OECD estimates to have grown from $340 billion 10 years ago to about $409 billion now for countries for which numbers exist, and probably half a trillion dollars in total. Such subsidies very significantly change the outcome of market interactions toward higher usage of energy resources. A good case in point is Saudi Arabia, which registered an absolute increase of its domestic consumption of oil of 1.2 mbd (million barrels per day) over the 2000–2010 decade, second only to China's (4.7 mbd) and higher than India (1.05 mbd), although the kingdom's population is below 30 million while it is well above one billion in India and China. The same European customers who pay high taxes to consume petroleum products often benefit from subsidies for coal-generated electricity. Altogether, producer rent and consumer subsidies play an essential role in shaping substitution effects across energy markets and in investment patterns, leading to patterns of resource allocation that differ profoundly from what would result from free market forces.

Resource Nationalism and Enlightened Resource Mix Tinkering: Omnipresent States

While supported and reinforced by all three features just reviewed, the fourth distinctive feature of energy investment and markets, namely omnipresent states, has strong roots of its own, notably in national security. The present rise of state power over energy decisions tends to be described under the heading of "resources nationalism," a cliché which assumes that some producer countries (notably OPEC members and Russia) treat energy as a political or geopolitical resource and constitute the major source of interference in energy markets (Gustafson 2012). Yet, as illustrated by the gamut of subsidies, taxes, renewable energy portfolios, feed-in tariffs, and other policy mandates, consumer states are also essential players in the energy arena. As observed by the IEA in its latest WEO report, "Renewable energy subsidies jumped to $88 billion in 2011, 24% higher than in 2010, and need to rise to almost $240 billion in 2035 to achieve the trends projected in the New Policies Scenario" (IEA 2012).

States take a direct interest in the consumption of energy as a condition to maintain support either from green-minded constituencies or from purse-constrained consumers, with the former weighting more in Europe and Japan and the latter more important in the US, Asia, and the Middle East. Europe as well as 28 States in the US and the District of Columbia conduct policies aiming to shift energy consumption away from what would result from the free operation of markets, by requiring a significant proportion of biofuels in transportation fuels and/or of renewable electricity in power brought to the market. While they often do it with the objective to "correct market imperfections" and notably to avoid a "market lock-in" of higher carbon energy sources, it remains true that resource allocations reflect a combination of market forces and policy mandates. This is true also of nuclear energy, as some countries set limits on it (France in 2012 decided to do so in ways still to be articulated) or ban it altogether (Italy, Germany).

With resource nationalism a poor guide, a typology of energy policy environments should be organized in terms of the combination of sovereign objectives and market objectives that exist in all countries (Bressand 2009). In the US, market objectives play an essential role, but nevertheless "energy independence" objectives and, increasingly, environmental policy mandates also have a profound impact. CAFE standards have proven powerful to reign in transportation fuels consumption. At the other extreme countries like Mexico and Saudi Arabia perceive energy as essential to their independence and national security. In between, countries like the Russian Federation attempt to strike a balance between sovereign objectives and market dynamics, with the development of national energy champions a major objective. The existence of international sanctions (most notably currently on Iran's oil exports and on investment in its gas resources) create another category, as do policies to reach energy independence through the facilitation of international investment in countries like Colombia. Altogether, six different groups of countries can be identified on matters of hydrocarbon policies and of investment and market conditions (Bressand 2009).

As a result, investment in energy and the working of energy markets differ significantly from what standard economic models would predict based on usual patterns of relatively free competition. Implications from the four specific features just reviewed include massive subsidies, enduring fragmentation of global gas markets, the fact that oil resources are being developed in deep water at costs 10–15 times higher than lower-cost resources still abundantly available in the Middle East, as well as the fact that some countries eagerly invest in large-scale power generation from non-competitive renewable technologies as an effort to create, over time, the set of market signals supporting the type of energy mix that their energy policies will have put in place.

A Northian Perspective on Resources, Institutions, Transactions, and Power

Altogether, energy is a domain in which some of the world's most liquid global markets and most capital intensive investments coexist with and are influenced by political, cultural, institutional, and geopolitical considerations. An analysis of energy investment and markets therefore calls for a fully fledged political economic perspective that can illuminate how economic relations are shaped by the factors we listed. In this vein, the analytical framework we see as most conducive to a comprehensive, forward-looking analysis of energy markets and energy investment consists of four interrelated components:

- A description of the *physical resources* and of the manner in which they are defined and relate to one another. In a traditional perspective, such resources are stocks of subsoil deposits and flows of water, wind, sunlight or heat. New resources such as stocks of carbon sink capacities may materialize from the world of "nature" to that of the economy (Heal 2000) while others may be subtracted at least temporarily from market interactions through nationalization or heavy-handed regulation;
- A set of *institutions and governance mechanisms* setting the regimes that apply to interactions within and about the energy system, including but not limited to markets and regulatory structures that govern the exercise or the trading of rights over energy resources;
- The *power structure* that determines which actors are in the position to exercise control over production or transit of energy assets and which actors can influence the governance of institutions setting the stage for energy market transactions. Such power may manifest itself, for instance, through the creation, design, and operation of energy markets. In oil, the power structure once revolved around a small number of international energy companies (the infamous Seven Sisters) before shifting in part to states bent upon challenging the existing international order, as OPEC did in the 1970s. Today, actors of influence include states, national energy companies, international energy companies, large-user groups, legislators, and political organizations such as the EU. Various civil society groups are also in a position to bring their values to bear onto the energy arena, such as Greenpeace and the Extractive Industry Transparency Initiative (EITI) among many others (Shell 2005).
- The *transactions* over energy assets and energy products and services themselves across the whole value chain from upstream to midstream and downstream. These transactions can happen in markets ranging from traditional commodities markets to esoteric futures markets or they can also unfold in state to state relations.

This four-dimensional framework, which emerges quite naturally from an informed look at the energy system, draws inspiration from the work of Nobel economic laureate Douglass North (North 1990). It is rooted in political economy, the branch of economics that defines its ambitions not in terms of what can be modeled but of what needs to be understood of the world in which politics, national sovereignty, conflicts, and power relations operate alongside the quintessential law of supply and demand. These four dimensions are relevant throughout the economy but, usually, a relatively stable and pacified power structure tends to support a relatively stable market-friendly institutional framework, making it easier to focus on two dimensions, namely resources (or "economic opportunities") and transactions. In the energy system, neither the institutional layer nor the power structure can be considered as having reached such levels of stability and market-friendliness; they tend to be contested and remain very significant. Appropriate for studying the emergence and working of energy markets, this Northian political economic framework will inform the rest of the present chapter. Each of the four dimensions in this analytical structure, however, is influenced by and influences all three other dimensions, with a tetrahedron a more relevant description of interdimensional relations than a linear sequence as the one just exposed. We shall begin with resources, then consider which institutional framework presides over investment and transactions in energy, notably when and how markets are the locus for investment decisions and transactions. The power structure and notably the role of states and the competition over rent will be discussed as a major influence on this institutional framework, before we turn to transactions over investment and in the market-place.

Energy Resources: The Paradigm Shift

Which resources can actually be mobilized is an essential consideration for energy markets, and so are substitution options among energy sources or "fuels." Resources however, in the Northian sense of economic opportunities, are not given once and for all – as the image of "oil in the ground" misleadingly suggests – but are a function of what existing technology and social practices render economically usable both on the production and the consumption sides.

Over the long term, the picture of energy resources is anything but stable. Forms of energy that had become archaic, such as the windmills of Europe's Middle Ages, can make spectacular comebacks as they did twice, first in the late nineteenth and early twentieth century when no less than 600,000 wind pumps provided US farmers with access to underground water (Gipe 1995) and today in the form of wind turbines. Some resources materialize, literally, out of the blue (but with major state involvement) as would be the case with carbon credits but also as is proposed in the Desertech project, whereby Northern African countries would supply Germany and Northern Europe with solar electricity from their desert lands. Resources that had been proclaimed nearing exhaustion (or at the "peak" preceding exhaustion) now appear plentiful, as is the case for natural gas for which no less than two centuries of reserves are now at hand; the IEA foresees the possibility of a golden age of gas that few would have predicted only a handful of years ago (IEA 2011b). Meanwhile investment in nuclear power generation was made uncertain in the wake of the Fukushima accident in Japan, with countries like China forging ahead after a few months of technical reassessment while Germany, Italy, and Switzerland lead a move out of nuclear energy. As a result, the present period is one of accelerated transition; a different energy system is taking shape, the main contours of which have begun to emerge. Uncertainties on the resources side have to do with geology (the peak oil debate), with technology (most vividly illustrated by the shale gas revolution in North America), and with the social and political acceptability of some resources, whether directly (nuclear power generation) or in light of the technologies needed for their production (tight oil and gas in continental Europe).

Peak Oil or Not Peak Oil?

An essential consideration throughout the energy system is whether oil resources have reached the limit beyond which investors need to run faster and faster on the Exploration and Production (E&P) treadmill without being able to offset a natural exhaustion in oil production. Views about "peak oil" are of importance not just for oil companies but for the overall structure of relative prices in the energy system, as declining oil production can be expected to go with rapidly increasing extraction costs for remaining oil reserves, which in turn should direct investment away from oil toward other energy sources – ideally renewables but in practice also coal even in green-at-heart countries like Germany. After decades of gradual decline that were widely interpreted as validating the views of Shell geologist M. King Hubbert when he coined the phrase "peak oil" (Hubbert 1956), the production of the 48 continental US States is on the increase again as a result firstly of a resumption of investment in US E&P after decades of underinvestment, and secondly of technology progress that reflects the experience gained with shale gas. Peak oil theory, further developed by energy economists in the ASPO,[1] is challenged, even if all can agree that stocks of in-ground resources are finite, arguing about "when" rather than "whether." In the words of a leading industry analyst, "the concept of peak oil is being buried in North Dakota, which is leading the U.S. to be the fastest growing oil producer in the world" (Morse 2012).

It is often overlooked that, for practical purposes, the economic resource on which transportation and the petrochemical industry depends is not "oil" – the general term in public discussions – but "liquids," a notion referring to a certain group of hydrocarbon molecules (such as middle distillates used to produce diesel, gasoline, and jet fuel) among those that are extracted from any given "oil" or "gas" reservoir. Crude oil, which comes in almost as many varieties as there are reservoirs on the planet, is the most accessible but not the only source of liquids. Liquids can be produced from gas (a process pioneered by Shell in its Bintulu plant in Malaysia and then brought to scale in the complex and expensive Pearl plant in Qatar) as well as from coal (as pioneered by Sasol in South Africa in the apartheid era) and even from biomass. According to IEA scenarios, the sum of gas-to-liquids and coal-to-liquids production should reach about 2 mbd in a couple of decades (IEA 2012: 106–107). Even within crudes, however, and leaving aside important considerations about sulfur content, three major distinctions must be made to distinguish "conventional oil" from, respectively, heavy oil, deepwater oil, and tight oil as these four types of liquid sources are developed and produced through quite different technical systems and, in part, by different economic actors.

- The first distinction is about viscosity (measured in API gravity degrees) with heavy crudes (lower than 15 API gravity degrees, or 20 in some other definition) and extra-heavy fuels (lower than 10 API) needing significant processing in upgraders to be made usable.
- A second important distinction also rooted in technical production systems is whether reservoirs can be exploited from fixed on-shore or shallow-water off-shore platforms, as opposed to having to be accessed from deep water thanks to very complex and costly floating structures that only a few energy companies are able to build and operate. The present limit for conventional platforms not requiring special design and capacities is about 400–500 meters.
- The third distinction is about the rock from which oil is produced, as it may either lend itself to the natural migration of hydrocarbon molecules to the wellhead (the conventional case) or, by contrast, require that such migration be caused by rock fracking and by the injection of massive amounts of sand and solvents to keep production flowing.

With these distinctions in mind, and observing that the term unconventional, a relative term, is not precisely defined and that the parameters mentioned here can be set slightly differently, "conventional oil" can be defined as crude oil of more than 15 degrees API found in rocks that do not require fracking, either on-shore or from waters no deeper than 400 meters.

An important additional consideration for students of energy investment and markets is that a reservoir categorized in terms of the dominant resource it contains (oil or natural gas) usually includes other types of hydrocarbons (condensates or natural gas liquids, in short NGL) that will be monetized on a different market. Presently, for instance, the economics of investment in US shale gas plays is very strongly influenced by such joint production consideration, leading to the development of plays that would not be economically viable at today's low US market prices for gas. While bankers will not hesitate to describe such "liquid rich plays" as producing "oil" as well as gas, the physical reality is different as the jointly produced resource is not oil but NGL, namely ethane, propane, and butane, each of which is sold on distinct markets at an average price influenced by specific regional considerations such as demand for ethane and butane in connection with the production of heavy oil, or the use of propane to power tractors and serve other fairly specific needs. Most economists writing about "the price of oil" or "investment in oil and gas" tend to ignore the importance, if not the existence, of these joint production features and distinct markets. The apparently irrational (from a perfect-market arbitrage perspective) price differential between the two major crude benchmarks, West Texas Intermediary (WTI) and Brent (itself consisting of three different crudes, the cheapest of which determines the Brent price at any given moment) has drawn attention to another aspect of the importance of the production system, in this case the need for complex and often customized midstream and downstream installations to create value out of the production of oil.

As these considerations suggest, the peak oil debate is in fact a peak liquid debate. In this vein the good news (from an ASPO perspective) is that King Hubbert has been proven right: conventional crude oil production did peak in 2005 and is on a 1.2% decline. While its share is expected to stabilize at around 65% of OPEC production thanks to large reserves in the Gulf producing states, conventional oil should represent less than half of non-OPEC production in the 2030s (Leonard 2012). Yet the three sorts of non-conventional crudes we just described and "liquids" produced from gas, coal, and biomass see their production rise fast enough to postpone by several decades the time of reckoning. Considering the existence of large untapped sources of hydrocarbons, notably methane hydrates floating deep in some oceans, it is too early to announce a production decline. Indeed, the planet may end up having too much hydrocarbons rather than too little as peak theory would suggest. A roundtable organized by Stephen Kopits in June 2012 led to a description of a long and sustained "oil plateau" extending from 2005 to probably the late 2030s or early 2040s thanks to successive peaks in various types of liquid resources, first in conventional oil (2005), then in deepwater oil (early 2020s), and lastly in tight oil (2030s?). The production of NGL is not an autonomous production process but the result of natural gas production (conventional gas or shale gas) as well as of the operation of refineries; it is expected to grow at about 3% per annum. Production of heavy oil is expected to keep growing throughout that period.

Another essential resource now determining economic opportunities in energy investment and markets is the earth's carbon sink capacities, a resource that no one even considered as such two decades ago. Unlike oil, carbon sink theory follows a pure "peak" logic to the extent that what really matters (in terms of human welfare and health) is a given envelope trajectory for Global Greenhouse Gasses (GHG) emissions compatible with relatively tame climate change (defined by diplomats as change of less than 2 °C on average). According to the IEA, GHG emissions under current and foreseeable policies (the New Policies scenario) are on a trajectory that will soon require the use of more carbon sink capacity than is available under this climate constraint and, notwithstanding a brave but globally irrelevant European effort, policies are not on a trajectory to modify this perspective (IEA 2011b).

The Electric Grid as the Energy Sources Integrator

A central element of the paradigm shift is the growing role of electricity and therefore of electricity markets in influencing demand for primary energy sources across markets.

Daniel Yergin, in his seminal history of the previous energy revolution (Yergin 1991), documented the substitution of King Coal with King Oil; we are now on the cusp of a new era in which King Electron presides over the restructuring of the energy system and of energy resources markets (Konoplyanik 2009).[2] One major implication is a greater substitutability between diverse energy sources.

Power can be generated directly from flows of energy (hydro, wind, sun, geothermic heat, and properly managed biomass) that are renewable in the sense that what is consumed now does not reduce flows available in the future or through the transformation of stocks of primary fuels (oil, gas, coal, and uranium). This diversity of power generation sources to produce the exact same resource (electrons) makes electricity markets powerful energy system integrators. Further electrification of the economy has begun to extend to transportation. Electric cars, like the Chrysler Volt or the Opel Ampera, and plug-in hybrid cars that can be powered from the grid are making their appearance, even if still timidly. The massive development of the Chinese car fleet now under way will probably be the major testing ground for the role of electric and hybrid cars (IEA 2011, 2012: 91). As a result, new substitution effects develop at the interface between the market for oil and markets for the fuels used in power generation, which today includes primarily markets for coal, natural gas, nuclear energy, and three main renewable power sources, namely hydro, wind, and solar energy. Whether such developments are part of the greening of the economy is a more open question as electric and plug-in hybrid cars can run on coal just as easily as on wind energy, depending on choices made by the power generation companies. Another less spectacular but important integration process is at work between the markets for power and for heat, namely cogeneration (the joint production of power and of hot water or steam), which has an essential role to play in increasing overall energy efficiency.

Looking even further ahead, scientists such as Klaus S. Lackner at the Columbia School of Engineering foresee a world where hydrocarbons will be produced from hydrogen extracted from the sea thanks to abundant, cheap solar energy and from the CO_2 that will, in their view, be in significant excess in the planet's atmosphere (Lackner *et al.* 2010; see also Lackner 2010). Should such a vision become true, energy markets will then rest squarely on the market for electricity. This will further exacerbate the role of what we call the Northian tetrahedron in shaping outcomes in energy investment and markets.

Energy Institutions: A Pre-WTO World Meets Post-Modern Synthetic Markets

A Northian perspective leads to ask how institutions facilitate, hamper or otherwise influence the definition of marketable rights and market processes over the resources we have just described. The institutional framework in which energy market interactions take place is far from having stabilized. In addition, as we have begun highlighting, this institutional framework sets limits on the role of markets that can be far more constraining than is the case in most other industries.

The capital intensive nature of the energy system means that the regime applicable to investment is at least as relevant to its overall evolution as is the regime applicable to trade in energy fuels and products. The failure to negotiate a Multilateral Investment Agreement at the OECD in the 1990s shows how far the world is from a multilateral, market-centric investment regime, with thousands of Bilateral Investment Treaties (BIT) providing only a very crude proxy.

In oil and gas, an essential aspect of competition between market participants is about access to upstream resources, namely "acreage" in which to explore for and, ideally, develop resources now that the lion's share of new discoveries happens in non-OECD countries – with a large part of the rest in Canada and the US. The investment regime prevailing in each resource-holding country is therefore an essential influence on the type of market opportunities that can be accessed and the merit order in which such opportunities will be developed. Major resource-holding countries regulate such access in a very discretionary manner. Except in a few places (most notably private land on the US mainland) such resources belong to the state. Hence the paradox whereby expensive, challenging resources are developed while lower-cost resources are not. A number of market processes are at work in the energy upstream, such as auctions for the development of oil and gas fields, but always within a policy-determined national energy framework. In many cases, access to investment opportunities is conditional to the creation of joint ventures with local partners, quite often with the national energy company. In some cases, international investors are excluded altogether and the development of energy resources is conducted under very close supervision by the state. Mexico, again, illustrates how far this approach can be pushed since the Mexican national energy company, Pemex, operates largely as an arm of the government, allocating a large, policy-determined share of its revenues to the national budget (significant reforms are likely to be introduced by the Mexican President-elect from 2013 onward).

In the US, an essential producer, the investment regime includes concessions, a term reminiscent of the Seven Sisters era but which applies in this case to private leases negotiated, usually through "land men," between investors and private landowners. Farmers in the US Northeast count on subsoil gas resources to take them through the challenging times many farms are going through. Resources on public land are accessed through a licensing and auctioning process which the Obama administration has endeavored to make faster and less litigation-prone (C-Span 2012).

Governance and the Uncertain Provision of Public Goods Essential to Market Operation

In all countries, even the most market-friendly ones, the production of private goods to be traded in markets (whether oil barrels or kilowatt-hours) depends in part on the availability of public goods reflecting the state of the governance of the energy system. Three public goods are most relevant to the working of the energy system, namely provision of a level playing-field in energy trade and investment (most noticeable by its very incomplete provision), energy security, and energy sustainability. As we review now, such public goods may or may not be present in whole or part of the energy system, with major implications for the existence and the operation of specific energy markets, as market participants often have to find alternatives through contractual arrangements.

Pre-Modern and Post-Modern Markets

Paradoxically, energy markets are both more "archaic" and more "modern" than the rest of the economy. More archaic because, together with armaments and water, significant parts of the energy field are among the few sectors not covered by the international trading regime (Keohane and Nye 1977) that emerged since the signing of the General Agreement on Tariffs and Trade (GATT) in 1946 and the commencement of the World Trade Organization (WTO), its successor since 1995. This exemption from WTO rules is true of natural energy resources such as crude oil that are not "products," as opposed to refined petroleum products (Desta 2003). Reinforcing this *sui generis* trade regime, the importance and the variability of economic rent that can accrue to investors, as observed by Honoré Le Leuch in Chapter 8 below, will lead most resource-holding countries to consider that "efficient taxation of upstream oil and gas projects cannot be achieved by using only the general taxation applicable to any economic activity." Developed countries like the UK are no exception and they do not hesitate to opportunistically modify tax schedules in light of evolving market conditions.

A telling illustration of this difference between parts of the energy sector and the rest of the economy was provided by two apparently contradictory actions that the US took in 2012: on March 14, the US joined Japan and the EU to sue China before the WTO for refusing to export enough of the "rare earth" metals for which China is the almost exclusive producer; yet earlier on February 14, US opinion and trade lawyers found nothing surprising or infringing on other countries' trade rights when ranking Democrat House Representative Edward Markey introduced a Bill that has to be cited as "The American Natural Gas Must Stay Here Act." The US, which for many years has banned the export of oil crudes, is far from alone in holding trade in energy to different standards from trade in other goods and services.

And yet, few markets are more complex, more open to design by regulators and business innovators than oil future markets and, even more strikingly, modern power markets. City dwellers switching on the light when returning home would hesitate to do so if they had to list, let alone understand, the various markets on which the timely and reliable supply of electrons depends. As described by Sally Hunt, to insure second per second matching of supply and demand, a fully deregulated power market requires the joint operation of half a dozen markets (Hunt 2002). The challenges of bringing higher-cost renewable energy sources into the electricity grid and the fact that wind and solar energy are intermittent in nature creates yet another layer of market design challenges that may lead to qualitatively different market structures, organization, and outcomes. Should markets, for example, deal only in power (kilowatt hours) or also in capacity (kilowatt installed) to reflect the need for back-up capacity for intermittent power generation? Should renewable resources always be used, even when fetching negative prices? This and many similar questions illustrate how power markets, at the heart of the overall energy transition, are open to innovators' and designers' imagination in ways that can only be compared with financial markets. The difference is that governments and international organizations rather than Wall Street strategists play an essential role in coming with market designs that they believe will produce outcomes better aligned with their political and value preferences.

The Power Structure Influencing Energy Investments and Market Relations

International investment relations over oil and gas have gone through four eras (Bressand 2009). First came the Concessions era in which many states outside of the US had only limited ways of monetizing their resources at a time when oil had not fully become the world's central energy commodity. The second era began with the seizure of power by OPEC in the context of two oil shocks, though it did not involve the much heralded emergence of a producer-controlled price formation mechanism that could sustainably substitute for the system of posted prices by which international companies used to appropriate most of the rent (Mabro 2005). The very success of OPEC countries then triggered a massive effort of energy efficiency and led international energy companies to prioritize OECD countries (North Sea, Gulf of Mexico) in the development of hydrocarbon resources. This third era was one in which market principles seemed to be gaining ground at the expense of the direct role of the state. A fourth era is now under way in which states (producer states but also those net importing states engaged in far-reaching energy policies) have reclaimed the central role. One reason in the shift to this new phase is that OECD energy resources have been dwindling, although the shale gas revolution in the US may change the game. In this fourth phase, large international oil and gas companies (referred to as IOCs or "energy majors") have had to rethink their positioning in light of the tighter control exercised by states over energy resources and of the competition by national oil companies (NOCs) for easily accessed hydrocarbon reservoirs. IOCs act therefore as a combination of technology companies and banks able to develop challenging resources beyond other companies' reach, to shoulder projects beyond many states' funding capacity, and to finance major technology development from their retained earnings. In short, IOCs are technology and finance houses that happen to specialize in energy. The importance of financial resources in preserving their role leads them to try to keep gearing ratios low enough to enjoy AA or A ratings. It also implies that they secure returns in the high teens or above to absorb project costs while strengthening their balance sheets and financial wherewithal.

In some circumstances, NGOs and local communities are also able to play a role in setting the framework for energy interactions. This can reinforce market mechanisms by reducing the scope for corrupt practices, as is the case of EITI when countries and companies agree to make public all payments they receive or make. The power of NGOs can also be harnessed in support of local content objectives and of broader sharing of the fruits of energy resources development (how broader depending on how independent and widely based such NGOs are, as captured NGOs may also be created by some politicians in a position to do so). In such cases, the role of NGOs in the power structure will end up facilitating departures from pure market price formation, for reasons that are politically and socially legitimate, a legitimacy dependent in part, again, on the integrity of the process. Ideally, tripartite relationships can develop in which investors, the host country, and local communities supported by NGOs can balance their various objectives toward sustainable development of oil and gas resources (Bressand 2011).

Transactions Over Energy Resources: Energy Value Chains and Energy Markets

Following up on this politico-economic perspective on energy decisions, we can now take a look at the bread and butter of energy economics, namely transactions over energy assets, products, and services, leaving detailed discussion for the relevant chapters in this volume.

Oil Markets, OPEC and the Oil Price Discovery Regime

The market for oil is home to the broadest market interactions subject, as we saw, to two essential caveats, namely the distinction between markets for crude oil and markets for petroleum products such as gasoline and diesel, and the role played by OPEC in the markets for crude oil.

Regarding the present price-setting process for crude oil, we shall follow Oxford's Robert E. Mabro's analysis showing how futures markets for crudes (Brent, WTI, and Dubai) have become central in setting the price for oil within the broad limits discussed above (i.e., the role of high-cost producers as marginal suppliers in spite of large undeveloped low-cost resources in the Gulf). The three types of crude grouped as Brent are traded on the London-based ICE electronic market; spot and future contracts for WTI are traded on the New York Mercantile Exchange (Nymex). OPEC then tries to influence these prices as they emanate from these liquid futures markets through the use of its quota system. This leaves OPEC as a price taker, able to influence market trends at the margin, most effectively when "sailing with the wind" through the direct use of production quotas and through the signals and expectations that OPEC quota decisions and policy debates generate. As stressed by Mabro, a cartel only exists when it is able to enforce price discipline including through oversupply by the core members to punish free rider strategies by cheating members. In this sense, paradoxically, OPEC operated really as a cartel only on a small number of occasions, the most remarkable of which was the enforcement of "netback prices" by Saudi Arabia in 1985. This episode led to a collapse in oil prices which Saudi Arabia was unable to transform into a rapid return to the higher prices it wanted to foster. Since then, and leaving aside the two or three slightly different price-setting regimes that existed after the collapse of the Posted Price regime, OPEC has given up setting prices, content to make its quota decisions based on adjustable, announced or non-announced, price bands. Within this framework, OPEC members are constrained by their own needs for revenues which have greatly increased since the Arab Spring and range from around $70 per barrel to as much as $100.

Structural shifts in demand are the strongest influence on oil price trends. Between 2000 and 2010, as shown by Christopher Allsopp and Bassam Fattouh in Chapter 5 below, oil demand growth in non-OECD consumption increased by around 13 mbd, while OECD oil demand dropped by 1.5 mbd (BP 2011). At the heart of this growth lies the Asia-Pacific region, which accounted for more than 50% of the incremental change in global oil demand during this period. As we discussed, structural changes on the supply side reflect the peaking of conventional oil in 2005 with investment redirected toward deepwater sources (which accounted for half of the new production in the past decade), heavy oil, and tight oil as well as toward liquid-rich natural gas plays in North America.

Natural Gas Markets

The market for natural gas is far from having achieved the same stage of globalization as the market for oil. Indeed, as we write, the same unit of natural gas is being traded at prices that oscillate between $2 and $3 per million btu ($/mbtu) in North America, $10/mbtu in Europe, and between $15 and $20/mbtu in the Pacific region. Not only are prices diverging but the price discovery mechanisms also differ profoundly, with the North American market the only one in which a fully fledged competition exists (referred to in industry parlance as gas-to-gas competition!), whereas natural gas is traded in Europe and the Pacific under longer-term contracts that include the significant element of indexation to oil prices. An essential question is how far further integration could proceed. The export of liquefied natural gas from the US would be a major factor of price convergence. In the institutional context we described – one outside of WTO rules – the possibility exist for various lobbies to prevent or limit such exports in the name of energy independence or simply to keep gas and electricity prices lower for large American industrial users.

Technological Innovation as an Essential Feature of Energy Markets, Notably for Renewable Energy Sources

As observed by Mel Horwitch and Mike LaBelle in Chapter 7 of this volume, "innovation is at the core of the global energy sector" in a manner consistent with the political economic perspective presented here since, in their words, "the choice of technologies is dependent on the cooperation of businesses, governments, and society" as part of a more general regulatory regime defined by "synergetic relationships, with private and public activities partially reinforcing each other" (Knill and Lehmkuhl 2002). This is especially visible in markets for renewables, whether biofuels – a pure product of agricultural policies parading as national security or environmental policies – or renewable power sources which, as said, depend on policy mandates and price support schemes like feed-in tariffs.

Evolution in all energy markets reflects directly (oil, gas, renewable electricity) or in larger part indirectly (coal) an ongoing flow of technological innovation commensurate with the need for major breakthroughs such as those associated with deep sea drilling or with photovoltaic production of solar electricity. In oil and gas, the innovation process is unrelenting; witness how better reservoir mapping and drilling techniques keep increasing the share of in-place reserves that can be economically produced. Game-changing innovations can be the result of entrepreneurial strategy on the part of smaller players deploying smaller levels of capital, a telling example being the adamant effort by US oilman George Mitchell to develop shale gas resources that were known to exist in large quantities, and had been exploited on a very small scale for one century, but were deemed uneconomic. While the breakthrough reflected entrepreneurial spirit and persistence in the face of repeated failures, a true inventory of the capital mobilized in this case should also include amounts spent at the initiative of the state. Mitchell Energy's breakthrough drew on the federally funded Eastern Gas Shales Project of 1976 and benefited from federal funding for the first horizontal well in the Barnett shale as well as from unconventional-gas tax incentives under Section 29 of the US tax code.

Conclusion

Altogether, an analysis of the contemporary global energy scene should shun black and white opposition between market and state, or good market-friendly states and bad resource-nationalist ones to capture instead the manner in which states and a plurality of actors interact throughout the energy value chain and across energy sources. How states, civil society groups, and investors interact is best apprehended from a political-economic perspective such as the one used here to assess economically relevant resources, the institutional framework in which they can be accessed and developed, the power structure governing the evolution of these rules, largely outside of the WTO framework, and the interaction of all players in markets for hydrocarbons, renewables, and power.

Notes

1. For views of the Association for the Study of Peak Oil (ASPO) founded by Colin Campbell see http://www.peakoil.net/. See also Deffeyes (2001). For the opposite view see Mills (2008).
2. For other work by Dr Konoplyanik see www.konoplyanik.ru.

References

BP. 2011. *Statistical Yearbook of World Energy*. London: BP.
Bressand, Albert. 2009. Foreign Direct Investment in the Oil and Gas Sector: Recent Trends and Strategic Drivers. In K. Sauvant, ed. *Yearbook on Investment Law and Policy*. Oxford: Oxford University Press.

Bressand, Albert, ed. 2011. *Getting It Right: Lessons from the South in Managing New Hydrocarbon Economies*. New York: UNDP Special Unit for South-South Cooperation.

C-Span. 2012. Interior Secretary Ken Salazar on US energy policy, April 24. Available from http://www.c-spanvideo.org/program/305624-1.

Deffeyes, Kenneth S. 2001. *Hubbert's Peak: The Impending World Oil Shortage*. Princeton, NJ: Princeton University Press.

Desta, Melaku Geboye. 2003. The Organization of Petroleum Exporting Countries, the World Trade Organization, and Regional Trade Agreements. *Journal of World Trade* 37, 3: 529–538.

Fattouh, Bassam, and Coby van der Linde. 2011. *The International Energy Forum, Twenty Years of Producer-Consumer Dialogue in a Changing World*. Riyadh: IEF.

Gipe, Paul. 1995. *Wind Energy Comes of Age*. New York: John Wiley & Sons, Inc.

Gustafson, Thane. 2012. *Wheel of Fortune: The Battle for Oil and Power in Russia*. Cambridge, MA: Harvard University Press.

Heal, Geoffrey M. 2000. *Nature and the Marketplace: Capturing the Value of Ecosystem Services*. Washington, DC: Island Press.

Hubbert, Marion King. 1956. Nuclear Energy and the Fossil Fuels. Paper read at Spring Meeting of the Southern District, American Petroleum Institute, March 7–9, at Plaza Hotel, San Antonio, Texas.

Hunt, Sally. 2002. *Making Competition Work in Electricity*. New York: John Wiley & Sons, Inc.

IEA. 2011a. *World Energy Outlook 2011*. Paris: OECD/IEA.

IEA. 2011b. *World Energy Outlook 2011 Special Report: Are We Entering a Golden Age of Gas?* Paris: OECD/IEA.

IEA. 2012. *World Energy Outlook 2012*. Paris: OECD/IEA.

Keohane, Robert O., and Joseph S.Nye. 1977. *Power and Interdependence: World Politics in Transition*. Boston: Little, Brown.

Knill, Christoph, and DirkLehmkuhl. 2002. Private Actors and the State: Internationalization and Changing Patterns of Governance. *Governance – An International Journal of Policy and Administration* 5, 1: 41–64.

Konoplyanik, Andrei. 2009. Gas Transit in Eurasia: Transit Issues between Russia and the European Union, and the Role of the Energy Charter. *Journal of Energy & Natural Resources Law* 27, 3: 445–486.

Lackner, Klaus. 2010. Washing Carbon Out of the Air. *Scientific American* 302 (June): 66–71.

Lackner, Klaus, Christoph Johannes Meinrenken, Eric Dahlgren, et al. 2010. *Closing the Carbon Cycle: Liquid fuels from Air, Water and Sunshine*. New York: Lenfest Center for Sustainable Energy, Columbia University.

Mabro, Robert. 2005. The International Oil Price Regime: Origins, Rationale and Assessment. *Journal of Energy Literature* 11, 1: 3–20.

Mills, Robin M. 2008. *The Myth of the Oil Crisis: Overcoming the Challenges of Depletion, Geopolitics, and Global Warming*. Westport, CT: Praeger.

Morse, Ed. 2012. Resurging North American Oil Production and the Death of the Peak Oil Hypothesis. *Commodities Strategy (Citi)*, February 15 (Citigroup).

North, Douglass. 1990. *Institutions, Institutional Change and Economic Performance*. Cambridge: Cambridge University Press.

Shell. 2005. *Shell Global Scenarios to 2025; The Future Business Environment: Trends, Trade-Offs and Choices*, ed. Albert Bressand. Washington, DC: Peterson Institute for International Economics.

Yergin, Daniel. 1991. *The Prize: The Epic Quest for Oil, Money, and Power*. New York: Simon & Schuster.

Chapter 2

The Entanglement of Energy, Grand Strategy, and International Security

Meghan L. O'Sullivan[1]

Introduction

Americans are pleasantly surprised about how their energy fate appears to have changed, in such a short time, with little notice or anticipation. Within the last five years, both actual US production of oil and gas and projections for future American production have changed dramatically. Whereas in the mid-2000s, experts predicted that the US should anticipate a future of severe dependence on imported natural gas, in 2012 Washington is debating the pros and cons of becoming an *exporter* of this resource.[2] Even more quietly, domestic production of oil has increased, in large part due to the development of the tight oil in the Bakken formation in North Dakota and the Eagle Ford in Texas.[3]

Innovation and technology deserve the credit for this transformation. The evolution of hydraulic fracturing and horizontal drilling has enabled the development of both oil and gas in locations where the resource was known to exist, but the prospects for extraction at commercial prices had previously seemed remote. These domestic advances are complemented by new energy developments among America's neighbors in the Western Hemisphere. Canada is set to double its production from its oil sands in the coming years, while Brazil is planning to embark on the development of vast sub-salt deepwater oil resources.

There is no question that these energy developments have major economic benefits to the United States. Inexpensive natural gas has spurred a revival in American manufacturing; diminished oil imports are shrinking the trade deficit and will strengthen the dollar; and both oil and gas revivals will bolster employment in direct and indirect ways.[4] But much has also been made of the security benefits that accrue and will continue to accrue to the United States on account of this boom. President Barack Obama has underscored the salubrious effects of these energy shifts for America's strategic position; former CIA director Jim Woolsey sees the energy revolution precipitating a rebalancing of America in the world.

But how exactly should we think about these domestic energy developments in the context of energy security? Much depends on how energy security is defined. Traditionally,

The Handbook of Global Energy Policy, First Edition. Edited by Andreas Goldthau.
© 2013 John Wiley & Sons, Ltd. Published 2013 by John Wiley & Sons, Ltd.

energy security referred to having access to sufficient supplies at a reasonable price. Energy security was largely perceived to be a notion of relevance to consuming or net importing countries. Access, supply, and affordability were the key concepts. By this traditional, basic definition, the recent revolution in domestic shale gas and tight oil most certainly make the US significantly more energy secure. Imports of oil and gas have dropped markedly, at least in part due to these developments.[5] Gas imports fell from 4.3 billion cubic feet (bcf) in 2005 to 3.5 bcf in 2011. Similarly net oil imports fell from their peak of 12.5 million barrels per day (mbd) in 2005 to 8.4 mbd in 2011, the lowest absolute value since 1995.[6] Moreover, the composition of these imports is likely to be comforting to Americans who examine it; 49% of crude oil imports and 89% of gas imports originate from America's neighbors.[7]

The concept of energy security has been modified in recent years to include additional dimensions.[8] Some rightly point out that consumers are not the only ones concerned with energy security; security of demand is as legitimate a preoccupation as security of supply to a country which reaps a significant portion of its national revenues from energy exports. Moreover, security of transit is a concept important to both consumers and producers. Secure infrastructure and transport routes are essential if accessible, affordable energy is to make it to its destination. Again, on these counts, America's energy boom seems to check the energy security box, and oil from Canada is certain to be less vulnerable to disruption than supplies that need to snake their way through the Bosporus or the Straits of Hormuz.

However, an even more sophisticated definition of energy security could go beyond these ideas to posit that being energy secure means having access to affordable energy *without* having to contort one's political, security, diplomatic, or military arrangements unduly. Is a country really energy secure if obtaining adequate energy supplies is dependent on a particular expensive, high risk, and limiting (in terms of opportunity cost) posture in the world? By this standard, recent energy developments may or may not make America fundamentally more energy secure. If such energy trends enable it to scale back its military presence in the Middle East, or allow it to be less vulnerable to political shocks in other parts of the world, or grant it a substantially freer hand in the pursuit of other foreign policy goals, then such energy developments would make America more energy secure in the fullest sense of the concept. While it is conceivable that the revolution in American energy could have these effects, it is not yet obvious this will be the case.

The purpose of this chapter is to look at the broader interplay between energy and security and the multiple ways in which the two concepts interact. It may be more useful to think about the "geopolitics of energy" rather than "energy security," with all its current associations. The chapter will offer a framework for thinking about the overlap between energy and security as it relates to "hard" national security issues, not economic or environmental ones. Because of the nearly infinite number of contemporary issues that inhabit this intersection, the chapter draws primarily from oil and gas in its illustrations, given the dominance of those fuel sources in the global energy mix. This is by no means to deny the importance of security issues associated with other energy sources, such as nuclear energy, or the possibility of complex interactions between security and renewable energies.[9]

Energy, Security, and the Grand Strategy of Countries

In exploring the interaction between energy and hard security, grand strategy is an appropriate place to begin. Such a lens offers a broader perspective on how energy and security are intertwined; rather than simply asking how countries meet their energy needs, a grand strategy framework reveals how energy factors into a whole host of interactions between countries, actors, and global institutions.

A grand strategy is an all-encompassing concept which guides a country in its effort to combine its instruments of national power in order to shape the international environment and advance specific national security goals. It is generally considered to have three parts: a vision of a desired outcome or set of objectives (ends); instruments or tools (ways) by which these goals are pursued; and the resources (means) available to apply to the effort. Good grand strategies offer a unifying vision to leaders, policy-makers, and citizens and help them prioritize inevitably scarce resources.

Not every country has a conscious grand strategy. Countries that do generally begin the formulation of such strategies by assessing the country's strengths, vulnerabilities, and the international or regional environment. As the basis for economic growth, energy will be at heart of virtually every country's evaluations. For instance, one of the drivers of a Japanese grand strategy would be the country's near complete dependence on external sources for energy; any subsequent political and military strategy would be crafted at least in part to secure this energy in the most reliable and least costly ways. In contrast, a country like Brazil would consider its energy resources to be a huge strategic asset and would likely develop a grand strategy which relied heavily on its energy for advancing other, non-energy objectives in the international and regional domains.

In short, energy can be and often is a key driver in each of the three components of grand strategy: ends, ways, means. Most commonly, people think about energy being the *ends* of a country's action; what does a country need to do to meet the objective of securing sufficient energy at affordable prices? But energy also has a profound influence on the *ways* of a grand strategy; energy is often the vehicle or tool by which a country achieves its non-energy objectives. And finally, energy – and the revenues it brings – can provide the *means* for pursuing non-energy goals. Examining how energy plays these very different roles in the formulation and execution of grand strategy helps us make better sense of the many ways in which energy is shaping the international landscape of today.

Energy as an End/Objective of Grand Strategy

Securing adequate energy supplies at affordable prices is often the end or objective of the grand strategies of consuming countries.[10] The energy imperative is so central to the prosperity and, as a result, the stability of countries that they will often use whatever instruments are at their disposal to ensure their energy security is met. For most countries, their tools are limited and they primarily need to rely on the global market to deliver them energy; their grand strategies may therefore revolve around generating enough foreign currency to purchase needed oil and gas or other forms of energy. But some countries have considerable tools and resources to extend to the pursuit of energy. For these countries, often elaborate political, diplomatic, economic, and military strategies are employed to ensure the energy goal is met.

Blood for Oil?

Many people think that countries often use military force or wage wars in order to meet their energy needs.[11] World War II offers many powerful examples, as the Axis powers believed that their energy needs would only truly be met by physical control of oilfields. This notion was behind Hitler's push to Baku and Japan's invasion of Borneo in southeast Asia.[12]

Since that time, however, relatively few international (as opposed to civil) wars have been fueled by the desire to gain *physical* control over energy resources.[13] Had the Soviets used Afghanistan as a launching pad to seize Iran's oilfields, as the US feared it would, that effort would be a notable exception. Iraq's invasion of Kuwait in August 1990 is perhaps the best contemporary example of a country waging war to control resources; Saddam Hussein wished to terminate the alleged Kuwaiti practice of slant drilling across the international border into the Rumaila oilfield.

Some, including many Iraqis, would attest that the two Iraq wars in 1991 and 2003 were motivated by America's desire to have physical control over Iraq's vast oil resources. Many allege that the US invaded Iraq either to directly control Iraq's oil or to bring in American companies to develop and dispose of Iraqi oil. Others deny that the Iraq wars had anything to do with oil.[14] In parsing these positions, it is important to distinguish between *commercial* and *strategic* interests, a distinction rarely made in the heated debates about "blood for oil."[15]

There is little evidence that the US was motivated by commercial interests – the desire to either directly control Iraq's oil or to bring in American companies to develop and dispose of Iraqi oil – in launching the first Gulf War in 1991. Had this been a primary driver, the US-led coalition would have not quickly relinquished the territory it gained in the push into southern Iraq, which is among the most oil saturated regions in the world.[16] Moreover, had commercial interests in oil been paramount, the US-led coalition might not have ceded sovereignty back to the Kuwaitis with virtually no conditions or understandings related to their vast oil resources. (With very few exceptions, Kuwait today still does not allow foreign involvement in the development of its oilfields.)

However, one can make a very strong case that *strategic* interests related to oil were a major factor in the decision to use military force to oust Saddam from Kuwait. Once Iraq had invaded Kuwait and seized its oilfields, Saddam had 19% of the world's oil reserves under his control. Had he continued his push into Saudi Arabia – as many suspected he might at the time – Saddam would then have had control of 44% of the world's oil reserves. The strategic implications of this situation for the US and the world would have been significant. Such control would shift the regional balance of power toward Iraq and away from other states more aligned with the US, at a time when the collapse of the Soviet Union was creating new and uncertain dynamics. It would also have given Saddam the ability to blackmail other Arab states, to threaten Israel, and to destabilize international oil prices, with all the consequent effects on the global economy. For these reasons, while the US may not have had commercial interests in securing plum positions for its companies in Iraq or Kuwait, it did have clear strategic interests in prying Saddam's grip off Kuwait and deterring Saddam from invading Saudi Arabia.

Similarly, commercial interests do not appear to have been a significant factor in the decision to invade Iraq in 2003. In fact, had access to Iraq's oil been America's primary preoccupation, there were many quicker, less costly ways to achieve it. For years before the 2003 war, Saddam had been negotiating lucrative production sharing agreements with national oil companies such as Russia's Lukoil and China's CNPC, "pending the lifting of sanctions." Given Saddam's demonstrated willingness to trade access to Iraq's oil wealth for the lifting of sanctions, the US could have leveraged its position in the UN Security Council and negotiated key oil agreements for US oil companies in exchange for acquiescing to an end to sanctions in the 1990s or early 2000s – as other members of the UN Security Council did.

The approach of the Coalition Provisional Authority (CPA), the US-led body governing Iraq under the occupation, to oil matters is further testimony to the negligible role that commercial interests played in spurring the 2003 invasion. While the CPA supported and oversaw the drafting of regulations by the Iraqi Governing Council (the CPA's Iraqi counterpart) to change laws on foreign investment and the ownership of private property, it steered clear of energy and oil, deciding that the disposition of these resources should await the decision of a legitimate, elected Iraqi government. This near allergy to aligning the US too closely to Iraq's oil policy is the outcome of the 2009 and 2010 bid rounds, which eventually did bring foreign companies into Iraq to develop oilfields under very strict terms; rather than "cleaning up," Chinese, Malaysian, European, and other companies fared much better than US ones.[17] This result is consistent with a US hands-off attitude toward Iraqi oil, but not necessarily proof of it; it is at least conceivable that US authorities simply failed in their efforts to exert influence.

Like the first Gulf war, there is stronger evidence that strategic considerations related to oil – as opposed to commercial ones – played a role in the decision to invade Iraq in 2003. But even here the linkage is not as strong as with the previous war. While in 2003, Saddam's oil resources were limited to those within Iraq's borders, they were still more than adequate to fuel what was perceived to be Saddam's nefarious and threatening behaviors, particularly the pursuit of weapons of mass destruction. As time progressed and international support for sanctions wavered, Saddam's ability to circumvent the sanctions heightened US concerns that Saddam was amassing sufficient resources to advance a destabilizing agenda in the region.[18] Moreover, while oil was not the impetus for war, its existence enabled war proponents to dismiss one possible impediment to the war: cost. Key figures in the Bush Administration downplayed the costs potentially associated with unseating Saddam by claiming that Iraq's oil wealth meant the country could pay for its own reconstruction.[19]

Conflict as a By-Product of Competition over Resources

The pursuit of energy as an end or an objective can also lead to conflict in less direct ways. In several contemporary cases, the quest to meet a country's energy needs is creating tensions which could be precursors to violent conflict. Here, war is not the strategy employed to actually secure energy resources, but armed confrontation is the possible by-product of other sorts of strategies countries employ to secure energy. In the terminology of today, conflict or even war could result from a global resource scramble, particularly if countries believe the world is closing in on the limits of its energy endowment.

The current situation in the South China Sea is a good case in point. This vast area is of strategic interest to the world and, in particular, to the six Asian countries of Brunei, China, Malaysia, Philippines, Taiwan, and Vietnam for two primary reasons. First, the South China Sea and the adjacent Straits of Malacca are key global oil transit waterways. In 2009, 30% of the world's shipped oil passed through the Straits of Malacca on its way to meet burgeoning Asian demand. Tankers move approximately two-thirds of South Korean energy supplies, nearly 60% of Japanese and Taiwanese energy supplies, and close to 80% of Chinese crude oil imports through the South China Sea.[20] Second, the South China Sea is believed to house significant oil and gas deposits, although territorial and other disputes have prevented good seismic studies which could yield more accurate data. Estimates for oil in the South China Seas range from 28 billion barrels to 213 billion barrels.[21]

The combination of these two strategic interests and longstanding territorial and maritime disputes in the area create a potentially dangerous stew. China claims the largest portion of the South China Sea and has used its military and paramilitary to intimidate other countries seeking claim over different areas of it. Each country has declared the

South China Sea a national priority and many are investing heavily in maritime military assets to advance and protect their interests. While China, the Philippines, and Vietnam have locked horns over competing claims since 1974, the intensity and the stakes associated have risen considerably since the prospect of significant hydrocarbon resources has come into play. In 2011, three Chinese vessels cut the survey cables of a Vietnamese ship exploring for oil and gas; in the same year, Chinese patrol boats allegedly challenged Filipino exploration vessels 250 km off the coast of the Philippines, with Manila responding by sending out two military aircraft.[22]

Some people look at these trends – and the burgeoning Asian demand for energy – and anticipate inevitable conflict. Robert D. Kaplan believes that a conflict with China is likely "if not a big war with China, then a series of Cold War-style standoffs that stretch out over years and decades."[23] Military confrontation or war, however, is not inevitable. No parties in fact have an interest in a military confrontation which would jeopardize the smooth passage of critical energy resources, not to mention significant and increasing amounts of trade between countries.[24] Moreover, given the long time lag to develop the energy resources, all parties have at least notionally an interest in resolving disputes in a way which allows for more aggressive exploration and development in the South China Sea. Conflict, however, may result not as a product of a deliberate strategy to assert dominance over resources, but as the upshot of miscalculation or miscommunication in an increasingly militarized arena.

Impediments to Preventive or Punitive Action

Finally, the pursuit of energy as an end of a grand strategy can have significant security implications by undermining or impeding the ability to address important, non-energy national security dilemmas. It is in this domain that China's approach to Africa is most relevant. Over the past 20 years, several factors spurred Beijing to reconsider its approach to securing energy. First, Chinese energy demand began to skyrocket, leading China to shift from being a net exporter of oil to a net importer in 1993. Since that time, China's dependency on foreign oil imports has increased, reaching 55% in 2011. Second, a souring Chinese relationship with the US in the 1990s, capped off by the failed Chinese bid to buy UNOCAL in 2005, led China to conclude that it could not rely entirely on the market mechanism to meet its energy needs, particularly in a time of future conflict.

Over the last decade, China developed and executed an approach which has come to be known as its "going out" strategy. Under this approach, China has sought to secure direct access to oil supplies through equity interests obtained by powerful Chinese national oil companies (NOCs). Unlike Western nations, whose official developmental and commercial interests are not coordinated, China has increased the appeal of commercial bids by Chinese NOCs by coupling them with government-to-government support packages. In many cases, granting Chinese NOCs the rights to an oilfield have at the same time brought the host government much-needed loans, financing, direct investment, and infrastructure development.[25]

This approach has generated significant controversy. Chinese engagement in Africa has fueled a construction boom of railroads, highways, and ports being built at unprecedented rates. But the "going out" approach has also raised legitimate concerns that China is undermining the efforts of Africans and the international community to promote governance, transparency, and accountability. Unlike the US and international institutions, China imposes no political conditions on the assistance which accompanies its commercial bids to develop Africa's resources. Not surprisingly, some African countries have foregone financial assistance from international institutions given the option of securing the same from China without any conditions: in 2004, the offer of generous, no-strings-attached Chinese aid undercut IMF efforts to advance economic and political reform in the wake of the civil war in Angola.[26] Other objections to China's "going out" strategy include complaints about the poor quality of Chinese infrastructure projects or the influx of Chinese labor to African countries.

But our real concerns lie in the security realm. Does China's "going out" approach in Africa create serious *security* challenges? Some critics worry that China is "locking up" Africa's natural resource wealth and that its aggressive approach will escalate US–China resource rivalries as both countries seek to develop more oil and gas reserves.[27] These fears are almost certainly exaggerated. First, China's absolute stake in Africa's oil is still comparatively small. According to a 2007 Wood Mackenzie study, China's oil companies are responsible for only 3% of total energy investment on the continent; Chinese production in Africa in 2006 was only one-third of that of Exxon Mobil's and only 7% of that of the Algerian NOC, Sonatrach, which is the largest energy producer on the continent. In addition, until recently, Chinese NOCs have tended not to compete for the same concessions that Western major oil companies were pursuing; rather, China has sought to develop oil in places that Western majors forewent due to high political risk or small anticipated payoffs.[28] The privileged financing that Chinese NOCs receive from their government has enabled them to develop resources that others would likely find too risky or insufficiently profitable. And, finally, oil developed by Chinese NOCs is not being spirited away to Beijing, but sold to China and many other buyers on the open market.[29] These realities suggest that competition for resources in Africa is unlikely to spur superpower conflict unless these patterns change significantly. Rather than "locking up" Africa's oil for itself, China is effectively adding oil to the global supply that would otherwise not be developed – a service to the tight international market and its customers.[30]

More valid are the security concerns that arise from the immunity that Chinese involvement gives some African countries from international pressure. Not only does China's mode of doing business provide countries with a way to skirt the governance conditions of international institutions, it also can neuter international efforts to address genocide, repression, and civil war. China's interests in maintaining its beneficial energy arrangements in certain countries have protected them from UN censure and in some cases sanctions and stricter action. For instance, in 2004, international efforts to condemn and penalize Sudan were significantly weakened at the insistence of the Chinese government, which many assumed was seeking to defend its significant investment in southern Sudan.

Chinese efforts to shield African countries from the censure of the international community are consistent with a longstanding Chinese policy of non-interference in the domestic affairs of other countries. This approach, thus far, has been a nice complement to an energy strategy and energy activities which sometimes have the effect of lending support to recalcitrant regimes. One can imagine, however, that over time, China's international energy strategy and its policy of non-interference could come into tension with one another. Political instability and civil conflict represent one of the greatest likely constraints on increased oil production in the years ahead. China could find itself, inadvertently, as having a major interest in the diffusion of domestic conflicts – and move to use its leverage inside a country to advance outcomes which facilitate the flow of oil.

Energy as a Way/Tool of Achieving Security Objectives

Countries endowed with substantial energy resources are bound to seek to use these assets in the quest to shape the global environment to their benefit. For many countries, energy is the main vehicle for promoting and protecting their interests; it is the way or instrument deployed in their grand strategies.

Energy as a Political Weapon?

Leaders, policy-makers, and consumers look back on history and fear the use of the "energy weapon" by producing countries. While the 1973 Arab oil embargo provides a sharp memory of such a case, a more dispassionate evaluation suggests that there are limited circumstances under which a country can successfully use energy as a political weapon today. The ability of a producing country or group of countries to exact political concessions by withholding energy is tempered by the state and nature of the international market for that commodity, specific characteristics of that country, and the willingness of the producer to also experience pain.

History is replete with instances in which producing countries have sought to advance specific political goals by terminating or curtailing oil sales to a particular country. In 1956, Saudi Arabia cut off sales to the UK and France, and continued its embargo of Israel, in response to these three countries seizing the Suez Canal after Egyptian nationalization of it; none of the countries, however, experienced supply shortages. In 1967, in an effort to get Western countries to cease their support for Israel in the Six Day War, Arab countries froze oil exports primarily to the US, Germany, and the UK. Again, none of the targeted countries suffered diminished oil supplies. However, in October 1973, at the start of the Yom Kippur War, the Organization of Arab Petroleum Exporting Countries (OAPEC) embargoed oil sales to the US, The Netherlands, Portugal, and South Africa in response to the US decision to restock Israel's depleted armoury. International prices rose dramatically and the global economy lurched into a severe recession which ushered in a period of pernicious "stagflation."

What accounts for the differences of effect in these three seemingly similar instances? In the first two instances, producing countries sought simply to deny particular countries access to their oil. But given that the market for oil is a global one, and that oil can be transported easily, countries denied oil from one source can generally replace it with oil from another as long as quantities in the overall market remain the same. Some difference in the quality of oil and the need to match oil with refineries may slow down or frustrate this process, but in general the global market redistributes oil efficiently. In 1956, moreover, the US was still a net oil exporter, further facilitating the redirection of oil to the UK, France, and Israel, once President Eisenhower had registered his disapproval of their intervention.[31] In contrast, in 1973, OAPEC coupled its embargo to specific countries with progressive cuts in production. Moreover, the state of the market in 1973 differed significantly from early years. Burgeoning demand and slower gains in supply were already contributing to a tightening oil market before the embargo. These combined factors forced the dramatic increase in global price from \$3.29 a barrel in 1973 to \$11.58 a barrel in 1974. Poor economic policies and efforts to ration gasoline are what drove the long gas lines in the US, not real shortages.

These episodes suggest the difficulty of targeting specific countries to undertake specific changes to their foreign policy or national security policies. In fact, in all three instances, there is little to suggest that the embargos did force the desired political changes.[32] Yet, they do underscore how producer countries can wield influence more generally by forcing up the global price of oil through production cuts. At the same time, however, they also caution that such a generalized attack on the global economy can boomerang on the instigators as it did in the 1970s. Rather than reaping huge profits from the sale of high-priced oil indefinitely, after several years of high oil prices the revenues of oil producers plummeted as the global economy was sent into recession, forcing down demand for energy and with it the price of oil. In addition, the oil crises of the 1970s spurred consumer countries to develop new institutions to mitigate the dangers which had been realized in the 1973 embargo and the subsequent Iranian revolution. In establishing the International Energy Agency (IEA) for greater coordination among consumers and launching strategic petroleum reserves, the OECD countries sought to – and to some extent did – inoculate themselves from future politically inspired oil crises.

These cases suggest that three factors seem to matter most in determining the prospects of using oil as a foreign policy or national security weapon: the state of the oil market, the willingness of producers to curtail production rather than just re-route it, and the risks producers are prepared to take to their own well-being. When we assess these factors in light of today, we can feel greater confidence that the scope for the successful use of energy as a political weapon is relatively limited. Oil markets could tighten in the foreseeable future, suggesting the potential for disruptions to cause sharp price spikes, but producer countries on the whole are sensitive to the notion that oil which is priced too high can be against their interests, if it drives the global economy into recession or galvanizes importing countries to find a substitute for oil. This realization not only mitigates the chances that oil will be used as a political weapon, but encourages Saudi Arabia and others to seek to moderate the global price of oil through calibrating their overall production. Middle Eastern producers are likely to be more sensitive to protecting their revenue streams today more than ever; political challenges within their own societies demand higher levels of social spending in order to placate domestic demands and protect regime stability.

One notable modern exception may be Iran, and its threat to wreak havoc on the global oil market in response to international pressure on it to abandon its alleged pursuit of a nuclear weapon. On various occasions over the last year, Iranian officials have claimed a willingness either to terminate its oil exports or to close the Straits of Hormuz, though which 17 mbd flowed in 2011, accounting for almost 20% of the global crude oil trade. While Iran could decide to take either course of action, both would come at significant economic costs to itself, with only a meager prospect that the pain endured would lead to a superior political outcome for Iran. In taking its 2.2 mbd of exports off the international market, Iran would be foregoing the revenues which make up 50–80% of its budget. At a time when Iran's economy is already suffering under international sanctions, the loss of this revenue could spur political unrest.[33] Closing the Straits of Hormuz would also effectively take Iranian oil (and most of the Gulf's oil exports) off the global market, with the same potentially destabilizing effects. In the first case, Iran's removal of its oil from global markets would only cause foreign countries significant pain (and stand the chance of exacting political change) if other producers, such as Saudi Arabia, were unable to increase production to substitute for the lost crude. In the Straits of Hormuz scenario, while a sharp increase in the price of oil would almost be inevitable, the US would likely clear the Straits in short order, making the spike relatively transitory and decreasing the chances that foreign countries would lift pressure on Iran as a result.

In contrast, producers of natural gas may be – at least theoretically – better positioned to use their resource as a tool for advancing political goals than their oil-producing brethren. Unlike oil, there is not yet a global market for natural gas.[34] International sales of natural gas are still conducted largely in three segregated markets (North America, Europe, and Asia), so a disruption in one market is less easily allayed by shifting resources from other global producers. Moreover, whereas oil can be transported via pipeline, ship, rail or even truck, the transport of gas requires huge infrastructure developments, either in terms of pipelines or equipment to liquefy and/or regasify natural gas, which adds greatly to the difficulty of switching seamlessly between suppliers of gas.

This theoretical ability to exercise energy clout is nevertheless limited by the interdependency between gas producers and consumers. Like oil producers, gas suppliers run the risk of damaging their own interests if the political use of energy leads their customers into economic depression, the establishment of alternative supply sources or the development of other resources to meet their energy needs. The major infrastructure requirements of gas trade not only make it hard for consumers to quickly find other avenues to meet their needs, they also limit the extent to which producers can shift gas to other customers. However, the growing trade of liquified natural gas and the rise of the spot market for gas will likely create greater opportunities for substitution, further weakening the ability of producers to use gas sales as a political weapon.

Although often seen as an example of the effective use of gas as a political weapon, Russia's cutoff of gas to the Ukraine actually demonstrates both the possibility of wielding political influence in the gas trade and the limits to doing so. Russia and the Ukraine began to head toward confrontation in 2005 after the Orange Revolution brought to power a westward looking government in Ukraine. Ukraine started to leverage its position as Russia's main transit country for gas exports to Europe and demanded higher transit tariffs.[35] At the same time, Ukraine's turn toward the West underscored the limited value Russia was getting from providing subsidized gas to Ukraine, leading Moscow to re-evaluate this policy and demand that Ukraine begin to pay full market prices. Subsequently, the two sides engaged in several tests of strength which reached a climax in January 2009 when Russian gas exports to Europe stopped for 15 days. The exact flow of events has not yet been fully revealed, with Russia accusing Ukraine of theft and Ukraine accusing Russia of cutting off gas supplies. The consequences were very real, as southeastern Europe found itself deprived of gas in the midst of winter. Lacking alternative supply sources, the countries were literally left out in the cold.

Whether this and other gas cutoffs served Russian interests is, however, debatable. In almost every instance, Russia was able to renegotiate the price at which the importing country bought Russian gas.[36] In the case of Belarus, Russia was also able to acquire greater interest in the country's energy infrastructure, giving it further control, in exchange for continued discounted gas. Almost by definition, the gas cutoffs also had an alleged desired political affect: inflicting punishment for revolutions which moved the countries' politics closer to the West. But in few cases did the gas embargoes reverse political changes and in one case – Georgia – the cutoff actually inspired the target country to diversify its suppliers away from Russia. The full effects of these episodes are still to manifest themselves, as the cutoffs to some extent undermined Russia's reputation as a reliable supplier and inspired greater European action to diversify Europe's supplies.

The use of energy as a foreign policy weapon, as mentioned above, is generally associated with producers. But, in some instances, consuming countries have also sought to leverage their buying power to induce behavior changes on the part of energy producers through the use of sanctions. In many respects, the ability of a country to successfully parlay its willingness to forego energy imports from a particular source into foreign policy influence is the flip side of a producer's embargo. Just as an embargoed consumer can find oil from other sources due to the global nature of the oil market, so too can a sanctioned producer generally find other outlets for its oil. In the 1990s, the US banned the import of oil from Libya, Iran, and Sudan; for the most part, these countries simply replaced the US customer with other buyers.[37] The state of the market – whether it is a buyer's or a seller's market – also affects the ease with which a producer can reposition itself in the global market. For these reasons, sanctions have only really seemed to bite when a large number of consumers band together to reject the purchase of oil from one source; even multilateral sanctions have generally been ineffective in stopping the flow of oil unless they include all the world's major consumers.

New sanctions innovations, however, are testing this age-old maxim that sanctions only work when they are multilateral. Sanctions on oil imports *do* appear to harm the producer, even if they are not universal, *if and when* they are coupled with other measures that frustrate the ability of non-sanctioning countries to purchase the target's oil. The complex array of UN, US, and EU sanctions do not amount to a comprehensive oil embargo on Iran, but the financial measures included in the sanctions regime have created real pressure on companies to stop purchasing Iranian oil, even if it is not official policy of their host country. "Extraterritorial" sanctions threaten to exact a price on third parties for doing business with Iran or entities like the Iranian Central Bank. When forced to choose between processing oil transactions through the Iranian Central Bank and losing some forms of access to the US financial system, many banks have steered clear of Iran. Others simply do not want the hassle, risk, or costs of ensuring they are complying with complicated restrictions, so opt out of handling trade or transactions with Iran altogether.

The Iran case also demonstrates how policy-makers in oil consuming countries are becoming more adept in using the leverage they have, without harming their own fundamental energy interests. Few worried about the impact on the global oil market of the sanctions placed on Iraq, Libya, Iran, and Sudan in the 1990s. With the price of oil dipping to a *real* price of under $10 a barrel in 1998, the prospect of taking some production off the markets was a positive one. In 2012, a tight market and high geopolitical risk has led policy-makers to seek ways of imposing sanctions on Iran *without* harming global oil supply and risking major price spikes. They have, as a consequence, constructed a sanctions regime under which – by limiting but not *eliminating* the number of countries willing to buy Iranian oil – prospective buyers can demand steep discounts on the price they pay per barrel. Ideally, Iran's oil stays on the market, but the Iranian government gets less revenue than it otherwise would.

Energy as a Cement in Alliances

Finally, energy is not only used as a tool of foreign policy in cases where it is wielded as a political weapon; it is also frequently used to shore up alliances and to build support for certain ideologies and national security positions. There are multiple instances in which an energy producing country has provided free or deeply discounted energy exports in order to keep another country in its orbit or sphere of influence. The Soviet Union provided huge energy subsidies to members of the Warsaw Pact during the late 1970s and early 1980s; similarly Russia continued the provision of subsidized energy to the Commonwealth of Independent States after the collapse of the Soviet Union. Saddam Hussein's Iraq also provided cheap oil to Jordan throughout the 1980s and even in the 1990s under the UN Oil-for-Food program. Current examples include Venezuela's provision of cheap energy to Bolivia and Cuba; such transfers are geared to shore up support for President Chavez's anti-American, revolutionary stance in Latin America.

While the motives of these suppliers in providing energy benefits to their near neighbors seem clear, their track record in cementing alliances is more questionable. While such energy largess appears to have been successful in buying support in international forums, it has a mixed record of securing the loyalty of countries to the supplier when national interests have been at stake. Jordan sat out the Gulf War, at least in part due to the fact that it benefited economically from Saddam's rule, while Yemen sided with Iraq against Kuwait in the first Gulf War *despite* the fact that Yemenis had received cheap energy from Kuwait and Saudi Arabia before Saddam's invasion.[38]

Energy as a Means/Resource for National Security Strategies

The final way in which energy can be a component of a grand strategy – and have major consequences on security issues as a result – is when it provides the means or resources with which countries can advance their foreign policy or national security interests. In short, energy sales in many instances provide large proportions of a country's revenues and national budget; the fact that there are rents associated with the extraction and sale of oil and gas makes these revenues particularly significant. Without these revenues, many countries would not have the financial resources to project power as they do today.

Of course, high revenues from energy exports are not in themselves a determinant of destabilizing international or domestic behavior. In 2011, Canada reaped 25.4% of its export revenues, and close to 7% of its GDP, from the energy industry, and remains a model democracy and international citizen. In Australia, mining and energy exports accounted for 72% of total export revenue in 2011, yet there are no allegations of adventurism abroad or repression at home against Australia. Norway, Brazil, the UK, Mexico, Angola, and China are all in the top 15 oil-producing nations of the world and not one of them is a major instigator of global disorder.[39]

Nor does a country have to have high energy revenues to repress its citizens or create international security woes. The annual *Country Reports on Terrorism* produced by the US Department of State listed Cuba, Iran, Sudan, and Syria in 2010 as state sponsors of terrorism under a designation developed in 1979. Cuba is dependent on external sources for energy; Sudan is now an energy exporter, but has been on the list since 1993, years before it began to export oil in 1999; Syria exports small amounts of oil, but is expected to become a net importer soon. Iran, of course, is a major exporter. North Korea, which produces virtually no energy, had been on the terrorism list from 1998 until 2008 when it was taken off in the context of negotiations about its nuclear program. Similarly, countries which have challenged the global nuclear non-proliferation regime are on the whole *not* energy powerhouses: North Korea, Pakistan, India, Israel, and – the exception – Iran.

While energy wealth is not necessarily a sole determinant of destabilizing behavior, there is clearly a subset of countries whose propensity to engage in such behavior is fueled by their energy wealth. What determines whether a country uses its energy revenues to fund such activity is still a matter of some debate. On the international side, what appears to matter the most is if the country has a revolutionary or expansionist ideology.[40] Without oil revenues, the current regime in Iran would still seek to extend its influence regionally and globally, but it would be far less successful in doing so. Estimates suggest that Iran has spent between $100 and $200 million a year supporting Lebanese Hezbollah and, until recently, an equal sum supporting Hamas.[41] No estimates exist about the level of support provided to Iraqi Shi'a militia, but it is likely to be substantial. Iran's suspected efforts to pursue a nuclear weapon may have cost it close to $1 billion thus far.[42] And, finally, Iran's oil revenues have helped the country weather international pressure and sanctions geared to sway the regime from attaining a nuclear capability. On a lesser scale, Venezuela's oil wealth has enabled Chavez to promote candidates sympathetic to his "Bolivarian" revolution across Latin America; he has provided financial support for the elections of like-minded politicians in Bolivia, Peru, Ecuador, and Paraguay. Oil revenues also have enabled Chavez to send financial support in the past to the FARC, the group seeking to destabilize neighboring Columbia.

There are also concerns that energy revenues enable and encourage the development of repressive regimes. Rentier states, as they are known in the academic literature, are ones which use the revenues from oil and gas (and other natural resources) to construct societies beholden to and/or repressed by the ruling group. Given that rents or energy revenues can absolve a government from imposing taxes on its population, the citizenry have little recourse to demand accountability from its rulers. Undemocratic regimes result, often with robust security apparatuses to ensure that any dissent which is not co-opted is crushed. Again, not every energy exporting country becomes a rentier state, but a subset of energy exporters clearly manifests some or many characteristics of one. Timing of institutional development, political culture, and diversity of economy seem to be some of the additional factors at play in determining whether an energy exporter also becomes a rentier state.[43]

Conclusion

This chapter has sought to provide a framework for thinking about the wide variety of ways in which energy and international security overlap, as well as to illuminate many contemporary issues inhabiting this intersection. It has explored how energy shapes and influences every component of a grand strategy that a country might develop. Energy may be an ends to a grand strategy, shaping political, military, diplomatic, and economic strategies created by leaders who seek to provide secure energy resources at reasonable prices to their economies and constituents. Energy can also provide the ways or the tools through which countries seek to advance their non-energy goals; countries use *either* their energy production or their energy demand to try to shape the international arena in a way which is most conducive to their broader national interests. And energy can also provide the means – or revenues – for countries to pursue particular foreign policy or domestic agendas which can have international security implications.

In spurring us to adopt a lens much broader than the traditional "energy security" one, this grand strategy framework has encouraged us to reconsider some conventional wisdoms about energy and international security. While energy resources can be the impetus for conflict, they may do so more by empowering expansionary states than by enticing others to invade. China's "going out" policy in Africa – an example of energy serving as an ends of grand strategy – may create security problems, but not the ones generally anticipated. Rather than sparking conflict over scarce resources, China's activities create potential security issues by insulating energy producing regimes from pressure to curb their own destabilizing behavior. Energy can be, and is, used as a political weapon, but the circumstances under which such an endeavor can be successful are fairly limited and specific, and the risks to the instigator quite high.

The grand strategy framework used also gives new meaning to what is construed as global energy policy. Nothing in this chapter diminishes the importance of international cooperation around policies related to managing oil stocks to buffer economies from price shocks, minimizing CO_2 in the environment, pursuing alternative sources of energy, or engaging in transnational energy projects from the Nabucco gas pipeline to the massive solar and wind effort envisioned by the conceivers of Desertec. This chapter, however, reveals how a country's "energy policies" go far beyond those generally considered to pertain to energy. Changes in political, military, diplomatic, and economic policies and strategies directly relate to the energy challenges and opportunities which shape the global energy landscape.

Given that oil and gas account for more than half of the energy used today in the world, it is no surprise that the predominance of energy related security issues connect to oil and gas. As this chapter has shown, the specifics surrounding oil and gas – their finite nature, their association with rents, the particular characteristics of oil and gas markets – are important factors in shaping the particulars in the interaction between energy and security. For instance, while gas producers may have better luck than oil producers in using their resource as a political weapon due to the lack of a global market and the massive infrastructure investments involved in gas trade, both are limited by mutual dependencies between producer and consumer.

Recognizing how the unique characteristics of oil and gas shape today's security issues should prompt us to think about the energy transitions which lie ahead. Although the shift away from fossil fuels in the global economy is and will continue to be slower than many desire, it is also inevitable. At some point in the future, likely out of both necessity and ingenuity, the dominance of fossil fuels will give way to one or more new energy sources. The shift to these new energy sources – whether solar, wind, biomass, nuclear or something we cannot yet imagine – will bring with it its own host of peculiar security issues. The challenges in anticipating this new landscape are further complicated by another related transition: the emergence of a more multipolar energy world, where emerging economies are now the drivers of energy demand and other global trends.

As oil and gas diminish in the global energy mix, the international security dilemmas associated with their use will gradually become less relevant to understanding global politics and security. In their place will arise a new set of factors, which will similarly shape the ends, ways, and means of grand strategies. Countries will continue to pursue energy as an end of their strategies, to utilize energy as a way or instrument of exerting influence on the international stage, and to use the resources that the energy trade provides to fund their foreign policy and national security strategies. Today, policy-makers, academics, businesses, environmentalists, and consumers are all focused on *how* the world can make the transition away from fossil fuels. While this question demands our urgent attention, we should also be anticipating how such a shift will change the world in which we live – in ways far beyond simply how we fuel our cars or make our electricity, but also in the sorts of international security challenges we will face.

Notes

1. I am grateful for the hard work and excellent insights of Kaweh Sadegh-Zadeh in helping me research this chapter. I also appreciate the comments of Andreas Goldthau on earlier drafts.
2. In 2005, energy companies were engaged in elaborate plans to construct more than 40 liquefied natural gas import terminals in the US, each with a price tag of $500 million to $1 billion. See Romero (2005).
3. American "tight oil" production has increased significantly in the last decade. Production grew from 10,000 bd in 2003 to 900,000 bd by the end of 2011. The Bakken formation in North Dakota and Montana, and the Eagle Ford in south Texas, account for 84% of total US tight oil production. The energy consultancy Wood Mackenzie expects tight oil production to hit 2.5 million bd by 2015.

4. See Morse *et al.* (2012).
5. The diminution of imports is also due to decreased economic activity in the last few years and increased efficiency over time. Yergin (2012) talks of the US (and other OECD countries) having reached "peak demand."
6. See Nerurkar (2011).
7. This 49% number includes imports from Venezuela, which make up 10% of overall imports. Put another way, the US receives 39% of its crude oil imports from friendly, stable neighbors.
8. See Yergin (2006).
9. See for instance, Evans and Kawaguchi (2009: 124–146); Lovins *et al.* (2008); Deutch *et al.* (2003).
10. Security of demand could also be an objective driving the foreign policy or national security strategy of producing countries. For instance, due to European efforts to diversify gas imports sources away from Russia, the Kremlin started to look to China as an alternative market. Energy is today one of the driving forces of cooperation between Russia and China. Although it is too early to determine, part of the motivation of Qatar in pushing for intervention in Libya could relate to Doha's need to secure markets for its gas in the future. Qatar is now involved in Libya's energy reconstruction and such involvement may give Qatar the opportunity to integrate some aspects of the two industries, helping guarantee continued Qatari access to European markets.
11. A total of 32.7% of 6,909 US respondents in Zogby interactive poll conducted January 16–18, 2007 said Iraq's oil was a "major" concern; 23.7% said it was not a factor. http://www.upi.com/Top_News/2007/01/25/UPI-Poll-Oil-seen-as-factor-for-Iraq-war/UPI-37491169740800/, accessed 04/06/2012.
12. For full accounts of these decisions, see Kershaw (2008: Chapters 2 and 8).
13. See Jaffe *et al.* (2008).
14. In a July 10, 2003 press briefing, Secretary of State Colin Powell stated: "We have not taken one drop of Iraqi oil for US purposes, or for coalition purposes. Quite the contrary . . . It cost a great deal of money to prosecute this war. But the oil of the Iraqi people belongs to the Iraqi people; it is their wealth, it will be used for their benefit. So we did not do it for oil."
15. Alan Greenspan wrote in his memoir, "I am saddened that it is politically inconvenient to acknowledge what everyone knows: the Iraq war is largely about oil." But even this statement obscures whether Greenspan saw commercial or strategic interests – or both – as the drivers behind the war (Greenspan 2007: 463).
16. Iraq's southeastern cluster of super-giant fields forms the largest known concentration of such fields in the world, accounting for close to 80% of Iraq's proven oil reserves.
17. Only two US companies (Exxon Mobil and Occidental) were part of consortium that successfully bid for 11 oil projects made available in 2009 and 2010 bid rounds. See *Bernstein Research* (2010).
18. A 2004 GAO report estimated that between 1997 and 2002, Saddam accumulated $10.1 billion in illegal revenues. Approximately $5.7 billion resulted from illegal sales, while the remaining $4.4 billion was due to surcharges and kickbacks imposed on those who bought the oil. See Christoff (2004).
19. Ari Fleischer, White House Spokesman, said in a February 18, 2003 briefing: "Iraq, unlike Afghanistan, is a rather wealthy country. Iraq has tremendous resources that belong to the Iraqi people. And so there are a variety of means that Iraq has to be able to shoulder much of the burden for their own reconstruction." Paul Wolfowitz, Deputy Defense Secretary, in speaking to the House Appropriations Committee on March 27, 2003, stated: "We're dealing with a country that can really finance its own reconstruction, and relatively soon."
20. Kaplan (2011).
21. See EIA (2008).

22. Hookway (2011).
23. Kaplan (2005). See also Lieberthal and Herberg (2006).
24. Trade volumes between China and the members of ASEAN are expected to hit a new record of U$350 billion in 2012 (Shan 2012).
25. For an in-depth analysis of Chinese energy strategy, see Kong (2010).
26. See Campos and Vines (2008) and Corkin (2011).
27. For different views on the future of US–Chinese relations, see Kissinger (2011) and Freidberg (2011).
28. This pattern, however, shows recent evidence of changing, as Chinese NOCs have competed with Exxon Mobil and other international oil companies in places like Ghana and Nigeria.
29. See Downs (2007).
30. According to an IEA study, there is no evidence that the Chinese government is imposing any quotas on Chinese NOCs for equity oil to be sold to China (Jiang and Sinton 2011). Andreas Goldthau (2010) argues that while Chinese oil investment in Africa and elsewhere is good for the global supply, it has the adverse effect of using capital less efficiently than if market mechanisms were to drive investments rather than political opportunity.
31. For an account of the 1956 Suez crisis, see Yergin (2008).
32. In 1956 France, the UK, and Israel did withdraw from the Suez Canal, but largely on account of diplomatic and financial pressure from the US, not due to the oil embargo. Even in 1973, despite significant economic pain, the US did not abandon its support for Israel.
33. Sanctions, however, can also have the opposite result, known at the "rally around the flag" effect, in which external pressure allows a government to solidify its support.
34. A variety of recent developments are putting pressure on gas markets to integrate. See Deutch (2011).
35. Around 80% of Russia's gas exports were transiting Ukraine before the Nord Stream pipeline came onstream and opened up a new route to Europe.
36. For more on these cases, see Stern (2005); Boussena and Locatelli (2005); and Rutland (2008).
37. See O'Sullivan (2003).
38. For a fuller treatment of this issue, see Crane *et al.* (2009: 35–37).
39. See Energy Information Agency data on top world oil producers in 2010. http://www.eia .gov/countries/index.cfm, accessed 04/06/2012.
40. Jeff Colgan argues that revolutionary petro-states are far more inclined to initiate military conflict than non-revolutionary petro-states (Colgan 2010).
41. See Bruno (2011).
42. Crane *et al*, (2009).
43. A vigorous debate also exists about whether resource rich countries are doomed to be undemocratic. See Ross (1999).

References

Bernstein Research. 2010. The Herculean Challenge of Lifting Iraq's Oil Production. August 3.

Boussena, Sadek, and Catherine Locatelli. 2005. Towards a More Coherent Oil Policy in Russia?" *OPEC Review* 29, 2 (June 1): 85–105.

Bruno, Greg. 2011. State Sponsors: Iran. *Council on Foreign Relations*, October 13. http://www .cfr.org/iran/state-sponsors-iran/p9362, accessed 04/06/2012.

Campos, Indira, and Alex Vines. 2008. *Angola and China: A Pragmatic Partnership*. Washington, DC: CSIS. http://csis.org/files/media/csis/pubs/080306_angolachina.pdf, accessed 04/06/ 2012.

Christoff, Joseph A. 2004. *Observations on the Oil for Food Program and Areas for Further Investigation*. Washington, DC: Government Accountability Office. http://www.gao.gov/new.items/d04880t.pdf, accessed 04/06/2012.

Colgan, Jeff. 2010. Oil and Revolutionary Governments: Fuel for International Conflict. *International Organization* 64, 4: 661–694.

Corkin, Lucy. 2011. Uneasy Allies: China's Evolving Relations with Angola. *Journal of Contemporary African Studies* 29, 2: 169–180.

Crane, Keith, Andreas Goldthau, Michael Toman, *et al.* 2009. *Imported Oil and US National Security*. Santa Monica, CA: RAND.

Deutch, J. 2011. Good News About Gas: The Natural Gas Revolution and Its Consequences. *Foreign Affairs* 90 (February): 82.

Deutch, J., E. J. Moniz, S. Ansolabehere, *et al.* 2003. *The Future of Nuclear Power*. Cambridge, MA: MIT Nuclear Energy Study Advisory Committee. http://web.mit.edu/nuclearpower.

Downs, Erica S. 2007. The Fact and Fiction of Sino-African Energy Relations. *China Security* 3, 3: 42–68.

EIA. 2008. *South China Seas Analysis Brief*, March. http://www.eia.gov/countries/regions-topics.cfm?fips=SCS.

Evans, Gareth, and Yoriko Kawaguchi. 2009. *Eliminating Nuclear Threats: A Practical Agenda for Global Policymakers*. International Commission on Nuclear Non-proliferation and Disarmament.

Friedberg, Aaron L. 2011. *A Contest for Supremacy: China, America, and the Struggle for Mastery in Asia*. New York: W.W. Norton & Co.

Goldthau, Andreas. 2010. Energy Diplomacy in Trade and Investment of Oil and Gas. In Andreas Goldthau and Jan Martin Witte, eds. *Global Energy Governance. The New Rules of the Game*. Washington, DC: Brookings Institution, pp. 25–48.

Greenspan, Alan. 2007. *The Age of Turbulence: Adventures in a New World*. New York: Penguin.

Hookway, James. 2011. Philippine Oil Vessel Confronted By China, Spurring New Dispute. *Wall Street Journal*, March 4. http://online.wsj.com/article/SB10001424052748703300904576178161531819874.html, accessed 04/06/2012.

Jaffe, Amy Myers, Michael T. Klare, and Nader Elhefnawy. 2008. The Impending Oil Shock: An Exchange. *Survival: Global Politics and Strategy* 50, 4: 61–82.

Jiang, Julie, and Jonathan Sinton. 2011. Overseas Investments by Chinese National Oil Companies. *IEA Information Paper* (February): 1–48.

Kaplan, Robert D. 2005. "How We Would Fight China." *The Atlantic*, June 2005. http://www.theatlantic.com/magazine/archive/2005/06/how-we-would-fight-china/3959/, accessed 04/06/2012.

Kaplan, Robert D. 2011. The South China Sea Is the Future of Conflict. *Foreign Policy* (October). http://www.foreignpolicy.com/articles/2011/08/15/the_south_china_sea_is_the_future_of_conflict?page=full, accessed 04/06/2012.

Kershaw, Ian. 2008. *Fateful Choices: Ten Decisions That Changed the World, 1940–1941*. London: Penguin.

Kissinger, Henry. 2011. *On China*. New York: Penguin.

Kong, Bo. 2010. *China's International Petroleum Policy*. Santa Barbara, CA: Praeger.

Lieberthal, Kenneth, and Mikkal Herberg. 2006. *China's Search for Energy Security: Implications for US Policy*. Seattle, WA: National Bureau of Asian Research.

Lovins, Amory B., Imran Sheikh, and Alex Markevich. 2008. Forget Nuclear. *Rocky Mountain Institute Solutions* 24, 1: 23–27.

Morse, Ed, *et al.* 2012. "Energy 2020: North America, the New Middle East?" Citi GPS: Global Perspectives & Solutions, March 20. http://fa.smithbarney.com/public/projectfiles/ce1d2d99-c133-4343-8ad0-43aa1da63cc2.pdf.

Nerurkar, Neelesh. 2011. *U.S. Oil Imports: Context and Considerations*. Washington, DC: Congressional Research Service. https://www.fas.org/sgp/crs/misc/R41765.pdf, accessed 04/06/2012.

O'Sullivan, Meghan L. 2003. *Shrewd Sanctions: Statecraft and State Sponsors of Terrorism*. Washington, DC: Brookings Institution.

Romero, Simon. 2005. Demand for Natural Gas Brings Big Import Plans, and Objections. *The New York Times,* June 15, sec. Business. http://www.nytimes.com/2005/06/15/business/15gas.html, accessed 04/06/2012.

Ross, Michael L. 1999. The Political Economy of the Resource Curse. *World Politics* 51, 2: 297–322.

Rutland, Peter. 2008. Russia as an Energy Superpower. *New Political Economy* 13, 2: 203–210.

Shan, He. 2012. Sino-ASEAN Trade Grows 36 Times in 20 Years. Xinhua Press Agency, February 20. http://cn-ph.china.org.cn/2012-02/20/content_4821179.htm, accessed 04/06/2012.

Stern, Jonathan P. 2005. *The Future of Russian Gas and Gazprom*. Oxford: Oxford University Press.

USGS. 2010. *Assessment of Undiscovered Oil and Gas Reserves of Southeast Asia*. Washington, DC: Department of the Interior, US Geological Survey. http://pubs.usgs.gov/fs/2010/3015/pdf/FS10-3015.pdf, accessed 04/06/2012.

Yergin, Daniel. 2006. Ensuring Energy Security. *Foreign Affairs* 85, 2 (April).

Yergin, Daniel. 2008. *The Prize: The Epic Quest for Oil, Money & Power*. New York: Simon & Schuster.

Yergin, Daniel. 2011. America's New Energy Security. *Wall Street Journal*, December 12.

Chapter 3

Sustainability, Climate Change, and Transition in Global Energy

Michael Bradshaw[1]

> The world's energy system is at a crossroads. Current global trends in energy
> supply and consumption are patently unsustainable – environmentally,
> economically, socially. But that can – and – must be altered; *there's still time to
> change the road we're on* [emphasis in the original].
>
> IEA (2008: 37)

Introduction

This chapter examines the various ways in which the current global energy system is
unsustainable and identifies the policy challenges that must be overcome to bring about a
transition to a more sustainable system. Other chapters in Part I focus on the economic,
security, and development aspects of the current energy system. Inevitably, this chapter
also touches on these issues, but the emphasis here is upon the environmental sustainability of the current energy system, with a particular emphasis upon climate change, which
is now unquestionably one of the major drivers of global energy policy.

The Rio+20 United Nations Conference on Sustainable Development that took place
in June 2012 marked two decades since the original Rio Earth Summit in 1992. Rio+20
identified seven key critical issues that need priority attention, of which energy was
one. In parallel, the UN Secretary-General Ban Ki-Moon declared 2012 International
Year of Sustainable Energy for All and set three interlinked objectives to be achieved
by 2030: universal access to modern energy services, doubling the rate of improvement
in energy efficiency, and doubling the share of renewable energy in the global energy
mix. This initiative is a somewhat belated acknowledgment that access to modern energy
services was missing from the Millennium Development Goals. The issue of energy and
development is the subject of the next chapter; however, as we shall see it is intimately
related to the relationship between energy and environment that is the subject of this
chapter.

The 1992 Earth Summit is also of significance as it opened the United Nations Framework Convention on Climate Change (UNFCCC) for signature. Today 195 states are

The Handbook of Global Energy Policy, First Edition. Edited by Andreas Goldthau.
© 2013 John Wiley & Sons, Ltd. Published 2013 by John Wiley & Sons, Ltd.

Parties to the Convention. The first World Climate Conference actually took place much earlier in 1979, the first Intergovernmental Panel on Climate Change (IPCC) met in 1988, and the first IPCC assessment report was released in 1990; the fifth assessment is due in 2014. The first Conference of Parties (COP-1) took place in Berlin, initiating a process that resulted in the Kyoto Protocol being adopted at COP-3 in 1997; however, it was not ratified until 2005. It expired in 2012 without a new agreement in place, but there remains a strong international commitment to reach a global agreement on greenhouse gas (GHG) emissions which is undoubtedly shaping global energy policy. The European Union (EU) has its own energy and climate change policy and has made a commitment to a 20% reduction in GHG emissions, a 20% increase in the share of renewable energy, and a 20% improvement in energy efficiency over 1990 levels by 2020. In addition, the so-called Copenhagen Pledges are the various unilateral commitments made by national governments in the absence of a post-Kyoto global agreement. Adding up the various pledges still leaves a substantial emissions gap in relation to the reductions needed to stand a good chance of reducing global warming to 2 °C relative to a pre-industrial level, This requires a peak atmospheric CO_2 of 450 parts per million (ppm), which was the commitment made in the Copenhagen Accord in 2009 and affirmed at Cancun in 2010 (Elzen et al. 2011). At the COP meeting in Durban in 2011 this emissions gap was noted with "grave concern," but all that could be agreed was the Durban Platform for Enhanced Action that seeks a new agreement by 2015, to be implemented from 2020. Thus, the global climate policy regime is now in the doldrums between the end of the Kyoto Protocol and a new global agreement on emissions reduction. At the COP-18 meeting in Doha in late 2012 the signatories to the Kyoto Protocol agreed to extend it until 2020. However, this leaves energy and climate change policy a matter for national governments and regional organizations.

From the above, it is evident that the contemporary policy concern about the development of a more environmentally sustainable global energy system dates back to the early 1990s; but from an energy industry perspective this was a period of plentiful supplies of oil and gas at low cost and it is fair to say that energy security was not then a major policy concern. Rather, it was the 1970s that had triggered concerns about the "secure and affordable" supply of energy to importing countries and consumers. The actions of the Arab members of OPEC provoked the developed economies of the OECD to create the International Energy Agency (IEA) and to coordinate their energy policies to increase their resilience to supply shocks. It also prompted them to develop their own production in more technologically difficult regions such as Alaska, the North Sea, and Gulf of Mexico, which brought with it many environmental challenges. The recent re-emergence of energy security concerns, the subject of Chapter 2, is the result of a number of drivers: the plateauing of non-OPEC production, a surge in resource nationalism that has placed the majority of oil and gas reserves under state control, the rising costs of sustaining production, and most importantly a surge in new demand from the non-OECD countries of the developing world.

The net result is that over the last decade or so the global energy system has increasingly struggled to match production growth with demand growth. Of course, there is a school of thought that this is because global oil output is close to peak production; this is a matter considered elsewhere in this volume. Whatever the underlying causes, the resulting "tight" energy market saw a rapid increase in price as the average annual oil price rose from $27 a barrel (in $2010) in the 1990s to over $56 a barrel in the 2000s (BP 2012a). In the summer of 2008 the oil price peaked at $147 a barrel, only to collapse to below $40 a barrel by early 2009 as a consequence of the global financial crisis; but it is now back around the $100 mark. This high oil price environment has resulted in massive profits for the oil and gas companies and has empowered oil-producing states and regions. It has also resulted in an enormous transfer of wealth from energy importing countries to energy exporting countries (often with negative consequences for both) and it has complicated global economic recovery (El-Gamal and Jaffe 2010). Previous recessions were associated with falling energy costs, today we are in uncharted territory in terms of the oil price and the economic cycle.

From the brief discussion above, it is easy to see why in 2008 the IEA reached the conclusion that current trends are unsustainable. As is discussed below, they are environmentally unsustainable because the current fossil fuel energy system is a source of widespread environmental degradation and because the GHG emissions associated with fossil fuel combustion are the major contributor to global climate change. They are economically unsustainable because high energy prices and price volatility pose a major threat to economic growth. Finally, they are socially unsustainable as the high cost of energy is a growing contributor to poverty for many and because billions remain without access to modern energy services, which are a prerequisite for poverty alleviation. Therefore, it is no surprise that there is now recognition of the need for a New Energy Paradigm (NEP) to set the world on a path to a more sustainable energy future.

There are various interpretations of the NEP. Some relate it to a move from a more state-based to a market-based system of energy relations, others to the need to accommodate the impact of climate change (Helm 2007; Goldthau 2012). In some contexts, such as the EU, the two are not mutually exclusive, but more generally we can think of the NEP as being associated with the need to address climate change through a transition to a low carbon energy system (Bradshaw 2011). However, I would also argue that a focus on global climate change policy increasingly foregrounds issues of social equity and the development agenda in relation to energy access. Thus, notions of energy security as the central goal of a nation's energy policy have now shifted away from the more traditional concerns of security of supply and affordability, to a more multidimensional approach that adds the issues of environmental sustainability and equity, amply demonstrated in this statement from the European Commission introducing its Energy 2020 strategy:

> A common EU energy policy has evolved around the common objective to ensure uninterrupted physical availability of energy products and services on the market, at a price which is affordable for all consumers (private and industrial), while contributing to the EU's wider social goals. European Commission (2010: 2)

To sum up the discussion so far, events over the last two decades have changed the focus of global energy policy such that current and future strategies must be aimed at delivering secure, affordable, environmentally sustainable, and socially equitable access to energy services. This is no easy task, as demonstrated by the discussion in each of these chapters of Part I that deal with a particular aspect of the global energy challenge. The remainder of this chapter is divided into two substantial sections. The first section examines global energy dilemmas and addresses two issues, the relationship between energy and climate change and the consequences of the globalization of energy demand. The second section examines global energy transitions; again two issues are considered, the nature of energy transitions and the lessons that can be learnt from history, and the specific challenges by the low carbon transition. The chapter concludes by identifying the key challenges that now face energy policy-makers.

Global Energy Dilemmas

The notion of the global energy dilemmas relates to a seemingly straightforward question: *can we have secure and affordable energy services that are environmentally benign?* (Bradshaw 2010). The term environmentally benign is used to stress the fact that the environmental impacts of the energy system are much more than just climate change. Events such as the Macondo oil leak in the Gulf of Mexico in 2010, the ongoing problems at the Fukushima nuclear power plant in Japan following the March 2011 tsunami, and the gas leak at the Elgin platform in the North Sea in 2012 are recent reminders of the environmental impacts of the energy industry. However, on a day-to-day basis the industry is responsible for widespread environmental degradation across the globe, be it in the Niger Delta in West Africa, the Tar Sands in Canada or the drilling for shale gas in the US; securing energy comes at a cost to the environment. In most instances, this is not a cost that is adequately passed on to the consumer. Equally, environmental degradation is not confined to fossil fuels. The pursuit of biofuels and biomass energy can be a major source of deforestation, the construction of large-scale hydroelectric schemes has long been recognized as having negative environmental impacts, nuclear power presents particular problems in relation to the disposal of spent fuel, and the large-scale construction of wind farms is seen by some as a source of both visual and noise pollution and a hazard to migrating birds. The key point is that all sources of energy supply – hydrocarbon or low carbon – have negative environmental impacts; in the end it is a matter of degree. However, most of the examples above are relatively local in terms of their impact and their management usually falls within a particular national jurisdiction. Problems such as acid rain that are also caused by the energy industry are transnational rather than global in terms of their impact. Even the radiation leak at Chernobyl in 1986 had a regional rather than global impact. The distinguishing feature of climate change is that it is global in impact, though the major GHG emitters may be concentrated in particular parts of the world. Thus, the reason that this chapter focuses on climate change in relation to sustainability is because this handbook is about global energy policy and at that scale climate change is the most important issue. However, that is not to understate the significance of all of the other environmental issues that are related to energy production and that are often ignored or understated.

Energy and Climate Change

Although there is now a consensus among politicians, policy-makers, and scientists that human actions are having a significant impact upon the Earth's climate system, it cannot be proved beyond doubt. That said, the industrial revolution and the more than 200 years of fossil fuel-based economic development that have followed coincide with a substantial increase in the levels of CO_2 concentration in the atmosphere. It is the next step that is the difficult one; can we prove beyond doubt that those emissions of CO_2 and other GHGs are a major cause of the warming trend that we see in the global climate records? In their fourth assessment, the IPCC reached the following conclusion: "Warming of the climate system is unequivocal, as is now evident from observations of increases in global average air and ocean temperatures, widespread melting snow and ice and rising global sea level" (Pachauri and Reisinger 2007: 30); they then went on to observe that: "There is a *very high confidence* (emphasis in original) that the global average net effect of human activities since 1750 has been one of warming" (2007: 37). On the same page, the report makes clear that global increases in CO_2 concentrations "are due primarily to fossil fuel use, with land-use change providing another significant but smaller contribution," and in 2004, according to the IPCC, CO_2 emissions from fossil fuel use accounted for 56.6% of total GHG emissions and fossil fuel combustion accounts for 80% of total CO_2 emissions (Pachauri and Reisinger 2007: 36).

Thus, for the IPCC at least, the changes in the global energy system that have taken place since the 1750s have been responsible for GHG emissions that have substantially increased from the pre-industrial level of 280 ppm to 393.53 ppm in February 2012 (the latest figure available from the US National Oceanographic and Atmospheric Administration). As noted earlier, climate change policy-makers have identified 450 ppm as a critical level at or below which emissions must be stabilized to reduce the chances of catastrophic climate change (the term they use). Thus, a key target for climate change policy is to substantially reduce the level of energy related GHG emissions. What does this mean for the energy system?

As explained in the section below on energy transitions, the current fossil fuel energy system has been constantly evolving since the advent of the industrial revolution and the harnessing of steam power based on coal as a raw material. Today the global primary energy mix is dominated by the three fossil fuels. According to BP (2012a: 41), in 2010 coal accounted for 29.6% of world consumption, natural gas 23.8%, and oil 33.6%. The remaining 13% was made up with nuclear energy, hydroelectricity, and renewables, but the latter was only 1.3% of global energy consumption. There are substantial regional variations in the energy mix that reflect differences in their energy endowment and their level of economic development. For example, in Europe and Eurasia natural gas accounts for 34.4% of primary energy consumption, while in the Asia-Pacific region coal still accounts for 52.1% of consumption. These differences have a significant impact on energy and climate change policy in the different regions. The energy mix is of significance because the different types of fossil fuels emit different levels of GHG when burnt. The IPCC fourth assessment provides information on the average amounts of CO_2 emitted by each fossil fuel. The figures that follow are averages and there is considerable variation depending on the characteristics of a particular coal deposit, oil well or gas field; there is also significant variation between conventional sources of oil and gas and unconventional sources such as tar sands and shale gas. According to the IPCC (Pachauri and Reisinger 2007), on average conventional coal produces 92.0 g CO_2/MJ of energy produced (grams of carbon dioxide emitted per megajoule of energy produced), conventional oil produces 76.3 g CO_2/MJ, and conventional gas produces 52.4 g CO_2/MJ (Metz *et al.* 2007: 264). Thus, conventional natural gas produces 43% less CO_2 per unit of energy produced than conventional coal. This is an important factor as the two fuels are competing inputs into electricity power generation. From these numbers it is clear why the energy mix of a particular country is an important factor influencing its carbon intensity of energy use (this is discussed in more detail below). Furthermore, a key element of energy policy aims at reducing GHG emissions from energy by changing the energy mix in favor of lower carbon fuels. For example, in the UK over the last 20 years the so-called "dash for gas" has replaced coal with natural gas as the dominant fuel in the energy mix and this has resulted in a substantial reduction in GHG emissions (Bradshaw 2012). Thus, replacing coal with natural gas is one way of decarbonizing the energy mix. The emergence of unconventional sources of oil and gas has complicated the situation and focused attention on the "full life cycle" emissions of different energy sources (Klare 2012: 100–127).

The statistics presented above tell us about the level of emissions at the point of combustion, but they do not tell us about the GHG emissions associated with the production and delivery of a particular fuel to a power station or household. This is important because it is the total GHG emissions that matter when it comes to climate change. It is clear that an unconventional oil like tar sands produces more GHG emissions per unit of energy than conventional oil, because a lot of energy has to be used to extract the bitumen and then turn it into a synthetic fuel that can be transported by pipeline and used in refineries. The situation with shale oil and shale gas is less clear and is the subject of considerable controversy, but it may be the case that such production involves higher GHG emissions than conventional sources of oil and gas (Hughes 2011). This is important because a shift from coal to shale gas power generation may result in a lower level of decarbonization than first thought, and according to some estimates may even increase GHG emissions. More generally, a switch from reliance on conventional oil and gas to unconventional gas, which may happen in North America, might actually increase the GHG emissions associated with fossil fuel combustion. This is an issue that is far from resolved, but it serves to illustrate the complexity and controversy surrounding life-cycle analysis of GHG emissions from different energy sources.

For the moment we can conclude that energy policy is central to climate change policy for the simple reason that the combustion of fossil fuels is the single most important source of the emissions responsible for anthropogenic climate change. Therefore, policies aimed at mitigating the causes of climate change must be aimed at the decarbonization of the global energy system.

The Globalization of Energy Demand and Carbon Emissions

The bulk of the anthropogenic GHG emissions currently resident in the atmosphere are related to the industrial development of what we know as the developed world, represented by membership of the OECD. As we shall see later, the relationship between energy and development in the OECD countries has evolved over time and they appear to have become increasingly efficient at turning energy into economic output; however, they are undoubtedly responsible for the majority of the historical emissions that are currently contributing to the warming of the climate. Country-level estimates of CO_2 emissions go back to 1850. Table 3.1 provides data on cumulative CO_2 emissions from fossil fuels. The table presents emissions from 1850 to 2002 and from 1990 to 2002; it also ranks countries during those two time periods.

A number of patterns can be discerned from the information in Table 3.1. First, the high degree of concentration of historical emissions in the industrialized world, and within that the US and Europe that together account for 55.8% of the cumulative emissions between 1850 and 2002. Second, there is evidence of a changing dynamic since 1990 that is the result of growing emissions in the developed world and falling emissions within the industrialized world. The share of the so-called BASIC economies (Brazil, China, India, and South Africa) has grown from 11.8% of historical emissions to 20.6% of emissions between 1990 and 2002. This is a trend that has actually accelerated over the past decade, during which China passed the US in 2007 as the world's top emitter. If one removes the BASIC countries, the share of the developing world has only increased from 12.6% up to 1990 to 18.6% between 1990 and 2002. The numbers show a rather different dynamic for those countries that were members of the Soviet economic system (Russia, Ukraine, and Poland), which have all experienced a reduction in their emissions. As a consequence of EU enlargement, some of this is also captured in the EU's reduction. Elsewhere I have disaggregated the world on the basis of a four-fold classification – developed, post-socialist, emerging, and developing – and used data from

Table 3.1   Cumulative emissions from fossil fuels, 1850–2002 (percentage of world total).

Country	1850–2002	Rank	1990–2002	Rank	% Change
United States	29.3	(1)	23.5	(1)	−20
EU-25	26.5	(2)	17.0	(2)	−36
Russia	8.1	(3)	7.5	(4)	−8
China	7.6	(4)	13.9	(3)	+83
Germany	7.3	(5)	4.0	(6)	−46
United Kingdom	6.3	(6)	2.5	(8)	−61
Japan	4.1	(7)	5.2	(5)	+26
France	2.9	(8)	1.6	(13)	−44
India	2.2	(9)	3.9	(7)	+79
Ukraine	2.2	(10)	1.9	(10)	−12
Canada	2.1	(11)	2.1	(9)	−3
Poland	2.1	(12)	1.5	(15)	−28
Italy	1.6	(13)	1.9	(11)	+17
South Africa	1.2	(14)	1.5	(16)	+26
Australia	1.1	(15)	1.3	(17)	+24
Mexico	1.0	(16)	1.6	(14)	+56
Spain	0.9	(20)	1.2	(20)	+31
Brazil	0.8	(22)	1.3	(18)	+58
South Korea	0.8	(23)	1.7	(12)	−127
Iran	0.6	(24)	1.2	(19)	+92
Indonesia	0.5	(27)	1.1	(21)	+110
Saudi Arabia	0.5	(28)	1.1	(22)	+116
Argentina	0.5	(29)	0.6	(30)	+16
Turkey	0.4	(31)	0.8	(24)	+82
Pakistan	0.2	(48)	0.4	(36)	+105
Industrialized	*75.6*		*60.8*		−20
Developing	*24.4*		*39.2*		+61

Source: Baumert *et al.* (2005: 115).

the World Bank's online Development Indicators database to analyze the recent dynamics of emissions. The results of this analysis are shown in Table 3.2.

This regional disaggregation of emissions, energy use, economic activity, and population makes it clear that we are in the midst of a significant shift in the locus of global emissions and energy consumption driven by the processes of economic globalization, population growth, and the consequences of the collapse of the Soviet economic system. The post-Socialist countries – the 27 economies that were under the original remit

Table 3.2   Key indicators by major world region (percentage of world total). (Some columns do not add up to 100 due to rounding errors in the original sources.)

	CO$_2$ Emissions		Energy Use		Gross National Income (PPP)		Population	
	1990	2007	1990	2008	1990	2009	1990	2010
Developed	41.7	39.0	48.6	42.2	58.6	47.6	16.1	14.1
Post-Socialist	18.8	9.1	19.7	10.7	8.9	7.6	7.8	5.9
Emerging	23.4	38.3	22.2	34.7	17.8	29.9	50.5	49.9
Developing	7.0	8.8	8.4	11.0	12.0	12.8	25.3	29.7

Source: Bradshaw (2013).

of the European Bank for Reconstruction and Development – have seen their share of global emissions and energy fall dramatically during this period, as their share of global economic activity has also declined together with their share of population. The developed economies have seen their global shares fall, but much more modestly. However, the emerging economies – which in this analysis is a larger group than just the BASIC economies – have seen their share of emissions, energy use, and economic activity increase significantly, while their share of global population growth has barely changed. Within this group China is by far the most significant source of emissions growth. In 2009 China accounted for 24% of global emissions of CO_2 and its emissions have tripled between 1990 and 2009 (IEA 2011a: 24). Finally, the developing world has shown little change in its relative share of emissions, energy use, and economic activity, but a more substantial change in its share of global population. The policy consequences of these dynamics are discussed later; for the moment we can note that a global shift is under way. This begs the question how much further are things likely to change between now and 2030–2040?

There are a number of organizations that produce projections and scenarios about future energy use and carbon emissions. This discussion is based on four recent publications: *World Energy Outlook 2011* (IEA 2011b), *International Energy Outlook 2011* (EIA 2011), *The Outlook for Energy: A View to 2040* (Exxon Mobil 2012), and *BP Energy Outlook 2030* (BP 2012b). What we are interested in is what these analyses have to say about the geographical dynamics of energy demand and the CO_2 emissions associated with that energy use. When it comes to four basic questions – by how much will global energy demand increase, what will be the role of fossil fuels, what will be the share of the OECD and non-OECD countries in global energy demand, and what will happen to carbon emissions – there is some degree of consensus. Starting with the EIA's (2011) reference scenario based on the polices currently in place, they project a 53% increase in world market energy consumption by 2035 and at that time fossil fuels will account for 79.1% of that demand. Between 2008 and 2035 energy demand in the non-OECD countries will grow by 85% and OECD demand by 18%. As a result of this differential growth, the non-OECD countries will account for 62.5% of global energy consumption in 2035, up from 51.6% in 2008. Overall, CO_2 emissions from fuels will increase by 43% and the level of non-OECD emissions will exceed the OECD by 100% in 2035 compared to 23.4% in 2008. Although the EIA provides no such assessment, this scenario is nowhere near restricting GHG emissions to a level that will keep global warming below 2 °C. The IEA (2011b) provides a range of scenarios, including a 450 ppm scenario that aims to achieve what is needed to restrain warming to the 2° C level. However, discussion here relates to the New Policies scenario that assumes that all the current policies in place are implemented; this is similar to the EIA's reference case as neither make assumptions about future policies that might be implemented. The New Policies scenario predicts a 40% increase in global energy demand by 2035, at which time fossil fuels will still meet 80% of that demand and the non-OECD share of global demand will be 64%. Over this period energy related CO_2 emissions will increase by 20%, a significantly lower level than the EIA forecast. Nonetheless, this scenario is on track to a 3.5 °C level of warming.

The next two forecasts come from international oil companies, which do not engage in the same range of scenario building exercises. Perhaps surprisingly, the Exxon Mobil (2012) forecast is the most optimistic and its time frame is to 2040. By 2040 they forecast that global energy demand will increase by 30%, that fossil fuels will satisfy 80% of that demand, and that non-OEDC energy demand will grow by 60%, while OECD demand will be flat. They provide no figures for emission growth, but they expect emissions to level off by 2030 and note that the non-OECD countries will account for 70% of CO_2 emissions from energy by 2040. The final forecast comes from BP (2012b) and is up to 2030. By that time they predict that total energy demand will have increased by 39% and that almost all of that growth (96%) will have come from the non-OECD countries where energy demand will be 69% above 2010 levels. Demand in the OECD in 2030 will only be 4% higher than it was in 2010. Overall, CO_2 emissions will increase by 28%, and they note that this is well above that required by the IEA's 450 ppm scenario.

Returning to the four questions posed above, we can conclude that these forecasts suggest that between now and 2030–2040 global energy demand will increase by about 40% and that fossil fuels will still dominate the global energy mix, accounting for about 80% of global energy demand. There is also consensus that the non-OECD will account for the vast majority of new energy demand growth and that this will bring about a re-balancing of the global energy system. Closer inspection of these analyses reveals that China, India, and the Middle East, in that order, are the major drivers of that demand growth. There is less agreement on the growth of CO_2 emissions, but this is related to underlying assumptions about technological change, energy efficiency gains, and the growth of low carbon energy sources. What is clear is that all of these scenarios fall well short of what is required to put the global energy system on a path that would constrain global warming to 2 °C. Returning to the earlier discussion of cumulative emissions, the IEA (2011b: 100) notes that "In the New Policies Scenarios, cumulative energy-related global emissions over the next 25 years are projected to be three quarters of the amount emitted over the past 110 years" and that the non-OECD countries will account for all of the emissions growth to 2035. Thus, if the industrialization of the OECD economies is largely responsible for cumulative emissions to date, it is clear that the economic growth and population increases in the non-OECD over the coming decades will be responsible for the majority of future emissions. This has significant implications for energy and climate change policy, and also suggests that the planet cannot afford for the non-OECD countries to develop in the same fossil fuel-intensive way that the OECD countries did. This requires a global transition to a more sustainable energy system.

Global Energy Transitions

From the discussion above it is clear that we are at the beginnings of a global shift in energy demand and that one aspect of a sustainable energy system is meeting the new demand for energy services that will come from population growth and improved living standards outside the OECD. However, to avoid catastrophic climate change, that demand must be met in ways that do not result in unsustainable increases in the emissions of GHGs. For that to happen both existing and new demand must increasingly be met from low carbon sources and that will require a transition in the nature of the system. This section examines how the fossil fuel system has transformed itself in the past and the lessons that can be learnt in terms of the low carbon transition.

Understanding Energy Transitions

The period of sustained economic development and population growth that the developed world has experienced since the industrial revolution has relied almost exclusively on access to cheap fossil fuel energy. However, within that period the fossil fuel system has experienced a number of transitions. According to Grübler (2004: 163) an energy transition is: "Change from one state of an energy system to another one, for example,

Table 3.3   Changes in the global primary energy mix, 1800–2008 (percentage of total energy demand).

Year	Coal	Crude Oil	Gas	Hydro-electricity	Nuclear	Biomass
1800	1.7					98.3
1900	47.3	1.5	0.5	0.1		50.5
1950	45.1	19.5	7.5	1.2		26.8
1960	38.1	27.1	11.0	1.7	0.0	22.0
1970	27.9	38.2	16.1	2.2	0.4	15.2
1980	27.4	37.8	17.8	2.1	2.6	12.3
1990	27.1	32.9	20.6	2.3	5.5	11.6
2000	23.0	33.7	22.6	2.5	6.4	11.8
2008	28.9	30.9	22.8	2.5	5.7	9.2

Source: Calculated from Smil (2010: 154).

from comparatively low levels of energy use relying on non-commercial, traditional, renewable fuels to high levels of energy use relying on commercial, modern fossil-based fuels." This kind of fundamental transition is represented by the industrial revolution. Prior to that the energy system relied on natural flows (sun, wind, and water) and animal and human power to provide the necessary energy services of heat, light, and work. This "somatic" energy system placed limits on the ability of human society to expand (McNeil 2000). According to Fouquet and Pearson (1998: 4), there have been three major changes in energy use: a dramatic change in per capita energy use in the second half of the millennium; a shift in the methods of supplying energy services from biomass to fossil fuels, from coal to petroleum products to natural gas, and from raw forms to more value added sources; and the replacement of direct methods or animate sources, wind, and water, by the use of heat to drive steam and combustion engines.

As Table 3.3 demonstrates, over the past 200 years or so when the world's population has increased sixfold and energy use has increased 20-fold, there have been three "transitions" within this fossil fuel system as new prime movers have created demand for new energy sources (Smil 2010). The first phase was dominated by coal and steam power, this was then replaced by the internal combustion engine that literally drove demand for petroleum, most recently we have seen the introduction of the gas turbine which has driven demand for natural gas. Along the way we have seen large-scale development of hydroelectric power and nuclear power, but these play a minor role relative to the three fossil fuels. The invention of electricity created continued demand for coal and now in the more advance economies oil is used as a petrochemical feedstock and for transportation, while coal and natural gas (and nuclear) compete against each other in the power generation sector. The majority of households in advanced economies use electricity and/or gas for heating and cooking. Thus, today we are locked into a fossil fuel-based energy system with a diversity of primary energy sources (Unruh 2000). As noted earlier, there are significant regional variations in the energy mix between them, dependent on the level of economic development of a particular economy and its indigenous energy endowment. The various forecasts discussed in the previous section see a relative balancing between the three fossil fuels with gas continuing to grow, oil stabilizing, and coal stagnating, but together the three remain dominant. This suggests that nuclear power, hydroelectric power, and renewable energy will only account for 20% of primary energy demand by 2030–2040, which can hardly be considered a low carbon transition!

The history of energy over the last 200 years suggests that significant changes in the structure of the primary energy supply take decades to occur. Smil (2010: vii) warns that the "inherently gradual nature of large-scale energy transitions is ... the key reason why – barring some extraordinary and entirely unprecedented financial commitments and determined actions – none of today's promises for a greatly accelerated transition from fossil fuels to renewable energies will be realized." One could argue that the actions implied in the IEA's (2011b) 450 ppm scenario represent such unprecedented action. However, a similarly cautious view is presented by Kramer and Haigh (2009) who note that there are physical limits to the rate at which new technologies can be deployed and that the experience of the twentieth century suggests that it takes 30 years for energy technologies to grow exponentially and become widely available. They suggest that given what they call the two "laws of energy-technology deployment" much greater emphasis needs to be placed on demand reduction and increased efficiency as demand-side solutions are subject to different laws and theoretically can be achieved within a shorter period of time. The final factor to consider is that the energy transitions within the fossil fuel system were driven by technological change that increased energy efficiency and reduced cost; they were also driven by innovation and competition. That fossil fuel system is now itself becoming less sustainable due to falling net returns on energy investment – in other words as the accessible conventional deposits are being used up more energy is being expended to extract energy from harder to access deposits and unconventional sources – and this is increasing the amount of energy expended to produce new energy (Murphy and Hall 2010). Nevertheless, the new renewable forms of energy are finding it difficult to compete against the incumbent fossil fuels and are requiring significant support and subsidy to gain market share (see Chapter 21). An alternative to direct subsidies, such as a feed-in tariff that guarantees a higher price for renewable electricity, is to set a cost for carbon that has the effect of increasing the price of fossil fuels relative to low carbon sources such as nuclear power and renewables (see Chapter 18). This is happening in some countries, but it does have the effect of increasing the cost of energy services, but that may be the price that society has to pay to accelerate the low carbon transition; higher energy costs also drive demand reduction and energy efficiency. Unfortunately, higher costs also aggravate the problems of energy poverty and access to energy services. Many of the chapters in this handbook discuss the technological and policy prescriptions needed to bring about a purposeful transition to a low carbon energy system. The irony is that many of the "new" technologies represent a scaling-up of energy sources that were prevalent before the industrial revolution, such as wind and water power. However, the problem remains that even when scaled up these renewable sources still have much lower energy equality and energy density than fossil fuels (Smil 2010: 113). In the final analysis, the transition to a more sustainable energy system is as much about changing the relationship between energy, economy, and society, as it is the means by which we secure our energy services.

The Challenge of the Low Carbon Transition

This final substantive section examines the key drivers that determine the level of CO_2 emissions from energy consumption. To aid in this discussion the Kaya Identity is used as a heuristic device to disaggregate the key elements that drive energy demand and emissions. This simple formula is named after its developer Yoichi Kaya, an engineer at Tokyo University.

The Kaya Identity has become very influential in the energy and climate change literature. It is used in the IPCC's models that forecast future carbon emissions, by the EIA (2011: 144–145) in its *International Energy Outlook*, and also in a number of scholarly and popular works (for example, Baumert *et al.* 2005; Jaccard 2005; Raupach *et al.* 2007; Pielke 2010). The EIA (2011: 144) describes the Kaya Identity as: "an intuitive approach to the interpretation of historical trends and future projections of energy-related carbon dioxide emissions." It is important to note that the Kaya Identity only deals with CO_2 from energy, thus it is only dealing with a particular cause of GHG emissions, but it is the most important cause. Baumert *et al.* (2005: 16) identify three elements within the formula: carbon intensity and energy intensity are treated separately, but GDP per capita and population are grouped together as activities. From a policy perspective this is an important distinguishing factor. Although many green groups advocate zero economic growth, this is not a viable policy option to reduce CO_2 emissions, though Jackson (2009) gives an alternative perspective. In the current context of global economic recession many politicians, especially in the OECD, are seeking to promote economic growth at any cost. However, the impact of the 2008 global economic crisis in reducing both energy con-sumption and carbon emissions does illustrate the importance of economic growth as a driver of energy demand. Since then economic recovery in Asia, the main location of new demand, has meant that emissions are once again climbing. Peters *et al.* (2012) report that GHG emissions fell by 1.4% in 2009, only to rebound and grow by 5.9% in 2010. Equally, while there are all sorts of benefits to be had from reducing the rate of population growth, especially in the developing world, managing population growth is not seen a policy lever when it comes to carbon emissions. That said, O'Neill *et al.* (2010: 17521) demonstrate that "slowing population growth could provide 16–29% of the emissions reductions suggested to be necessary by 2050 to avoid dangerous climate change." Thus, measures like improving female education and workforce participation, both of which are known to limit population growth, could also have positive environmental conse-quences. This leaves the two remaining elements of *energy intensity* and *carbon intensity* as the key levers for global energy policy-makers.

Energy intensity is a measure of the amount of energy used to produce a unit of output in a particular national economy. If it is reported in terms of GDP produced per unit of energy consumed, a higher number means more output per unit of energy consumed; equally it can be expressed as energy consumed per unit of GDP produced in which case a lower number means less energy per unit of output. It is also expressed in a variety of different units in terms of economic output, normally GDP, and energy consumed. Therefore, it is important to be clear about the specific measure of energy intensity that is being used. In this discussion an improvement in energy intensity simply means that less energy is being consumed per unit of economic output. Energy intensity is not the same as energy efficiency, although improvements in efficiency do usually have the effect of improving energy intensity. The World Energy Council (2010: 13) has identified a number of factors that influence a country's energy intensity: economic structure, the primary energy mix, the climate, the level of economic development, the organization of the transport sector, and technical energy efficiency. There is not the space to delve into them all here and most are self-explanatory; however, economic structure is of particular significance in terms of the relationship between economic development and energy intensity. Economies at an early stage of industrialization are typified by a concentration in heavy industry and manufacturing that are energy intensive activities. As an economy matures it tends to shed those sectors and move more into the service sector that results in a reduction in energy intensity. Of course, this post-industrial economy still consumes the outputs of heavy industry and manufacturing, but it imports them from elsewhere, rather than producing them domestically (Peters *et al.* 2011). Thus, while the levels of energy intensity, and with it carbon emissions, in the post-industrial economy appear to decrease, the reality is that they have just been exported elsewhere. This pattern describes the recent relationship between the OECD and China. Rühl *et al.* (2012: 20) explain that there is a general pattern of rising and then falling energy intensity at the scale of the global economy and that there is also a process of convergence under way in terms of levels of energy intensity. They suggest that it is globalization that is accelerating these long-term trends, "because all tradable fuels can now be traded across all international borders, because technologies are becoming increasingly shared internationally, and because even consumption baskets (determining the end-use of energy) are becoming standardised and similar across formerly very different countries and cultures." This may well be the case for the developed and emerging economies, but it is not apparent that the developing world is part of this convergence. At any rate, while this process of convergence and declining energy intensity may slow the growth of energy consumption, it will not do it quickly enough to mitigate climate change. From a policy perspective, measures that reduce energy demand and improve energy efficiency have the positive effect of reducing the level of energy consumption and, if that energy is derived from fossil fuels, the level of associated CO_2; they also improve the energy security of import dependent countries as less has to be imported.

The second measure is the carbon intensity of energy use. This is usually expressed as a physical measure of CO_2 per unit of energy consumed, for example as tons of CO_2 per ton of oil equivalent energy consumed or metric tons per billion Btu. Whatever the measure, the higher the level of carbon intensity the more CO_2 is produced per unit of energy consumed. For obvious reasons, the energy mix is the key factor influencing the carbon intensity of energy use in a particular economy. The greater the reliance on more carbon-intensive primary energy sources, such as coal (and potentially unconventional sources of oil and gas), the higher the level of carbon intensity. One possible solution is to deploy technologies to capture and store the CO_2 emission from coal- and gas-fired power stations and other large industrial facilities; at present this Carbon Capture and Storage (CCS) technology is not commercially proven, but it is a priority area of development that needs comprehensive policy support (Watson 2012). Of course, CCS does not remove the other environmental impacts and sustainability concerns related to the fossil fuel energy economy.

As noted earlier, the OECD economies tend to have a more diverse energy mix with a higher reliance on natural gas and low carbon sources such as nuclear power and increasingly renewable energy. If we return to the BASIC economies, China, India, and South Africa are heavily reliant on coal; Brazil is different as it relies on hydroelectric power and biofuels. In the case of China, when its coal reliance is combined with its current stage of development with a recent emphasis on heavy industry and manufacturing, it is easy to see why its carbon intensity of energy use has remained high and its absolute emissions of carbon have surged (Guan *et al.* 2009). Returning to the policy consequences of Kaya, to reduce the carbon intensity of energy use an economy needs to decarbonize its energy mix. This means to reduce its reliance on high carbon fossil fuels such as coal, to replace it with gas in power generation, or better still low carbon sources as nuclear power and/or renewable energy.

Conclusions

This chapter has examined the challenges facing a transition to a more sustainable global energy system. The discussion has focused on the issue of climate change. The scientific evidence suggests that the emissions of GHG from fossil fuel combustion are the most significant source of the emissions that promote anthropogenic climate change. Therefore, to reduce the risk of catastrophic climate change it is necessary to significantly reduce the amount of fossil fuels consumed in the delivery of energy services; this requires a transition to a low carbon energy system. The net result of this low carbon transition is to reduce the amount of carbon produced per unit of energy consumed. Such an approach is the focus of energy policy in the OECD, particularly in the EU, where energy demand is stable and there is the policy imperative and economic resources to finance low carbon energy and CCS. In the context of the emerging economies where demand is growing very rapidly, and sustained economic growth is essential to meet the demands of the population, it is a matter of harnessing all available sources of energy regardless of their carbon consequences. In the developing world the imperative is to improve access to energy services, high carbon or low carbon. Thus, the immediate policy challenge is to find ways to control and reduce the carbon intensity of energy use in the rapidly developing economies such as China, India, and in the Middle East, at the same time as significantly reducing the carbon intensity of energy consumption in the developed world. This will also allow for universal access to energy services without increasing GHG emissions. The Kaya Identity makes clear that the most likely policy prescription to promote a more sustainable energy system is: *demand reduction, improved energy efficiency*, and the *decarbonization of energy use*. This is well recognized by policy-makers and these elements are evident in both the EU's 20-20-20 and the UN's targets related to a sustainable energy system for all, which include doubling the rate of improvement in energy efficiency and doubling the share of renewable energy in the global energy mix. The policy prescriptions to create a more sustainable global energy system may be clear, but the progress to date is limited. The 2011 edition of the IEA's *World Energy Outlook* starts with the observation: "There are few signs that the urgently needed change in direction in global energy trends is underway" (IEA 2011b: 39). Returning to the quotation at the beginning of this chapter, also from the IEA, we can only hope that there is still time to change the road that we are on and that we can implement global energy policies that provide universal access to secure and affordable energy services that are environmentally benign.

Note

1. This chapter draws on a larger piece of work entitled Global Energy Dilemmas: Energy Security, Globalization and Climate Change that was funded by a Leverhulme Major Research Fellowship and that will be published as a book by Polity Press.

References

Baumert, K. A., T. Herzog, and J. Pershing. 2005. *Navigating the Numbers: Greenhouse Gas Data and International Climate Policy*. Washington, DC: World Resources Institute.

Bradshaw, M. J. 2010. Global Energy Dilemmas: A Geographical Perspective. *The Geographical Journal* 176: 275–290.

Bradshaw, M. J. 2011. In Search of a New Energy Paradigm: Energy Supply, Security of Supply and Demand and Climate Change Mitigation. *Mitteilungen der Österreichischen Geographischen Gesellschaft* 152, 9–26.

Bradshaw, M. J. 2012. *Time to Take the Foot off the Gas? Gas in UK Energy Security*. London: Friends of the Earth (UK).

Bradshaw, M. J. 2013. *Global Energy Dilemmas: Energy Security, Globalization and Climate Change*. Cambridge: Polity Press.

BP. 2012a. *Statistical Review of World Energy 2011*. London: BP.

BP. 2012b. *BP Energy Outlook 2030*. London: BP.

EIA (US Energy Information Administration). 2011. *International Energy Outlook 2011*. Washington, DC: EIA.

El-Gamal, M. A., and A. M. Jaffe. 2010. *Oil, Dollars, Debt and Crisis: The Global Curse of Black Gold*. Cambridge: Cambridge University Press.

Elzen, M. G. J., A. F. Hof, and M. Roelfsema. 2011. The Emissions Gap Between the Copenhagen Pledges and the 2 °C Climate Goal: Options for Closing and Risks That Could Widen the Gap. *Global Environmental Change* 21: 733–743.

European Commission. 2010. *Energy 2020: A Strategy for Competitive, Sustainable and Secure Energy*. Brussels: European Commission.

Exxon Mobil. 2012. *The Outlook for Energy: A View to 2040*. Houston, TX: Exxon Mobil.

Fouquet, R., and P. J. G. Pearson. 1998. A Thousand Years of Energy Use in the United Kingdom. *The Energy Journal* 19, 4: 1–41.

Goldthau, A. 2012. From State to the Market and Back. Policy Implications of Changing Energy Paradigms. *Global Policy* 3, 2: 198–210.

Grübler, A. 2004. Transitions in Energy Use. *Encyclopaedia of Energy* 6: 163–177.

Guan, D., G. P. Peters, C. L. Weber, and K. Hubacek. 2009. Journey to World's Top Emitter: An Analysis of the Driving Forces of China's Recent CO_2 Emissions Surge. *Geophysical Research Letters* 36: L04709.

Helm, D. 2007. The New Energy Paradigm. In Dieter Helm, ed. *The New Energy Paradigm*. Oxford: Oxford University Press, 9–36.

Hughes, J. D. 2011. *Lifecycle Greenhouse Gas Emissions from Shale Compared to Coal: An Analysis of Two Conflicting Studies*. Santa Rosa, CA: Post Carbon Institute.

IEA (International Energy Agency). 2008. *World Energy Outlook 2008*. Paris: OECD/IEA.

IEA. 2011a. *CO_2 Emissions from Fuel Combustion: Highlights*. Paris: OECD/IEA.

IEA. 2011b. *World Energy Outlook 2011*. Paris: OECD/IEA.

Jaccard, M. 2005. *Sustainable Fossil Fuels: The Unusual Suspect in the Quest for Clean and Enduring Energy*. Cambridge University Press: Cambridge.

Jackson, M. 2009. *Prosperity Without Growth? The Transition to a Sustainable Economy*. London: Sustainable Development Commission.

Klare, M. 2012. *The Race for What's Left: The Global Scramble for the World's Last Resources*. New York: Metropolitan Books.

Kramer, G. J., and M. Haigh. 2009. No Quick Switch to Low-Carbon Energy. *Nature* 462: 568–569.

McNeil, J. 2000. *Something New under the Sun: An Environmental History of the Twentieth Century*. London: Penguin Books.

Metz, B., O. R. Davidson, P. R. Bosch, et al., eds. 2007. *Reports of the Intergovernmental Panel on Climate Change*. Cambridge: Cambridge University Press.

Murphy, D. J., and C. A. S. Hall. 2010. Year in Review – EROI or Energy Return on (Energy) Invested. *Annals of the New York Academy of Sciences* 1185: 102–118.

O'Neill, B. C., M. Dalton, R. Fuchs, et al. 2010. Global Demographic Trends and Future Carbon Emissions. *Proceedings of the National Academy of Sciences* 107, 41: 17521–17526.

Pachauri, R. K., and A. Reisinger, eds. 2007. *Contribution of Working Groups I, II and III to the Fourth Assessment Report of the Intergovernmental Panel on Climate Change*. Geneva: IPCC.

Peters, G. P., G. Marland, C. Le Quéré, et al. 2012. Rapid Growth in CO_2 Emissions after the 2008–2009 Global Financial Crisis. *Nature Climate Change* 2: 2–4.

Peters, G. P., J. C. Minx, C. L. Weber, and O. Edenhofer, 2011. Growth in Emission Transfers via International Trade from 1990 to 2008. *Proceedings of the National Academy of Sciences*, early view, doi: 10.1073/pnas.1006388108.

Pielke, Jr., R. 2010. *The Climate Fix: What Scientists and Politicians Won't Tell You About Global Warming.* New York: Basic Books.

Raupach, M. R., G. Marland, P. Ciais, *et al.* 2007. Global and Regional Drivers of Accelerating CO_2 Emissions. *Proceedings of the National Academy of Sciences* 104: 10288–10293.

Rühl, C., P. Appleby, J. Fennema, *et al.* 2012. Economic Development and Demand for Energy: A Historical Perspective on the Next 20 Years. Paper published to coincide with *BP Energy Outlook 2030.* http://www.bp.com/liveassets/bp_internet/globalbp/STAGING/global_assets/downloads/R/reports_and_publications_economic_development_demand_for_energy.pdf, accessed May 3, 2012.

Smil, V. 2010. *Energy Transitions: History, Requirements and Prospects.* Denver, CO: Praeger.

Unruh, G. C. 2000. Understanding Carbon Lock-in. *Energy Policy* 28: 817–830.

Watson, J., ed. 2012. *Carbon Capture and Storage: Realising the Potential.* London: UK Energy Research Centre.

World Energy Council. 2010. *Energy Efficiency: A Recipe for Success.* London: World Energy Council.

Chapter 4

The Development Nexus of Global Energy

Gilles Carbonnier and Fritz Brugger

Introduction

Energy has been one of the key drivers of human development since prehistoric times. The domestication of fire and the subsequent ability to use it for smelting iron ore boosted the development of new civilizations. The ability to use animal traction for transport and agriculture with the invention of harnesses and stirrups, windpower for sailing, and hydropower for milling all contributed to increasing productivity.

Yet, these traditional sources of energy did not allow for any sustained period of economic growth. Human standard of living did not progress much from antiquity to the "thermo-industrial revolution" of the early nineteenth century. According to Maddison (2001), the world's economy actually declined during the first millennium and then grew at an average annual rate of 0.05% from 1000 to 1820, slightly above the world's demographic growth. Annual global economic growth remarkably increased to 1% between 1820 and 1950, and reached almost 3% between 1950 and 1973 (Severino and Ray 2011). This has been made possible thanks to an increasingly massive consumption of cheap fossil fuels and, to a lesser extent, fossil materials.

Figure 4.1 shows the increasing consumption of coal in the nineteenth century and oil thereafter, in particular since the end of World War II, resulting in a steep takeoff of the global emissions from burning fossil fuels between 1800 and 2007. The thermo-industrial revolution in Europe and North America, sustained by military and civil engineering skills, opened new frontiers for industrialization, speeded up mobility, and boosted trade (Carbonnier and Grinevald 2011). Modern energy gave rise to a unique and sustained demographic and economic growth while traditional, renewable energy sources have been marginalized in the energy mix of industrialized and emerging economies. The exploitation of cheap oil, with its exceptional properties (liquid, easy to store and to transport), has spurred a massive leap in the development process, which has been all too often overlooked by the traditional evolutionary paradigm that considers a linear relationship between the progressive command over different energy sources and the development of civilizations.

The Handbook of Global Energy Policy, First Edition. Edited by Andreas Goldthau.
© 2013 John Wiley & Sons, Ltd. Published 2013 by John Wiley & Sons, Ltd.

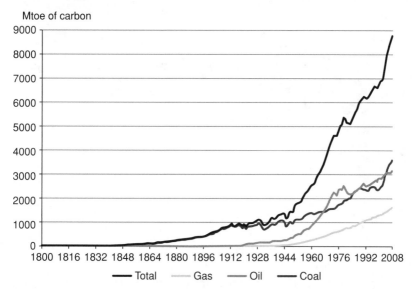

Figure 4.1 Global CO$_2$ emissions from fossil-fuel burning, 1800–2008.
Source: Boden *et al.* (2011).

Today, the energy–development nexus encompasses a complex set of direct and indi-rect relations. In this article, we focus on three interrelated clusters: first, the relationship between access to energy and development; second, energy security, food, and water policies; third, the impact of energy resources on development prospects in producer countries. From a development perspective this is of particular relevance since a number of developing countries recently joined the ranks of energy producers (e.g., Chad, Ghana, Uganda) while others see major exploration activities and may follow suit (Mozambique, Tanzania, Puntland (Somalia), Ethiopia, Mauretania, Kenya). This opens a window of opportunity for development but the cases of Angola, Nigeria, Gabon, and Equatorial Guinea also illustrate the risks associated with revenue windfalls.

The concluding section discusses the lack of coherence and coordination between these distinct policy fields. Environmental externalities of energy production and consumption obviously have a major impact on development but are discussed elsewhere in this volume.

Energy Poverty, Inequality, and Development

Use of energy per capita data, measured in kilogram oil equivalents (kgoe), across coun-tries shows a correlation with human development as measured by the Human Develop-ment Index (HDI) that combines monetary income with health and education indicators. Figure 4.2 illustrates that the relationship between energy and human development is not linear.

Martínez and Ebenhack (2008) observe a steep rise in human development relative to energy consumption for energy poor nations; a moderate rise for transitioning nations; and no rise in human development for countries consuming large amounts of modern energy. More concretely, a threshold of annual energy consumption of close to 4000 kgoe per capita seems to be required to achieve an HDI value of 0.8 for oil importing countries, but only 0.7 for oil exporting nations. Energy consumption above 4000 kgoe does not lead to any significant improvement in the HDI. Oil exporting countries, marked by the black

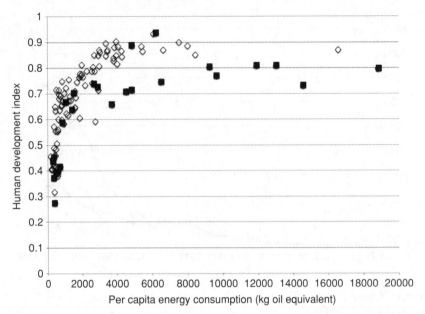

Figure 4.2 Per capita energy consumption versus HDI, 2008. Oil and gas producing countries are in black.
Source: Martínez and Ebenhack (2008).

squares in Figure 4.2, tend to have a higher energy consumption rate relative to their HDI, which may reflect the well-known "resource curse" phenomenon (see below). Accounting for the share of biomass dependence, the correlation between modern energy sources and HDI is even stronger: According to Martínez and Ebenhack (2008) countries with more than 60% biomass energy dependence systematically score below 0.6 on the HDI.

Access to modern energy is very unevenly distributed. The least developed countries, with 10% of the world's population, account for 1% of global energy consumption (and 1% of the world's GDP). The US, with less than 4.5% of the world's population, consumes 24% of the world's energy and accounts for 30% of its GDP. According to the International Energy Agency, about 1.4 billion people do not have access to electricity, 85% of them living in rural areas. Besides, 2.7 billion have to rely on traditional biomass for cooking and heating (IEA *et al.* 2010: 9).

The consequences for economic and human development are far-reaching. Using firewood as cooking fuel causes close to 1.5 million premature deaths every year due to indoor air pollution, more than deaths caused by tuberculosis or malaria (IEA *et al.* 2010: 14), putting children and pregnant women at particular risk (Gaye 2007: 7). Collecting firewood occupies a large part of women and children's time, and accelerates deforestation. Using mainly kerosene for lighting, those without access to electricity spend nearly $40 billion a year, which is almost 20% of all global lighting expenditures (Mills and Jacobson 2011: 524). Low efficiency of lanterns and high prices paid for buying small quantities of kerosene make such lighting 200 times more expensive than for grid users.

Africa's energy situation is paradoxical in that the continent requires better access to energy for poverty reduction and economic development, yet it is a net exporter of commercial energy (IAE 2008: 355–376). Africa produces about 8% of the world's commercial energy but accounts for more than 12% of world exports while it consumes only 3% of global commercial energy. The data in Table 4.1 shows that, in the 10 top

Table 4.1 Africa's energy exporters and their populations' access to energy.

| | Oil | | | Gas | | | Population | | | | |
| | Reserves | Production | Export | Reserves | Production | Export | Total 2006 | Without Electricity | | Cooking with Wood/Charcoal | |
	bn b	mbd	mbd	bcm	bcm/yr	bcm/yr	mio	mio	%	mio	%
Angola	9.0	1.70	1.64	270	0.8	–	16.6	14.6	88	15.7	95
Cameroon	0.2	0.09	0.06	135	–	–	18.2	14.2	78	14.2	78
Chad	1.5	0.14	0.14	–	–	–	10.5	10.1	97	10.2	97
Congo Rep	1.6	0.21	0.21	91	1.7	–	3.7	2.9	78	2.9	80
Côte d'Ivoire	0.1	0.06	0.03	28	1.3	–	18.9	11.6	61	14.7	78
Eq. Guinea	1.1	0.36	0.36	37	1.3	–	0.5	0.4	73	0.3	59
Gabon	2.0	0.23	0.22	28	0.1	–	1.3	0.9	70	0.4	33
Mozambique	–	–	–	127	2.7	2.7	21.0	18.6	89	16.9	80
Nigeria	36.2	2.35	2.03	5207	29.3	18.9	144.7	76.6	53	93.8	65
Sudan	5.0	0.47	0.39	85	–	–	37.7	26.9	71	35.2	93
Total	56.8	5.61	5.09	6008	35.9	21.6	273.1	176.9	65	204.0	75
% in world	4.3	7.0	12.1	3.4	1.2	5.2					

Source: Data compiled from IEA (2008: 356 and 358).

oil and gas exporting countries of Sub-Saharan Africa, 65% of the people do not have access to electricity and 75% rely on firewood and charcoal for cooking.

Improving access to energy for people in rural areas should be a development priority. Assuming that annual investment in improving access to energy averages $14 billion per year (up from $9.1 billion in 2009), an additional $34 billion per year are required to achieve universal access to modern energy sources by 2030, or the equivalent of 3% of global investment in energy infrastructure from 2010 to 2030 (IEA 2011: 20). Achieving universal access to modern energy sources by 2030 would increase global electricity generation by 2.5%. Demand for fossil fuels would grow by 0.8% and CO_2 emissions go up by 0.7% (IEA 2011: 7).

This would result in an increase of 239 million tons of CO_2, or 60% of the CO_2 currently emitted by the flaring of 150 billion cubic meters of "excess" natural gas associated with oil production every year (IEA 2011: 27; World Bank 2011). Avoiding that waste of gas in flame is a question of political will in the first place. While Norway or Brazil have declared gas flaring illegal (which in the case of the Brazilian pré-sal ultra-deep offshore deposits requires massive investments in pipelines to capture associated gas), most oil producing countries have not implemented effective regulatory frameworks to enforce use of associated gas, which leads companies to prefer flaring the gas instead of investing in required infrastructure.

Government policies in developing countries play a key role in setting a national energy agenda. For example, the Dominican Republic's decision to subsidize natural gas in the 1990s released pressure on wood as primary source of energy. The forest coverage in the Dominican part of Hispaniola started to recover while in neighboring Haiti, where charcoal still covers over 85% of energy needs, forest cover is below 2%, as can easily be noticed on Google Earth. The difference is also clearly reflected in the countries' ranking in the Energy Development Index (IAE 2010) that tracks the transition of developing countries to the use of "modern" fuels.

Some countries like Brazil, Ghana, India, South Africa, and Zambia have launched national electrification programs that faced difficulties in providing reliable access to electricity in the absence of competent utility companies or because of politically moti-vated subsidies that provide perverse incentives. While state-owned utilities often operate under politically imposed rules that hamper their ability to run operations efficiently, out-sourcing to the private sector or public–private partnerships proved to be no blueprint either, raising issues of accountability and responsiveness. Strong corporate governance of utilities, ring-fenced from political influence, is key for sustainable service provision (Andrés *et al.* 2011).

Energy Security, Food, and Water

This chapter outlines the linkages between an energy security and geopolitics, food and agriculture policy, and water policy. While there is considerable literature for each policy area, they are usually examined in isolation. Not surprisingly, policy coherence is lacking. Today, energy policy must integrate social and sustainable development and vice versa. More than ever, our own energy security lies in the security of others.

Geopolitical Reconfiguration: The Scramble for African Resources

With raising global demand for energy, in particular from Asian emerging economies, energy security policies result in a heightened scramble for energy resources with Asian extractive industries rushing to invest abroad. In this context, African countries (re-) specialize in primary commodity exports and have gained renewed geostrategic significance. The heightened interest in Africa's energy resources opens a window of opportunity for African governments to redefine their own development path.

The Gulf of Guinea has become an important source of oil supply, in particular for the US that expects to import a quarter of its oil supplies from this region by 2015, which remains valid despite the shale gas production boom in the US. After 9/11, policymakers and oil companies under the African Oil Policy Initiative Group lobbied in Washington to increase US security focus on West Africa (AOPIG 2002), contributing to the establishment of AFRICOM in 2008. But instability in producer countries along the Gulf of Guinea does not affect the US alone. Some experts estimate the geopolitical risk premium between $15 and $20 per barrel (Blas 2011). Minor events can have a direct upward effect on the price of oil on world markets. This has for instance been the case when Gabon's National Organization of Oil Workers went on strike on March 31, 2011 in an attempt to force the government to reduce the number of foreigners in the oil industry down to the promised 10% level.

Parallel to the expansion of the US hegemony, Chinese and to a lesser extent Indian, Korean, and Malaysian engagement in Sub-Saharan Africa increased significantly over the past two decades. China considers Africa not only as a source of raw materials but also as a market for its products and services. With this broad agenda, China became a significant investor and aid donor, maintaining friendly relationships with the ruling elites and adopting a policy of non-interference in domestic politics.

Food and Energy

The nexus between food and energy has grown stronger as a result of rising energy intensity in agriculture and increasing land use competition between food and energy production. Both trends have pushed food prices upward, which has affected the (urban) poor in developing countries harder to the extent that they spend a larger share of their income on food than the rich in industrialized countries.

Concerns over peak oil (a debate that is far from being conclusive, as discussed in Chapter 11) and climate change have led to the development of biofuels as an alternative source of energy, supported by tax and regulatory incentives (tax breaks, mandated use, and subsidies) in Europe and North America. The first generation of biofuels, which are made primarily from sugar, starch, and vegetable oil, has a perverse "crowding out" effect whereby agricultural land and crops are diverted toward the production of biofuels, which results in a reduction in global crop supplies and cereal stocks (Mitchell 2008).

The International Food Policy Research Institute estimates that the increased demand for biofuels between 2000 and 2007 has contributed up to 30% of the increase of weighted average grain prices (Rosegrant 2008). However, the longer-term effect on food prices is expected to decrease with the transition to the so-called second generation of biofuels that put less pressure on agricultural land and are meant not to compete with food production (OECD 2008). Chapter 16 of this volume discusses in more detail the prospects and perils of biofuels and biofuel policies.

Besides the pressure coming from biofuels, modern agriculture heavily depends on fossil fuels. As a result, there is a strong correlation between oil and food prices, as Figure 4.3 illustrates. Both direct energy use for crop management and indirect energy use for fertilizers, pesticides, and machinery production have contributed to the major increases in food production since the 1960s (Woods *et al.* 2010).

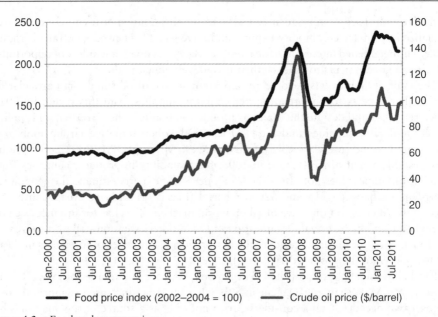

Figure 4.3   Food and energy prices.
Source: FAO Food Price Index; US Energy Information Administration; FAOSTAT.

Nitrogen fertilizer production uses large amounts of natural gas (90% of the fertilizer price is the cost for gas) and can account for more than 50% of total energy use in commercial agriculture (Abram and Forster 2005). Fertilizer expenditure has increased disproportionately since fossil fuel prices skyrocketed. As long as agriculture remains highly dependent on fossil energy, food prices will be coupled with energy prices, and food production will remain a significant contributor to anthropogenic greenhouse gas emissions. High oil and gas prices further render access to fertilizers often too expensive for farmers in developing countries.

Increasing the productivity of smallholder farms in developing countries and closing the yield gap are critical to improve food security. Changing diet patterns toward more meat and dairy products in emerging and developing countries following urbanization and the emergence of a growing middle class may lead to a more energy intensive agriculture, with an increasing use of synthetic inputs (fertilizers, chemicals) and mechanization. Some argue that genetically modified crops could reduce the reliance on such inputs. But starting by reducing food losses could save much energy, reduce CO_2 emissions and increase food security. Globally, about one third of the food produced for human consumption is lost or wasted, which amounts roughly to 1.3 billion tons per year (Gustavsson *et al.* 2011: 4). In developing countries, this results mainly from a host of financial, managerial, and technical limitations in harvesting techniques, storage and cooling facilities, and infrastructure, packaging, and marketing systems. In upper middle- and high-income countries food is more often wasted at the consumption stage.

Energy and Water Policy

While the food–energy nexus has gained prominence with the food price riots in 2008 and the debates around biofuel, water issues have been somehow neglected in the wider policy-making circles. Water deserves special attention since it is required for energy production and it depends on energy for transport, treatment, purification, desalination, pumping, etc. "Thermoelectric power plants running on coal, natural gas, oil and uranium are water-cooled, withdrawing trillions of cubic meters of water from rivers and streams, consuming billions of gallons of water from local aquifers and lakes, and contaminating water supplies at various parts of their fuel cycle" (Florini and Sovacool 2011: 67). But the required water might not be always available where needed. By 2025, more than 60% of the world's population is expected to live in countries – mainly in Asia, Africa, and Latin America – where water supplies no longer meet requirements (Feeley *et al.* 2008: 1).

Today, over 50 countries rely primarily on hydroelectric dams to generate power and large dams produce 19% of the world's electricity (Zehnder *et al.* 2003: 5). While Europe and North America use more than half of their potentially commercially viable hydropower resources, 60–80% of the potential in developing countries remains unexploited (Zehnder *et al.* 2003). But this is about to change as dozens of dams are planned in South Asia over the next few decades mainly for the sake of producing energy (*Economist* 2011a). Upstream countries like Nepal or Bhutan could benefit from selling electricity to India.

However, large dams for hydropower and irrigation are also a source of major tensions. They can cause large displacement of population and negatively impact on the hydrology of the catchment area. They make downstream water users more dependent on upstream users, which can add stress on already tense relations between neighboring countries. Pakistan for example may feel threatened by the Baglihar dam on the upstream Indian part of the Indus. Some reports warn that the "cumulative effect of [many dam] projects could give India the ability to store enough water to limit the supply to Pakistan at crucial moments in the growing season" (US Committee on Foreign Relations 2011: 5). Water diplomacy will be required to keep cooperation prevailing over confrontation.

Impact of Resource Extraction on Development in Producing Countries

Changing Landscape in Oil Production

The number of oil and gas producing countries has increased since more investment went into exploration over the last decade; this was triggered by growth of demand, renewed interest in nuclear energy, and rising prices. A large portion of newly discovered reserves (oil, gas, uranium) and unexplored territory are located in developing countries. Former net importing countries are becoming exporters like Sudan and Chad a few years ago or Ghana and Uganda in 2010 and 2011 respectively. Most of the increase in demand originates from the fast growing Asian economies, mainly China (which became a net importer of oil in 1993), India, and Korea. All three have formulated explicit internationalization strategies to secure access to energy resources through their state-owned extractive industries venturing abroad (Jiang and Sinton 2011; Jung-a 2006; Lewis 2007; Paik *et al.* 2007; PRC Ministry of Foreign Affairs 2006; Suh 2006). This strategy is particularly noteworthy for Korea, which has no significant domestic oil production.

The "going out" of Asian state-owned energy enterprises has changed the extractive sector in several ways. First, it has contributed to increased foreign direct investment (FDI) in upstream projects. Second, it has increased competition. Third, the fact that these new Asian consumer heavyweights have decided not to rely only on buying oil on the world's markets but also strive to get a share of the production raises new energy security issues. Finally, there is a debate on whether Asian state-owned companies initiate a race to the bottom in the extractive sector regarding environmental standards and respect for human and labor rights, particularly in countries where governmental regulatory capacity is limited; the absence of transparency, the policy of non-interference, and reluctance to cooperate with non-governmental actors might also change the prospect for governance initiatives such as the Extractive Industries Transparency Initiative.

Turning Energy Extraction into Development

The nexus between primary resource extraction and development has come to be known under the "resource curse" label, capturing the paradox that many resource-rich economies in the developing world suffer from lower economic growth than resource-poor countries (Auty 1993; Collier and Goderis 2007; Karl 1997; Sachs and Warner 1997). This has been explained by focusing on high corruption levels (Dietz *et al.* 2007), on the authoritarian nature of the regimes in oil producing countries (Ross 2001), and on prolonged civil war and severe human rights abuses correlated with high natural resource dependence (Collier and Hoeffler 2004; Le Billon 2003). Yet there is little consensus about the transmission channels through which resource wealth causes negative development outcomes (Papyrakis and Gerlagh 2004) and on the relative incidence of rentier state dynamics (Beblawi and Luciani 1987; Mahdavy 1970), of Dutch disease and revenue volatility (Auty 1993) or of governance and the quality of institutions (Torvik 2009).

Also, as Ross (1999: 307) noted, all these theories fail to explain why the governments which play an exceptionally large role in the resource sectors of almost all developing countries fails to take corrective measures. At least in theory, they have the policy tools to mitigate each of these hardships.

Against the theoretical and practical shortcomings of the Dutch disease, rentier state, and rent-seeking versions of the resource curse, Jonathan Di John (2011) finds that the resource impact on a country – the extent to which mineral and fuel abundance generate developmental outcomes – depends largely on the nature of the state to appropriate mineral rents, the ownership structure in the export sector, and the ability of the state to implement a dual-track growth strategy where a ring-fenced sector promotes the emergence of a competitive industry to drive the diversification of the economy.

It is critical to understand the root causes for significant variations in the performance of resource-rich developing countries. One can distinguish three main clusters of arguments: a first cluster finds that rents deteriorate institutions; a second cluster points to the importance of the quality of institutions prior to windfall revenues; the third cluster disentangles the notions of governance and institutions and puts the role of "impartiality-enhancing institutions" center stage (Acemoglu and Johnson 2005; Kolstad and Wiig 2008). Recent research highlights the importance of effective checks and balances that constrain the power of the executive (Carbonnier *et al.* 2011b). The quality of institutions and the existence of checks and balances that limit the discretionary power of the government are key to avoid resource curse dynamics (Stoever 2012), a finding that can be traced back to the silver windfall from Spain's South American colonies in the seventeenth and eighteenth centuries that eroded nascent Spanish democratic institutions (Drelichman and Voth 2008; North 1973). This explanatory variable focuses on the output side of the political system. The decisive question for the performance of resource dependent economies is: how are (policy) decisions being made and how is power exercised with impartiality as the organizing principle (Rothstein and Teorell 2008: 15).

Policy Options and Development Impact

In most jurisdictions, ownership of subsoil assets rests with the state, which defines the conditions under which national and international extractive industries can extract and sell energy resources (for a discussion of the issue of resource ownership from a legal and ethical perspective, see Viñuales 2011 and Schaber 2011). National resource extraction policies determine critical issues for development purposes such as the pace of extraction, local content requirements, and the allocation of the resource rent. Policy-making implies domestic power relations and pressures from external actors such as the international financial institutions (IFIs), multinational extractive firms, and the major consumer states, which all tend to lobby for a liberal, unimpeded access to energy resources.

Schematically, domestic policies have swung over time between a nationalist, state-led model and a neoliberal approach, which influenced the way concessions or production-sharing agreements (PSA) have been crafted. While there is a mix of both in the real world, considering the two extremes helps clarify the debate. The neoliberal model focuses on attracting FDI, minimizing taxes and royalties, granting generous tax exemptions, and no conditions with regard to local content. Extraction is seen as just another industrial activity. To reward investors' risk, royalties should be abandoned and taxes should be limited to corporate income tax. While this approach, promoted by the IFIs under the "Washington consensus," has been relatively effective in bringing in foreign investors, the development outcome of extraction has been limited (Di John 2011: 177; UNCTAD 2009). The opposite approach considers the allocation of the extractive rent between the host state and foreign firms as a critical element, where the rent should accrue to the resource owner. This implies a strong role for domestic state-owned companies or even full nationalization. Under this approach, international oil companies (IOCs) may be invited to invest for the sake of technology transfer under the direction of national oil companies (NOCs).

Over time there has been a swing between the two poles with nationalization in the 1970s followed by rounds of liberalization in the 1980s and 1990s. Interestingly, these trends were closely associated with upward and downward trends in oil prices. By 2003/4, resource nationalism came back to the fore in Latin America, led by Venezuela, Ecuador, and Bolivia (UNCTAD 2006). Many developing countries introduced more stringent fiscal arrangements following record prices of extractive resources (UNCTAD 2007). In Brazil, the discovery of the large pré-sal deposits led to a shift from the former liberal system whereby concessions were to be sold at auctions in which any company, Brazilian or foreign, could bid equally. A new state enterprise, Pré-Sal Petróleo, shall now own all pré-sal deposits and be able to veto projects it deems not in the national interest. Future pré-sal concessions will be auctioned to consortia which must include Petrobras as the operator. And once a consortium will have pumped enough cost oil to cover its initial investment, profit oil will have to be shared with the Brazilian state, and winning bids shall be those that hand over more of profit oil to Brazil (*Economist* 2011b).

Revenue Management

Sound revenue management by producer states is a critical element of any policy seeking to avert the resource curse. Price volatility has traditionally contributed to making this more difficult. Over the past years, volatility has increased since oil has become a financial asset class. About 70% of today's investors in the oil trade are not active in the oil sector, in essence speculators who increase price volatility (Mabro 2005). This increase in commodity trading has been encouraged by the US Commodities Futures Modernization Act of 2000, which removed legal constraints on speculative trading in over-the-counter (OTC) derivatives (Fattouh 2010; Medlock and Jaffe 2009). Some provisions of the Dodd–Frank Wall Street Reform and Consumer Protection Act signed into law in 2010 aim at turning back to tighter regulation by restoring legal limits on speculative derivatives that are traded outside of a clearing house. However, a number of exemptions may limit the effectiveness of its enforcement (Greenberger 2011).

The World Bank has actively promoted national revenue funds to compensate for price fluctuations and save for future generations, once the fuel or mineral reserves are exhausted. Since the 2008 financial crisis, these funds have come under mounting critique and their performance in poor countries is found to be weak (Luciani 2011). Collier *et al.* argue that resource-rich developing countries should focus investing in assets yielding high social returns (basic infrastructure, public health) instead of investing in low-yielding global assets (Collier *et al.* 2010: 85–86). We agree with the principle from a capital perspective, whereby the exploitation of exhaustible natural resources is a liquidation of an inherited asset and corresponds to the depletion of a producer country's natural capital. Investing the ensuing rent in infrastructure, capital goods, and human capital is a prerequisite for continued prosperity beyond resource extraction. Yet investment opportunities are often limited by a weak absorption capacity that can result in low-quality but overpaid infrastructure and poor return on investment (Dabla-Norris *et al.* 2010; Haque and Kellner 2008). In fact, many resource-rich developing countries show negative genuine savings rates, which means that they have squandered part of their national wealth and have not reinvested the proceeds of natural capital depletion into other forms of capital (Carbonnier *et al.* 2011b).

Governance Responses

For net importing countries, energy security understood as the availability of cheap energy and access to energy resources for their extractive industries is a top foreign-policy priority. Concerns about doing business with authoritarian regimes are rather the exception than the rule, and very few instruments have been developed so far to address the resource curse. This is the case with the Extractive Industries Transparency Initiative (EITI) and the Voluntary Principles on Human Rights and Security, which were introduced as a response to civil society pressure and are structured as voluntary governance mechanisms. The eventual effectiveness and outcome of such voluntary multi-stakeholder regimes remains to be evaluated in a rigorous manner. For the time being, they suffer from the fact that Asian states and companies stay outside and that local civil society organizations in producer states are too weak to effectively discharge monitoring, advocacy, and whistle-blowing functions (Carbonnier *et al.* 2011a). Voluntary regimes may be complemented by Section 1504 (also called the Cardin–Lugar provision) of the Dodd–Frank Act of 2010 that requires all extractive firms registered with the US Security and Exchange Commission (SEC) to disclose their payments to producer states on a country-by-country basis. Depending on the eventual enforcement modalities, this may clearly improve transparency on the industry side, but cannot help to improve accountability on the allocation of revenues by producer states, which, in the end, is the decisive factor for development. The only attempt so far to regulate revenue allocation was tried out by the World Bank in the case of the Chad–Cameroon pipeline. It fell victim to the traditional obsolescing bargain problem once investments were made and the Chadian government enjoyed a steady flow of revenues (Gould and Winters 2011; Pegg 2009).

Conclusions

In tracing the links between energy and development we have crossed a number of policy fields, including food security, water governance and management, trade and investment, security, and environmental policies, all having local, national, and international dimensions. Energy is the backbone of our industrialized society, which has largely hung on the consumption of cheap and abundant fossil fuels over the past four centuries. But the twin constraints of a looming peak of oil production and climate change radically call into question this development model, just when billions of people in emerging economies can reasonably expect – and aspire to – join the club of wealthy industrialized countries. Prices will increasingly provide strong incentives to move to a less carbon-intensive and a less energy-intensive development pathway. But price incentives cannot reverse the depletion of exhaustible resource stocks, the concentration of greenhouse gases in the atmosphere, and the loss of biological diversity.

Domestic and international development efforts should focus on granting access to modern energy to the 20% of the world's population that are still deprived of it and to reduce the current dependence of the 2.7 billion people that depend on biomass for cooking. Another priority is to reduce energy waste and increase energy efficiency in the booming urban centers in the global South as in the North. Faced with the prospect of nine billion people on planet Earth by 2050, the nexus between energy, food, and water urgently requires more attention from scholars and more coherence from policy-makers.

Energy security remains all too often confined by a nationalist perspective that overrides other priorities such as development and peace-building objectives. Engaged in a race to secure access to energy resources, Western and emerging economies overlook that sustainable development in producer countries and increased cooperation between exporting, transit, and importing countries is a cornerstone of any serious attempt to improve energy security and avert the multiplication of conflicts over energy resources.

References

Abram, Aleksander, and D. Lynn Forster. 2005. A Primer on Ammonia, Nitrogen Fertilizers, and Natural Gas Markets. *Agricultural, Environmental and Development Economics* 53: 50.

Acemoglu, Daron, and Simon Johnson. 2005. Unbundling Institutions. *Journal of Political Economy* 113, 5: 949–995.

Andrés, Luis Alberto, José Luis Guasch, and Sebastián López Azumendi. 2011. *Governance in State-Owned Enterprises Revisited. The Cases of Water and Electricity in Latin America and the Caribbean.* Policy Research Working Paper 5747. Washington, DC: World Bank.

AOPIG. 2002. *African Oil: A Priority for U.S. National Security and African Development.* Research Papers in Strategy 14. Washington, DC: Institute for Advanced Strategic and Political Studies.

Auty, Richard M. 1993. *Sustaining Development in Mineral Economies: The Resource Curse Thesis.* London: Routledge.

Beblawi, Hazem, and Giacomo Luciani. 1987. *The Rentier State.* London: Croom Helm.

Blas, Javier. 2011. Oil Hits $120 After Strike in Gabon. *Financial Times*, April 4.

Boden, T. A., G. Marland, and R. J. Andres. 2011. Global, Regional, and National Fossil-Fuel CO2 Emissions. http://cdiac.ornl.gov/by_new/bysubjec.html#carbon, accessed January 12, 2012.

Carbonnier, Gilles, and Jacques Grinevald. 2011. Energy and Development. *Revue internationale de politique de développement* 2: 9–28.

Carbonnier, Gilles, Fritz Brugger, and Jana Krause. 2011a. Global and Local Policy Responses to the Resource Trap. *Global Governance* 14, 2: 247–264.

Carbonnier, Gilles, Natascha Wagner, and Fritz Brugger. 2011b. Oil, Gas and Minerals: The Impact of Resource-Dependence and Governance on Sustainable Development. *CCDP Working Paper* 8: 36.

Collier, Paul, and Benedikt Goderis. 2007. *Commodity Prices, Growth, and the Natural Resource Curse: Reconciling a Conundrum*. CSAE Working Paper 15.

Collier, Paul, and Anke Hoeffler. 2004. Greed and Grievance in Civil War. *Oxford Economic Papers* 56, 4: 563–595.

Collier, Paul, Michael Spence, Frederick van der Ploeg, and Anthony J. Venables. 2010. Managing Resource Revenues in Developing Economies. *IMF Staff Papers* 57, 1: 84–118.

Dabla-Norris, Era, Jim Brumby, Annette Kyobe, *et al.* 2010. *Investing in Public Investment: An Index of Public Investment Efficiency*. IMF Working Paper WP/11/37.

Di John, Jonathan. 2011. Is There Really a Resource Curse? A Critical Survey of Theory and Evidence. *Global Governance* 14, 2: 167–184.

Dietz, Simon, Eric Neumayer, and Indra de Soysa. 2007. Corruption, the Resource Curse and Genuine Saving. *Environment and Development Economics* 12: 33–53.

Drelichman, Mauricio, and Hans-Joachim Voth. 2008. *Institutions and the Resource Curse in Early Modem Spain*. In Elhanan Helpman, ed. *Institutions and Economic Performance*. Cambridge, MA: Harvard University Press.

Economist. 2011a. South Asia's Water: Unquenchable Thirst. *The Economist*, November 19.

Economist. 2011b. Brazil's Oil Boom: Filling Up the Future. *The Economist*, November 5.

Fattouh, Bassam. 2010. *Oil Market Dynamics Through the Lens of the 2002–2009 Price Cycle*. Oxford: Oxford Institute for Energy Studies.

Feeley, Thomas J., Timothy J. Skone, Gary J. Stiegel Jr., *et al.* 2008. Water: A Critical Resource in the Thermoelectric Power Industry. *Energy* 33, 1: 1–11.

Florini, Ann, and Benjamin K. Sovacool. 2011. Bridging the Gaps in Global Energy Governance. *Global Governance* 17, 1: 57–74.

Gaye, Amie. 2007. *Access to Energy and Human Development*. Occasional Paper 25. New York: UNDP, Human Development Report Office

Gould, John A., and Matthew S. Winters. 2011. Petroleum Blues: The Political Economy of Resources and Conflict in Chad. In Päivi Lujala and Siri Aas Rustad, eds. *High-Value Natural Resources and Post-Conflict Peacebuilding*. Washington, DC: Environmental Law Institute.

Greenberger, Michael. 2011. *Will the CFTC Defy Congress's Mandate to Stop Excessive Speculation in Commodity Markets and Aid and Abet Hyperinflation in World Food and Energy Prices? Analysis of the CFTC's Proposed Rules on Speculative Position Limits*. University of Maryland Legal Studies Research Paper 20.

Gustavsson, Jenny, Christel Cederberg, Ulf Sonesson, *et al.* 2011. *Global Food Losses and Food Waste. Extent, Causes and Prevention*. Rome: Food and Agriculture Organization of the United Nations (FAO).

Haque, Emranul, and Richard Kellner. 2008. *Public Investment and Growth: The Role of Corruption*. Discussion Paper Series 098. Manchester: Centre for Growth and Business Cycle Research, Economic Studies, University of Manchester.

IEA (International Energy Agency). 2008. *World Energy Outlook 2008*. Paris: IEA.

IEA. 2010. *World Energy Outlook – Energy Development Index*. Paris: IEA.

IEA. 2011. *Energy for All: Financing Access for the Poor. Special Early Excerpt of the World Energy Outlook 2011*. Paris: IEA.

IEA, UN Development Programme, and UN Industrial Development Organization. 2010. *Energy Poverty: How to Make Modern Energy Access Universal? Special Early Excerpt of the World Energy Outlook 2010 for the UN General Assembly on the Millennium Development Goals*. Paris: IEA.

Jiang, Julie, and Jonathan Sinton. 2011. *Overseas Investments by Chinese National Oil Companies. Assessing the Drivers and Impacts*. Paris: IEA.

Jung-a, Song. 2006. Posco in Talks Over $10bn Rail Project in Nigeria. *Financial Times*, November 2.

Karl, Terry Lynn. 1997. *The Paradox of Plenty: Oil Booms and Petro-States*. Berkeley, CA: University of California Press.

Kolstad, Ivar, and Arne Wiig. 2008. *Political Economy Models of the Resource Curse: Implications for Policy and Research*. Bergen: Chr. Michelsen Institute.

Le Billon, Philippe. 2003. *Fuelling War: Natural Resources and Armed Conflicts*. Oxford: Oxford University Press.

Lewis, Steven W. 2007. *Chinese NOCs and World Energy Markets: CNPC, Sinopec and CNOOC*. Houston, TX: Baker Institute for Public Policy, Rice University.

Luciani, Giacomo. 2011. Price and Revenue Volatility: What Policy Options and Role for the State? *Global Governance* 17, 2: 213–228.

Mabro, Robert. 2005. The International Oil Price Regime. Origins, Rationale and Assessment. *Journal of Energy Literature* 11, 1: 3–20.

Maddison, Angus. 2001. *The World Economy, a Millennial Perspective*. Paris: OECD.

Mahdavy, H. 1970. Patterns and Problems of Economic Development in Rentier States. The Case of Iran. In M. A. Cook, ed. *Studies in the Economic History of the Middle East*. Oxford: Oxford University Press.

Martínez, Daniel M., and Ben W. Ebenhack. 2008. Understanding the Role of Energy Consumption in Human Development Through the Use of Saturation Phenomena. *Energy Policy* 36: 1430–1435.

Medlock, III, Kenneth B., and Amy Myers Jaffe. 2009. *Who Is in the Oil Futures Market and How Has It Changed?* Houston, TX: Baker Institute for Public Policy, Rice University.

Mills, Evan, and Arne Jacobson. 2011. From Carbon to Light: A New Framework for Estimating Greenhouse Gas Emissions Reductions from Replacing Fuel-Based Lighting with LED Systems. *Energy Efficiency* 4: 523–546.

Mitchell, Donald. 2008. *A Note on Rising Food Prices*. Policy Research Working Paper 4682. Washington, DC: World Bank.

North, Douglas. 1973. *The Rise of the Western World: A New Economic History*. Cambridge: Cambridge University Press.

OECD. 2008. *Biofuel Support Policies. An Economic Assessment*. Paris: OECD.

Paik, Keun-Wook, Valerie Marcel, Glada Lahn, *et al.* 2007. *Trends in Asian NOC Investment Abroad*. London: Chatham House.

Papyrakis, Elissaios, and Reyer Gerlagh. 2004. The Resource Curse Hypothesis and Its Transmission Channels. *Journal of Comparative Economics* 32, 1: 181–193.

Pegg, Scott. 2009. Briefing: Chronicle of a Death Foretold: The Collapse of the Chad–Cameroon Pipeline Project. *African Affairs* 108, 431: 311–320.

PRC Ministry of Foreign Affairs. 2006. *China's African Policy*. Beijing: Ministry of Foreign Affairs.

Rosegrant, Mark W. 2008. Biofuels and Grain Prices: Impacts and Policy Responses. Testimony for the US Senate Committee on Homeland Security and Governmental Affairs.

Ross, Michael L. 1999. The Political Economy of the Resource Curse. *World Politics* 51, 2: 297–322.

Ross, Michael L. 2001. Does Oil Hinder Democracy? *World Politics* 53, 3: 325–361.

Rothstein, Bo, and Jan Teorell. 2008. What Is Quality of Government? A Theory of Impartial Government Institutions. *Governance: An International Journal of Policy, Administration, and Institutions* 21, 2: 165–190.

Sachs, Jeffrey D., and Andrew M. Warner. 1997. Natural Resource Abundance and Economic Growth. In *Leading Issues in Economic Development*, 7th edn. Oxford: Oxford University Press.

Schaber, Peter. 2011. Property Rights and the Resource Curse. *Global Governance* 17, 2: 185–196.

Severino, Jean-Michel, and Olivier Ray. 2011. *Le Grand Basculement: La question sociale à l'échelle mondiale*. Paris: Odile Jacob.

Stoever, Jana. 2012. On Comprehensive Wealth, Institutional Quality and Sustainable Development – Quantifying the Effect of Institutional Quality on Sustainability. *Journal of Economic Behavior & Organization* 81, 3: 794–801.

Suh, Moon Kyu. 2006. Oil Security and Overseas Oil Development Strategy in Korea. Paper read at Toward Regional Energy Cooperation in Northeast Asia: Key Issues in the Development of Oil and Gas in Russia, Korea Energy Economics Institute's International Symposium, September 15, Seoul.

Torvik, Ragnar. 2009. Why Do Some Resource Abundant Countries Succeed While Others Do Not? *Oxford Review of Economic Policy* 25, 2: 241–256.

UNCTAD. 2006. *World Investment Report. FDI from Developing and Transition Economies: Implications for Development*. Geneva: UNCTAD.

UNCTAD. 2007. *The Least Developed Countries Report 2007. Knowledge, Technological Learning and Innovation for Development*. Geneva: UNCTAD.

UNCTAD. 2009. *The Least Developed Countries Report 2009. The State and Development Governance*. Geneva: UNCTAD.

US Committee on Foreign Relations. 2011. *Avoiding Water Wars: Water Scarcity and Central Asia's Growing Importance for Stability in Afghanistan and Pakistan*. Washington, DC: US Committee on Foreign Relations.

Viñuales, Jorge E. 2011. The Resource Curse: A Legal Perspective. *Global Governance* 17, 2: 197–212.

Woods, Jeremy, Adrian Williams, John K. Hughes, *et al.* 2010. Energy and the Food System. *Philosophical Transactions of the Royal Society B* 365: 2991–3006.

World Bank. 2011. Global Gas Flaring Reduction. http://go.worldbank.org/R54GVV2QD1, accessed January 12, 2012.

Zehnder, Alexander J. B., Hong Yang, and Roland Schertenleib. 2003. Water Issues: The Need for Action at Different Levels. *Aquatic Sciences* 65: 1–20.

Part II   Global Energy and Markets

Chapter 5

The Oil Market: Context, Selected Features, and Implications

Christopher Allsopp and Bassam Fattouh

Introduction

A focus on oil hardly needs justification. Oil is a political commodity (Penrose 1976) and has always assumed center stage in key international affairs. Moreover, oil is a global commodity: crude oil together with its refined products is the most widely traded physical commodity measured either by volume or value (Stevens 2005). On a weight or volume basis, oil has the highest energy content compared with other fuels such as gas or coal. Despite a reduction in its relative share within the global energy mix in the last decade, oil remains the largest source of primary energy (34.4% in 2010), followed by coal (29.1%) and natural gas (23.4%) (see Table 5.1). As a liquid fuel, it is highly convenient, and exhibits large technical economies of scale at the various stages of production and transportation (Frankel 1969). The transport and aviation sectors, the lifelines of a modern economy, are still almost totally reliant on refined products from crude oil.

From a producers' perspective, oil is a crucial resource. Despite efforts to diversify their economies away from hydrocarbons, the oil sector remains the engine of economic growth and development in most producing countries. Oil exports generate the bulk of the foreign currency needed to meet import requirements. They also generate the bulk of the government revenues needed to implement key developmental and social projects and to diversify and industrialize their economies to achieve sustainable economic growth and to create employment opportunities for the hundreds of thousands of workers entering their labor markets each year. Given the dominance of the oil sector in their economies, producers are vulnerable to episodes of price instability and to protracted declines in oil prices, especially as compared with the more diversified OECD economies.

Moreover, since the large oil price shocks of the 1970s, it has been widely argued that these shocks have had large effects on the global economy, with nine of the ten post-World War II recessions in the US preceded by episodes of sharply rising oil prices (Hamilton 1983). However, recent studies indicate that the sharp oil price rises during 2002 and 2008 did not have the anticipated adverse effect on the global economy (Kilian 2009; Segal 2011) or core inflation (Cecchetti and Moessner 2008). For instance, Rasmussen

The Handbook of Global Energy Policy, First Edition. Edited by Andreas Goldthau.
© 2013 John Wiley & Sons, Ltd. Published 2013 by John Wiley & Sons, Ltd.

Table 5.1   Share of fuel in the world's primary
energy balance, 2010.

Oil	34.40%
Natural Gas	23.42%
Coal	29.09%
Nuclear Energy	5.40%
Hydro electricity	6.48%
Renewables	1.21%

Source: BP (2011).

and Roitman (2011) show that oil price shocks have generally not been associated with a contemporaneous decline in output, with the exception of the US, and although there is evidence of lagged negative effect on output especially for OECD economies, the effect has been relatively small. Nevertheless concerns about the impact of high oil prices on the global economy continue to dominate the public and academic debate, with Christine Lagarde, the managing director of the IMF, describing the recent oil price rises as "a new threat that could derail the recovery"[1] while Ali Naimi, the Saudi Oil Minister, described high international oil prices as "bad for Europe, bad for the US, bad for emerging economies and bad for the world's poorest nations."[2]

Increasingly, oil market developments need to be seen in a broader context for more than one reason. To start with, energy policies interact with other political and economic policy agendas, particularly energy security and climate change. Another reason is that long-term analyses of the position of oil, usually based on extrapolation, have frequently turned out to be spectacularly wrong. Is extrapolation of relatively high oil prices today justified or will oil market dynamics take a different turn? The recent sharp oil price movements have added another dimension related to the drivers of oil prices, the role of fundamentals versus speculation in the oil price formation process, and the design of regulation of commodities derivatives in the aftermath of the 2008 financial crisis (Fattouh *et al.* 2012; Turner *et al.* 2011). Another reason for analyzing oil in the broader international context is the issue of rent distribution and how competition to capture a higher share of the rent in the oil supply chain shapes producer–consumer relations (Fattouh and van der Linde 2011).

The objective of this chapter is to focus on issues surrounding international oil markets within the wider context of international energy, energy security and climate change policies, the global economy, and producer–consumer relations. The following sections cover the position (and uncertainties) of oil in the energy mix, energy security, climate change, taxation and subsidies, pricing in the international oil market, and the absence of anticipated feedbacks. The concluding section reverts to some of the big questions and contradictions in the current energy discourse and how they might be resolved.

Projections and Uncertainty about Policy

Long-term projections of oil prices, usually based on extrapolation, have frequently turned out to be spectacularly wrong. Extrapolation from the 1950s and 1960s led the Club of Rome to predict unsustainable growth and unaffordable prices. High prices were indeed a feature of the 1970s and early 1980s. But two world recessions (often regarded, rather simplistically, as oil induced), the spectacular substitution of natural gas for oil in power generation and in space heating, as well as technological innovation and developments on the supply side (e.g., the North Sea, Alaska) confounded the conventional wisdom. Instead, oil prices effectively collapsed from 1985 into 1986 (sometimes described as the "counter shock") ushering in nearly two decades of low oil prices and low investment in all segments of the oil sector.

The dangers of extrapolation were, again, spectacularly illustrated by the *Economist* newspaper's (1999) prediction that "the world is awash with the stuff, and it is likely to remain so" and that "$10 might actually be too optimistic" and oil prices might be heading for $5 per barrel. Only nine years later, in 2008, the price of dated Brent reached its historic high of $144.2 per barrel on July 3, and at the time of writing (November 2012), with recovery from the "great recession" on its way and geopolitical uncertainty abounding, Brent is again trading at around $110 per barrel.

Is extrapolation of high oil prices justified now? The conventional wisdom among oil analysts is that structural changes have tightened market fundamentals and have placed oil prices in an upward path. One of the most important shifts in oil demand dynamics in recent years has been the acceleration of oil consumption in non-OECD economies. Between 2000 and 2010, oil demand growth in non-OECD outpaced that of the OECD in every year. During this period, non-OECD oil consumption increased by around 13 million barrels per day (mbd) while that of the OECD dropped by 1.5 mbd (BP 2011). At the heart of this growth lies the Asia-Pacific region, which accounted for more than 50% of the incremental change in global oil demand during this period.

According to the conventional wisdom, this demand can only grow further as economic development proceeds, household incomes improve, and car ownership increases. Evidence from countries with long time-series data such as the US, Japan, and European countries shows a slow growth of car ownership at early stages of economic development. As income per capita reaches a certain threshold, growth in car ownership is twice as large as the growth in income. At high levels of income, growth in car ownership tends to slow down but will continue to grow as fast as income (Dargay *et al.* 2007). Although many expect OECD economies to reach a saturation point very soon, the evidence of such a saturation effect is not yet very strong. This stylized fact also applies across countries: countries with relatively higher income per capita tend to have higher car ownership.

On the supply side, despite the sharp rise in the oil price between 2001 and 2008 and the price rebound since 2009, the response of non-OPEC supply outside the Former Soviet Union has been muted. Between 2000 and 2010, non-OPEC production added only around 4 mbd to world oil supplies (BP 2011) with the bulk of the increase accounted for by Russia. This slow growth can be explained by a number of factors including sharp decline rates in mature oil fields and the increasing cost of exploration and development of new reserves. According to OPEC (2009), the weighted average annual observed decline rate, over the period 2000–2008, stood at 4.6% per annum, implying that 1.8 mbd of non-OPEC supply needs to be replaced each year simply to prevent oil supply from declining. A combination of high oil prices, hardened fiscal terms, and limited access to reserves has also pushed non-OPEC producers to explore new frontiers. These include the exploitation of oil reserves in deep and ultra-deep waters in places such as the Gulf of Mexico and shale oil in the US, Angola and Nigeria in Africa, and Brazil in South America. In addition, oil companies have turned to developing unconventional resources such as oil sands, bitumen, extra heavy oil, and shale oil, as well as biofuels, coal to liquids (CTL), and gas to liquids (GTL).

These changes on the supply side have a number of important implications. In effect, the world has entered the phase of substituting a relatively cheap-to-extract barrel with a relatively expensive-to-extract one. It has become technically, financially, and managerially much more challenging to extract oil in new areas. In addition, maintaining stable decline rates in mature fields requires the use of advanced and more costly technology. Moreover, the production of unconventional resources such as oil sands and shale oil raises serious issues about the environmental costs, including the effect on greenhouse gas (GHG) emissions. Finally, because of the risks and the higher costs involved in development and production, non-OPEC supply has become more sensitive to oil price cycles. Specifically, there seems to be an asymmetric response to oil price changes. A sharp rise in the oil price induces a modest investment response in non-OPEC countries, while a decline in the oil price generates a sharp fall in investment in the oil industry, especially in those segments with relatively high marginal costs.

With these changes in the dynamics of oil supply and demand, combined with the fact that oil supply and demand responses to prices are low especially in the short run, it is easy to produce scenarios that lead to higher prices into the future. An extremely important aspect of the story is that oil will continue to be required in the transport sector where substitution is extremely difficult. Producers of oil, so the story goes, can be relatively complacent because oil is "special."

But history suggests the need for caution. There have been enormous changes in the relative prices of different sources of primary energy. In the 1980s, a comparable change in relative prices led, in the industrial countries, to a massive substitution of gas (and coal) in power generation and in space heating, essentially eliminating oil from the mix except in niche sectors such as stand-by generation capacity. It is argued that the easy substitutions (the low hanging fruit) have already occurred. But is this true? And will it remain true? There are substitutes for crude oil in its main usage. One, obviously, is biofuels. They are small as yet. But Brazil, perhaps a special case, has substituted around 50% of its gasoline consumption in the transport sector with ethanol, and it happened over a relatively short period of time. Corn ethanol in the US is highly controversial, given its costs, its carbon footprint, and its effect on land use and food prices. But technological breakthroughs are possible. As with other technological developments, they are, perhaps, difficult to predict, but the economic incentives are already in place. Ethanol, even at present, makes a significant contribution to US fuel supply.

The possibility of substitution for crude oil in its main use is even more obvious from another direction. Different fossil fuels, coal, oil, and gas, can be converted into each other, at a price and with costs in terms, for example, of thermal losses and of carbon emissions. At current prices, especially in North America, both GTL and CTL appear to be economic. The capital costs are high, but in the longer term there are major implications for a study such as this. Within the fossil fuel sector, backstop technologies should limit the price of oil and oil products, in much the same way as the prospective costs of developing more difficult oil reserves such as oil sands and shale oil should limit prices at the margin. It is possible to imagine a transport sector, technically much like it is today, with little dependence on crude oil as such. Of course, low cost oil, for example from the Middle East, could still compete, but not at any price! One implication is that the idea of "peak oil," usually based on some idea of the physical availability of oil, needs to be critically evaluated. If oil, gas, and coal are considered together as potential sources of liquids for the transport sector (and feedstock for the chemicals industry) the problem is not shortage but, from a climate change point of view, abundance. The consequences of using what is available without some way of dealing with CO_2 emissions would, according to climate scientists, be catastrophic.

A rather different potential channel is the substitution of gas for oil at the point of use. There is still oil used in power generation, especially in non-OECD countries, which is increasingly uneconomic. More radically, compressed natural gas (CNG) or liquefied petroleum gas (LPG) could be used directly in the transport sector. Public sector transport in Delhi, for example, is almost entirely based on CNG, which can compete with diesel even at relatively high Asian gas prices (Jain and Sen 2011). The potential for such substitution is very large.[3] Clearly, too, there is the possibility of substitution through transformation, especially through the use of electricity to replace end use fossil fuel burning applications, such as electric cars and heat pumps. The processes are in their infancy, and the lags are likely to be long even with favorable technical developments, especially in battery technology. But, increasingly, the different elements in the complex international energy system are likely to come together via the electricity sector. Many observers strongly believe that hybrid and electric cars are destined to play a key role in the future. Deutsche Bank (2009), for instance, predicts that in the US, hybrid and electric cars will account for around 25% of new vehicles by 2020 and 8-9% of the vehicles on the road. For China, it predicts that about two thirds of new light vehicle sales will be highly efficient and that half of all light vehicles will be electric or hybrid by 2030. In its reference scenario, the EIA (2010) expects the market share of "alternative" vehicles to increase to 49% of new vehicle sales by 2035 in the US, from the 2008 level of 13%.

The "shale gas revolution" is a clear example of how technological innovation could prove to be a game changer. The conventional wisdom had it that the US would become a major importer of LNG within the Atlantic basin, competing with Europe and Asia. Instead, its demand for imports of LNG effectively disappeared. Reflecting this, domestic gas prices in the US fell sharply, and at the time of writing are a fraction of Asian LNG prices. Some analysts now see the US as a substantial potential net exporter of gas to the rest of the world, with arbitrage working the other way. Hydraulic fracturing and horizontal drilling technology are also helping to unlock billions of barrels of oil, with some analysts predicting the US supply from tight oil will reach 3 mbd by 2020 (*Wall Street Journal* 2012). Others go further, claiming that recent developments in the US energy sector have "unexpectedly brought the United States markedly closer to a goal that has tantalized presidents since Richard Nixon: independence from foreign energy sources" and predict wide repercussions which "could reconfigure American foreign policy, the economy and more" (Krauss and Lipton 2012).

The main message is that, though oil is special, it is not that special, and the possible technical and economic substitutions against oil could, in the longer term, be very great, especially if electricity makes major inroads into the transport sector. A crucial part of the story is price: both the general level of energy prices relative to other goods and services, and the relative prices of competing fuels within the overall mix. There have been large increases in the price of energy over the last decade and large changes in relative prices. If they persist, the future is unlikely to be like the past.

The wide divergence in views about the future position of oil in the energy mix is mainly about the role of technology and policy: what policies will be adopted and how effective these policies will be in shaping the oil market. Two types of policies stand out: energy security and climate change.

Energy Security and Investment

Much oil is concentrated in the Middle East and other politically unstable areas of the world economy, distant from the main concentrations of consumption. This gives rise to security concerns among oil importing countries, which fear disruption to the regular flow of oil supplies. Such disruptions can occur at any segment of the very long oil supply chain, which includes refining, international and local transport, storage, and delivery facilities. Disruptions can be caused by a large number of factors such as technical failures, weather events (hurricanes and storms), terrorist attacks on oil facilities, civil strife in producing countries, wars involving oil exporters, revolutions and regime changes that restrict the export capability of some producers, closure of oil trade routes, and a deliberate action by one or a group of exporters to restrict their oil supplies to certain consuming countries (sometimes referred to as the "oil weapon").

During the 1990s, the availability of large spare capacity and the willingness of key OPEC member countries to fill the gap in case of disruption meant that concerns about physical disruptions received little priority in consumer countries' policy agendas. This however has changed in recent years. A decline in the size of spare capacity and a series of supply shocks in key producing countries such as Iraq, Venezuela, Nigeria, and Libya have brought to the fore the issue of energy security. Iran's recent threat to use the oil weapon, the European embargo on Iranian oil imports, and the US sanctions on financial institutions engaging in direct dealings with Iran's Central Bank have further elevated the geopolitical risks as well as fears of a major supply disruption (El Katiri and Fattouh 2012).

Depending on the nature of the disruption and the availability of spare capacity, the market often adjusts to disruptions through sharp price increases. In such events, consumers who are concerned about securing oil supplies tend to increase their precautionary demand, causing prices to jump higher than what is justified by the reduction in supplies due to the disruption. Sharp adjustments in the oil price often impose high economic and social costs on oil importing countries. In order to mitigate the impact of such supply shocks, many governments hold strategic oil stocks, with the US, through its Strategic Petroleum Reserve, holding the largest stockpile. At the core of the International Energy Agency (IEA), established in 1973 in the wake of OAPEC's decision to restrict oil exports to the US and selected industrial countries, are the requirements that IEA members maintain emergency oil stocks equivalent to at least 90 days of net oil imports and participate in oil allocation among members in case of emergency disruption. India and China have also embarked on ambitious plans to build their strategic reserves, with recent heightened geopolitical risk providing a strong impetus for accelerating such plans and expanding the size of strategic stocks. In the short term, maintaining strategic stocks remains the most concrete and effective instrument available to oil importers to deal with physical disruptions.

In the long run, governments can pursue policies aimed at diversifying energy sources, reducing oil dependency, and encouraging oil substitution policies through regulations, incentives, subsidies, taxation, moral suasion, and/or combination of these instruments. There is much uncertainty as to whether these various policies will be implemented and the potential impact of such policies on long-term oil demand. The large sums of government investment in research and development, and financial incentives for alternative forms of energy and for reducing dependency on oil, are not new on the political agenda. Comparable investment pledges and incentives have been made in the past century with few tangible results. Furthermore, these policies and debates are very much influenced by economic developments and by oil price behavior. Economic recessions, combined with low oil prices, might dampen enthusiasm for some expensive alternative energy projects and carbon taxes, while high and volatile oil prices can speed up efforts for alternative energy projects. That being said, the pressure to restructure the energy mix away from oil will not disappear. However, the effects of policies on oil demand, even when widely implemented, will not be disruptive to the oil market. Nevertheless, the impacts of these policies are cumulative and most probably irreversible and hence cannot be ignored in the long term.

The dynamics of supply and demand may also result in market dislocations. In the worst possible scenario, global oil supply may not grow fast enough to meet the expected demand growth due to insufficient investment in new productive capacity. In such a scenario, given the long gestation lags in investments in the oil sector, most of the adjustment occurs through price increases. New oil supplies, the entry of competing fuels, and/or the development of efficiency measures cannot act as immediate adjustment mechanisms. Declines in oil demand associated with economic slowdowns and high prices can resolve investment bottlenecks and may even create spare capacity in the system. Spare capacity and an environment of low oil prices can, in turn, discourage investment in the oil sector. The disincentive to invest then creates the roots of the next oil price shock once oil demand recovers. In other words, the adjustment mechanism in the oil market is far from smooth: the oil market can witness long periods of large surplus capacity followed by periods of tight capacity. These alternating states of the oil market affect investment decisions and, hence, future supply availability.

This feature of cyclicality is common to other industries as well, but there are three special features that distinguish the oil industry from other industries. First, in countries where proven oil reserves are highly concentrated the decision to extract and develop these reserves is in the hands of governments or state actors. This has important implications, as decisions about whether and how much to invest are affected by economic and political factors and by events both inside and outside the oil market. The oil price is one of the various determinants of investment. Other determinants include political impediments such as sanctions, civil strife or internal conflicts; the nature of the relationship between the owner of the resource and the national oil company responsible for exploiting these reserves; the technical and managerial capability of the national oil company; the degree of access to reserves to foreign investors; and the petroleum regime and the fiscal system that govern the relationship between national and international oil companies. One factor that has received special importance in the consumer–producer dialogue is long-term oil demand uncertainty. Oil producers often argue that the policies of consuming governments, both implemented and announced, play an important role in inducing uncertainty and thus, in the face of calls for security of supply, they have coined the concept "security of demand."

Second, oil projects have long gestation periods and can be subject to delays. These delays do not only occur because of the size of the projects and the large capital outlays involved but can also be due to issues such as access to reserves and the complexity of the negotiations between international oil companies, national oil companies, and the owner of reserves in both the pre- and post-investment stages. The relationship between the international oil companies and the owner of the reserves (the government or state-owned enterprise) is affected by oil price developments, but equally importantly, it affects oil price behavior through the investment channel.

Finally, producers' investment decisions affect the market structure in a fundamental way. High oil prices do not necessarily induce governments of producing countries to increase investment and productive capacity. In contrast, a combination of high oil prices and limited access to reserves has pushed many international oil companies to explore new frontiers. The effect is that the cheapest oil reserves are not necessarily developed first, allowing for the coexistence of both high-cost and low-cost producers, with important consequences for the process of oil price formation.

The interactions between energy security and geopolitical issues, on the one hand, and more straightforwardly economic issues such as investment, on the other, account for part of the complexity surrounding discussions of international oil. Within such a framework, the main issue is how the oil market and its different players adjust to ensure the market does not suffer from a serious dislocation. But such a framework already appears out of date, due to the developing climate change agenda, which now interacts with nearly all aspects of the energy debate.

The Climate Change Agenda

The climate change agenda starts with an imperative, essentially to control cumulative GHG emissions (especially CO_2) in order to limit the risk of global temperature rise above some dangerous or very costly level. Needless to say, the science is subject to considerable uncertainty (there remain climate change skeptics), which is an essential part of the policy problem. Put thus starkly, the issues may appear mainly scientific and technical rather than economic, except in the rather trivial sense that whatever needs to be done should be done in the least costly way.[4] This methodology is apparent in the widespread use of scenarios or "backcasts" illustrating possible ways of meeting particular specified targets for emissions at some future date: see, for example, the IEA's 450 scenario (IEA 2010), intended to limit global temperature rise to 2 °C. It is not surprising that analysts in the economic realist tradition, as well as many climate change economists, are highly skeptical about the usefulness of such exercises. The output of such exercises is the delineation of a gap between what is likely under a business as usual (BAU) scenario and what is required.[5] The typical response has been the widespread adoption by governments of targets for GHG emissions or for renewable energy use, essentially a declaration of intent rather than a clear, worked-out policy response.

But the existence of the climate change agenda markedly alters the nature of the debate over international energy, including oil. First, the largest and fastest growing source of GHGs is coal, not oil (the largest users of coal are China, followed by the US, India, Indonesia, Russia, and Germany). For a given amount of energy, coal is roughly twice as polluting as gas, with oil roughly in between. Obviously there are potentially huge benefits to GHG emissions from the widespread substitution of gas for coal, largely in the power sector. To go further, however, would require substitution of non-carbon sources of primary energy, such as hydro and nuclear, renewables, and/or the development of carbon capture systems allowing the continuing use of hydrocarbon fuels.

Second, some potential substitutions for oil such as CTL and GTL are themselves extremely bad from the point of view of carbon emissions. Thus, they are a potential solution to a shortage of oil, but not to wider climate concerns.

Third, a potential solution for the transport sector via electrification would be no use at all if the electric power was produced by burning coal. (There is an irony in the enthusiasm for electric vehicles by the US and China, the two largest coal-burning countries in the world.) Most climate scientists believe that meeting targets such as those of the IEA would require the more or less complete decarbonization of the power sector as well as substantial reductions (compared with BAU) in hydrocarbon use in transport.

Finally, climate change policies induce wide uncertainty about the future position of oil in the global energy mix. As suggested in the literature on irreversible investment under uncertainty, the large investment outlays in oil projects and the irreversible nature of these investments have the effect of increasing the value of the option to wait. There is thus a case for producers to delay their investment decision until there is clearer picture of how climate change policies will affect their core product, at what pace this will occur, and what impact these policies will have on the oil price. The lack of a credible global approach aimed at putting a universal price on carbon only adds to this uncertainty.

Clearly, the climate change perspective on international oil issues is very different from the conventional wisdom among many oil industry analysts of expected tight market fundamentals. At the extremes, different assumptions account for the difference between worries over "peak oil," on the one hand, and worries about "stranded oil" on the other. Essentially, the differences hinge on technical and economic assessments (e.g., about efficiency developments and substitutions) and about the policies that are likely to be adopted. Thus forecasts for oil depend crucially on assumptions, or forecasts, about the policies that will be adopted and about how effective they will be. Integrating potential policy responses in projections, however, creates wide uncertainty about the future evolution of oil market fundamentals.

It might be thought that assessments or forecasts based on "likely policies" are the most "reasonable" in the circumstances. But the contradictions remain. BAU, as climate change analysts often point out, does not work, and the same may be true for "likely policies." The message then becomes essentially a contradiction: the forecast will not materialize since, if it did, something else would have to happen in response to the increasingly urgent need for climate change mitigation strategies.

Taxation, Subsidies, and Rent Distribution

The energy security and climate change agendas interact with another core feature of the oil market: the distribution of rent among the various players in the supply chain and between consuming and producing countries. As put by Fattouh and van der Linde (2011), "oil creates large economic rents, which are contested between producing and consuming countries, and among the various other players active in parts of the value chain, each wanting to capture a share. The sizable economic rents have been a prize deemed worth fighting for, far beyond the normal competition among market players. They have guaranteed persistent involvement by governments everywhere, either as producers or tax collectors."

The failure to appreciate this special dimension of the oil market has often rendered analyses of oil and its role in the political and economic order incomplete and misguided. Consumer governments, not surprisingly, would prefer to capture the rents involved via domestic taxation or equivalently by cap and trade systems, such as the EU's Emissions Trading System. Demand for petroleum products is highly inelastic due to limited substitutes, while its consumption is associated with negative externalities such as air pollution and adverse health effects. Thus, taxes on petroleum products are also perceived to be an efficient way to raise revenues for consuming countries and as a way to correct for negative externalities.[6] Since taxes represent a large portion of the price of petroleum products at the pump, a given rise in international crude oil prices is associated with less than proportionate increase in the price of petroleum products. Thus, in many countries, taxation weakens the demand response of petroleum products to changes in crude oil prices in international markets. In recent years, many OECD and non-OECD countries are stimulating the use of renewable energies, often through a combination of subsidies and taxes, to change the composition of their energy mix to one with lower carbon content. Such policy measures can have large impacts on the demand and supply for certain fuels.

From the oil producers' perspective, taxes on petroleum products are seen as discriminatory, tending to dampen oil demand growth, and reducing the producers' export share in the energy mix in the long term. Equally importantly, they raise a distributional issue since, through taxation, consuming countries can capture part of the rent, and in many cases more than the share extracted by producers themselves.

The counterpart of producers' complaints about taxation of petroleum products is concerns about energy subsidies. The policy of maintaining tight control of domestic energy prices has characterized the political and economic environment in many producing countries for decades. The objectives behind such a policy range from overall welfare objectives such as expanding energy access and protecting poor households' incomes to economic development objectives such as fostering industrial growth and smoothing domestic consumption, and to political considerations, with energy subsidies constituting one of the various channels through which resource-rich countries can distribute oil and natural gas rents to their population (see El Katiri *et al.* 2011 in the context of Kuwait). While energy subsidies may be seen as achieving some of a country's objectives, they are a costly and inefficient way of doing so. Energy subsidies distort price signals, with serious implications on efficiency and the optimal allocation of resources. They also impose enormous fiscal burden on state budgets. Over time, energy subsidies may also strain the export capabilities of producers, when demand continues to rise faster than supply. Subsidies may also undermine the climate change agenda and sustainable development.

Oil Price Movements: Speculation versus Fundamentals

During the period 2002–2008, the oil market experienced the most sustained increase in prices in its recent history, with the annual average price rising for seven consecutive years. This sustained rise in the oil price occurred in the absence of shifts in the power structure between producers and consumers (as happened in 1973 when OPEC took control over the pricing system) or big supply shocks (such as the disruption that followed the Iranian revolution in 1979). The oil price boom ended with a spectacular collapse toward the end of 2008, which saw the oil price decline by more than $100 per barrel in December 2008 from its July 2008 peak.

The sharp swings in oil prices in 2008 have polarized views about the key drivers of oil prices. On the one hand, some observers within the oil industry and in academic institutions attribute the recent behavior in prices to structural transformations in the oil market. According to this view, the boom in oil prices can be explained in terms of oil demand shocks, low price elasticities, rigidities in the oil industry due to long periods of underinvestment, and structural changes in the behavior of key players such as OPEC.

Within this tradition, some argue that price instability is an intrinsic feature of the oil market since there is a wide range within which the oil price can clear (Mabro 1991: 23). The lower boundary of the range is set by the cost floor of oil production in key OPEC countries, while the upper boundary is set by the potential entry of oil substitutes and, more recently, by the anticipations and behavior of participants in the financial markets. When the market is characterized by excess capacity, as it was in 1998, the oil price tends to move toward the lower boundary. When the market is characterized by excess demand (ex-ante), potential substitutes and adjustments in demand patterns cannot place a cap on the oil price in the short term. Instead, in the absence of spare capacity, most of the market adjustment is likely to occur through sharp increases in oil prices.

An opposing view is that changes in "fundamentals" or even expectations about these fundamentals have not been sufficiently dramatic to justify the extreme cycles in oil prices.

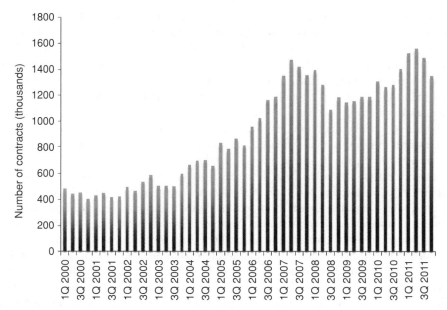

Figure 5.1 Average daily open interest in crude oil futures on US exchanges
Source: US Energy Information Administration website.

According to this view, the oil market has been distorted by the entry of speculators and financial players and particularly by substantial and volatile passive investment by index investors in deregulated or poorly regulated crude oil derivatives markets (Masters 2010).

While financial institutions have been the largest traders of oil since 1985, banks have become more involved in bridging the gaps between producers and a more diverse set of customers. In the last few years, other financial players such as pension funds, hedge funds, and retail investors increased their exposure to the oil market, a process referred to sometimes as "financialization" of oil markets (Tang and Xiong 2010). Financial innovation provided an easy and a cheap way for various participants, both institutional and retail, to gain exposure to commodities through a variety of financial instruments such as futures, options, index funds, exchange-traded funds, and other bespoke products. Figure 5.1 shows that between 2000 and 2011 the quarterly average of the number of outstanding oil futures contracts at the end of each NYMEX trading day increased dramatically from around 400,000 contracts to more than 1.4 million contracts (each representing 1000 barrels).

Many factors have been suggested to explain why financial players have increased their participation in commodities markets. Tight market conditions encouraged the entry of active money funds and institutional investors into commodities markets, including the crude oil market. Tight market conditions increase the upside potential for financial investments and speculative bets, especially in the presence of shocks originating from various sources. The historic low correlation between commodities in general and other financial assets such as stocks or bonds increased the attractiveness of holding commodities for portfolio diversification. Because commodity returns are positively correlated with inflation, many investors entered the commodities market to hedge against inflation risk and a weak dollar. Furthermore, expectations of relatively high returns to investment in commodities compared to other financial assets motivated many investors to increase their exposure.

Due to data limitations, issues of causality and endogeneity, and lack of a clear definition of concepts such as speculation and financialization, the academic literature has not provided convincing evidence that speculation or "excessive speculation" has been the key driver of oil prices during the period 2002–2008 (see Fattouh *et al.* 2012 for a recent review). Quite to the contrary, evidence from structural models indicates that oil demand shocks driven by global economic expansion account for the bulk of price increases. Some studies such as Lombardi and Van Robays (2011) find some evidence of destabilizing speculation, but its importance is very limited, especially in the long run.

This does not imply that the large entry of financial players had no impact on oil markets. However, the literature has so far failed to provide an answer as to whether increased financialization has been detrimental to the functioning of the oil market and whether it has resulted in the improvement or the deterioration of social welfare. For instance, some argue that financialization contributed to an increase in price co-movement between oil and financial assets such as stocks and bonds, enhancing volatility spillover effects from financial markets to the oil market (Tang and Xiong 2010) and eroding the diversification benefits in commodity markets (Silvennoinen and Thorp 2010). Hamilton and Wu (2011) provide evidence of large shifts in the risk premium after 2005 while Büyüksahin *et al.* (2009) find that increased financialization is associated with more efficient derivatives pricing methods. Pirrong (2011) shows that greater financial market integration increases liquidity, reduces the market price of risk, and increases the level of inventories, with the effect of lowering of probability of future price spikes.

Nevertheless, concerns about the impact of "excessive" speculation and financialization have pushed many governments to tighten regulation of commodity derivatives markets. In 2010, President Barack Obama signed the most sweeping financial rules since the Great Depression into US law (the Dodd–Frank Wall Street Reform and Consumer Protection Act). The Commodities Futures Trading Commission (CFTC), the agency with the mandate to regulate commodity futures and option markets in the US, has identified 30 areas where rules will be necessary. These reforms will have direct impact on the functioning of commodities' derivatives markets. Some of the rules relate to clearing and trading mandates, the purpose of which is to move "standardized" derivatives onto clearing houses to reduce risk. Other rules relate to data reporting requirements. These include the establishment of swap data repositories (SDRs) and the requirement that both cleared and uncleared swaps be reported to SDRs registered with the CFTC. Another important area is the imposition of limits on aggregate positions and the amount of positions (other than bona fide hedging positions) that may be held by any person with respect to physical commodity futures. The European Commission has also issued many rules concerning commodities. These include the Regulation on Energy Market Integrity and Transparency (REMIT) which covers all wholesale energy trading in the EU, including contracts and derivatives for the supply and transportation of natural gas and electricity, and the European Market Infrastructure Regulation (EMIR) aimed at mandating central clearing of eligible derivatives contracts and improving reporting by requiring over-the-counter trades to be reported to electronic trade repositories.

Increased Uncertainty and Limited Feedbacks

Fattouh (2010) proposes an interpretation of the long-term behavior of oil prices based on a shift to a regime of increased uncertainty about oil market fundamentals and lack of feedbacks. Until recently, expectations about short-term oil price behavior rested on the assumption that changes in oil prices would induce responses or feedbacks from supply, demand or policy, or a combination of all three, which would prevent prices from rising above a certain ceiling or falling below a certain floor. On the demand side, the feedbacks from high oil prices to demand operated through two main channels. High oil prices would have an adverse impact on oil demand through a price effect, an income effect, and changes in consumer behavior. Furthermore, high oil prices would eventually slow economic growth and induce recessionary pressures, with a detrimental effect on global oil demand. On the supply side, high oil prices encourage investment in the oil sector, inducing a supply response, but with a multi-year lag. High oil prices also encourage substitution at the margin by increasing the price of oil relative to other energy sources. One important factor that allowed the conventional framework to persist for a long period of time was the availability of large spare capacity. Spare capacity effectively increased the elasticity of oil supply and generated strong feedbacks, even when the market endured strong shocks that resulted in large supply disruptions.

As oil prices rose sharply during the boom years, perceptions of strong feedbacks were replaced by perceptions of limited feedbacks. Uncertainty as well as difference of opinion about the existence, the size, and the timing of feedbacks from prices to oil supply and demand increased markedly. In particular, four key feedbacks that were expected to put a ceiling on the oil price were not strong or visible enough: high oil prices did not trigger inflationary pressures and a subsequent recession; high oil prices did not induce an immediate strong growth in supply at the margin; during the upswing, producer power did not appear to be the cause of the price rises but nor was it used to limit the rise in prices; and the gradual erosion of spare capacity has had the effect of steepening an already highly inelastic supply curve. The lack of feedbacks affected the way in which expectations were formed, with important implications for oil price determination. The market entered into a phase of indeterminacy of beliefs, where market participants (including oil companies and oil producers) did not know where to anchor the anticipated oil price that would balance supply and demand in the short and long run. In effect, prices in the short and long run became jointly determined (Fattouh and Scaramozzino 2011).

In a framework of difference in opinions, heterogeneity of traders and uncertainty about feedbacks, higher order beliefs (players' beliefs about other players' beliefs: players' beliefs about other players' beliefs about other players' beliefs, and so on) can play an important role in influencing short-term price movements (formalized by Allen *et al.* 2006). This captures some of the intuition provided by Keynes's "beauty contest" metaphor where traders are motivated to guess other traders' guesses to benefit from short-term movements in oil prices, and offers useful insights that help to explain the sharp rise in oil prices in the first half of 2008. One such insight is the importance that public information or publicly observed signals acquire in the context of beauty contests, even if these public signals do not necessarily reflect large changes in underlying fundamentals or provide new information to the market. Since public signals can affect a player's guess about other players' guesses, they could have a disproportionate impact on the oil price (Morris and Shin 2003). The framework can also account for another interesting feature of the oil market. While there is an endless stream of public news and information, traders often seem to limit their attention to a few signals that they consider important. For coordination games to work in practice, market participants should only consider signals that are public and are thought to affect the expectations of other participants at a particular point in time. After all, it is impossible to coordinate on a large number of public signals. To what extent do these features that characterize price behavior in equity markets also play out in the commodities market? This is yet to be determined and is need of further research.

Conclusions

This chapter has emphasized a particular view of oil markets and their instabilities, stressing the uncertainties about the fundamentals, indeterminacies resulting from a lack of feedbacks, and the need for the market to coordinate on some consistent view (or story) in a situation where there is a wide range of plausible future possibilities (at least quite a wide range). But what are some of the policy implications if this kind of account is true? It is certainly of little help to those whose job it is to forecast the oil price, since it suggests intrinsic uncertainty. To an extent, it may justify the use of scenarios which focus on what goes with what, rather than unconditional projections. It may help in throwing some light on the debates over speculation versus fundamentals, since, if the situation is as described, it would seem unlikely that regulatory measures such as position limits would make much difference. On a more positive note, it may suggest more research and analysis on the factors that may, if perceived by the players in the market, limit the range of possibilities. An obvious example would be the dissemination of research on backstop sources of liquid fuels, and backstop technologies.

One obvious point is that, if the problem arises from a lack of feedbacks, and uncertainty about the fundamentals, then there are policies that could help. These range from simple dissemination of information about fundamentals, to policy moves to strengthen the perception that there will be responses, on the supply side or the demand side, to commitments by certain players to act in particular (contingent) ways. For instance, if market perceptions are wrong about the extent and the timing of feedbacks (for instance, if the market believes that there are no feasible instruments while in fact these exist), then policy could play a role in preventing sharp price movements by increasing the visibility of these feedbacks and policy responses.

The analysis points to the importance of a transparent and cooperative international approach to the issues involved, within institutional forums such as the International Energy Forum, the G20, and other bodies. This appears especially important when oil market issues are seen in the wider context of other agendas such as security and climate change, since developments in these areas can clearly influence oil market outcomes.

In fact, a lot is known in broad terms about the drivers of oil prices. For example, it is a safe bet that the anticipation of a marked slowdown in non-OECD growth (on the basis of news about the data, or because of new analysis from the IMF or other bodies) would lower the oil price. News about potential supplies from Iraq, or about deep-water technical problems in the Gulf of Mexico or offshore Brazil, feeds into the general picture with usually predictable effects. The list of potential impacts is endless. The uncertainty about many of these means that it is no surprise that forecasts and assessments often turn out to be wildly wrong. When the world changes, so does the oil market.

A particular uncertainty, which is especially important for the future of oil, concerns technical change and the diffusion processes involved in large transitions. Despite the lack of a coherent international climate change mitigation program, there is in fact a great deal going on, with different approaches and different policies in different countries responding to local conditions and local market failures. Much of this is irreversible and cumulative. Could it turn into a major transition with large effects on the future demand for oil? It is very hard to quantify. Such a large-scale transition certainly appears necessary if climate change mitigation goals are to be realized. We have already noted that, for something like the IEA's 450 scenario to be realized, this would require more or less complete decarbonization of the electricity sector, and substantial decarbonization of the world's transport sector as well. This is a scenario in which oil demand could fall substantially, with stranded assets in the longer term.

There are some paradoxical aspects to this. The deployment of renewables on a large scale, which would almost certainly require subsidies, could lead to low oil prices, leading to the need for even higher subsidies. An alternative, within consumer countries, would be higher taxes on conventional fuels or on their carbon content to make alternatives economic, unless there were spectacular falls in the costs of alternatives and technologies. Countering the problem with higher taxes would have to be international, otherwise cheaper oil would flow to areas where fossil fuels were not taxed. Another, equally unlikely, solution would be for the producers to receive compensation for not producing fossil fuel on the lines of "set aside" schemes often applied in agriculture. There are already a few schemes of this type designed to protect carbon sinks such as rain forests. Such issues are very complex and highlight a core feature of the oil market: the competition over rents between producers and consumers. Consumer governments would prefer to capture the rents involved via domestic taxation or equivalently by cap and trade systems. Producers would rather claim the rents for themselves by maintaining increasingly stringent constraints on investment and supply.

Finally, we revert to the apparent dissonance or disconnect between analysis of oil and other energy markets from an economic realist perspective and the climate change mitigation agenda. Can this be resolved? Essentially the difference between the two positions is about technology and policy – what policies will be adopted and how effective they will be. (This includes policy about technology and efficiency.) If climate change mitigation is taken as an imperative and policy-makers are expected to succeed, then the climate change perspective becomes also the realist perspective. But if the appropriate policies are not expected to be introduced, or are expected to be ineffective, then the dissonance re-emerges. One response is to ignore it. Another is to ask what feedbacks should be anticipated. One is to estimate the effects on global warming and the economic consequences and to include these within the realist perspective. Clearly, given the uncertainties, this would not be easy but, depending on the base case, they could be substantial. A second would be to anticipate policy change, as the situation develops. This is also not easy. The upshot is that it is extremely difficult to produce a consistent picture. This helps to account for the very wide range of views about the future of oil and international energy that seem to coexist. It also helps to account for the way in which changes in policy (and policy credibility) are likely to feed into energy markets and affect, amongst other variables, the oil price.

Notes

1. Address by Christine Lagarde at the China Development Forum 2012, Beijing, March 18, 2012, http://www.imf.org/external/np/speeches/2012/031812.htm, accessed November 28, 2012.
2. Saudi Arabia Will Act to Lower Soaring Oil Prices, *Financial Times*, March 28, 2012.
3. The use of natural gas in the transport sector may have many advantages (large availability of gas reserves, environmental impact) and the technology is well established. But CNG cars are still likely to make limited penetration in the transport sector due to infrastructure issues, size and weight of natural gas tanks, purchase cost, just to mention a few difficulties. There is also the issue of duplication of infrastructure costs and whether it is more effective to encourage a transition to a single type of technology, such as the electric/hybrid car.

4. For instance, in the UK's Stern Review of 2006, the authors tried to put price tags to the climate change problem, arguing that mitigation is cheaper than adaptation and that the cost of non-action tends to rise exponentially.
5. In a recent analysis of global energy trends by BP, based not on BAU but on a judgment of the most likely path of development to 2030, the base scenario has CO_2 emissions about 35% above the IEA 450 scenario in 2030. A more aggressive policy scenario still ends up with a substantial gap of about 21% (BP 2011).
6. Taxation of petroleum products and oil substitution policies are not only limited to the OECD. For many net importers in developing countries, taxes on petroleum products constitute a main source of government revenue, ranging from 7% to 30% of total government revenues (Gupta and Mahler 1995).

References

Allen, F., S. Morris, and H. S. Shin. 2006. Beauty Contests and Iterated Expectations in Asset Markets. *Review of Financial Studies* 19: 719–752.

BP. 2011. *BP Statistical Review of World Energy*. London: BP,

Büyüksahin, Bahattin, Michael S. Haigh, Jeffrey H. Harris, *et al.* 2009. *Fundamentals, Trader Activity, and Derivative Pricing*. Working Paper. Washington, DC: CFTC.

Cecchetti, S., and R. Moessner. 2008. Commodity Prices and Inflation Dynamics. *BIS Quarterly Review*, December.

Dargay, Joyce, Dermot Gately, and Martin Sommer. 2007. Vehicle Ownership and Income Growth, Worldwide: 1960–2030. *The Energy Journal* 28, 4: 163–190.

Deutsche Bank. 2009. The Peak Oil Market: Price Dynamics at the End of the Oil Age. October 4.

Economist. 1999. Drowning in oil. *The Economist*, March 4.

EIA. 2010. *International Energy Outlook 2010*. Washington, DC: Energy Information Administration.

El-Katiri, L., and B. Fattouh. 2012. Oil Embargoes and the Myth of the Iranian Oil Weapon. *Energy and Geopolitical Risk* 3, 2 (February).

El-Katiri, L., B. Fattouh, and P. Segal. 2011. Anatomy of an Oil-based Welfare State: Rent Distribution in Kuwait. In David Held and Kristian Ulrichsen, eds. *The Transformation of the Gulf: Politics, Economics and the Global Order*. London: Routledge.

Fattouh, B. 2010. *Oil Market Dynamics through the Lens of the 2002–2009 Price Cycle*. Working Paper 39. Oxford: Oxford Institute for Energy Studies.

Fattouh, B., and P. Scaramozzino. 2011. Uncertainty, Expectations, and Fundamentals: Whatever Happened to Long-term Oil Prices? *Oxford Review of Economic Policy* 27, 1: 186–206.

Fattouh, B., and C. van der Linde. 2011. *The International Energy Forum: Twenty Years of Producer–Consumer Dilaouge in a Changing World*. Riyadh: IEF.

Fattouh, B., L. Kilian, and L. Mahadava. 2012. *The Role of Speculation in Oil Markets: What Have We Learnt So Far?* Working Paper 45. Oxford: Oxford Institute for Energy Studies.

Frankel, P. 1969. *Essentials of Petroleum*. 2nd edn. London: Routledge.

Gupta, S., and W. Mahler, 1995. Taxation of Petroleum Products: Theory and Empirical Evidence. *Energy Economics* 17, 2: 101–16.

Hamilton, James D. 1983. Oil and the Macroeconomy since World War II. *Journal of Political Economy* 91, 2: 228–248.

Hamilton, James D., and J. Cynthia Wu. 2011. *Risk Premia in Crude Oil Futures Prices*. Working Paper. University of California at San Diego.

IEA. 2010. *World Energy Outlook 2010*. Paris: International Energy Agency.

Jain, A., and A. Sen, 2011. *Natural Gas in India: An Analysis of Policy*. Working Paper NG 50. Oxford: Oxford Institute for Energy Studies.

Kilian, L. 2009. Not All Oil Price Shocks Are Alike: Disentangling Demand and Supply Shocks in the Crude Oil Market. *American Economic Review* 99, 3: 1053–1069.

Krauss, C., and E. Lipton. 2012. U.S. Inches Toward Goal of Energy Independence. *The New York Times*, March 22.

Lombardi, Marco J., and Ine Van Robays. 2011. *Do Financial Investors Destabilize the Oil Price?* Working Paper. European Central Bank.

Mabro, R. 1991. *A Dialogue Between Oil Producers and Consumers: The Why and the How.* Oxford: Oxford Institute for Energy Studies.

Masters, Michael W. 2010. Testimony before the US Commodity Futures Trading Commission, March 25.

Morris, S., and H. S. Shin. 2003. Global Games: Theory and Applications. In M. Dewatripont, L. Hansen, and S. Turnovsky, eds. *Advances in Economics and Econometrics (Proceedings of the Eighth World Congress of the Econometric Society).* Cambridge: Cambridge University Press.

OPEC. 2009. *World Oil Market Outlook 2009.* Vienna: OPEC.

Penrose, E. 1976. Oil and International Relations. *British Journal of International Studies* 2, 1: 41–50.

Pirrong, S. Craig. 2011. *Commodity Price Dynamics: A Structural Approach.* Cambridge: Cambridge University Press.

Rasmussen, Tobias N., and Agustin Roitman. 2011. *Oil Shocks in a Global Perspective: Are They Really That Bad?* Working Paper WP/11/194. Washington, DC: IMF.

Segal, P. 2011. Oil Price Shocks and the Macroeconomy. *Oxford Review of Economic Policy* 27, 1: 169–185.

Silvennoinen, Annastiina, and Susan Thorp. 2010. *Financialization, Crisis, and Commodity Correlation Dynamics.* Working Paper. Sydney: University of Technology.

Stevens, P. 2005. Oil Markets. *Oxford Review of Economic Policy* 21, 1: 19–42.

Tang, K., and W. Xiong. 2010. *Index Investing and the Financialization of Commodities.* Working Paper. Princeton, NJ: Princeton University Department of Economics.

Turner, A., J. Farrimond, and J. Hill. 2011. The Oil Trading Markets, 2003–10: Analysis of Market Behaviour and Possible Policy Responses. *Oxford Review of Economic Policy* 27, 1: 33–67.

Wall Street Journal. 2012. Market Talk Roundup: Updates from IHS CERA Energy Conference. *The Wall Street Journal*, March 6.

Chapter 6

Natural Gas Going Global? Potential and Pitfalls

Matthew Hulbert and Andreas Goldthau

A Brave New Gas World

These are interesting times for international gas markets. Fueled by soaring unconventional gas production in North America, depressed demand in economically struggling consumer regions such as Europe, policies aimed at decarbonizing or denuclearizing the global energy mix, and strategic decisions taken by key gas producers, a perfect storm was created – an enormous liquidity in gas volumes available across the world. As a result, after decades of long term contracts (LTCs) firmly tying producers and consumers into bilateral relationships, markets have started to move toward new models. Gas-to-gas competition is in the making, and spot market trade has become a significant part of global gas arrangements, challenging oil indexation, the mechanism that traditionally pegged the price of gas to the price of oil. In short, we are starting to see the first serious signs of global gas prices based on *actual* gas fundamentals. As a corollary, the logic for international gas price parity has been set in motion across diverse geographic locations. What happens in one part of the world – be it the Arab Spring, nuclear catastrophes, new resource finds or intractable economic woes – has an influence on gas fundamentals, pricing, and outlook far beyond the immediate geographic region. In a nutshell, natural gas is about to "go global." By some accounts, this comes close to a revolution and has the potential to fundamentally alter the geopolitics and geo-economics of natural gas (Butler 2011; Yergin and Inieson 2009).

Natural gas is widely considered the fuel of choice for the decades to come. As the IEA stresses, gas demand has grown at twice the pace of oil over the past decade, with further consumption growth of 50% expected over the next 20 years. If one was to believe the IEA's analysts, the "golden age of gas" has just started (IEA 2011a). This is for various reasons. One, natural gas is a comparatively clean fossil fuel, whose carbon impact is some 50% lower than coal, the competitor fuel in the electricity and heating sector (Worldwatch and Deutsche Bank Climate Change Advisors 2011). Two, natural gas is abundant, with soaring unconventional gas production having recently boosted global reserves-to-production ratios to 200 years or more (EIA 2011; IEA 2011a; MIT

The Handbook of Global Energy Policy, First Edition. Edited by Andreas Goldthau.
© 2013 John Wiley & Sons, Ltd. Published 2013 by John Wiley & Sons, Ltd.

2010). The outlook therefore sounds promising: switching from coal to natural gas may simultaneously help decarbonize the global energy mix, serve as a bridge fuel to new and clean energy technologies, and provide for energy security, a key policy concern. Yet, while observers were quick to proclaim an all-too-bright future for global gas, we are not there yet. As anywhere else, transitions are bumpy processes, so also in natural gas markets. They are bound with uncertainty; they create winners and losers; and it may in fact take longer than expected to find a new and stable institutional equilibrium. In light of this, the current transition to new market structures will likely not be a quick or easy process for anyone concerned – neither producer states yearning for the "demand securities" of old, nor for consumers who will have to bear the heavy burden of high gas prices in some parts of the world, compared to cheap prices in others.

Furthermore, natural gas is not just another fuel. It is and has always been a strategic commodity, and its international trade is subject to state intervention. Whether gas is indeed going global is therefore not only a question of market fundamentals and technological progress, but equally depends on whether governments will eventually allow it to happen. If the world were to function according to the "iron law of economics," gas trade should reflect price arbitrage opportunities. Liquefied natural gas (LNG) should therefore start linking regional markets in Europe, Asia, and the US. Physical assets should simply go from low price markets (in this case, the US at some $2.5/MMbtu) to high yield plays (Asian spot at some $20/MMbtu) (*Financial Times* 2012). Over time the $1bn a day arbitrage spread gets whittled down, with Europe occupying the middle (geographic *and* price) ground. Gas on gas competition would eventually morph the Atlantic and Pacific Basins into a single gas "Pangaea." This would certainly mark a dramatic shift for a commodity previously deemed to hold insufficient "calorific clout" to make it worth sending halfway round the world and back (Stevens 2010). Yet the key question is what happens if gas fundamentals start to tighten and global liquidity literally dries up? Here, economic theory could easily crash on the harsh rocks of political preferences to keep gas as a regional, if not national affair based on oil fundamentals, rather than pushing toward efficient global allocation of the "blue stuff." Russia, the biggest gas producer in the world, clearly aims at keeping prices tied to old formulas. In the US, a key future export market for LNG, a tortuous political debate has emerged as to how much gas it intends to put onto global markets, despite sitting on a vast oversupply of cheap domestic gas – coined the "American gas must stay here" debate by an author in this book (see Chapter 1). Political preferences may therefore severely impact on market developments.

This chapter discusses the political and economic drivers behind the current transition in international gas markets, and sketches the various contingencies characterizing this process. It starts by looking at the "climatic conditions" that have made global gas convergence a serious debate. What factors contributed to triggering a transition from old market structures to new – even if yet largely undefined – ones? As we will argue, reduced OECD demand *and* meteoric unconventional gains in the US has freed up vast tonnages of LNG tankers from the Middle East that should have hit US ports to find their way to European hubs instead. As a consequence, European utilities contracted to expensive Russian pipeline gas have been bleeding customers and cash ever since, constantly being undercut by new market entrants using spot purchases to good effect over term. As a result, Europe – more specifically European wholesale hub prices and Russo-German border prices – started to emerge as the main battleground for new pricing models, with as yet no conclusive winner, given the conflicting "fundamentals" in play. But the war over pricing models is not just being waged in Europe, it increasingly divides Asia as well.

We therefore turn to look at Asia, and particularly the China nexus. A key Asian growth market for natural gas (imports), China will have a crucial role to play in determining future pricing models. Here, the question emerges whether producers can continue to sell gas at oil indexed prices to China, or whether they have to shift toward gas prices based on gas fundamentals. As we argue, it is the "vital supply side relationship" between Russia and Qatar that will affect how these two worlds eventually play out.

To be sure, all upstream players will keep pushing for security of demand (and term prices) to sink significant capital investment into new upstream developments. Most producers still want long-term contracts to get fields developed, infrastructure built, pipes welded, and even LNG tankers filled to make sure someone covers the costs. This historical legacy is not going to lose contemporary resonance overnight – particularly as 90% of gas is still traded on a regional, pipeline basis. Hence the real question is not whether long-term bilateral supply contracts will be struck, but what is used as the pricing reference point within them: spot market prices based on supply and demand fundamentals or oil indexation. With this in mind, we next take a specific look at the options gas producers play in shaping the new international market landscape. Obviously, producer countries are not a coherent block. The mere prospect of additional North American LNG supplies hitting the market has the potential to create major pricing problems for other producer states trying to stick to incumbent formulas – not only Russia and Qatar but also Australia. Since US developments tend to retain links to Henry Hub, the main American trading spot, traditional oil indexation pricing methods could be additionally challenged. We therefore move on to discuss a much neglected aspect in this regard: political risk in developed markets. As we argue, the "US energy independence" narrative traditionally enjoys strong traction in America and may constitute a serious threat to large-scale LNG exports in the future. In Europe, regulatory (in addition to geological) risks are coupled with strong environmental concerns against domestic shale gas extraction. In addition, vested interests may play in favor of importing Russian gas rather than developing unconventional European reserves.

Turning to the future, we address a critical question during any transition and ponder: who would ultimately profit most from a globalized gas market? In the short run, we argue, gas on gas pricing should benefit consumers, driving competition and efficiency gains, not to mention far greater energy security by fostering diversification of supplies. But the long-term conclusion might actually tilt toward greater supply side collusion – a potential gas cartel in which producers coordinate prices and volumes. As we argue, a single price point could well lend itself to a core set of swing producers. That would certainly be an ironic twist in a fascinating gas convergence tale – out of a supposed existential crisis could come the biggest opportunity gas producers ever had. We conclude by drawing some implications for global energy policy.

Perfect Storm: European Eye

We first need to appreciate how the gas world has undergone a fundamental transformation in the past few years from a seller's market to a buyer's bonanza. As recently as 2008, producers found themselves in a comfortable market environment. Pricing preferences were under little pressure, oil prices were soaring, pipeline projects were signed off based on traditional models linking producers to consumers in regional markets, and those playing the emerging global LNG game were getting even better returns. Burgeoning demand and supply had producers dictating prices and politics to consumer states – all of whom were desperate to secure their gas supplies.

Two major trends significantly altered this environment by 2010. The first was that global gas demand took a battering from the economic crisis that is still yet to fully recover. Demand was cut by 3% in 2009, with the EU seeing a 7% slide in 2010/11 that has since plummeted to 9.9% into 2011/12 (BP 2012). Furthermore, a swathe of new gas was all coming onstream at exactly the wrong time for producers – be it pipeline gas, LNG from Qatar or elsewhere, or more critically the breakthrough in unconventional gas production in North America. In fact, the scale and impact of shale developments has been widely underestimated. As with most "revolutions" this was not achieved by accident, but by years of development spanning back to the 1970s, with fracking technologies tying into deep and liquid US markets and lots of capital (ironically from high oil prices). The result was the development of massive Marcellus, Haynesville, Barnett, and Utica plays (unconventional gas "fields"), helping the US to catapult its production to 651 billion cubic meters (bcm) in 2011 (EIA 2012). That makes the US the largest single gas producer in the world, accounting for 20% of global share, while shale gas now makes up a third of all US consumption. Shale developments and technological advances have been so successful that they have driven gas prices to under $2/MMBtu (the standard unit of measurement) in Henry Hub (the main US gas trading point) as the quintessential example of gas on gas competition. What is more, the EIA's latest estimate is that US unconventional recoverable reserves now stand at some 13,65 trillion cubic meters (EIA 2012).

What happened in the domestic American market has had a profound global impact. Most importantly, it has thrown existing international pricing regimes into turmoil. Not only was global demand down, exporters also lost their "LNG banker of choice," the United States of America. Since the turn of the millennium, the US had been busy building regasification terminals in anticipation of a tightening domestic market. But with the shale boom, the US market closed down to LNG imports. This tilted the market in favor of the consumers. To the surprise of unprepared producers, pesky consumers started negotiating down long-term Gas Purchase Agreements toward lower spot prices in flooded markets. All of a sudden, Russia, Central Asia, Africa, MENA (Middle East and North Africa), and Australian suppliers all found themselves fiercely competing for market share wherever they could. Expensive ($200m) tankers carrying "–162° C cargo" literally had to find a new port in a pricing storm. And European hubs were the harbors of choice.

As a consequence, a "hybrid" model has emerged in Europe, with oil indexed pipeline gas averaging around $12–14/MMbtu against spot traded prices at $8–10/MMbtu (Bloomberg 2012; see also Stern and Rogers 2011). In practical terms, that largely means the UK National Balancing Point (NBP), as the most mature and liquid market, sets European wholesale benchmarks via its interconnection to the mainland (TTF, Zeebrugge). LNG imports to the UK amounted to 22 bcm in 2011, 85% of which was unsurprisingly sourced from Qatar, the largest single LNG producer in the world, responsible for 30% of global supply. Smaller European markets increased spot deliveries as well (ironically often for re-export purposes), which added to a raft of other suppliers ensured that close to 50% of all physically traded gas in Europe was done so on a spot market basis in 2011 (Energy Intelligence 2011). For incumbent European utilities, these developments meant trouble. They had hard times pushing into the market even the minimum quantities they were obliged to buy according to their off-take agreements, given an uncompetitive price determined by a relatively inflexible formula. They also found it hard to roll-over contracted volumes as economic recovery in Europe was highly uncertain. It was particularly the inability of incumbents to pass through costs to re-claw the weighted average cost of gas from consumers – rather than the price per se – that has rendered the situation rather dramatic for them. Soon, Germany's E.ON Ruhrgas, Italy's Eni or France's GDF Suez saw their decade-old business model being fundamentally put in question, endangering margins and profits. In short, incumbent European utilities felt being left out of the money.

This, however, should not have come as a surprise to them. In fact, incumbent utilities failed to prepare for competition coming with market liberalization measures pushed through by Brussels. Several "energy packages" put together by the European Commission over the course of a good decade had opened the door for second tier players to take market share from incumbents, bypassing traditional wholesalers and going straight to large end users from spot (Talus 2012). Incumbents instead remained unable to retain market share by offering discounted supplies. As a consequence, and unsurprisingly, E.ON and its like started pushing for more flexibility in their LTCs with producers around 2010, triggering a wave of pricing disputes and arbitration cases (RBC 2010; Reuters 2010). Some eventually managed to renegotiate their contracts. E.ON now sources from Russia's Gazprom by indexing a minority share of contracted volumes to a new formula (RBC 2012; see also Konoplyanik 2011). Likewise, Norway's Statoil now sells up to 30% of its gas contracted to European utilities at spot prices (RIA Novosti 2010) and even signed supply contracts with Centrica directly linked to NBP prices. More cases are still pending across Europe, with virtually all major European utilities desperate to carve out a larger spot component.

As much as European utilities seek to add more flexibility to existing contracts, many gas producers aim at keeping the status quo at present. Particularly with regard to Russia's Gazprom, it appears highly unlikely that some more flexible middle ground can be found on oil-indexed and independent spot benchmarks. For one, Gazprom has a vital interest in retaining its position as the main and often exclusive supplier of the European market, given the crucial margins the company makes through its European exports. By some estimates, and in addition to exports depressed by more than 10% in 2009 (*Moscow Times* 2010), Gazprom has already lost an estimated $2 billion in 2010 alone thanks to discounts granted to European utilities (RIA Novosti 2010). Second, as the major Eurasian resource holder, Russia seems to think it can sit through the current soft market environment, make minor tactical concessions, wait for fundamentals to tighten again, and eventually return to business as usual. Meanwhile, policy decisions such as Germany's nuclear phase-out have certainly helped to give Russian pipeline projects such as the 55 bcm Nord Stream more hope, as have slower LNG developments in Qatar. But Russia also has indicated it has no problem playing rough to safeguard its perceived interests. Upstream asset sweating has been one argument on the Shtokman and Sakhalin developments; bypassing incumbent wholesalers and selling directly to traders, second tier players, and more end users is another. The idea of pegging gas to expensive European renewables has also been floated, more to "prove the point" that playing around with pricing formulas can "cut both ways" (UPI 2012). Gazprom also raised the structural reality that without Russian gas as the backbone to European supply, no spot market would ever be able to exist (Komlev 2012). Alexei Miller, Gazprom's CEO, pondered over opportunities to look for attractive markets elsewhere. He allegedly quipped that he would get out of bed and decide which way to send his gas that day depending on price: to Europe or Asia. The logic behind this plain-spoken threat and Russia's "Eastern Strategy" is very clearly to maintain the status quo – long-term contracts, oil price peg, and bilateralized business relationships.

In all, since 2010 Europe finds itself in the eye of a perfect storm in natural gas markets, with producers and consumers pulling toward opposite ends. Yet the question who will win the ongoing price war will be not be answered in Europe. For this, we need to turn to China.

A (Very) Complex Pacific Game

The market that really matters in Asia is China. The Middle Kingdom already consumes 155 bcm/y of gas, a figure that many analysts think could easily double by 2030 given 15% year on year growth since 2000. LNG growth shot up 31% as soon as China's fifth import terminal went online, with a dozen more terminals being planned. To put this into perspective, a 1% increase in Chinese gas consumption equates to around 25 bcm of gas, one quarter of Germany's annual consumption, the largest European market (*Economist* 2012). China is *the* growth market that no producer can afford to miss. Past practice would almost certainly have seen Beijing put security of supply ahead of price and sign up for whatever it could get. But this dynamic is changing, precisely because China has opened up sufficient scope from strategic investments over the years to start hedging price risk far more effectively. Throughout the 2000s Beijing signed numerous governmental memoranda of understandings with major reserve holders for prospective supplies, while sourcing *actual* resources from Central Asia. Turkmenistan has been critical to this, with 30 bcm of Turkmen gas expected to flow into the Chinese mainland by 2015. Additional agreements toward 65 bcm are in place, with Uzbekistan and Kazakhstan added to China's Caspian ranks (Downstream Today 2012). China has on its doorstep Australia, that could well become the largest LNG producer in the world by 2018 (80 mt/y) (Forster 2012). Burma also holds considerable resource potential, as do MENA states still actively seeking supply agreements with Beijing. Significant East Africa LNG plays currently developed in Tanzania, Mozambique, and Kenya will eventually plug into the Pacific Basin as well. On top, China's own domestic unconventional potential could be enormous. Beijing extracted 10 bcm of coal bed methane in 2011 and has set ambitious shale gas targets for a 100 bcm expansion by 2020 (IEA 2011a, b). Whether they are reached is debatable, but China has brought Chevron and Shell in to provide technological edge, not to mention financing their own national champions to acquire shale assets (mainly in the US) to learn their "unconventional trade" and bring it back to the Chinese mainland. The upshot is hardly surprising; China is playing increasingly hard to get with core suppliers on oil-indexed pricing – both in liquid and pipeline form. On the important Chinese market the two worlds are increasingly blurring into one under a single rule: Beijing is not going to pay oil-indexed rates for either of them.

Let us take pipelines first. Russia initially signed a framework agreement to supply China with up to 70 bcm/y of gas in 2009 (*China Daily* 2009). Where the deal has since fallen down is Moscow's insistence that gas is sold at oil-indexed rates at $350–400 per thousand cubic meters (roughly the European price), compared to China's preference of $200–250 – more or less the domestic coal benchmarks. China can also handily point to independent (and in fact much lower) gas benchmarks in Europe as credible price points for "Sino-Soviet" deals. As a response to its failure to secure an oil-indexed LTC with Beijing, Moscow has voiced an inclination to develop gas to liquids in its eastern fields and sell LNG into jurisdictions such as Vietnam and Thailand, beyond Chinese markets. This however does not seem a viable strategy. In fact, Russia seemingly fell for its own press that it could sell expensive Siberian pipeline gas into China, which in turn would be used as leverage over other Asia-Pacific consumers for fringe LNG, and more importantly over its core European demand base. Russia currently does not have serious

LNG capacity. In addition, China has alternatives to Russian piped gas imports. In this, Moscow's "Eastern Strategy" looks very much like an empty threat.

That brings us directly back to LNG, and more specifically to Qatar as the main primer of European liquidity. Tiny in size but a big player in natural gas, the Middle Eastern state is playing a very strategic game that is not just about maximizing receipts, but enhancing its long-term global potential. Qatar's medium-term strategy is to keep feeding European spot markets as a transitional step toward an Asian future. It therefore stays in the European market despite netbacks on Qatari spot into Asian ports hovering about $14/MMBtu, twice the figure achieved on UK and Northwest Europe deliveries. Many industry insiders think Qatar would even need to see Asian spot prices hit $25/MMbtu before it comprehensively exited European markets. For Qatar, the step into Asian markets is both promising and risky. Qatar is trying to place up to 50 million tonnes of LNG into Asian markets over the next few years, ramping up the 34 million tonnes it already ships out east (under half its total 77 mt production), mainly to India, South Korea, Taiwan, and Japan, all relatively "easy" recipients. But for Qatar the real prize lies elsewhere: China. It is yet to reach conclusive agreements over contract duration and pricing formulas for full oil parity, providing a measly 2.1 mt/y to China. In total, Qatar has only sold around 2 mt/y at significant discounts into Asia, making clear that its vast riches (both in geological and paper form) do not need to instantly rake in the RMB to stay afloat. As long as it keeps feeding European markets, the underlying hedge is that China will have to pay a decent price on decent terms for Qatar to turn most of its tankers east. Yet this entails a very delicate balancing act for Qatar to get right, and it is one that ultimately points toward discounts on long-term gas agreements with Beijing. As much as Qatar wants a good price for its gas, it still needs to head off a full-scale Pacific Basin pricing war, with over 50 mt/y expected to come online in the region in the next few years (in effect, not that far off a new Qatar coming online). Leave things too late, and it risks losing out on the Beijing market altogether as others may serve it earlier and at lower prices. Beijing has a raft of supply options to draw on in the next five years, and would presumably not also take too much Qatari gas at premium prices. Rather, and closing the circle to Russia again, Beijing might try and keep Qatar involved in the European market. Allowing European spot prices to burgeon on the back of Qatari gas is in fact coming in remarkably handy for China, specifically because it knows in the longer term it will have to start sourcing large amounts of Russian gas. From a strategic point of view, China would much rather forego relatively small quantities of Qatari supplies to maintain spot prices on European hubs now, in order to drive a harder bargain for procuring larger quantities of Russian gas in future. It is telling that China refuses to touch Russian West Siberian supplies, precisely because it is worried Moscow could simultaneously supply Beijing and Brussels with the same fields. For Russia this means that if it wants entry to the Chinese markets, it has to develop fresh East Siberian fields – another blow to its Eastern Strategy.

Obviously, the Pacific part of the globalizing gas story is a complex one. What makes the Russia-Qatar-China nexus so important is the possible effect on future pricing models. Assuming Beijing forced Russia's hand to agree to spot dynamics in their pricing, China will then place further downward pressure on future Central Asian supplies as a "domino" effect. It has already revised price offers downward to Turkmenistan (to around $200 per thousand cubic meters) in return for infrastructure loans (Energy Intelligence 2011). Such pressures would also be used against Australia for long-term Chinese contracts by using Qatar as a potential hedge to check Australian LNG price ambitions. Without us getting bogged down in any more price points, the bottom line is that Qatar provides China far more arbitrage options with Central Asian, Russian, Australasian, and MENA producers by keeping them in the spot game rather than bringing them under Beijing's wing. If there is a race to the bottom on price, soaking up European liquidity is a no brainer for China, at least at this stage.

Producer Competition over Markets and Models

In a nutshell, the Qatari-Russian intrigue rests on the same debate: can producers continue to sell gas at oil-indexed prices, or do they have to shift toward gas prices based on gas fundamentals? Clearly, a trend has already been set in motion. Of the 330 bcm of LNG gas shipped globally, 25% of it is now done on a genuinely spot basis. With another 250 mt/y of LNG potentially coming to market over the next decade from every point on the compass – Nigeria, Angola, Israel, PNG, Mozambique, Equatorial Guinea or other – LNG growth *should* continue to erode old market rules and structures. What will become a decisive factor, however, is how one key market will develop: North America. In fact, after having reached self-sufficiency by now, North America promises to emerge as one of the largest export markets of all (EIA 2012: 3). In Canada, Shell, PetroChina, Kogas, and Mitsubishi are lining up 12 mt/y exports from British Columbia for Asian markets. That follows export licenses already agreed for BG Group, and Apache through Kitimat LNG, as well as the Alaskan North Slope plumping for LNG to monetize its 706 million tonnes of recoverable reserves (EIA 2012). And since it can hardly place LNG into its neighboring US market, Canada is likely to aim at selling 30 mt/y of stranded gas to Asia to 2020. Likewise, US producers have started to look for export markets in Asia. Because of the phenomenal breakthroughs of its shale developments, the US now ironically risks becoming a victim of its own success in terms of Henry Hub prices dropping so low that full-cycle economics for US shale gas plays have become negative. Unless prices organically firm, or US producers artificially restrain supply, current output levels will be difficult to maintain or enhance for US consumers alone. Companies will fold, fields will be mothballed, with the woes of Chesapeake, arguably the front runner in unconventional gas, providing the best case study example. The quick fix option to get Henry Hub back at a sustainable $4–7/MMbtu level (and by far the most lucrative for some of the mid-cap players involved), is to sign up international LNG contracts and seek export markets. That is exactly what is being done, with some of the larger international oil companies such as Royal Dutch Shell and Exxon Mobil also aggressively pushing for LNG exports to capitalize on the huge spreads. In total, the US Federal Energy Regulatory Commission (exercising oversight over natural gas pricing and LNG terminals) has around 125 bcm/y of LNG applications currently awaiting approval; some 40–50 bcm exports should therefore be feasible by 2020. That would make the US the third largest LNG player in the world.

As a matter of fact, US LNG could be the straw that breaks oil indexation's back. The mere prospect of North American LNG hitting the market is creating major pricing problems for producer states trying to stick to old formulas in Asia – and not just Russia and Qatar. Australia is facing serious cost inflation with coal bed plays (a form of unconventional gas) looking more costly than originally thought. International players are still investing down under (ironically as a double hedge against the risk of US LNG flopping), but given that Australian LNG docks into Asian ports for around $17–18/MMbtu any softening of prices could leave current (and prospective) LNG projects in the red. That might sound problematic for future supply prospects, but it is also extremely interesting when we consider how recent US supply agreements to Asia are being brokered:

Henry Hub is the underlying price point. Cheniere Energy's export terminal at Sabine Pass, Louisiana, will sell LNG into South Korea at $9–10/MMBtu (*Economist* 2012). The "general" formula is to set a minimal $3/MMBtu (i.e., Henry Hub) capacity leasing charge as default payment if gas is not lifted, with a 115% mark-up to bridge differentials on actual deliveries over a 20-year period (3.5 mt/y). Indian outfit GAIL brokered a very similar deal, while European off-takers from Sabine Pass, most notably BG Group (5.5 mt/y) and Gas Natural Fenosa (3.5 mt/y), have pegged leasing charges even lower at $2.25–2.5/MMBtu. Other planned projects such as Excelerate Energy's floating LNG plants off the Texas coast or ConocoPhilip's 10 mt/y Freeport, Texas LNG project may work on similar pricing models.

Even if planned US LNG export terminals such as Cove Point (Maryland), Lake Charles (Louisiana) or Jordan Cove (Oregon) retain only notional links to underlying Henry Hub prices (plus mark-ups), then traditional oil-indexation pricing methods could be in deep trouble. Put simply, the fact that Cheniere Energy has inked Henry Hub deals with Japan's Mitsui and Mitsubishi provides a very explicit price link between the cheapest market on earth and the most expensive. Given the enormous margins to play with between the US and Asia, shipping costs (with or without Panama Canal tariffs) all of a sudden become more of a "rounding error" rather than a core cost. US players will have no problem taking market share, providing far cheaper gas and more flexible contracts, a development that should create more space for excess LNG supplies to be placed onto wholesale hubs internationally. At the very least, long-term contracts will be far more dynamic in terms of pricing, while forcing the likes of Qatar, Russia, Australia, and even Canada to be more flexible on contractual terms if they want to secure Asian markets. It is still very early doors to call structural decoupling from indexation toward some future "Shanghai spot" driving competition between LNG, pipeline gas, and domestic production – the Shanghai Petroleum Exchange has launched a gas spot trading platform only very recently and is not a liquid trading spot as of now. But the fact that Singapore is eagerly expanding its LNG capacity in order to maintain its leading energy hub role in Asia gives a fair indication of where important market players may expect things to go.

Europe should be watching this space very closely as well, as European hub prices typically sit mid-way between the US and Asia. With new LNG plays predominantly eyeing high yield Asian markets to 2020, fierce price competition in the Pacific Basin might ironically leave Russia with a relatively free European hand until LNG tankers find their way back to European ports. The key question is: how quickly will Brussels see price convergence between Europe and Asia to make sure they don't get dragged back down Russian indexation preferences? Until recently the answer would have been several years, but Japan has already filled most of its nuclear gaps, marking a dramatic 25% drop in Asian spot prices in one month to $13/MMBtu (June 2012). On that trajectory, it is very possible more and more spot cargoes could dock in Europe pretty soon. While an Asian premium on the gas price might persist given strong demand side fundamentals, analysts have already noted notional Henry Hub five-year curves look rather similar to the NBP plus liquefaction and shipment costs. In effect, the market seems to be preparing for the US to send tankers directly across the pond. Intra-Atlantic Basin spreads are on their way to glorious long-term convergence.

Headwinds: Grabbing Defeat from the Jaws of Victory?

Or are they? Overall, current trends seem to point toward truly globalizing gas markets, and fundamental shifts in contractual models and new pricing structures coming with them. Yet, as dicey as things look for traditional pricing methods, our discussion so far fails to consider the big elephant in the room: *political risk* in developed markets. Unless that is addressed, the kind of liquidity needed for truly independent gas prices to become dominant is likely to remain absent. Arguably, the US is the number one risk to take the froth out of the spot market cappuccino. Despite the very clear economic logic of converting its shale plays into LNG, it is politics in Washington that will ultimately decide how much gas is allowed to leave US shores. As a matter of fact, a rather motley crew of actors in US energy debates has the potential to form a strange but effective alliance against LNG exports. On the one hand, the "energy independence" narrative has a strong voice and is traditionally represented by national security proponents (Jaffe 2011; Morse 2012). On the other hand, environmental campaigners happily rejoice in significant shifts from dirty coal to cheap gas – clipping US emissions by 450 million tonnes over the past five years. At the same time, industry is profiting immensely from low natural gas prices – the petrochemicals industry is getting its feedstock close to free by switching to gas – all of which is good news for broader US economic output. On top, a debate over gasification of the US transportation fleet has emerged recently. On paper this looks an interesting prospect, as gas at $2.5/MMbtu is about $15/b in oil terms and converting shale to compressed natural gas, or LNG, or putting it into gas to liquid form is all possible. Given only 3% of natural gas is being used in the US transport sector, there clearly is ample room for growth. Taken together, all of this provides for a neat argument to sell to the US electorate the idea of restricting LNG exports – at the expense of US companies suffering from low price environments and the 30 or so US States that currently benefit from hydrocarbon royalties. In short, if for political reasons the US gas market remains a dislocated island, with Washington "capping" Henry Hub prices to $5/MMbtu (Levi 2012), many of the pricing pressures in the Asia-Pacific region discussed above will rapidly abate.

Elsewhere, political risks characterize shale gas prospects from different ends. In Europe, regulatory risk couples with environmental concerns and a generally difficult business environment for shale gas (Stevens 2010). On most counts, Europe's unconventional gas industry is at best a nascent sector: European rotary rig counts only number around 120 compared to 2,000 in the US, 700 in Canada, and 450 in Latin America; Europe has no serious in-house fracking expertise and generally a rather short history in mineral extraction on European soil; subsoil resource laws often remain unclear while fiscal regimes fail to capture unconventional gas altogether; and "nimbymania" is a widespread phenomenon on the densely populated European continent. Yet, it is in the political realm that most concern rests to get shale going across EU member states. Despite sitting on 5100 bcm of shale, France recently put a ban on shale. The Netherlands seems uninterested in shale given they still have conventional gas fields such as Groningen in operation. Germany decided to split its energy mix between beefing up wind and solar power and – at least in the medium term – even more lignite coal, using Russian gas to fill any residual gaps. That does not leave much room for North Rhine-Westphalia shale plays. In the UK political backing for shale gas rests with a few Conservative parliamentarians who think shale might offer a British version of the US shale revolution in Northwest England. Sweden has not shown inclination to pursue the prospects of some estimated 1162 bcm of reserves. Spain remains plugged into Algerian production, while Italy has been oversupplied with Russian and ongoing Libyan deliveries.

Central, Eastern, and Southeastern Europe (CEE/SEE) offers yet another lesson for the complex interplay between politics and economics in the natural gas sector. In most countries of the former communist bloc, import dependency on Russian gas fosters strong political desire to develop domestic shale gas deposits. Poland or Hungary therefore made major political and fiscal efforts to enlist US majors to replicate the shale revolution in the US. Yet, as quickly as most of them came, many of them have since left. First, geology remains a challenge. While studies indeed suggest promising prospects, the overall technical and economic viability of shale plays remains deeply circumspect in Poland, Austria, Romania, Hungary, and Ukraine (EIA 2011). Second, Russia still casts a long geopolitical shadow over CEE and SEE states. As Gazprom has no interest in seeing domestic competition emerging in Europe, its traditional and most important export market, it will likely do everything in its power to prevent shale developments for those close to its borders. It is telling that no sooner had Exxon Mobil signed agreements to develop West Siberian tight oil plays in Russia, it exited from Polish shale exploration. Bulgaria has put a ban on any shale developments, officially due to domestic opposition. Bucharest has actually followed "Sofia's suit" and imposed an outright moratorium on any further shale developments within its borders. Although Hungary remains open to shale exploration (with US Irish company Falcon maintaining its activities), the Carpathian-Balkan Basin (538 bcm) straddling joint reserves of Bulgaria, Hungary, and Romania is not likely to be tapped any time soon. In Prague, finally, the Czech upper house is in the process of pushing through a shale ban having canceled previous overseas contracts.

To be sure, the benefits of developing European domestic shale gas plays would certainly not lie in making the continent "energy independent" from Russian imports. Rather, by adding an additional (domestic) resource base, additional gas volumes would continue feeding spot market liquidity, enhance gas on gas competition, and reinforce the current process undermining incumbent pricing models. Developing indigenous shale could contribute to developing far deeper, liquid, and more mature wholesale hubs across Western Europe, reaching all the way to the Baumgarten hub in Austria and NCG in Germany. In conjunction with enhanced state-funded infrastructure to foster interconnections and back-flow capacity, CEE and SEE states would be in a better position to negotiate "cost reflective" contracts with Russia.

Overall, while much attention is placed on the geological potential of shale, it is very much the politics involved that matter (Butler 2012). IEA analysts suggest that if all goes well with shale, gas should take around 25% of the global fuel mix over the next 20 years (IEA 2011a). Shale has doubled the gas resource base and could potentially be five times larger than conventional supplies, leaving reserves-to-production ratios of 200 years or more of gas supplies. It has tilted the traditional 8:10 ratio (80% of oil and gas sitting in OPEC and Russia with 10% in OECD states and 10% in China) in favor of the OECD world. Yet, if politics in Europe and the US remain on the current track, the long-term supply outlook appears far less certain.

Who Gains? A Risky Transition

As stressed at the beginning, transitions are anything but smooth processes and create winners and losers. This leaves our discussion with a final question: who would ultimately profit most from a globalized gas market? Considering the contingencies first, it remains unclear whether shale is going to pan out on a global basis, either on geological *or* geopolitical grounds. Asia *might* develop its own domestic shale in India and China; Australia *might* come good on its LNG potential to make sure Qatari LNG can keep supplying European hubs. Europe *might* become home to liquid gas trading hubs thanks to developing its domestic resource base. East Africa *might* add extra depth to LNG supplies; West Africa *might* up its game directly in response. Latin America *might* pull back from the resource nationalism abyss, a process that has gained strong traction again lately. The Southern Corridor *might* deliver more gas to Europe from the Caspian Basin than many currently think, and from multiple sources. The US *might* develop serious LNG export potential, sending tankers across the world to put gas benchmarks on a "fast track" convergence. And the enormous (private) capital required to make all this happen *might* come to the fore – possibly a function of how long the current economic crisis will last. If shale gas becomes a global play, earth is cracked, LNG trains loaded, then incumbent producers of unconventional gas may find themselves in deep commercial and political trouble. But there are too many *mights* to make a clear bet on unconventional plays, and by implication to make a definitive conclusion as to how price convergence does or does not play out – ultimately the determining factor of who wins and who loses.

Contingencies aside, gas market developments so far have clearly put traditional oil price indexation in question. In fact, it is highly questionable whether oil can be credibly deemed as a sensible way of pricing gas, a commodity no longer in direct competition. The principle that gas markets should be based on demand and supply fundamentals gains traction, and the "gas on gas" genie is clearly out of the bottle. True, the way to liquid international markets and global price convergence is not a straightforward one. We will probably continue to see a hybrid model playing out over the next decade until fundamentals tips the balance one way or the other. Many traders will continue to hedge gas price exposure against heavily traded oil markets, some might even point out the utility of long-term indexed contracts to renegotiate terms, just as upstream players will still demand long-term contracts to develop fields, built infrastructure, lay pipes, and even load LNG tankers. But if gas market liquidity continues to hold up, the contracts will probably not only be subject to far more dynamic price points, they could also be considerably shorter. The first conclusion to draw would therefore be that things do not look particularly good for traditional producers. Incumbents such as Russia might come under strong pressure to compete with newcomers such as Qatar on both their traditional European customer base and on emerging markets in Asia, or strike a deal on supply side cooperation. Under a "business as usual" scenario Russia will therefore have to make difficult decisions between what volume-to-price balances it is happy with. Consumers in the Atlantic and Pacific Basins, in turn, expect lots of LNG to keep coming their way at preferential prices. They would come out as the winners of globalizing gas markets and profit from suppliers being stuck in fierce competition over market shares.

On second thoughts, however, not all of this is bad news for producers. Most con-sumers (especially continental Europeans) seem to be laboring under the illusion that spot markets mean cheap prices, an assumption clearly also underpinning the European Commission's energy packages (see European Commission 2011). Setting gas prices *based on gas fundamentals* however has got nothing to do with being cheap: it is purely about achieving a cost reflective price for whatever market fundamentals suggest. Gas on gas competition might well have positive medium-term effects on price, given that marginal costs of production are generally cheaper than oil. But there are never any guarantees. If anything, prices could initially be far more volatile than those associated with piped gas given the cyclical nature of the business, not to mention adapting to new upstream investment regimes unable to fall back on the oil "certainties" of old. The UK provides a vivid example of increased volatility during gas market liberalization and thereafter.

But even assuming these initial hurdles are jumped and gas markets are politically allowed to bed in, additional pitfalls may be looming. In short, an expanding market share traded on spot rather than bound in LTCs also means an expanding market share open to cartelization. As much as import dependent consumers tend to think they have taken the political sting out of gas producers' tails, spot markets could actually give producers far more leverage to manipulate prices, either on a collective or bilateral basis. A quick look at the map reveals that supply side dynamics are essentially oligopolistic in Europe, a position that Russia might decide to capitalize on. Much would depend on pricing pressures involved and how far convergence has gone, but the lower that prices go, the more compelling supply side collusion would become in order to recalibrate markets back toward producer interests. The emerging Gas Exporting Countries Forum (GECF) may become the organization of choice for this purpose. The GECF currently represents some 70% of world natural gas production. Major producers and reserve holders such as Russia or Iran have shown revived interest in the Forum, which has now moved toward a more formalized organizational structure (see Stern 2010: 4; Stern and Rogers 2011).

A gas cartel would obviously face similar challenges as OPEC, in that someone would have to shoulder initial opportunity costs and absorb likely free riding, enforce quotas, and restrict new market entry at the fringe. Moreover, the cartel would also need to find a swing producer, that many have long thought would be Qatar, but actually this may flag up an interesting opportunity for Russia. At present, and despite sitting on over 30% of global gas supplies, Russian LNG production accounts for less than 5% of global share. In short, Russia – the gas equivalent to Saudi Arabia for oil – has let itself become a fringe player in a global gas world. Developing Shtokman, Sakhalin, and indeed Bazhenov and Achimov fields will undoubtedly not be to everyone's liking, but given that Russia's own unconventional reserves are estimated to be ten times larger than the whole of Europe, it still has the time and potential to take the lead back on volume to dictate long-term prices. If global gas benchmarks are the way of the future, then Russia has potential to play a pivotal role as the swing LNG producer of the world. Not only could Russia lean far heavier on Qatari, Australian, Algerian, East African, and burgeoning Latin American LNG production to align short-term prices, it would set the stage for a serious approach toward a gas cartel that would be the logical conclusion of independent global gas prices. On top of all, Russia's swing status would be built on the shoulders of a well-supplied but largely isolated US market. If the US goes "energy independent," Russia can start applying its own logic of gas on gas competition. Ironically, gas convergence – a supposed existential crisis for traditional producers – could eventually emerge as a tremendous opportunity to tilt the gas market in their favor.

Overall, the transition toward global price benchmarks should in the short run benefit consumers by driving competition and efficiency gains, in addition to diversifying supplies. In the long run, however, greater supply side collusion may be looming, both in terms of volumes and price. Ignoring a potential gas cartel could prove to be a costly mistake, just as it was to ignore the world's largest oil producers aligning their interests in OPEC in the 1960s.

Conclusion

This chapter has covered a wide range of political and economic drivers characterizing an ongoing fundamental shift from the old world of natural gas (characterized by regionalized markets and oil-indexed LTCs) to a new and admittedly rather uncertain one (liquid spot trading replacing LTCs, and global price convergence). Whether that new gas world will be a brave and bright one will however depend on a number of things. One is economics. Overall liquidity is both the core driver and the core challenge in globalizing gas markets, which means shale options need to be globally developed and LNG trains set in motion. This obviously comes with immense investment needs, which, according to IEA estimates, amount to some $8 trillion until 2035 (IEA 2011a: 8). Public funds will not do the trick here, and private capital will only find its way into upstream projects and infrastructure if the expected return on investment looks promising enough, if access is granted by reserve holders, and if risks can be hedged. As discussed, however, crucial markets in North America may fail to provide for an attractive pricing environment in the future, while geology may prevent shale plays from coming to fruition in Europe, in addition to non-business factors negatively impacting on investment decisions. Natural gas markets will keep on globalizing only if the underlying economics allows for it. This gets us to a second important insight: politics remains key in this process. Whether domestic reserves are made available to free up volumes for export, new plays are licensed out, and pricing mechanisms continue going global are, in the end, questions for political decision-making in key producer and consumer countries. Gas markets do not operate in a vacuum. More than other less politicized markets, they are created by states and ordered by institutions. Economics clearly do not trump politics in gas. To the contrary, the future direction natural gas markets take fundamentally depends on strategic decisions taken in Washington, Brussels, Beijing, Doha, and Moscow. Europe can shun domestic shale developments and bank on global gas fundamentals going their way. This, however, leaves them with a strategic supply game played out between Qatar and Russia, and makes them rule takers with regard to their own destiny. The US has a strategic decision to take on whether it becomes a self-sufficient island or continues shaping the new gas market order. Russia can stick to a failed "Eastern Strategy" or get to grips with the fact that US shale has made the energy heavyweight a price taker in Europe (and Asia) and start developing LNG prospects to reclaim control of global gas fundamentals. In fact, as discussed, the brave new gas world may have much to offer producers – a single price point could reduce price volatility, while far broader and flexible markets may offer better opportunity than a single consumer at the end of a pipeline where the price is set by oil, meaning OPEC. Markets will not take the decisions here, but governments.

One thing is for sure: gas is here to stay. It is widely regarded as key to fueling emerging markets, alleviating energy poverty, decarbonizing the fossil fuel energy mix, and building a bridge to a low carbon future. It simply is the fuel of choice for decades to come, and will claim higher shares in the global energy mix. While coming with adjustments costs for both producers and consumers, there is much to gain from globalizing gas markets, from greater liquidity to a pricing mechanism based on actual market fundamentals. The trend toward new, more market-based models in natural gas trade seems irreversible; *status quo ante* no longer seems an option. Producers and consumers had better get the policies right to make the inevitable transition a success.

References

Bloomberg. 2012. NBP Pricing Data http://www.bloomberg.com/quote/NBPGWTHN:IND, accessed July 26, 2012.

BP. 2012. *Statistical Yearbook of World Energy*. London: BP.

Butler, Nick. 2011. How Shale Gas Will Transform the Markets. *Financial Times*, May 8.

Butler, Nick. 2012. Prepare to Celebrate OPEC's Demise. *Financial Times*, May 21.

China Daily. 2009. China, Russia Sign 12 Agreements During Putin's Visit. *China Daily*, October 13.

Downstream Today. 2012. Turkmengas Signs Framework Agreement with CNPC. Downstream Today (.com), June 7.

Economist. 2012. Special Report, Natural Gas, Shale Shock. *The Economist*, July 14.

EIA. 2011. World Shale Gas Resources: An Initial Assessment of 14 Regions Outside the United States. Washington, DC: US Energy Information Agency.

EIA. 2012. *Annual Energy Outlook*. Washington, DC: US Energy Information Agency.

Energy Intelligence. 2011. Natural Gas Week. Energy Intelligence Group.

European Commission. 2011. *Energy Roadmap 2050*. Brussels.

Financial Times. 2012. Japan Pushes Asia Gas Price Close to High. *Financial Times*, May 17.

Forster, Christine. 2012. US LNG Exports Loom But Australia's Santos Keeps Watch on Qatar. Platts, The Barrel, April 3. http://blogs.platts.com/2012/04/03/us_lng_exports/, accessed November 15, 2012.

IEA. 2011a. *World Energy Outlook 2011 Special Report: Are We Entering a Golden Age of Gas?* Paris: International Energy Agency.

IEA. 2011b. *World Energy Outlook 2011*. Paris: International Energy Agency.

Jaffe, Amy Myers. 2011. The Americas, Not the Middle East Will Be the New World Capital of Energy. *Foreign Policy*, August 15.

Komlev, Sergei. 2012. European Gas Market Reforms Undermine Security of Supply. *European Energy Review*, May 7.

Konoplyanik, Andrei. 2011. Gazprom's Concessions in Oil-Indexed Long-Term Contracts Reflect "Forced Adaptation" to New Realities. *Gas Matters*, April.

Levi, Michael. 2012. A Strategy for US Natural Gas Exports. Washington, DC/New York: Brookings Institution / Council on Foreign Relations.

MIT. 2010. *The Future of Natural Gas: An Interdisciplinary MIT Study*. Cambridge, MA: MIT.

Morse, E. 2012. *Energy 2020, North America, The New Middle East*. New York: Citi.

Moscow Times. 2010. E.ON Asks for Lower Gas Prices. *Moscow Times*, August 23.

RBC. 2010. Gazprom's European Partners to Get Gas with Discount. RosBusiness Consulting, February 27.

RBC. 2012. E.ON, Gazprom Iron Out Price Dispute. RosBusiness Consulting, July 3.

Reuters. 2010. Gazprom Adjusts Gas Pricing to Defend Market Share. Reuters, February 19.

RIA Novosti. 2010. E.ON Ruhrgas Presses Gazprom for Discounts Again. Russian International News Agency, August 20.

Stern, Jonathan. 2010. Continental European Long-Term Gas Contracts: Is a Transition Away From Oil Product-Linked Pricing Inevitable and Imminent? *Oil, Gas & Energy Law Intelligence*, June.

Stern, Jonathan, and Howard Rogers. 2011. *The Transition to Hub-Based Gas Pricing in Continental Europe*. Oxford: Oxford Institute for Energy Studies.

Stevens, Paul. 2010. *The "Shale Gas Revolution": Hype and Reality*. London: Chatham House.

Talus, Kim. 2012. Winds of Change: Long-Term Gas Contracts and the Changing Energy Paradigms in the European Union. In C. Kuzemko, M. Keating, A. Goldthau, and A. Belyi, eds. *Dynamics of Energy Governance in Europe and Russia*. Basingstoke: Palgrave Macmillan.

UPI. 2012. Gazprom Weighs Renewables as Price Peg. United Press International, June 7.

Worldwatch and Deutsche Bank Climate Change Advisors. 2011. *Comparing Life Cycle Greenhouse Gas Emissions from Natural Gas and Coal*. Frankfurt am Main: Deutsche Bank AG.

Yergin, Daniel, and Robert Inieson. 2009. America's Natural Gas Revolution. *Wall Street Journal*, November 2.

Chapter 7

The Breakout of Energy Innovation: Accelerating to a New Low Carbon Energy System

Michael LaBelle and Mel Horwitch

Introduction

Innovation is at the core of the global energy sector. The choice of technologies is dependent on the cooperation of businesses, governments, and society. Innovating toward a new low carbon energy system requires technological breakthroughs and general agreement by all actors. Global technologies emerge from research and development (R&D) and become localized through research and innovation (R&I) tightly connected to partnerships with social and business stakeholders. Providing the right political and economic environment is essential for the success of new energy technologies, broadly referred to as Cleantech. However, new technologies must contend with actors' risk avoidance, resulting in stymied efforts to introduce the most efficient technologies.

Climate change and its global implications require drastic reductions in greenhouse gas (GHG) emissions in an attempt to control the increase in global temperatures. The year 2050 is viewed as a common goalpost for action to alter the current course and reduce GHG emission levels in the atmosphere, with the European Union (EU) aiming for an 80–95% reduction by this date (European Commission 2011). Innovative solutions are necessary to quickly alter carbon-based trends for meeting growing global energy demand. Demand and supply must occupy a new low carbon trajectory – a sustainable technological pathway.

This chapter will highlight how innovation in new technologies and partnerships can disrupt and distort the lingering fossil fuel energy regime by creating a more sustainable energy system. Even in this scenario, carbon-based energy technologies remain. Coal, gas, and oil continue to play a prominent role in the future global energy mix; however, they become part of an overall diversified energy mix. Centralized electrical power systems also remain, but smaller and distributed generation offers the potential for life without an electrical grid. Overall, as the rising population and economic activity continue to stress the Earth's capacity beyond replacement levels, an energy regime will emerge that utilizes a wider variety of technologies and solutions to level off (and possibly decrease)

The Handbook of Global Energy Policy, First Edition. Edited by Andreas Goldthau.
© 2013 John Wiley & Sons, Ltd. Published 2013 by John Wiley & Sons, Ltd.

GHG emissions; therefore, fossil fuels and centralized systems remain, but are part of a larger diversified energy delivery system.

The two arguments in this chapter rest on the concept that the energy sector operates within a buffered regulatory and policy environment. This means that the choice of technology and the deployed scale, including everything from power generation to energy saving technologies, are strongly mediated by government regulations, public perceptions, and political decisions. In the race to reduce GHG emissions, action by these actors is as pivotal as inaction. High infrastructure investment costs take decades to be fully paid off. When these costs are combined with significant technical rules and highly tailored regulations the status quo is kept; it takes technological generations to implement a new energy system.

This chapter, in the first section, will argue that technological lock-in is perpetuated by state institutions within a broader regime of actors; a barrier is formed that prevents a faster roll-out of technologies able to shepherd in a low carbon era. The second section argues that a greater use and awareness of different types of innovation provide the potential to eventually break this technological lock-in and could, if more fully acknowledged by current regime actors, disrupt and displace the current carbon-based energy regime. The third section is structured around three types of innovation: disruptive, discontinuous, and sequential. Brief case studies highlight how renewable energy technologies (RET), smart grids, and shale gas represent different types of innovation and pose challenges to the current energy regime. In the final section, these case studies inform how innovation and the diffusion of energy technologies occur at a global scale then become embedded in the local.

Locking Out Risks

Universal access to electricity underpinned the growth of the electricity (and gas) sector in developed countries in the early twentieth century. Electric utilities rolled out power lines within communities, and as technology developed the systems became larger and more interconnected. In the US, as each geographic and political scale was reached – local, state, and interstate – various regulatory bodies stepped in to oversee and shape market and technical standards. The Great Depression, set off in part by the fictitious bookkeeping of utilities, resulted in a tighter regulatory environment where vertically integrated utilities were given monopolistic status under tight regulatory conditions, enabling them to own everything from production to distribution (Hirsh 1999).

Universal service and monopolistic practices began to falter in the 1990s. The effort to introduce competition between electricity generation companies gave a huge boost to electricity generated by independent power producers using natural gas. The twenty-first century is now marked by a network infused with competitive players overseen by energy regulators. The rise of RET allows the opportunity to be disconnected from the grid. Just as competitive players emerged to compete against former monopolists, innovative technologies and businesses may further weaken established players. Regulators and dominant market actors still attempt to retain their monopolistic niche. They try to maintain their position and protect investments by citing the sunk costs (invested amount into system infrastructure) or risks that endanger the centralized system. This risk aversion prevents a more robust technical and regulatory response where greater innovation in technology and regulations are fully embraced.

The relations that exist in the current energy regime are an expression of a deeper structure. The regulatory regime is marked by a "configuration of policies and institutions" structuring relations between social interests, state, and economic actors in the economy (Eisner 1993: 1). For a new regime to emerge, "new regulatory policies are initiated in several regulatory issue areas ... and [are] combined with significant institutional change" (Eisner 1993: 3). Since the 1990s there has been a new regulatory regime in energy; this is defined by "synergetic relationships, with private and public activities partially reinforcing each other" (Knill and Lehmkuhl 2002: 42). This provides further opportunity for new types of business and technological relationships to form and shape how carbon-based energy can be replaced with cleaner energy technology.

The growth of the Cleantech sector offers the opportunity to change long-established relationships. Cleantech involves a range of technologies including "energy generation, energy storage, energy infrastructure, energy efficiency, transportation, water and wastewater, air and environment, materials, manufacturing/industrial, agriculture, and recycling and waste" (Horwitch and Mulloth 2010: 25) to meet a range of needs, coupled to a diverse range of business relationships and practices as new solutions require embedding new technologies into societal practices.

Technologies like smart meters, solar power, and energy efficiency measures all require direct interaction with consumers. Just as mass communications through television and radio were altered through improvement in communication technologies, like the internet, energy consumption is experiencing its own interactive communication revolution, altering traditional relations between energy providers and consumers. Smart technologies redefine and reshape how consumers mange their energy demands.

There are three ways technological change can be encouraged or resisted. First, technologies require return on investments. Regulations and policies are used to encourage investments within a buffered regulatory environment; within the energy sector, government activity affects deployment of technologies. Due to high capital costs and buffered market environment in the energy sector (Islas 1997; van der Vleuten and Raven 2006), regulations can ensure incremental change becomes the norm and obsolete technologies are thereby maintained. The perpetuation of less than optimal technology is called *technological lock-in* (Unruh 2000).

Second, institutional inaction may discourage innovation in energy technologies thereby locking in high GHG emission levels. "Capital investments thus may have longterm implications for GHG emission levels, particularly in the energy sector, where investments are typically long-lived and require long lead times" (Buchner 2007: 13). Fostering technological evolution through state institutions prevents locking in older technology through institutional inaction. Unruh (2000, 2002) develops the concept of the Techno-Institutional Complex (TIC) as a means to explain the interlinked nature of technological progress and institutional evolution. *Institutional lock-in* occurs when institutions fail to encourage this uptake and old inefficient technologies remain the norm (Unruh 2000: 824; for further discussion on carbon-based TIC, see Calvert and Simandan 2010). Institutions, with an eye on long-term development of the energy sector, need to prompt efficient technological change, thereby ensuring policies and regulations fulfill public goals that encourage new technologies.

Third, longstanding relations between current actors, coupled with a proactive stance by companies and institutions, holds the potential for a more robust adoption of new technologies. Barriers do not have to remain in place. A faster and more measured shift to a more technically advanced energy sector is possible. This may be preferable because a "superior institutional arrangement" which is compatible with existing laws may set about significant technological *and systemic* change (Woerdman 2004: 74). In business–government relations, it is shown that a business holding an "insider" status by providing consistent and credible information facilitates regulatory policy-making. Here, in the regulatory arena, "reputation and goodwill become key components" (Coen 2005: 377). Joint initiatives for new types of regulatory structures based on new technologies can be developed together. For the EU, rolling out a pan-European smart grid (discussed below) is reliant on cooperation between sectoral actors.

There is potential to drive GHG emissions lower by fostering closer relations with energy sector actors, who are more proactive and willing to utilize new technologies along with new business practices. Coupling climate change mitigation goals with innovation in new energy technology has resulted in higher levels of innovation and Cleantech applications (Jänicke 2012: 51). However, resistance and a perpetuation of the current regime may be more of a reality. This means Cleantech backers seek to go around, below, and through the current institutionalized barriers. Described next is how types of innovation emerge on the other side.

Innovation and Collaboration

Technological innovation can progress along a continuum to provide gradual improvements, or it can disrupt and/or shift technologies onto a different track, rendering certain technologies obsolete. This section will define different types of innovation that hold the potential to alter an energy regime. This includes how R&I, through Cleantech, is central to deploy a broad range of sustainable technologies. Partnerships hold the possibility to foster new energy technologies, shifting the broader energy regime toward a low carbon path.

Three types of innovation are identified: disruptive, discontinuous, and sequential (or incremental) (Walsh 2012). Based on these types different strategies are necessary to ensure a return on investments when these are deployed. "The choice of commercialization strategy is influenced by the type of innovation and the related commercial risk (cost risk, product risk and market risk) associated with introducing that innovation in the marketplace" (Walsh 2012: 33). The links between risk assessment, regulatory regime, and types of lock-in all play a part in the extent of technological innovation. According to Walsh (2012: 33) the three types of innovation are:

- *Disruptive:* radical innovation providing for the introduction to the market of a product or service that is distinct from any other provided in the marketplace.
- *Discontinuous:* shift of an existing technological learning curve in order to enhance the value of an existing product or service.
- *Sequential (or incremental):* when the product or service undergoes continual improvement in order to maintain its uniqueness and superiority and thus its competitive advantage.

The land of energy technology is filled with various innovative products that disrupt, improve, and maintain their edge. The case studies below describe in greater detail how particular technologies correspond to types of innovation. The pairing of innovation definitions with technology demonstrates the growing momentum for technological (and regime) change.

The following connections will be made in the case studies. (1) *Disruptive innovation* describes RET by allowing disconnection from the centralized grid system. However, as Walsh argues, "RETs vary by countries and technologies, which can alter the classification of innovation type" (2012: 34) thus reflecting the localized nature of energy technology which is dependent on local energy sources and the policy and economic environment. (2) *Discontinuous innovation* emerges for smart grid technology. Transmission and distribution grids are boosted by inserting communication technologies, enabling centralized power plants, distributed generators (small power plants), and even home appliances to talk to each other, enhancing the overall system. (3) *Sequential innovation* describes shale gas fracking technology. Extracting unconventional gas reserves is a significant improvement over past extraction technologies, boosting gas supplies, lowering prices, and improving product penetration. Technological innovations offer the opportunity for a more efficient and lower carbon energy regime.

Collaboration in business and between global and local actors emerges as a key ingredient to sustainable innovation. "The only way to create and speed up sustainable innovation is to collaborate openly and actively with a large number of organizations" (World Business Council for Sustainable Development 2011: 3). Transforming how and where energy technologies are rolled out must evolve if companies and society target both economic growth and improvement in living standards. New Cleantech products and services from energy generation, storage, and energy efficiency methods to infrastructure become embedded in the local communities. Innovation and community action, by leveraging social entrepreneurship, become two constituting and significant forces for sustainable development (Horwitch and Mulloth 2010; Seyfang and Smith 2007).

Advancing innovation under the auspices of climate change and the need for new technologies to reduce GHG emissions is reliant on an R&D pipeline. Increasing R&D expenditure in GHG mitigation technologies can be increased by leveraging both private and public investments. "Firms innovate because it helps them become more profitable, but they innovate less than would be socially desirable because they know that other firms and consumers will reap some of the benefits" (Weyant 2011: 675). A range of solutions are possible to leverage private investments. Increasing public financial support for innovation in the energy sector can make a difference. For example, rather than just taxing GHG emissions, increasing the public financial support for investment projects can double the "social rate of return" when leveraged with private investments (Weyant 2011: 675).

R&I emerges as a means for the EU and the US to aim for longer-term development and deployment of new energy technologies (Hervás Soriano and Mulatero 2011). This means fostering cooperation between social actors, companies, and governments to leverage the human and market capital necessary for technology to take root. Rolling technologies out of laboratories is reliant on relationships and matching high skilled labor with social/user engagement. Institutional efforts are necessary to increase cooperation across political boundaries, reduce market uncertainty, promote open grid access, and educate consumers (Ewing Marion Kauffman Foundation 2010). Through this strategy, R&D is supplemented with knowledge and expertise and, when applicable, attached with a social component: forming the basis of R&I. Business and social practices are central in devising GHG mitigation techniques; technology supplements this relationship and over time supplants embedded technologies.

Case Studies

Cleantech can challenge and eventually compete against carbon based energy sources but requires cooperation by a broad spectrum of society, businesses, and government. The case studies and innovation typology presented here (Figure 7.1) are reflective of this upswelling of innovation that attempts to challenge the current energy regime. The

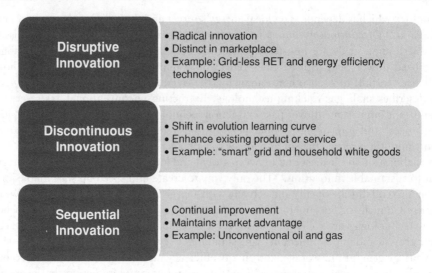

Figure 7.1 Types of energy innovations and examples.
Source: Categories developed by Walsh (2012).

categories of innovation discussed above provide the foundation to understand how technologies emerge to alter the trajectories of current practices.

The case studies are based on three innovative categories and three types of technologies. (1) The use of RET demonstrates the challenge non-centralized generation (distributed generation) holds for the current energy system. The emergence of a global market for solar power accelerates the innovation process and creates a new basis of technological competition – through *disruptive innovation*. (2) The building of a smart electricity grid holds the potential, through *discontinuous innovation*, to create new business opportunities and serves as a platform for economic development by developing the backbone to a non-carbon-based energy system. (3) Despite the rise of Cleantech, fossil fuels will remain dominant and continue their *sequential innovation*. In the third case study, the role that shale gas is playing to accelerate the energy market toward a *lower* carbon energy regime is examined. However, even this technology, which is more evolutionary rather than revolutionary, faces resistance by society.

Disruptive Innovation: Renewable Energy and Energy Efficient Technologies

The concept of disruptive innovation is based on breaking the stasis of established technologies. In the realm of energy, within the buffered regulatory walls of the sector, the best or most innovative technology may not win. Solar photovoltaic (PV) and LED lighting technologies hold the potential to disrupt the present business practices of electric utilities, by dramatically dropping the amount of electricity withdrawn from the grid – holding the potential to also dramatically drop company revenues. The current regulatory system that structures the utility business and profit levels must consider evolutions in technologies. The market and regulatory environment in which new technology is launched becomes essential for understanding long-term "disruptor credentials."

Establishing these credentials requires asking the question: a disruptor for whom and what? Since the social and political acceptance of vertically integrated monopolies in the early years of the twentieth century, the stranglehold over the type of generation and electricity distribution has remained largely intact. Only natural gas has recently emerged as a competitive production technology to coal, nuclear, and hydroelectric (with the development of these technologies supported by government efforts). However, the emergence of RET and energy efficiency efforts may radically, and disruptively, alter how utilities operate: not by going through, but by going around utilities, thereby disrupting, distorting, and possibly dismantling a century-old business model.

"A major characteristic of disruptive technologies is that they are rarely directly employed in established markets, but change the architecture of the market in the medium and long term" (Christensen and Bower 1996; cf. Richter 2011: 9). New technologies are often deployed in niche markets. But the question should also be asked: what if, with the rise of India and China and other quickly developing countries, the niche market is them? The market is not just high value technical markets in developed countries, as the early semiconductor industry was for the US military (Alic et al. 2010), but the technology is engineered and geared to developing markets. Electricity is expensive in these countries, above rates in developed countries. In addition, the price paid for energy in developing countries is highest among rural and poor consumers. The use of batteries, kerosene, and candles costs as much as $4 per kilowatt hour – 66 times the price in Manhattan (Pearson et al. 2012). Therefore, the global market for solar PV technology is truly global. While consumers in developed OECD countries operate under subsidized government regimes for solar PV (in an attempt to foster domestic Cleantech industries and demand), consumers in developing countries live under high priced oil-based energy systems. Solar technology is funneled through different paths, thereby gaining wider deployment.

Developing countries have the chance to skip centralized power systems, resulting in a "second great leapfrog" (Pearson et al. 2012), which describes the process of moving from kerosene to solar power. "The price of solar PV modules dropped from more than $4 per Wp in 2008 to just under $1 per Wp by January 2012, and the global installed capacity increased from 4.5 GW to more than 65 GW today" (Aanesen et al. 2012: 3). This global ramping up of the solar PV market has a broad impact. In developed countries, the installation of systems in homes or for industry influences the price along with the interconnected infrastructure and services. Grid parity, the rate at which solar PV becomes competitive with grid electricity, is foreseen by 2020 (Buijs 2012: 44), if not sooner. If solar can operate competitively without the backbone of a grid then this can have a disruptive effect for utilities, including how they invest and operate their grids. More fundamentally, "the growing differential between the rising costs of the old fossil fuel energies and the declining cost of renewable energies is setting the stage for an upheaval of the global economy and the emergence of a new economic paradigm for the twenty-first century" (Rifkin 2011: 31). The global solar PV market is an essential part of this new economic paradigm.

Coupled with the rise of renewable energy is the awareness that energy efficiency measures are also essential for reducing costs and also reduce the need for new power plants. However, reducing energy demand can be at odds with established businesses and their operations. Energy efficiency emerges as a co-benefit, usually quantifiable for a secondary reason (rather than strictly saving the planet). "Although more a driver than a barrier, co-benefits beyond efficiency improvement are often observed for industrial energy-efficient techniques. They may result from waste reduction, reduced material consumption, lower maintenance needs, lower emissions or improved reliability and better product quality" (Fleiter et al. 2011: 3101).

The innovation and deployment of new technologies can significantly alter how factories and homes are powered. The introduction of LED street lighting, besides being more energy efficient, also has the benefit of lasting 15 years compared to 3–4 years for current sodium lights. Once the costs of replacing sodium lights are considered, LEDs emerge on the cusp of being competitive. However, utilities are reluctant to switch based on a decline in revenue and cite the risk of using a technology with a limited track record (Linebaugh 2011, 2012). Disruptive technologies hold tremendous potential to create clean energy while reducing the amount of energy consumed. Figuring out how to break apart old business models and technological lock-in, due to risk factors, are key steps in the innovative process. Secondary benefits can act as the primary drivers to technological change.

The long-term timeline that disruptive innovation operates along means the "selling points" for these technologies must become apparent before their widespread adoption. The flag that has been planted in 2050 looms on the horizon for RET, requiring broad adjustments to regulations, operations, and business practices. Altering the business model of the centralized electricity system in developed countries requires supporters of the secondary benefits to push institutions to alter how some regulated businesses operate. For consumers in developing countries, open to market forces, innovative business solutions must be deployed that allow the high upfront costs of RET to be spread out over time. The medium and long-term horizons allow the momentum of RET to build, with the wave finally breaking over the sunk costs of the carbon-based technologies of today.

Discontinuous Innovation: Smart Grids

Shifting to new products and services from an existing technology requires accelerating to a new technological platform beyond the current technology. Smart grids and overall smart technology in the electricity system provide the basis for discontinuous innovation in the energy sector. The communication elements of the smart grid transforms the established network of high and low voltage electricity lines into an Internet-based system that enables each component, whether a power plant or a washing machine, to communicate. While various levels exist within this infrastructure, the scope of change is demonstrated by understanding how drastically new communication technologies have already altered the information sphere. The energy sphere, as a basis for economic activity, can also become revitalized through communication that will accelerate the use of new products and services.

The energy infrastructure is a dynamic space that structures economic activity, therefore innovation in this field ripples through the broader economy. "Infrastructure is an organic relationship between communications technologies and energy sources that, together, create a living economy" (Rifkin 2011: 27). Building smart grids serves as the basis for new energy technologies, economic activity, and even the survival of utilities. The "smart grid will act as a backbone infrastructure, enabling a suite of new business models, new energy management services and new energy tariff structures" (Accenture 2009: 3). How these business models are developed and the services provided will be dependent on the types of technologies deployed.

Overall, the aim of a smart grid is to create a better balanced electrical system. From the feed-in of renewable energy technology to smoothing peak demand periods, communicative grids act to balance supply and demand. The smart grid is reliant on in-home technology as much as consumers, as stated by an ICT expert, "Smart grids and smart users must go together … this interaction will result in dynamic pricing and dynamic response" (LaBelle and Sajeva 2010: 66). The dynamics of a smart grid also must be managed, as the technology does not disrupt grid operations, rather the communication technology supplements the energy network, not significantly reducing the use of electricity. "Smart metering enables more accurate billing to occur but will not reduce the demand for electricity, it should be viewed as a load balancing method for supply side management, rather than benefiting consumers like energy efficiency measures would" (Warren 2009). Therefore, while peak periods of energy use can be reduced, consumers can benefit by turning off appliances at high-cost periods in the day, furthering energy and financial savings.

Smart grids rolled out at a broad scale can provide load balancing that renewable energy requires, like wind and solar, which have varied production during the day and night. At this broad scale, state and transnational organizations need to facilitate building the infrastructure and the applied technical standards. China, performing a "double leap-frog" by skipping over a static grid, requires the grid to balance the increasing load for renewable energy. Since 2006 wind power has doubled every year, while solar power grew 44% per year (State Grid Corporation of China 2010: 6; Buijs 2012: 27), therefore prioritizing and balancing RET generation over coal generation in the grid can drive carbon emissions lower. In order not to exclude further technological innovation from being connected to the grid, establishing common standards for technology and how the system operates as an integrated but open system is important for long-term success (State Grid Corporation of China 2010: 20).

EU efforts to build a smart grid as the backbone of a European electricity system start with a top-down institutional response. With the price tag of a European smart grid at 150 billion euros (Smart Grid Today 2009) significant regulatory and stakeholder cooperation must occur. This can be seen in the new institutional architecture established by the EU and instituted by the European Commission. The establishment of the Agency for Cooperation of Energy Regulators in 2009 is meant to foster greater cooperation between national energy regulators, thus improving cross-border cooperation on infrastructure projects and market opening. Within this framework, the EU has established working groups for smart grid technologies that will work with national regulators to facilitate technical standards, enabling a faster roll-out of smart meters for 80% of consumers by 2020 (see LaBelle 2012).

Innovation in smart grid technology therefore rests not just on the availability of the technology, but how widely it is accepted by governments and companies and deployed across borders and then connected to homes. At the core of innovation in the smart grid is the concept of cooperation and coordination of stakeholders. The organic technological platform encourages the growth of new business models by fostering greater interaction between producers and consumers. Discontinuous innovation, marked by an improved learning curve and new products and services, matches the developments around a smart grid. R&I becomes reliant on technological developments *and* how consumers choose to integrate the technologies into their daily lives.

Sequential Innovation: Gas

The terms sequential innovation and technological risk go together for fossil fuels. For the natural gas sector, dwindling conventional gas reserves require greater innovation but with consideration for environmental risks. The risks hold the potential to prevent the use of new technologies and thereby prevent further extraction of unconventional gas. Shale gas is now emerging as an important unconventional source that is reliant on technology which holds perceived high environmental risks. Citing environmental risks, France and Bulgaria have moratoriums in place. Nonetheless, the impact in the North American gas market is demonstrated in an oversupplied market with low prices. Natural gas is now competitive against coal in the US with a double benefit. First, being a key low carbon fuel it is a bridging fuel toward a zero carbon economy. Second, the technology has earned its popular title of "game changer" for the North American gas market. Low prices are now boosting manufacturing competitiveness as energy intensive industries find cheaper power markets in the US (Boxell 2012). However, the perceived environmental risk surrounding the technology has spread globally. The success in the US must contend with the environmental risk perceptions in Europe, where moratoriums are in place and opposition exists against the technology.

The opposition to shale rock fracking techniques demonstrates that technologies do not have to be disruptors in order to experience resistance in the energy landscape. Environmental concerns have been raised over the techniques used to extract gas from shale formations. These concerns stem from the large amount of water, sand, and chemicals used, along with the release of other chemicals and gases, through the fracturing process. Together these can contaminate aquifers and the air. Hydraulic fracturing of shale formations uses water mixed with chemicals and pumped under intense pressure to crack the shale, releasing gas. Afterwards, the injected water, which returns as "flowback," must be reinjected into the ground and/or treated to isolate pollutants.

The emergence of shale gas technology has transformed the energy landscape in the US. In the late 1970s a moratorium was placed on the construction of gas fired power plants, due to the understanding that the US would run out of gas. While the moratorium was lifted in the 1980s, the view of dwindling supplies persisted into the 2000s, with LNG import facilities constructed to continue to fuel the demand for gas. The impact has been dramatic, with shale gas expected to surpass conventional gas as a main supplier on the US market. The dramatic impact on the US market is possible because of significant technological innovation. Shale gas wells use newly developed horizontal drilling techniques. New technologies give the ability to more fully understand the geology and the best way to extract gas. Microseismic tools and 3D imaging made these advances possible, and emerged after years of government supported research and joint work with the oil and gas industry (Breakthrough Institute 2011; Shellenberger and Nordhaus 2011; Trembath 2011).

R&D into unconventional gas leading to R&I by the increased cooperation between companies and government research units, the application of technical resources, human know-how, and government policy support, unrolled the shale gas revolution. In the example of shale gas, it was after a business merger in 1997 that the "slick water" extraction technique was applied to horizontal drilling, dropping the price per well from $375,000 to $85,000. Later in 2006 and 2007 work done by the Federal Bureau of Economic Geology utilized technology from metallurgy and found nano-sized pores that contained gas, further contributing to improving extraction techniques (Breakthrough Institute 2011).

Reflecting on innovation in shale gas technology, and echoing the continuous innovation process in the gas sector, "What needs to be done is continuous innovation. Innovation, technology and know-how developed in the US are available for the world" (Rashid Khan, Natural Gas Europe 2012). Reflecting on this perspective, and the role that shale gas has played in altering the energy landscape, is a chief business manager involved in shale gas technology. "Gas has bought us the time to develop the other things . . . Government has to be looking down the road. We really cannot wait to develop those other energies. Industry doesn't look as far down the road as the government should" (Dan Steward, Breakthrough Institute 2011). Innovation in the energy sector is dependent on a continuous path where government and industry work together to provide expertise and support for new energy technologies.

The development of new technological solutions for extracting oil and gas also pushes against environmental risks with the possibility that technologies will be rejected by society and governments – even in the case of sequential technologies. Broader institutional and societal risks create boundaries. Engagement with social considerations and the right policy environment are essential for allowing the utilization of technologies. Comparing the game changing market analysis of US shale gas to the resistance in Europe exposes how the risks of technological and institutional lock-in affect the industry. Examining the risks for the European shale gas industry exposes a range of constraints that impact the growth of the industry. Despite some moratoriums against the technology in Europe, other countries are proceeding slowly, largely in part due to industry and infrastructure limitations, but also in consideration to broader environmental questions.

Extraction technology for oil and gas is based on sequential innovation, thereby as demand and costs rise for conventional reserves, unconventional technologies can make a difference. Unconventionals, as they become financially viable, help maintain supplies and can stabilize prices. They may also increase market share, as is the case in the US. However, conventional extraction techniques may win the day if the concerns put forward by society, politicians, and regulators cannot be addressed. Sequential innovation may extend technological boundaries, but as these boundaries are pushed, they also test the boundaries of society. R&I is based in cooperation between government, companies, and societies, and as the energy challenges become more pressing due to failing and depleted fossil fuel reserves, working through new energy options requires a new consensus on acceptable technologies.

Discussion: Innovating to Unlock Technology

Energy innovation is reliant on broad-based support that allows disruption of the embedded technological regime. Transitioning toward low carbon technologies in a rapid and consistent manner is based on globally deployed innovative solutions. Unlocking new technologies, which mitigate carbon emissions and provide access to electricity, rests on innovative partnerships. Cooperation enables a reduction for various risks associated with new technologies. R&I is a means to shape the global deployment of energy technologies that are transformed and transfixed to local environments. This can be seen in the three types of innovation and the related case studies.

First, disruptive innovation emerges as the biggest challenge to the current energy system. RET and solar PV will also take the longest to implement. Global penetration of solar power accelerates as both developed and developing countries seek to use this free resource with technologies that are rapidly declining in cost. Second, smart grid technology boosts the learning curve and opens a new space for products and services. Discontinuous innovation can have broad economic and social effects. The fusion of power plants to home appliances, through a communicative grid, accelerates decentralized power production, while boosting capacity for RET and low energy technologies. Third, sequential innovation can be seen in the gas output from fracking technology in shale formations. The gradual development of the technology and movement from vertical to horizontal wells has prevented the US from becoming an importer of natural gas. However, while shale gas is established in the US, in Europe environmental and social opposition may prevent wide deployment. Each of these technologies reduces GHG emissions but is reliant on collaboration with social actors, governments, and companies. R&I emerges as an essential element to disrupt and reorganize the global carbon energy regime through Cleantech.

Breaking through the institutional and regulatory barriers erected by dominant actors around the legacy technology can take a variety of paths for new technologies. Promotion of the technological switch is reliant on external and secondary reasons, whether reduction of GHG emissions or the financial savings that can be had from new technologies. Byproducts emerge as the game changers, either because of pricing that is equal or less than current technologies or through collaboration that provides the opportunity to enter a new economic paradigm. Institutional and technological lock-in becomes harder to justify as price and performance between technologies becomes equalized or is improved. LED lighting and RET demonstrate that as prices fall, the uptake improves, and over the long term gradual replacement of older technologies can be transformed into a new system. R&D and R&I are reliant on cooperation, with government institutions playing a central role; moving technologies from laboratory to the field must overcome these technological, institutional, social, and financial barriers.

Conclusion: Global Technology and Local Markets

Initiatives of R&I reduce carbon emissions through a strategic approach of long-term technological regime change. Technology is global, but how a technology competes against other sources is local. Innovative technologies and relationships emerge as key means for RET to go around and supplant entrenched carbon technologies and business practices. Working within a local context enables global technologies to become entrenched and compete against established technologies. Various factors of price, government support, social acceptance, and even entrepreneurship affect how new technologies infiltrate and are locally deployed.

Identifying the type of innovation of a particular technology enables a greater understanding of where resistance may emerge. Disruptive, discontinuous, and sequential innovations pose unique risks for institutions and businesses which must lead the way in adjusting to societal choices and technological advances. The integrated nature of R&D between the private sector and government entities means that exploiting short-term commercial interests must be matched with long-term strategic visions.

The global implication of climate change requires the rapid roll-out of low or zero carbon technologies. Solutions emerge locally as technology is deployed in different markets and adapted to an existing pricing and technological regime. The global demand for fossil fuels will remain constant, as economic development continues. However, the limitations of the Earth's resources require greater technological know-how as economic development and energy demand continues to rise. Decoupling this growth from the use of fossil fuels and GHG emissions is essential for maintaining a habitable planet. Cleantech emerges as a combination of technologies and partnerships. Innovative energy solutions are represented in technology, but require society and governments to work with the private sector to build an energy system of the future.

Overhanging the possible success of a wide roll-out of new technologies is the marketplace. Deployed with little government initiative for change, technology emerges into a treacherous landscape. Within the confines of the buffered regulatory arena of electricity markets, regulatory approval and government legislation can dissuade and inhibit the significant roll-out of new technologies. Technological and institutional lock-in can prevent more efficient solutions for supplying electricity and reducing demand. However, as

global niche markets emerge, and off-grid options for RET offer alternatives to centralized power systems, new technologies are taken up by consumers, improved further, and
over time can destabilize established business practices. Breaking the lock-in of carbon
energy technology only becomes a matter of time – and innovation.

References

Aanesen, Krister, Stefan Heck, and Dickon Pinner. 2012. *Solar Power: Darkest Before Dawn.*
McKinsey & Company, eBook.
Accenture. 2009. Accelerating Smart Grid Investments. Geneva: World Economic Forum / Accen-
ture.
Alic, John, Daniel Sarewitz, Charles Weis, and William Bonvillian. 2010. A New Strategy for Energy
Innovation. *Nature* 466 (July): 316–317.
Boxell, James. 2012. US Ahead of Europe on Energy Policy. *Financial Times*, May 13.
Breakthrough Institute. 2011. The Breakthrough Institute: Interview with Dan Steward, Former
Mitchell Energy Vice President. http://thebreakthrough.org/blog/2011/12/interview_with_dan_
steward_for.shtml.
Buchner, Barbara. 2007. *Policy Uncertainty, Investment and Commitment Periods.* Paris: Interna-
tional Energy Agency.
Buijs, Bram. 2012. *China and the Future of New Energy Technologies: Trends in Global Compe-
tition and Innovation.* The Hague: Clingendael International Energy Programme.
Calvert, Kirby, and Dragos Simandan. 2010. Energy, Space, and Society: A Reassessment of the
Changing Landscape of Energy Production, Distribution, and Use. *Journal of Economics and
Business Research* 16, 1: 13–37.
Christensen, Clayton M., and Joseph L. Bower. 1996. Customer Power, Strategic Investment, and
the Failure of Leading Firms. *Strategic Management Journal* 17, 3: 197–218.
Coen, David. 2005. Business-Regulatory Regulations: Learning to Play Regulatory Games in Euro-
pean Utility Markets. *Governance* 18, 3 (July): 375–398.
Eisner, Marc Allen. 1993. *Regulatory Politics in Transition.* Baltimore, MD: Johns Hopkins Uni-
versity Press.
European Commission. 2011. A Roadmap for Moving to a Competitive Low Carbon Economy in
2050. http://ec.europa.eu/clima/documentation/roadmap/docs/com_2011_112_en.pdf.
Ewing Marion Kauffman Foundation. 2010. *A Clean Energy Roadmap: Forging the Path Ahead.*
Kansas City, MO: Ewing Marion Kauffman Foundation.
Fleiter, Tobias, Ernst Worrell, and Wolfgang Eichhammer. 2011. Barriers to Energy Efficiency in
Industrial Bottom-up Energy Demand Models. *A Review. Renewable and Sustainable Energy
Reviews* 15, 6 (August): 3099–3111. doi: 10.1016/j.rser.2011.03.025.
Hervás Soriano, Fernando, and Fulvio Mulatero. 2011. EU Research and Innovation (R&I) in
Renewable Energies: The Role of the Strategic Energy Technology Plan (SET-Plan). *Energy
Policy* 39, 6 (June): 3582–3590. doi: 10.1016/j.enpol.2011.03.059.
Hirsh, R. 1999. *Power Loss: The Origins of Deregulation and Restructuring in the American
Electric Utility System.* Cambridge, MA: MIT Press.
Horwitch, Mel, and Bala Mulloth. 2010. The Interlinking of Entrepreneurs, Grassroots Movements,
Public Policy and Hubs of Innovation: The Rise of Cleantech in New York City. *Exploring
Technological Innovation* 21, 1: 23–30.
Islas, Jorge. 1997. Getting Round the Lock-in in Electricity Generating Systems: The Example of
the Gas Turbine. *Research Policy* 26, 1 (March): 49–66. doi: 10.1016/S0048-7333(96)00912-2.
Jänicke, Martin. 2012. Dynamic Governance of Clean-Energy Markets: How Technical Innovation
Could Accelerate Climate Policies. *Journal of Cleaner Production* 22, 1 (February): 50–59. doi:
10.1016/j.jclepro.2011.09.006.
Knill, Christoph, and Dirk Lehmkuhl. 2002. Private Actors and the State: Internationalization and
Changing Patterns of Governance. *Governance* 15, 1: 41.

LaBelle, Michael. 2012. Constructing Post-Carbon Institutions: Assessing EU Carbon Reduction Efforts Through an Institutional Risk Governance Approach. *Energy Policy* 40 (January): 390–403. doi: 10.1016/j.enpol.2011.10.024.

LaBelle, Michael, and Maurizio Sajeva. 2010. Pathways for Carbon Transitions: Workpackage 4. Transition Towards Post-Carbon Society. University of Turku, Finland / REKK, Budapest. http://www.pact-carbon-transition.org/delivrables/D-4.2.pdf.

Linebaugh, Kate. 2011. Cities, Utilities Are Poles Apart Over Streetlights. *Wall Street Journal*, December 24, sec. Business.

Linebaugh, Kate. 2012. LED Streetlight's Price Cut in Half. *Wall Street Journal*, April 9, sec. Technology.

Natural Gas Europe. 2012. Unconventionals: All About Innovation. http://www.naturalgaseurope.com/unconventionals-all-about-innovation-5855.

Pearson, Ben Sills, Natalie Obiko, and Stefan Nicola. 2012. Farmers Foil Utilities Using Cell Phones to Access Solar. Bloomberg. http://www.bloomberg.com/news/2012-04-11/farmers-foil-utilities-using-cell-phones-to-access-solar.html.

Richter, Mario. 2011. *Business Model Innovation for Sustainable Energy*. Lüneburg: Leuphana University, Centre for Sustainability Management. www.leuphana.de/csm/.

Rifkin, Jeremy. 2011. *The Third Industrial Revolution: How Lateral Power Is Transforming Energy, the Economy, and the World*. New York: Palgrave Macmillan.

Seyfang, G., and A. Smith. 2007. Grassroots Innovations for Sustainable Development: Towards a New Research and Policy Agenda. *Environment and Politics* 16, 4 (August): 584–603.

Shellenberger, Michael, and Ted Nordhaus. 2011. A Boom in Shale Gas? Credit the Feds. *The Washington Post*, December 16, sec. Opinions.

Smart Grid Today. 2009. European Commission Smart Grid Task Force Sets Goals. Smart Grid Today. http://www.smartgridtoday.com/public/939.cfm.

State Grid Corporation of China. 2010. SGCC Framework and Roadmap for Strong and Smart Grid Standards. http://collaborate.nist.gov/twiki-sggrid/pub/SmartGrid/SGIPDocumentsAndReferencesSGAC/China_State_Grid_Framework_and_Roadmap_for_SG_Standards.pdf.

Trembath, Alex. 2011. History of the Shale Gas Revolution. Breakthrough Institute, December 14. http://thebreakthrough.org/blog/2011/12/history_of_the_shale_gas_revolution.shtml.

Unruh, Gregory C. 2000. Understanding Carbon Lock-in. *Energy Policy* 28, 12 (October): 817–830. doi: 10.1016/S0301-4215(00)00070-7.

Unruh, Gregory C. 2002. Escaping Carbon Lock-in. *Energy Policy* 30, 4 (March): 317–325. doi: 10.1016/S0301-4215(01)00098-2.

van der Vleuten, Erik, and Rob Raven. 2006. Lock-in and Change: Distributed Generation in Denmark in a Long-Term Perspective. *Energy Policy* 34, 1) (December): 3739–3748. doi: 10.1016/j.enpol.2005.08.016.

Walsh, Philip R. 2012. Innovation Nirvana or Innovation Wasteland? Identifying Commercialization Strategies for Small and Medium Renewable Energy Enterprises. *Technovation* 32, 1 (January): 32–42. doi: 10.1016/j.technovation.2011.09.002.

Warren, Andrew. 2009. European Alliance of Companies for Energy Efficiency in Buildings. *Personal interview*, July.

Weyant, John P. 2011. Accelerating the Development and Diffusion of New Energy Technologies: Beyond the "Valley of Death." *Energy Economics* 33, 4 (July): 674–682. doi: 10.1016/j.eneco.2010.08.008.

Woerdman, Edwin. 2004. *Path Dependence and Lock-in of Market-based Climate Policy*. Developments in Environmental Economics 7. Amsterdam: Elsevier.

World Business Council for Sustainable Development. 2011. *Collaboration, Innovation, Transformation Ideas and Inspiration to Accelerate Sustainable Growth – A Value Chain Approach*. Geneva: World Business Council for Sustainable Development.

Chapter 8

Recent Trends in Upstream Petroleum Agreements: Policy, Contractual, Fiscal, and Legal Issues

Honoré Le Leuch

Introduction and Overview on Upstream Petroleum Contracts

Upstream petroleum[1] agreements and the associated fiscal regimes are the cornerstone of the relationships between petroleum countries (the host countries) or their national oil companies (NOCs) and foreign investors (the international oil and gas companies, IOCs). They are indeed the instruments implementing the specific *petroleum policy* decided by each sovereign country. Their terms directly reflect the conditions under which IOCs accept to invest in an upstream venture.

Upstream petroleum agreements are often called *petroleum exploration and production contracts* (E&P contracts). They govern, under the law of the country and in particular the *petroleum law*, the obligations and rights of the petroleum investors. These are quite long-term contracts concerning a specific contractual area awarded to the investor on an exclusive basis by the host country, based on the applicable legislation and taxation. In particular, they define, in case commercial petroleum discoveries are developed and exploited, how the production, incomes, and risks will be allocated between the government and the investor over the field's producing life as a direct consequence of the *fiscal regime* associated with E&P contracts.

E&P contracts and the associated fiscal regimes may be of several types depending on the legal system used in the country and its selected petroleum policy. While the main types of E&P contracts and fiscal regimes have been established a long time ago, the terms of each type of upstream petroleum contracts and fiscal systems have considerably changed in the last decades. The objectives of this chapter are to review the features of each main type of upstream petroleum agreements and their associated fiscal systems, and to outline their major trends and evolution. In particular the chapter seeks to answer the following questions. How are a country's petroleum policy and E&P contracts interrelated? What are the main differences between each type of E&P contract and fiscal regimes? What main changes in E&P contracts have recently occurred? How to define "a fair government take" under a fiscal system? When may a fiscal stabilization clause be justified? What adjustments are required to account for unconventional oil and gas?

The Handbook of Global Energy Policy, First Edition. Edited by Andreas Goldthau.
© 2013 John Wiley & Sons, Ltd. Published 2013 by John Wiley & Sons, Ltd.

Evolution of Upstream Petroleum Policies and Their Impact on E&P Contracts

In most countries the state is the sole owner of the underground resources (ownership of which is generally distinct from the rights vested in surface landowners) and exercises sovereign rights over their exploration and exploitation. It decides on the petroleum policy it wishes to apply in the long term and in particular on the role it selects to grant to private investors, mostly foreign companies, in the exploration and production of petroleum in the country. The sovereign state may decide to open or not selected areas within the country and its exclusive economic zone for further upstream activities by foreign companies; when to award E&P contracts and how to select the winning companies among qualified applicants for a given area; or whether to encourage, and at what extent, direct investments by foreign companies in its territory.

An upstream petroleum policy represents a balance between the interests of the country and those of investors. This balance depends on many factors and varies over time. Today, only some countries remain closed to IOCs (see Box 8.1), considerably less than

---

Box 8.1   Countries open or closed to foreign direct investment in petroleum exploration and production.

Since the opening of E&P to foreign investors in the former Soviet Union in the early 1990s, most countries are now largely open to direct investments by foreign petroleum companies, with the exception of the following:

- *Mexico*, since the nationalization of the oil industry in 1938. However, risk services contracts (RSCs) are now authorized in selected areas. Several contracts of this type were signed since 2005, mostly related to mature fields to be redeveloped.
- *Kuwait*, since the entire nationalization of the oil industry in 1976. However, technical assistance contracts were entered into with IOCs. Other types of contracts were considered, such as RSCs, but the law enacting them is not yet promulgated.
- *Saudi Arabia*, since full nationalization of the industry in 1976. Four concession contracts were however signed with IOCs since 2003 but only for non-associated gas exploration and exploitation, excluding oil.

In some other exporting countries where the petroleum industry was previously nationalized selected areas are now open to foreign direct petroleum investment, while the national or local companies continue to carry out a significant share of the petroleum activity, in particular from the petroleum fields producing before the reopening to foreign investors. This is the case in many OPEC countries, Russia, and China. Today over 50% of world petroleum production continues to be directly carried out by national or local oil companies without any involvement of IOCs. Globally NOCs have access to over 80% of worldwide discovered reserves.

Access to new exploration and production areas has become a key priority for IOCs in the last decade in order to renew their reserves. Technological progress and higher prices now allow exploration in new zones or deeper horizons for unconventional oil and gas resources.

two decades ago. In all other countries which desire foreign investments the pace for the awards of E&P contracts and the location of the contract areas remain at the sole prerogative of the state, which often associate companies in the pre-selection of the areas of interest to the industry.

To implement its upstream petroleum policy, each country promulgates an *upstream petroleum law* which defines in particular how to authorize IOCs to carry out exploration and exploitation operations and how to define the terms and conditions to be met by specific E&P agreements. The applicable petroleum tax regime is often governed by a special chapter added at the same time to the general tax law. The petroleum law deals with the E&P provisions regarding exploration for both oil and natural gas, and in case of a commercial discovery, its development and production until field decommissioning and site restoration operations at the end of the exploitation. Exploration of a specific contractual area may exceed 10 years and is performed in phases with distinct work commitments. Petroleum exploration remains a quite risky activity because no commercial discovery may be found. When a commercial discovery is demonstrated and approved by the state, the contract-holder will develop and then produce it during a period which may exceed 30 years, subject to the applicable petroleum tax regime.

The terms of new upstream petroleum contracts and tax systems have considerably changed throughout the last decades, following closely the respective political evolution in developed and developing countries and the many changes in the international oil and gas markets.[2] Undoubtedly the role and policy of the Organization of Petroleum Exporting Countries (OPEC) since 1960 was a catalyst in the evolution of the relationships between petroleum host countries and investors, resulting today mainly in higher government take in upstream profits and higher state control and participation in upstream activities, especially in the most promising producing countries in terms of reserves. Meanwhile, other countries not yet at this stage of petroleum maturity have adopted, or were induced to adopt, more attractive policies and fiscal incentives designed to encourage companies to search for and exploit petroleum in their territory.

The accelerated volatility in international oil and gas market prices in the last decades, reflecting the changing world oil supply and demand balance, had a major impact on the terms of E&P contracts and fiscal regimes, requiring for example the introduction of more progressive fiscal systems, robust enough to maintain a fair government take in case of large price fluctuations. In addition, both petroleum producing countries and IOCs are giving more and more priority to the environmental and socio-economic issues, because petroleum resources, being non-renewable riches, should benefit during their relatively limited exploitation life all the stakeholders and foster a sustainable development for the country and the local communities concerned by the activity. To that end, new clauses are now included in E&P contracts.

The Main Forms of Upstream Agreements and Licenses and the Related Fiscal Regimes

E&P contracts concluded between host countries and qualified investors may take different forms in relation to the legal system used by the country and more specifically to the policy selected for enacting the applicable petroleum law. Box 8.2 summarizes the different forms of E&P contracts or licenses awarded by states. These may consist in the award of *oil and gas licenses* (e.g., an exploration license followed by, in case of commercial development, a production license), without the signing of a distinct agreement, when the country's petroleum law and its regulations define in great detail the conduct

Box 8.2   Differences between E&P petroleum contracts and licenses.

The term "E&P contract" in this chapter refers to any agreement or legal instrument authorized under the applicable law between a host country (or its NOC) and an oil or gas company (or more frequently a consortium of companies) selected by the country to conduct exploration within a specific area, on an exclusive basis and at its own risk, and in the event of commercial discovery, the development and production of the discovered oil and natural gas fields.

The petroleum law enacted in a country may provide for different legal instruments, generally known as *E&P contracts*, depending on the petroleum policy decided by the country. Two main approaches are followed.

1. The award of specific petroleum licenses under terms entirely defined in the national legislation and regulation. The license may be (1) an *exploration license* (or *permit*) over an exploration area authorizing exploration, or (2) a *production license* (or *lease* or *concession*) over a restricted development area authorizing the development and production of a commercial field. In this case, no E&P contract per se is signed because all the terms are fixed by the law and detailed regulations, as in the US, Canada, or Australia, etc. Even if such licenses are not governed by a distinct concession contract they are categorized from a legal point as pure *concession agreements*. Sometimes, the award of a license under the country legislation is subject to the execution of a short license (or concession) agreement providing for a few specific terms regarding exploration work obligations (such as in the UK, Norway, and Denmark) and sometimes selected tax terms (e.g., related to an additional profits tax).

2. In countries which do not have yet issued comprehensive legislation and regulation, as currently in many developing countries, there is the conclusion of a *detailed E&P contract* under the *national law*, which provides for the terms and conditions applied to exploration and production not covered under the legislation and regulation, including for certain tax aspects when the law does not fix them. In most cases, the contract deals with the subsequent award of administrative *exploration and exploitation licenses* or *authorizations* as an automatic consequence of its signing and satisfactory performance of each period of the contract.

In this chapter any reference to "E&P contracts" covers the two above situations.

of petroleum activities and the related fiscal regime.[3] Alternatively, when the petroleum law only provides for the main principles and when the fiscal system is not fully defined by national law, the award of E&P rights may result from the effective signing of a quite detailed *E&P contract*, which corresponds to the ad hoc investment agreement for petroleum. The contract deals both with exploration and exploitation provisions consistent with the petroleum law and defines the elements of the fiscal regime not fully fixed under the law. The requirement for signing a detailed contract is often the situation in developing countries or in countries without a long petroleum history. The law generally states that as a consequence of the signing of contracts, petroleum licenses or specific exclusive authorizations, which are quite short administrative documents, will then be issued for authorizing exploration and exploitation pursuant to the procedure set forth in the law or its regulation.

Each E&P contract is associated with a *fiscal regime* (also called *system* or *package*) which consists of the set of economic and tax provisions applicable to the contract-holder. That fiscal regime is defined by the tax and petroleum laws of the country. It is however often supplemented by a few specific terms provided for in the E&P contract when so authorized by the country law. Nevertheless, there is a new tendency to apply in a country, at a given time for new contracts, the same tax regime to all newcomers (except for one or two fiscal terms) in order to facilitate the implementation of many contracts and develop transparency between investors.[4] The fiscal regime has a paramount importance both for the government and the company because it directly governs the allocation of the production and revenues between the country and the contract-holder. It determines the aggregate of the share of production and the different taxes payable to the country, referred to as *the government revenues* and often expressed in percentage as *the government take* in the petroleum profits derived from a given contract area or project. In any country, the petroleum fiscal system consists of several sources of government revenues. Their nature directly depends on the type of E&P contracts, as explained below. This is the reason why in this chapter both the evolution of E&P contracts and fiscal regimes are highlighted when presenting each type of upstream contract.

The award of E&P licenses or contracts more often results today from competitive biddings organized by the country pursuant to a *transparent tender procedure*. In some cases however a contract may be awarded following direct negotiations when the petroleum law so provides and the conditions justify it, for example when the competition would be insufficient for the concerned area. In both systems of awarding E&P licenses, only a few numbers of terms, including for the fiscal regime, are today subject to bidding or negotiation. The contract has also to be established on the basis of the *model E&P contract or license* elaborated by the country before the tender or the negotiation.[5]

Evolution of the Main Types of Upstream Agreements and Associated Fiscal Regimes

When a country wishes to attract new investors, the types of E&P petroleum contracts and related fiscal regimes available today in the world are so-called *modern concession agreements* and the *production sharing agreements*; *risk services agreements* are sometimes also used but less frequently. Indeed, in the last decades no new types of upstream agreements were introduced, but the terms of each type of agreement and their related fiscal systems changed significantly, adapting to the new volatile environment of international oil and gas markets. Depending on its petroleum policy and the characteristics of the acreage to be tendered, the country selects the type of E&P contracts that it considers the most appropriate for the tender or negotiation.

In all these arrangements, the state remains owner of the petroleum resources when in the ground and the contract-holder is obliged to undertake and fund, at its sole risk, all the petroleum operations, only being remunerated if the exploration is successful and leads to the exploitation of commercial fields. The main differences between the main types of upstream petroleum contracts arise indeed from the remuneration scheme for the contract-holder, namely (1) its share in the production extracted from the fields, varying from 100% under concessions to significantly lesser percentages under the other types of contracts, and (2) its specific tax obligations under the national law. The total government take under a contract integrates the different sources of revenues for the state or its NOC, both the tax payments made in cash and the revenues derived from the access to a share of production, if any, directly taken by the country. The main characteristics of each type of E&P contracts and associated fiscal regimes are summarized below.

Concession Agreement

This is an old legacy from the regime used in the mining industry in the nineteenth century, whereby the government grants oil and gas mining rights (generally named *exploration license* or *production lease*) giving exclusive rights to all the petroleum extracted by the concessionaire from the licensed area. The company becomes the owner of the entire oil and gas production when extracted at the wellhead – or at another agreed point of transfer of title – and markets it. The original concession regime was criticized by many developing countries and has gradually evolved toward what is known today as the *modern concession agreement* updated to better safeguard the legitimate interests of the host country. The concessionaire, also called the licensee or lessee, is subject to different tax obligations depending on the country policy implemented in national law, namely:

- *Ad valorem royalty on production*, payable in cash or in kind at the election of the country and equal to a percentage of the monthly (or quarterly) petroleum revenues. In some cases, the concessionaire is exempted from royalty, as in many European countries for the North Sea offshore operations. The royalty rate is generally fixed by the law either as a unique rate or a progressive royalty scale based on different technical or, as in Canada, economic parameters.
- *Corporate income tax* (CIT), corresponding to a percentage of the annual net incomes or profits computed after deducting eligible expenses, costs, and capital allowances. The CIT rate may be the generally applicable corporate tax rate stated in the tax code of the country or the higher tax rate specific to upstream petroleum operations fixed by law. The corporate tax is often determined on a consolidated country basis for all the upstream activities of a company, and not per contract, concession or field, unless the tax law provides explicitly for smaller tax ring-fencing, for example per concession.
- *Additional profits tax* (APT), which may have different names, is an annual tax payable, in addition to the CIT, only when some conditions of profits or petroleum price are met. That type of supplementary tax on profits was introduced by several countries since the 1970s and 1980s when the oil price was abruptly raised. The APT is assessed in many producing countries on an adjusted cumulative cash flow basis determined per company or per concession, such as in the UK, Norway, Denmark, The Netherlands, Australia, etc. APTs are now applied under different mechanisms in more and more countries with the objective of achieving a more progressive fiscal scheme when the effective profitability of projects exceeds predefined levels.
- *Miscellaneous taxes or quasi-taxes*, including bonuses, rental fees, and training fees; withholding taxes on dividends paid by the taxpayer to its shareholders, on interest paid to the lenders for loans, and on the remuneration of foreign subcontractors for services; stamp duties; and in a very few countries export and import duties.

Those tax payments constitute the total petroleum government revenues under concession contracts. The amount and the timing of payments depend on the terms of the petroleum legislation and the applicable concession contract. One of the major drawbacks of concession contracts was their lack of flexibility, as in many countries most of the components of the fiscal package are entirely fixed by the tax law, except for some fiscal parameters.

In some countries, the state or its NOC may also benefit under a concession contract from an *option to participate* at a given percentage in the event of a commercial discovery, receiving a proportionate share in the production. The obligation of the participating state is generally assimilated by the investors to a tax obligation when the state is *carried* by the licensee during exploration and sometimes during development. This means that the state or its NOC holds as co-investor no funding obligations during those periods and therefore does not directly bear the exploration risks, being however, but only in case of production, subject to reimbursing the investor under the agreed terms its pro-rated share in the past investments.

Production Sharing Contract (PSC)

First introduced in 1966 by Indonesia, its use has spread rapidly in many developing countries, for political and economic reasons and above all for its fiscal flexibility. Indeed, the PSC provides in the contract itself for specific progressive production sharing percentages when the law does not fix them. Another frequent advantage of PSCs, especially when there is no efficient additional profits tax used by the country under concession agreements, is to allow the country to get a higher government take in the profits from the first years of production.[6]

Under the terms of PSCs, the company does not directly hold the petroleum rights related to the area concerned but is legally appointed to conduct on an exclusive basis the petroleum operations *as a contractor* to the state or its NOC, by virtue of its contract. As any concessionaire, it is committed to undertake and finance all the work stipulated under the PSC to search for and exploit the oil and gas which may be trapped in the area.

In compensation for its activity, and in the sole event of commercial production, the company is repaid in kind for the recovery at cost of its eligible expenses and capital expenditures incurred in exploration, development, and production by being allowed to market a portion of the total oil (or gas) produced, called the *cost petroleum* (or *cost oil* and *cost gas*), up to a maximum annual percentage of total production, called the *cost petroleum stop* or *ceiling*. As an incentive to invest, the company also receives a profit element from the portion of the remaining amount of petroleum produced after deducting the cost petroleum, which is called the *profit petroleum* (or *profit oil* and *profit gas*), shared between the contractor and the state according to the terms tendered or agreed upon when signing the PSC, prior to the commencement of exploration. The state's share in the profit petroleum remains with the host country (or its NOC) and may be directly marketed and sold by the latter. This access by the state to a share of the production is the major difference with the concession contract.

The contractor under a PSC is also subject to several tax obligations dealt with in the law and, when authorized by the law, in the applicable contract. Those obligations may include the payments of royalty (if the law so provides), corporate income tax (CIT) on profits, and various taxes or quasi-taxes (such as bonuses, surface rentals, and social fees). To render the PSC simpler in its understanding and implementation, the profit petroleum sharing may be agreed on an *after corporate tax* basis, as it was in the original PSCs, under which the income tax is deemed included in the state's share of profit petroleum allocated to the host country. Under that *after tax* sharing basis, obviously the host country receives a higher share in the profit petroleum than under a *before corporate tax* sharing basis – another system used in other countries where the PSC provides that the country receives two separate revenues: first, a lower state share in profit petroleum; second, a payment in cash from the contractor for its corporate tax liability. Globally, the two petroleum sharing schemes result in similar takes for the state, but the first one grants to the country a higher access to the production as the corporate tax in this scheme is paid in kind and not in cash.

Risk Service Contract (RSC)

This was first implemented around 60 years ago by large producing countries. It has today a number of variations. Under RSCs, the company does not hold direct access to a share of the petroleum produced, but in compensation for its work and investments performed *as contractor* to the state or the NOC it receives, only in case of commercial production, a monetary remuneration which cannot exceed a portion of the market value of the petroleum production extracted from the contract area. RSCs remain today a considerably less common form of E&P contracts between host countries and IOCs. It originated in exporting countries such as Iran, Iraq, Qatar, and Venezuela that nationalized their petroleum industry and gave their NOC the monopoly of exploration and production. They gave to their NOC the right of using companies as service contractors in some specific cases for two main reasons: firstly, their technical capacity to use the most advanced technology, and, secondly, their financial capacity to obtain third-party funds and loans for investment under better terms that those to which the host country would have access. More recently it developed in other exporting countries such as Bolivia, Ecuador, and Mexico.

The main difference between the RSC and the PSC is the fact that, according to the terms of the former, the company does not benefit from direct access to a share of the petroleum produced. It receives during the contract's duration a monthly or quarterly remuneration, generally called a *service fee*, paid in cash, which is determined under a formula designed with two objectives: (1) to reimburse over several years, sometimes with interest, the eligible investment and operating costs incurred, by allocating up to a maximum percentage of the annual production value, and (2) to provide for a profit element, such as a fee per barrel, which may be variable with parameters (for example levels of production). The service fee is generally subject to the payment of CIT on profits. The company is often entitled under the RSC to purchase at market price and lift a share of the oil produced. This *buy back clause* authorizes the contractor to buy a share of production equal in value to the payable service fee. It provides security to the contractor for receiving its remuneration in case the country would encounter problems with timely paying in dollars of the service fee.

Analyzing over time trends in the type of upstream petroleum agreements selected by countries, an increasing use of PSCs in emerging and developing countries is observed worldwide. Concessions continue however to remain largely selected, especially in most developed countries, but also in some developing countries where a concession contract is often associated with state participation rights. Meanwhile, some emerging and developing countries used to concession agreements decided to introduce PSCs, for example Brazil in 2010 considered PSCs more consistent with its petroleum policy for its new promising pre-salt province. Only a very few countries abandoned PSCs in the last decade in favor of other types of E&P contracts while keeping in force the existing PSCs already signed, namely the Russian Federation and Kazakhstan – where the change was caused by difficulties in implementing a quite specific sharing mechanism as a result of considerably delayed development projects[7] – and Algeria. RSCs have been signed in the last decade in certain exporting countries, mainly for redeveloping producing fields, such as in Iraq (with the exception of the Kurdistan region which continues to prefer PSCs for E&P projects), Mexico, and Ecuador.

Evolution in E&P Fiscal Regimes: What Could Be a Fair Government Take?

The main purpose of any E&P petroleum fiscal system consists in determining how the profits are shared between the host country and the IOC, because it directly governs, first, the expected return on the future investment the company is evaluating, and consequently its decision on whether or not to make such investment, and, second, the amount and timing of government petroleum revenues for the country.[8]

The fiscal terms for new E&P contracts awarded by countries, whatever their type, facilitate their implementation and correct any identified loopholes. These fiscal terms evolved in the last decades to reflect the changing international petroleum environment, and at any given time greatly depend both on the international oil and gas price environment at their date of design, and the petroleum attractiveness of the country and area to be licensed.

Since the turn of the twenty-first century, while OPEC countries were offering limited acreage, the combination of rising oil and gas prices with the priority for most IOCs to renew their naturally declining reserves has increased the *competition* between IOCs for obtaining the most promising open areas elsewhere. The direct consequence was companies offering to countries more favorable fiscal and contractual terms than before for recent upstream petroleum agreements.

The higher petroleum market price environment and the favorable fiscal terms achieved in some countries induced the other producing countries either to demand renegotiation of their fiscal terms or to impose new fiscal terms less favorable to the investors but more representative of the terms generally accepted in other places. Such requests were sometimes accepted and in other cases rejected by investors, leading to arbitrations still ongoing.

By contrast, in the two preceding decades, under the then declining oil and price environment, especially after the price drops of 1986 and 1998, the opposite trend was observed. For continuing their operations or for deciding on new development projects, a number of IOCs were asking host countries to improve the fiscal terms applied in existing contracts or to amend the petroleum taxation in order to provide better terms to investors than previously agreed. Many countries updated their petroleum policy and eventually accepted such requests when duly justified as a means to foster upstream investments and attract new E&P investors under low petroleum prices.

The huge impact of the petroleum market price on the economics of any E&P project, which by experience is *highly uncertain and unpredictable*, explains the importance of designing fiscal systems associated to E&P contracts with terms *sufficiently progressive* for maintaining a fair sharing of the profits between the state and the contract-holders when the circumstances become different from those expected, namely, when the oil or gas price significantly rises or decreases, or when the field characteristics are considerably different from those assumed when the petroleum fiscal regime was designed. Therefore, new fiscal schemes were introduced to achieve these mutually beneficial objectives.

Indeed one of the most striking constraints in designing a fiscal package for E&P activities in a country, in addition to price uncertainty, is that the profitability of upstream projects varies in quite large proportions depending on their characteristics, such as the location of the project (onshore, shallow offshore, deep offshore), the petroleum prospectivity of the area, the chance of success, the expected reserves and production profiles, the type and quality of petroleum to be extracted, the availability of infrastructure in the region to transport the production, etc.[9] As a consequence, the *petroleum economic rent* for a project – equal to the difference between the realized gross incomes from the production and the total technical costs incurred for exploration, development, production, and abandonment of the field – varies widely in the world. For example, oil prices rose from less than $20 to over $125 per barrel in the last decade, and technical costs today may range from less than $10 to over $40 per barrel, depending on the field characteristics, with the result that the petroleum rent could now fluctuate between $10 and $85 per barrel. The range for rent will indeed continue to increase in the future with the anticipated price rise.

Therefore, *efficient taxation of upstream oil and gas projects* cannot be achieved by using only the general taxation applicable to any economic activity. Petroleum resource taxation requires sector-specific contractual and fiscal schemes. The challenges in designing an efficient fiscal package for upstream contracts, which are entered into for long periods of validity, are to meet the two following objectives at any time while remaining simple enough to be understood: (1) to encourage the company to perform further investments, and (2) to grant to the country a fair share in the rent, not fixed but progressive enough when the project economics and the price improve.

Government revenues and their timing during the production period depend directly on the type of E&P contracts and their fiscal terms. Under *concession contracts,* the country only receives taxes, namely royalty on production, corporate tax on profits, more frequently an additional profits tax, and other taxes. Royalty is paid from the commencement of production on gross incomes, while corporate tax is only paid from the date profits are generated. Additional profits tax (APT) is generally paid from a later date when a predefined profitability criterion is triggered, which may be several years later. A drawback for the national fiscal system is to accurately forecast its future annual revenues, as they depend not only on future prices but also on the schedules of future expenditures and depreciation of past investments allowed as deductions for determining taxable profits. One way to limit the variability of annual corporate tax payments when the IOC holds several contracts at different stages of exploration and production in the country is to introduce a *ring fence* per contract for the determination of corporate tax. This limits the amount of deductible expenditures from other contract areas in the country, with the drawback for the investor of reducing the attractiveness of doing supplementary investment outside the already producing contractual area. For the same reason, most APT mechanisms provide for a ring fence per commercial field in order not to delay APT payments.

Under *PSCs,* the country keeps from the commencement of production a share of production which can be sold at market prices, by the country or its NOC, and also receives payments of agreed taxes. The main fiscal terms governing PSCs are: (1) the cost petroleum ceiling and ring-fencing rules which control the pace for cost recovery, (2) the progressive profit petroleum sharing mechanism under which the country receives a higher percentage in production when the economics improves, and (3) taxes, such as corporate income tax and bonuses.

The original simple PSC system of a fixed percentage of profit petroleum sharing is no longer used. It was replaced by a *sliding scale* based on different progressive systems, such as: *increments of daily production or cumulative production*; the *R-factor* – a profitability criterion equal to the effective ratio between the accumulated revenues and the accumulated costs for the period from the contract signing date to each date of production sharing – related to a field or a contract area; or the *effective rate-of-return* (ROR), another well-known profitability criterion related to the field or the contract area achieved for the same period up to each date of production sharing. In the last decade, the R-factor became the profitability criterion most used for triggering progressive sharing mechanisms in the world. The reasons are mainly its simplicity, making it easily understood by policy-makers and investors, and its demonstrated effectiveness. Moreover, in case of cost overruns or project delays, those risks are theoretically less transferred to the country under an R-factor scheme than under ROR systems, as ROR computation integrates by definition the timing of annual cash flows whereas R-factor determines a ratio between cumulative revenues and costs at the date of assessment without any time value consideration. In Angola the ROR system with a ring-fencing per field is however used with great success for deep offshore fields, probably because they are rapidly developed and produced.

The *government take* (GT) in the *petroleum economic rent* – the average weighted share of the country's revenues in the total profits generated by a field along its production life – varies considerably from one country to another. This depends more on the applicable fiscal and contractual terms agreed in each country or region than on the type of contract. In fact, a concession or a PSC can theoretically produce similar economic results and GT when their terms are appropriately selected, but the timing of revenues will be different due to the nature of each type of contract. Today GTs in the world may range from around 45% for the state (corresponding to 55% for the company) to 70–90% for the state (10–30% for the company), when excluding exceptional cases of lower or higher takes and the possible impact of state participation. The weighted average GT in the world is now estimated at around 65–75%, bearing in mind that this average integrates many different situations. In the last decade the general trend everywhere in the world was a continuing increase in GTs. For example in the UK, where there is no fiscal stabilization clause protecting the investors, the effective corporate income tax rate for E&P activity was increased in several steps from 30% in 2000 to 62% in 2011 in order to give a higher GT in the profits, justified by the significant rise of petroleum prices over this period.[10]

Fiscal terms greatly vary in the world between countries. Globally, the most advantageous fiscal terms to governments occur in countries or basins where the petroleum geology is considered as the most promising and where technical costs and risks are deemed the lowest. Within a given country, fiscal terms may vary depending on the type of E&P projects, the exploration maturity and risks of the contract area, the field characteristics, the category of conventional and unconventional petroleum, and above all the date of award of the relevant E&P contract. This complicates the comparative analysis of petroleum fiscal systems between countries, which should be always done with sound assumptions and interpreted with great care, especially when they are only performed on a simple stand-alone field basis, a scenario which generally does not correspond to the reality as companies are generally investing in a series of projects within a country.

Indeed no magic and unique percentage exists for fixing GTs fairly. For example, fiscal terms leading to a given GT being considered too lenient for a specific field in a country may become too onerous for other less profitable fields located in the same country, explaining why there is no single fair GT. Though there is no exact science for determining the fair GT that should be applied by a country or used in a given E&P contract, there are recognized skills, expertise, and advisory capacity available in the world for recommending, after detailed analysis and rigorous economic modeling, what could be *the possible range of a fair GT* after taking into consideration the petroleum potential of the acreage and all the specificities of the investment, including the selected way to award the contract, the competition, and whether a fiscal stabilization clause applies under the contract.

Box 8.3 Illustrative examples of progressive fiscal schemes.

Progressive fiscal schemes in concession contracts: royalty and additional profits tax

Regarding royalty, many countries have adopted *sliding scales* based on daily production and more recently on petroleum price or one economic criterion (such as the payout time of a project, allowing a lower royalty rate up to the payout time and a higher rate thereafter; or more frequently linked to an effective profitability criterion of the project). Royalty rates are quite variable in the world, from 0% in case of exemption up to 40% in exceptional cases.

The province of Alberta in Canada is today the country applying the most elaborate and relatively complex royalty system, with specific regimes for *conventional oil* (with rates between 0% and 40% depending on oil price and daily production per well), *natural conventional gas* (with rates between 5% and 36% depending on gas price and daily production per well), *shale gas* and *coal bed methane* (with rates similar to conventional gas except during the first 36 months of production), *oil sands, horizontal oil or gas wells*, and *enhanced oil recovery projects*. In the US, the royalty rates are generally constant, but they rose in the last decade above the traditional 12.5%, up to 18.75% in federal offshore waters and around 25% in the most favorable onshore shale gas private lands.

More and more countries (such as Algeria, Denmark, Ghana, Ireland, Namibia, Israel, Kazakhstan, The Netherlands, Senegal) have adopted an *additional profits tax* (APT) of different types, payable in addition to corporate tax, taking into account the experience of supplementary tax systems gained by the UK, Norway, and Australia and based on an adjusted cumulative cash flow. In another fashion, some countries, instead of introducing a distinct APT, decided to use *variable income tax rate* schemes for E&P activity based on daily production, or more recently on an economic criterion (such as the oil price or the R-factor, or a combination of these two parameters), a solution which works efficiently when appropriately defined.

Progressive fiscal schemes in PSCs: cost petroleum and profit petroleum

First, some countries have introduced a sliding scale for determining *cost petroleum ceiling* percentage (instead of a single percentage) in relation with a parameter, such as the type of petroleum (oil, gas), daily or cumulative production, petroleum price, etc.

Second, a progressive split for *profit petroleum sharing* may be agreed, based on daily or cumulative production, and more and more frequently in the twenty-first century on the effective profitability achieved by the E&P project, measured at each date of sharing with a given profitability criterion. The retained criterion is either, in a few cases, the rate-of-return on investment, as in Angola and Kazakhstan, or in many more cases the R-factor, as in Algeria, Azerbaijan, Cameroon, India, Libya, Malaysia, Qatar, Tunisia, etc.

The differences in the appreciation of any upstream project explain why in case of competitive bidding for E&P contracts, which is becoming the customary practice, the offers submitted on the biddable contractual or fiscal terms may greatly vary. The offers made by interested companies integrate not only the uncertainties on many technical and economic factors but also the differences between the bidders regarding their long-term strategic priorities and assessment of risks. There is however a general tendency to design for upstream contracts more progressive fiscal systems when profitability changes. In terms of sound fiscal policy, the petroleum economic rent sharing in order to remain fair to all parties when circumstances change has to be *progressive* in relation to the effective profitability achieved by petroleum projects: the greater the profitability of the project, the higher the GT. Illustrative examples on how to introduce fiscal progressivity are summarized in Box 8.3.

Fiscal progressivity is generally more accepted today because it leads to *a win-win situation encouraging investments and sanctity of contracts*: granting better fiscal terms to the IOC for encouraging the development of small, costly or risky projects, while obtaining higher but still reasonable GT in case of quite profitable projects. A direct consequence of such progressive schemes is that countries increasingly bear the petroleum price and cost risks, but in exchange they benefit from a higher take when the profitability of the project rises. On the contrary, *regressive* fiscal systems leading to a smaller GT when profitability rises are no longer sustainable in the long run by a country.

One fiscal evolution of importance in the world is the longer experience now accumulated in implementing each type of E&P contracts and upstream fiscal regimes, leading today to more clarity on how each party interpret contracts and fiscal regimes and how to reduce identified loopholes. Nevertheless, more thoughts remain to be developed to mitigate aggressive IOC tax planning when only designed to reduce petroleum government revenues, as such an objective may be contrary to fostering the necessary long-term cooperation between the parties. Some key issues requiring the greatest care when designing upstream fiscal regimes are summarized in Box 8.4.

Why Do Some E&P Contracts Contain a Fiscal Stabilization Clause?

Any petroleum contract may include, in addition to many technical, operational, and economic clauses, a set of legal clauses, such as the applicable law, dispute resolution, force majeure, stabilization, transfer of interest, liability and indemnity, termination, etc., when those issues are not dealt with in the petroleum law itself. The reason why IOCs may request a stabilization clause in an E&P contract is explained below. Stabilization clauses have considerably evolved in the last decades and today, when they have to be accepted by countries for allowing investments, their scope has been reduced to only some fiscal issues.[11]

As petroleum E&P contracts are long-term agreements by nature, sometimes for terms exceeding 40 years, the overall equilibrium of the contract resulting from the agreed fiscal scheme may be deeply affected by the occurrence of unexpected circumstances and above all by the impact of petroleum price volatility. It is unusual under E&P contracts to have an *adaptation provision* which would automatically change the fiscal terms toward restoring the original equilibrium (except under the new progressive fiscal schemes described above which try to achieve this goal), or a provision that would oblige the parties to negotiate a change if one party suffers hardship.

To limit political risks resulting from a possible change in law by the sovereign host country, IOCs often demand the protection of a *stabilization provision*. While IOCs

Box 8.4 Toward more detailed E&P fiscal systems to increase clarity and mitigate aggressive tax planning.

Many issues have to be dealt with in petroleum tax laws and E&P contracts, concerning the contract-holders themselves but also their shareholders, employees, subcontractors, and lenders. Thus, a contract-holder is generally constituted by more than one entity, each one subject to tax.

The experience in producing countries demonstrated than even when the tax system is quite simple, as under royalty and corporate tax regimes, the administration of such tax clauses in the E&P sector when not properly drafted is often more complex than expected and requires a highly professional *petroleum taxation office* working in close cooperation with the minister or agency responsible for petroleum.[12]

Such difficulties should not be an excuse for rejecting additional profits tax or progressive profit petroleum sharing schemes, when their principles are properly designed to be fully understandable and not too complex and the rules are clear enough.

As in any fiscal system, loopholes may exist, and they have to be progressively eliminated to give clarity to the interpretation of the upstream fiscal regime and E&P contracts and to allow in the long run a fair application by the parties of the principles governing the selected fiscal scheme and contract.

Key petroleum fiscal issues to be addressed in sufficient detail in tax law may concern: (1) the comprehensive definition of the eligible costs and deductions for tax or cost recovery purposes, (2) the tax treatment of direct and indirect transfers of interests in E&P contracts and how to tax the resulting gains in case of cash considerations, or (3) how to mitigate the unexpected reduction in tax liabilities derived from possible tax planning and double taxation treaty shopping. For example, many countries have still not clarified their law for taxation of capital gains in case of direct or indirect transfers of interests under E&P contracts, which continues to raise great uncertainty in terms of tax liabilities resulting from such transactions, which are customary in the petroleum industry.

New producers in developing countries are in an especially asymmetrical position with IOCs in terms of resources for defining with sufficient detail their petroleum taxation and regulation, because the domain is relatively new for them and they are not yet holding sufficient experience. This is the reason why countries should prepare with the greatest care and with the help of external advice their petroleum law and model E&P contract, with the necessary amendments to their tax laws and how to administer them.

In the same way, companies should recognize this state of fact and give the highest priority to help in clarifying in a fair manner the implementation of the upstream fiscal and contractual regime, with the objective of encouraging long-term cooperation with the producing country.

accept entering into E&P contracts in developed countries without any stabilization clause, they largely require the benefit of such a clause in most developing countries. Enforceability of stabilization clauses in E&P contracts often raised difficulties in the past. Considering this experience, the wording of stabilization provisions in new upstream contracts has recently evolved under the following principles.

The applicable law in most new E&P contracts and their associated fiscal regimes is no longer the law in force at the date of its signing – the *freezing type stabilization clause* often used in the past – but the law applicable *at any time* during its term. The reason is that a sovereign state may always change its laws or regulations and can generally apply them to previously issued licenses or contracts, and therefore the enforceability of freezing type stabilization clauses may be problematic.

Stabilization is now limited to a few specific fiscal issues or rates, strictly listed in the contract, the country having the right to change non-fiscal aspects such as environmental regulations and the other fiscal rules not stabilized. With the rapid evolution of the world and techniques, a sovereign country holds the unilateral right to issue new regulations applicable to any contract-holder, for example to take into account new environment, health, labor, and safety priorities, new techniques, more advanced conservation of petroleum, or for performance of operations in more stringent conditions. Moreover, fiscal issues applicable to any industry in the country, such as labor laws and taxes, and non-discriminatory by nature are generally no longer stabilized.

In the event that a change in a fiscal term listed in the stabilization clause has a material impact on one of the parties, which may be the government or the contract-holder, such party may ask for the benefit of the so-called *economic-equilibrium stabilization provision* of the contract, designed to restore the economic benefit prevailing at the signing date of the contract by adapting the economic or fiscal terms of the contract in an appropriate way, subject to mutual agreement. There is also a trend to limit the duration of fiscal stabilization to a period shorter than the entire validity of the contract itself.

Most developed countries do not provide for fiscal stabilization clauses, keeping at any time the unilateral right to amend their petroleum taxation when they consider, under changing circumstances, that the fiscal regime leads to unfair sharing of the profits, and all their existing license-holders become automatically subject to such fiscal changes. This was the case for example in the UK, which increased its petroleum corporate tax significantly in the last decade, as mentioned above.

When a fiscal stabilization clause applies under an E&P contract, as in many developing countries, the government does not have the same flexibility to adjust its petroleum tax policy in case of changing circumstances. Therefore, it is of paramount importance that the fiscal scheme associated with E&P contracts should have to be designed in such a way to automatically protect the parties when the petroleum market price, or more generally the profitability, considerably varies over the duration of the contract. The only way to achieve this goal in such countries is by implementing a sufficiently progressive fiscal system, based ideally on a profitability criterion as presented above.

Each party to an E&P contract containing a fiscal stability clause bears a special responsibility in designing a progressive fiscal system. There is no other way in the long term to foster sanctity of contracts and minimize political risks for investors in the developing countries concerned.

Necessary Adaptation to Upstream E&P Contracts for Unconventional Petroleum

In the last two decades new categories of petroleum, called *unconventional petroleum*, have begun to be explored and exploited in more and more countries, in addition to what is called *conventional petroleum*. Their share in global petroleum production, relatively small today, will progressively increase both for oil and gas and may become significant, following the striking example of their successful development in North America.

Generally speaking, in most countries, the upstream petroleum legislation and contracts dealing with conventional petroleum also apply to the most recent unconventional petroleum activity, unless it is explicitly provided for under special legislation enacted by the country. Some countries have discovered with surprise that E&P contracts awarded for searching for conventional petroleum are now used by their holders for unconventional prospecting. The legal reason is that in many petroleum laws the definition of petroleum is only based on its *chemical composition* and as a consequence the petroleum rights granted may indeed cover both conventional and unconventional petroleum, except when the contrary was clearly intended.

Up to now, most upstream petroleum laws, regulations, contracts, and fiscal regimes were drafted having in mind only conventional petroleum activity, ignoring the specificities of unconventional petroleum, which was not of commercial interest at that time. The main categories of unconventional petroleum and their differences with conventional oil and gas are summarized in Box 8.5. The differences between conventional and unconventional petroleum activity originate mainly from the techniques used for their exploration and exploitation, their respective risks, costs, and possible impact on the environment, in particular when the extraction of unconventional petroleum requires using hydraulic fracturing, a technique known in the industry but not yet approved in some countries.

Box 8.5 The different categories of unconventional oil and gas.

Unconventional petroleum has basically the same chemical composition as *conventional petroleum* but their exploitation may result in significant differences in terms of techniques used, possible impact on the environment, costs, production profiles, and economics. The main technical differences consist both in the location of oil or gas underground, often deeper and in more compact rocks for unconventional petroleum, and in some specifications of the petroleum extracted, such as density or viscosity, leading to use of different techniques for their extraction. Unconventional petroleum includes today the following categories.

Unconventional natural gas

Tight gas produced from compact and deep reservoirs; *shale gas* which is natural gas directly extracted from shales where it was generated; *coal bed methane* (CBM), also called *coal seam methane* (CSM) in Australia, gas existing in coal beds which is extracted by drilling wells up to such beds. Due to the extremely low porosity and permeability of shales, the production of shale gas requires the hydraulic fracturing ("fracking") of the rock from horizontal wells, a technique also used in tight gas reservoirs for the same reasons.[13]

Unconventional oil

Oil extracted from *deep offshore* reservoirs below the sea; *heavy oil* and *oil sands* which require specific techniques for their recovery; and more recently *shale oil* extracted from compact shale formations requiring hydraulic fracturing of the rock.

Where the host country awarded a conventional license or contract, today its holder may wish to use it to conduct unconventional petroleum activity, leading to possible conflicts in the interpretation of laws or contracts when they are silent on unconventional petroleum. For example, what are the regulations applicable to unconventional petroleum activity when no specific regulation was issued? Is CBM extracted from coal seams covered by mining law or by petroleum law? In the future any petroleum legislation, regulation, and taxation will have to contain, in addition to the provisions dealing with conventional petroleum, special clauses concerning the specificities of each category of unconventional oil and gas activity.

Deep offshore activities are an exception because the existence of those zones of interest are known before awarding E&P contracts. This explains why most countries have already introduced special tax incentives under their petroleum legislation and contracts for deep offshore to take care of higher costs, such as lower royalty rates under concession agreements and a higher-cost petroleum ceiling or more favorable profit petroleum sharing under PSCs. Special regulations concerning deep offshore operations have also been issued.

For the other categories of unconventional petroleum, a few countries have already started to introduce special laws or clauses dealing with it. The most illustrative case is probably the province of Alberta in Canada regarding oil sands, gas shale, and CBM. An ad hoc legal, fiscal, and contractual regime is provided for oil sands activity under the Oil Sands Conservation Act and the Oil Sands Tenure Regulations, providing for specific oil sands permits and leases, along with the Oil Sands Royalty Regulations which define a more attractive royalty scheme in favor of companies holding oil sands leases than standard conventional petroleum leases. Regarding shale gas and CBM, the standard gas royalty framework applies in Alberta, but supplemented by a specific incentive limited to the first three years of production of unconventional gas from a well. However, most countries have still to promulgate special rules for unconventional petroleum activity and the new E&P contracts concerning such activity will be adapted accordingly.

Conclusion and Suggestions for Fostering Future Upstream Developments

This review of the evolution of E&P contracts shows that, while no new types of contract were introduced in the last two decades, their terms significantly changed. The main changes deal with their associated fiscal regime and concern the introduction of higher and more progressive GT, the level of which is closely related to the petroleum attractiveness of the contract area and also of the country, the degree of competition, and the expected future petroleum price.

Depending on their petroleum policy and pursuant to the law, host countries may authorize upstream concession contracts or, more and more frequently in developing countries, PSCs, while in specific cases RSCs may be selected by large exporting countries. New terms may apply to unconventional petroleum. State participation depends on the national petroleum policy and is part of the overall fiscal package.

The fiscal regime and the overall government take resulting from the applicable upstream tax law and contract terms are of paramount importance both for the country and the investor when looking to future upstream investments. Only E&P contracts which are built with a reasonably progressive fiscal scheme designed to allow in the long run, first, the host country receiving a fair GT whatever the changing circumstances and, second, the international petroleum company achieving its profitability criteria, can be sustainable, encourage fiscal stability, and foster the pursuit of new investments.

Notes

1. Petroleum means oil and natural gas.
2. For a comprehensive review on the evolution of legal, fiscal, and contractual issues, see Duval *et al.* (2009).
3. For country examples of licensing policies and methods, see Cameron (1984), Daintith and Willoughby (2000), and Lucas and Hunt (1990).
4. Transparency is becoming a key objective in the upstream sector for its sustainable development. For a review on how to achieve better transparency in relation with contracts, taxation and revenues, see IMF (2007).
5. For more information on upstream contracts and policy insights, see the following source book freely available on the Internet and regularly updated: *The Extractive Industries Source Book for Oil, Gas and Mining* at http://www.eisourcebook.org.
6. Theoretically, the economics for the investor from a decision-making point of view should be neutral whatever the type of E&P agreements. It is theoretically possible to design fiscal packages for concession contracts and PSCs giving relatively similar economic results for both the country and the investor. Nevertheless, in reality a PSC will generate under most contracts higher revenues to the host country during the first years of production because the *cost petroleum ceiling* mechanism indirectly allows the country to take a higher share of production under the PSC than the royalty payable under a concession contract, when the CIT is not yet payable.
7. In those two countries, due to exceptional cost overruns and delays in the carrying out of a project, the use of a production sharing scheme based on rate-of-return increments with substantially high threshold rates led from the point of view of the state to unbalanced production and cost overruns sharing. This defect could however be easily eliminated in new PSCs when the rate-of-return system contains appropriate safeguard clauses to better protect state interests.
8. For a more detailed analysis on petroleum taxation, see Daniel *et al.* (2010), Kemp (1987), Nackle (2008), and Tordo (2007).
9. For details on petroleum costs, see Favennec and Bret-Rouzaut (2011).
10. When the 50% supplementary petroleum tax is applicable to a field, the UK marginal GT rose from 65% in 2000 to 81% in 2011, a marginal take close to the maximum of 78% applied in Norway, excluding state participation.
11. For more information on stabilization, see Cameron (2006).
12. The most illustrative example concerns royalty in the US where after a long implementation many practical issues have still to be resolved, in particular for agreeing on the market price at the wellhead.
13. Hydraulic fracturing has been used for over 50 years by the petroleum industry. However, this technique is more massively used in shale gas wells, with the use of larger quantities of water containing a tiny percentage of associated chemicals. Local communities are voicing more and more concern against this technique. This new situation requires such communities to be given more information on the quite limited risks of the technique and for host countries to develop ad hoc regulations and contractual clauses on how to use hydraulic fracturing safely.

References

Cameron, Peter. 1984. *Petroleum Licensing Comparative Analysis*. London: Financial Times Business Information Ltd.

Cameron, Peter D. 2006. *Stabilisation in Investment Contracts and Change of Rules in Host Countries: Tools for Oil & Gas Investors*. Research Paper. Houston, TX: Association of International Petroleum Negotiators.

Daintith, Terrence C., and Geoffrey D. M. Willoughby. 2000. *United Kingdom Oil and Gas Law*. 3rd edn. London: Sweet & Maxwell.

Daniel, Philip, Michael Keen, and Charles McPherson. 2010. *The Taxation of Petroleum and Minerals: Principles, Problems and Practice*. London and New York: Routledge.

Duval, Claude, Honoré Le Leuch, Andre Pertuzio, and Jacqueline Weaver, eds. 2009. *International Petroleum Exploration and Exploitation Agreements: Legal, Economic and Policy Aspects*. New York: Barrows Company Inc.

Favennec, Jean-Pierre, and Nadine Bret-Rouzaut. 2011. *Oil and Gas Exploration and Production: Reserves, Costs, Contracts*. Paris: Éditions Technip.

IMF. 2007. *Guide on Resource Revenue Transparency*. Washington, DC: International Monetary Fund.

Kemp, Alexander. 1987. *Petroleum Rent Collection Around the World*. Halifax, Nova Scotia: Institute for Research on Public Policy.

Lucas, Alastair R., and Constance D. Hunt. 1990. *Oil and Gas Law in Canada*. Calgary: Carswell.

Nackle, Carole. 2008. *Petroleum Taxation: Sharing the Wealth*. London and New York: Routledge.

The Extractive Industries Source Book for oil, gas and mining at http://www.eisourcebook.org.

Tordo, Silvana. 2007. *Fiscal Systems for Hydrocarbons: Design Issues*. Working Paper 123. Washington, DC: World Bank.

Chapter 9

National Oil Companies: Ensuring Benefits and Avoiding Systemic Risks

Charles McPherson

Introduction

The past several years have seen a significant increase in policy and academic interest in the role of state participation in natural resource sectors and national resource companies in particular. This is especially so in the case of national oil companies (NOCs), reflecting their continuing, and growing, economic, social, and political importance, both domestically and internationally. The activities of the NOCs influence domestic development through their impact on macro-economic management, investment, and overall governance and stability. How well they perform has major implications for global energy supply and security.

NOCs now exist in almost all countries identified by the International Monetary Fund as being petroleum-rich (IMF 2007: Appendix 1; McPherson 2010). The first NOCs were established in the 1920s and 1930s in Argentina and Mexico. Numbers grew rapidly in the 1960s and 1970s with the formation of the Organization of Oil Exporting Countries (OPEC) and widespread nationalism and enthusiasm for state ownership of the "commanding heights" of the economy (Yergin and Stanislaw 1998). Their global and domestic relevance is undeniable: NOCs control 73% of world oil reserves, 61% of production, and account for a very high percentage of international trade flows in oil. The numbers for natural gas are similar (Wood Mackenzie group as quoted in Victor *et al*. 2012). At the domestic level, these numbers translate into financial flows that frequently account for 50% and even up to 90% of government revenues or export earnings and high percentages of gross domestic product (IMF 2007).

Both the performance of and policy support for NOCs have gone through cycles. In the 1960s and 1970s, high oil prices and substantial cash flows led to expectations that the NOCs could successfully address not only commercial challenges, but a considerable range of non-commercial objectives as well. Subsequent oil price and revenue collapses, coincident with global moves toward liberalization and privatization, led to a critical

The Handbook of Global Energy Policy, First Edition. Edited by Andreas Goldthau.
© 2013 John Wiley & Sons, Ltd. Published 2013 by John Wiley & Sons, Ltd.

re-examination of NOCs and to the introduction of petroleum sector restructuring and market-oriented reforms in many countries.

The return of high oil prices restored national government confidence in the role of NOCs. Virtually every newly emerging petroleum producer has opted to establish an NOC or consolidate and expand the responsibilities of an existing NOC.[1] The NOCs show every sign of being an enduring phenomenon.

A considerable literature on NOCs now exists, rich in both breadth of research and depth of analysis.[2] This chapter reflects and comments in a summary fashion on the central themes and findings of that literature.[3] The next section reviews the several roles and responsibilities commonly assigned to, or appropriated by, NOCs. Each of these has raised issues in the past, and led to recommendations for reform. The following section examines some of the systemic governance risks that have been associated with NOCs and which have jeopardized looked-for benefits. The final section provides a summary assessment and outlook.

Roles and Responsibilities

The non-exhaustive list of NOC roles considered below comprises those of: commercial participant and revenue generator; sector regulator and overseer; promoter of national capacity; development agency; fiscal and financial agent; and instrument of foreign policy.

Commercial Player

An early and continuing expectation for NOCs was that they would emulate and offset, or even replace the commercial roles played by the international oil companies. They were expected to run efficient and profitable operations and to produce for government a substantial revenue stream through taxes, other payments, and dividends. Ideally, that revenue stream would contribute importantly to sustainable economic and social development in the home country.

It has not proved easy to access the data required to make an assessment of NOCs' commercial performance. That said, theoretical analysis, limited empirical work, and anecdotal evidence (see respectively Hartley and Medlock 2008, Victor 2007, and McPherson 2003) suggest that NOCs have not done well under this heading, either in absolute terms or in comparison with private sector operators. The poor results have been attributed to a variety of factors. Prominent among them have been: weak institutional capacity; lack of competition; and, perhaps especially, the broad range of non-commercial objectives, discussed in the next subsections, that quickly became a standard part of their terms of reference. Funding, governance, and the overall political context have proved to be major challenges to performance. Funding is considered below, and governance and the political context in the next main section.

Reforms introduced to improve NOC commercial performance represent a response to these causal factors and include: capacity building and corporatization programs; benchmarking against private sector performance; introduction of competition; divestiture of non-core functions; joint venturing with the private sector; partial privatization through listing on stock exchanges; and sector-wide reform and restructuring. The jury is still out on how successful these initiatives have been, but initial results are encouraging. It is noteworthy that Petrobras, Brazil's NOC, reflects many of these reforms and has been counted in one survey as among the world's 100 most respected companies, the only petroleum company to make the list.[4]

One of the central issues raised by an NOC's commercial participation in the petroleum sector is that of funding. While equity participation may generate handsome revenues for the state, it also implies a major demand on public funds. The oil and gas industry is notoriously capital intensive. A commitment to NOC equity participation, if met, will inevitably divert funds from other worthwhile, and typically urgent, budget priorities. Funding for social and physical infrastructure – health, education, roads, telecommunications, etc. – may suffer. The competition for funds may also mean that the NOC gets squeezed, resulting in costly delays or interruptions to its projects and to government revenue. This has been a major problem in Nigeria.

Additional drawbacks to state equity funding of an NOC's commercial roles include the sizeable risks to which it exposes public funds, and the single sector focus which it fosters, to the detriment of economic diversity and possibly of social and political stability (Dutch Disease: loss of labor-intensive sectors).

NOC borrowing may represent an alternative to equity funding, but it has proven difficult to arrange and is likely to be expensive, available only at rates greater than those charged to private sector companies. It may involve either government guarantees, in which case it constitutes a fiscal risk to government, or costly pledging of future production or income.

Given these concerns, NOC equity participation in the oil sector ought to be considered carefully, especially since good alternatives are usually available. If a country's resource base is attractive, private sector firms are likely to be more than willing to take the place of the NOC, and accept the related technical, financial, and operational risks involved. Where an efficient fiscal regime is in place, the revenue loss may be minimal or even putative (McPherson 2010).

Many countries have now opted for sector participation arrangements that reduce or defer NOC funding obligations or eliminate them altogether. Joint ventures with the private sector reduce NOC funding exposure. So-called "carried interests," where private sector partners spend on behalf of the NOC until either development or production begins, defer funding exposure. Finally, production sharing agreements with investors require no expenditure at all by the NOC, but do provide for NOC involvement in investment and operational decision-making. (See McPherson 2010 for a summary description of the various possible forms of state participation.)

Regulator

Based on their familiarity with industry operations, superior technical capacity, and preferential access to information, NOCs are often assigned the role of industry regulator, or this role may be taken on by default in situations where the nominal regulator does not have the capacity to do the job.

Responsibilities under this heading can include several, or all, of the following: ensuring industry compliance with sector legal, contractual, and regulatory obligations; input to the drafting of regulations; collection and maintenance of industry technical and contractual data; oversight of procurement; and cost or "value for money" audits. Critical policy functions, including input to draft sector legislation or licensing procedures, may also migrate, at least in part, to the NOC. For example, Nigeria's NOC, the Nigerian National Petroleum Corporation, has been heavily involved in the drafting of new omnibus petroleum legislation.

The scope for serious conflicts of interest is clear. Taking on any of these responsibilities, while at the same time playing a commercial role, presents the NOC with huge opportunities to favor its interests either vis-à-vis other oil sector participants or jointly on behalf of itself and its sector partners.

The standard policy recommendation in face of these potential conflicts is, not surprisingly, to separate out any regulatory or related functions from the NOC and place them in, or restore them to, more appropriate arms-length agencies or ministries, at the same time building necessary capacity in those agencies (sometimes referred to as the Norwegian model or "trinity," see Al-Kasim 2006). In practice, however, this may not be the preferred option. Where administrative capacity is traditionally and chronically weak, regulatory functions may be better left with the NOC, ideally with "firewalls" built between those functions and commercial operations (Thurber *et al.* 2011 make this case). Angola has adopted this approach. In the face of manifest weaknesses in the Ministry of Petroleum, a number of critical policy and regulatory functions have been assigned to, or left with Sonangol, Angola's NOC. Units responsible for these functions, at least in theory, have an arms-length relationship to other units within Sonangol.[5] This seems an unfortunate second best choice. It underscores, however, the profound importance of building institutional capacity outside the NOC, if the NOC itself is not to be burdened with non-core responsibilities.

Promoter of National Capacity

NOCs are expected to promote and develop a broad national oil sector capacity by building their own internal capacity, and by encouraging building of capacity in related outside activities. Developing a full suite of managerial and technical skills within an NOC has proved difficult in many countries where the requisite educational backgrounds and facilities are missing. Under these circumstances, joint venturing with experienced international investor partners, and implementation of the mandatory training programs that are now commonly provided for in legislation or contracts, can prove critical. Care needs to be exercised to ensure that the resulting skills transfers are appropriate to needs and are applied, rather than ignored. To be effective, skills development needs to be complemented by a corporate culture that values those skills and provides opportunities and incentives to exercise them.

It is worth noting that national capacity can be very effectively built up in the private sector. Shell's Nigerian subsidiary, the Shell Petroleum Development Corporation (SPDC), is a highly sophisticated company that prepares and manages a wide array of projects from modest to world scale. Yet SPDC's staffing at all levels, through to and including its chief executive and chairman, is almost entirely Nigerian.

In recent years, the development of national industry capacity has gone beyond something internal to the NOC. It now encompasses the promotion of "backward" and "forward" linkages to the exploration and production activities of the NOC and private investors. Backward linkages refer to activities that supply or service core petroleum operations, from catering services to the fabrication of drill pipe. Forward linkages are meant to add value to local production by encouraging or facilitating its further domestic processing to produce petrochemicals or supply domestic gas distribution networks. So-called "local content" policies focus on growing these linkages, and with them the domestic economy. If policies are not to be abused, building capacity in the linked activities is critical. Again, as the public locus of oil industry expertise, the NOC is often expected to play a role.

Development Agency

Relatively few governments stand back from their NOCs and allow them to concentrate solely on commercial efficiency and revenue maximization.[6] Non-commercial assignments have been and still are common. They may range far afield from the need for petroleum industry expertise, but depend instead on perceptions of the NOC's managerial strengths and access to funds. Social and physical infrastructure obligations – schools, hospitals, roads, and bridges – are typical.

One of the most typical and demanding social obligations of NOCs in oil-rich countries is income redistribution, through the supply of petroleum products at below – often well below – market prices. Product price subsidies are expected almost as a matter of right in major producing countries.[7] A popular means of garnering political support, they can be extremely expensive, and very difficult to remove once in place. Past subsidies by Indonesia's Pertamina have imposed costs of up to US$3 billion annually, or 15% of the national budget. The costs associated with similar policies implemented by SOCAR in Azerbaijan and NIOC in Iran have been estimated at in excess of 10% of GDP (Gupta *et al.* 2003). Efforts to remove or phase out subsidies in Nigeria have regularly met with violent protests and strikes.

These roles all properly belong with government, and for that reason are referred to as "quasi-fiscal functions" (QFAs). Their assignment to NOCs and exclusion from formal budgeting processes seriously blurs lines of accountability and complicates macroeconomic, fiscal, and budgetary management. In some cases, this resulting obfuscation, rather than any comparative advantage of the NOC in terms of implementation, may be a motivating factor for QFAs. Channeling funds through an NOC avoids unwelcome scrutiny by the legislature or other agencies and may be preferred by political elites.[8]

Capacity permitting, these functions related to social and physical infrastructure should all be transferred or restored to the government agencies, departments or ministries where they belong. Care should be taken to provide for their continuation during any transition period. Responsibility for subsidies, if they persist, belongs with the finance ministries. Where these transfers are not or cannot, for capacity reasons, be effected, QFAs should be explicitly recognized in government budgets and subjected to regular budgetary and planning procedures.

Fiscal and Financial Agent

Based either on legislation, contractual provisions, or simply practice, NOCs commonly perform a number of fiscal and financial functions. They may be responsible for the assessment and collection of royalties, especially where they have been assigned or have acquired regulatory roles. As government's representatives in production sharing contracts, NOCs assess and collect profit oil shares, a role which entails detailed cost audits. As equity participants in operations, they are responsible for managing the financial flows involved and the calculation and payment of dividends to government. Participation agreements more often than not have a fiscal dimension, the benefits of which go to the NOC, or are at least administered by the NOC (Daniel 1995: McPherson 2010).

Royalty provisions may, and production sharing contracts, by their very nature, do call for payments to be made in kind. NOCs are responsible for the marketing or commercialization of in-kind payments. Very substantial amounts of money are associated with these several functions. Under production sharing regimes, the sums involved could easily amount to 50% or more of government revenues from the sector. Assignment of these functions to the NOCs can, once again, be traced to their technical expertise and proximity to operations. In acting as fiscal agents, however, the NOCs are effectively usurping authority traditionally granted to ministries of finance or revenue collection agencies. This division of responsibility adds to the challenges of revenue collection and to the difficulties of audit and control of fiscal and financial flows – areas where performance is essential to avoid mistakes, waste, abuse, and corruption.

Fiscal roles assigned to NOCs are likely to remain intact for the near term. Where this is the case it becomes critical that the NOCs cooperate closely with fiscal agencies and comprehensively share information and data on a timely basis. Unfortunately this level of cooperation is often lacking. The frustration experienced in many finance ministries and revenue authorities at being, in practice, excluded from an important part of their traditional mandate has led them to seek support from agencies like the IMF or World Bank to build small strategic petroleum units within the finance ministry or revenue authority. The objective is to create the internal capacity to dialogue more effectively with the sector ministry and NOC and win a greater say in matters with clear relevance to fiscal and financial management.

Foreign Policy Representative

A number of oil-importing countries with established NOCs have encouraged their NOCs to acquire petroleum assets abroad. These include, notably, China, India, and South Korea. Support may be given in the form of subsidized loans, parallel loans to the country selling or licensing the asset, and/or direct foreign aid and technical assistance, often to infrastructure construction. The ostensible reason for these moves is national energy security. Access to petroleum resources becomes a dimension of foreign policy.[9]

It is not clear that this is a rational policy given the existence of a well-developed international and competitive oil market as an alternative to purchased access. Buying oil in the international market is likely to prove less expensive and less risky in both technical and political terms than negotiated country-to-country deals. From a global governance perspective, the policy has a negative feature in that multi-dimensional access deals are rarely transparent.

Of course, NOCs can be found outside their home countries for other reasons – attractive investment opportunities, diversification, and acquisition of vital competitive skills among them. These reasons are unrelated to foreign policy, but closely aligned with NOCs' commercial objectives.

Governance and Transparency

Growing interest in the resource sectors and their national companies has sparked a complementary interest in critical issues of governance (Benner and Soares de Oliveira 2010: see also Chapter 15 below for an extended discussion of these issues). Given the massive financial flows they control or oversee and the influential role they play in their respective economies, NOCs have proved to be irresistible targets for control by local elites in pursuit of personal or political gains. Not surprisingly, they have been lightning rods for corruption (McPherson and MacSearraigh 2007). This has had two consequences.

First, it has seriously undermined governance in many NOCs. Controlling elites have politicized NOC boards and executive management, shown little interest in meaningful corporatization, and promoted secrecy. Whatever formal governance structures might exist for NOCs, they are chronically diminished by informal channels of governance based on personal connections and political persuasion (Hults 2012). NNPC in Nigeria provides a classic illustration of these governance problems (see Chapter 29 below). Executives of NOCs often spend more time managing and/or lobbying political interests than they do on pursuing commercial efficiency. All of this runs directly counter to the central tenets of good governance for state-owned enterprises (OECD 2005).

Second, where it occurs, the capture of NOCs by elites has undermined governance at the level of government itself. With non-transparent access to significant funds and economic influence, ruling authorities are likely to have little need for, or interest in accountability. Oil wealth, rather than delivering expected benefits, may instead negatively impact economic performance, human development, and social and political stability – outcomes collectively referred to as the "Paradox of Plenty" or the "Resource Curse" (Humphreys *et al.* 2007; Karl 1997).

These outcomes should not be seen as inevitable. They have been avoided or reversed in a number of countries. The political context is critical (Eifert *et al.* 2003; Warshaw 2012). Adverse outcomes are more likely in "predatory autocracies" or "factional democracies" characterized by few checks and balances on ruling authorities and by very short political time horizons. Good outcomes depend on the opposite – adequate checks and balances and longer time horizons. Political will at the highest level and public support are equally important. Ironically these qualities may emerge when negative outcomes have reached intolerable levels, creating a space for broad-based support of reform.

Transparency has been widely referenced as being a powerful force in support of improved governance.[10] Once credible information on management of the oil sector, on sector financial flows, and on NOC operations is placed in the public domain, waste, mismanagement, and corruption are harder to conceal and opportunities to press for accountability are enhanced.

The formation of Publish What You Pay (PWYP), a coalition of concerned non-governmental organizations, and the launch of the Extractive Industries Transparency Initiative (EITI) marked the opening of a sustained multi-stakeholder drive for greater transparency in the extractive industries with an initial focus on oil.[11] NOCs have attracted special attention. Their inclusion in the transparency criteria of the EITI was regarded as a *sine qua non* for that initiative. The transparency movement has gained considerable traction over the past decade. EITI has been endorsed by 37 petroleum-rich countries, 16 of which are now deemed compliant, having satisfied the EITI criteria. PWYP has recorded similar successes, including incorporation in recent US legislation of extractive industry payment transparency as a requirement for company listing on the New York Stock Exchange. The European Commission is considering similar requirements for European Union companies. Successive G8 and G20 summits have committed to the extractive industry transparency agenda, as have many governments on a bilateral basis.

While all this must be seen as encouraging to transparency, results on the ground, at least to date, have been disappointing. Extractive industries transparency initiatives have mostly been concerned with publication of audited numbers on payments made (by companies) and revenues received (by governments). Translation of these numbers into better sector and NOC governance depends on those receiving the information having the capacity to interpret it, and to use it to demand accountability and reform. Further, it depends critically on having the political space to make such demands. So far, this has not come as easily on the heels of transparency as one might have hoped. Countries that have signed up to the EITI, and even met its criteria, have in many cases simply enjoyed the "points" this garners in the international community, without pursuing the governance reforms such transparency is intended to produce (Benner and Soares de Oliveira 2010 cite, e.g., Azerbaijan). Further, oil-consuming countries and investor home countries have been reluctant to press as hard as they might have to avoid jeopardizing either access to oil or their companies' investments.

Assessment and Outlook

As noted in my introduction, NOCs have shown remarkable resilience through cycles of popularity and criticism. They have persisted when state-owned enterprises in other sectors around the world have succumbed to privatization or undergone profound reform. This section closes the chapter with three observations on the persistence of NOCs and their differing implications for sustainable development of domestic and global energy policies, and by implication global energy supply and security.

First, almost any review of the evolution of NOCs will expose and underscore the fundamental importance of institutional capacity. NOCs cannot hope to improve performance and grow without internal capacity. The development of administrative capacity in other agencies and sectors is equally important, if NOCs are not to be burdened with non-core objectives and activities that get in the way of performance. The persistence of many of the negative attributes and functions of NOCs, and of NOCs themselves, is a direct result of capacity failures elsewhere. Leadership in this area must begin at home, but the international community – aid agencies, bilateral donors, and investors – can play a major role by stepping up support. Without improvement, the promise the oil sector ought to hold out for sustainable development will be slow in coming.

Second, NOCs continue to be popular because they offer controlling elites the prospect of easy, non-transparent access to substantial funds, patronage opportunities, and political and economic influence. Unless these abuses of governance are corrected, any promises of sustainable development will be seriously set back, not just deferred. Entrenched interests in mature oil states and motivated opportunists in new oil states will resist reform vigorously. These obstacles will be all the more difficult to overcome because they are often symptomatic of countrywide governance failures. Transparency, if persistently and broadly pursued, has the potential to be a powerful instrument in support of reform, but only if it is complemented by meaningful assistance to local civil society groups to make use of it, and by sustained high level, unambiguous international pressure on oil country elites to pay attention.

Third, some NOCs, hopefully a growing number, will persist because they and their governments have made a serious commitment to reform and development. This chapter and the more comprehensive surveys it has referenced have exposed a wide range of NOC issues and abuses. These are now well understood, and appropriate policy responses have been identified. Implementation of these responses can be expected not only to reduce the risks of adverse outcomes but also to substantially increase the probability of achieving anticipated benefits. Global support to reformers, at both policy and technical levels, will play a critical role.

In summary, NOCs that perform well will enhance oil sector performance and, with accommodating domestic economic policies and governance, contribute importantly to domestic sustainable development and stability. These outcomes, in turn, bode well for global energy supply and security. Avoiding the apparent systemic risks associated with NOCs and ensuring their better performance will require both domestic and international political will – commitment at the highest levels to the support of institutional capacity, transparency, and accountability.

Notes

1. Timor-Leste, Mauritania, Ghana, Uganda, and Mozambique provide examples of this trend, as does Argentina's recent decision to renationalize its flagship oil company, Yacimientos Petrolíferos Fiscales.
2. See Victor *et al.* (2012) and Tordo (2011) for excellent and comprehensive treatments of the topic. Both the Natural Resources Charter, Precept 6, and the Extractive Industries Source Book summarize the issues.
3. The chapter also draws on two previous papers by the author (McPherson 2003 and 2010).
4. Based on the annual survey by Reputation Institute as reported in *Upstream*, June 11, 2012.
5. See Heller (2012) and Soares de Oliveira (2007) for insightful analyses of Sonangol and its roles.
6. Stevens (2007) points to Saudi Arabia as an example.
7. Jaffe (2010) notes: "most NOCs in major oil-producing countries provide subsidized fuel to both consumers and local industry."
8. McPherson (2010) observes, in the context of Venezuela: "channeling funds through the NOC made it easier to target favored recipients and gave the Presidency and executive branch a competitive advantage over Congress in the control of funds."
9. See Goldthau (2010) for a full discussion of the role of NOCs in the conduct of foreign policy.
10. Karl (2007), Gillies (2011), and McPherson (2005) underscore the relevance of transparency to extractive sector governance. See also Natural Resources Charter, Precept 2, Level 3, and Extractive Industries Source Book.
11. See www.publishwhatyoupay.org and www.eiti.org.

References

Al-Kasim, Farouk. 2006. *Managing Petroleum Resources: The Norwegian Model*. Oxford: Oxford Institute of Energy Studies.

Benner, Thorsten, and Ricardo Soares de Oliveira. 2010. The Good/Bad Nexus in Global Energy Governance. In Andreas Goldthau and Jan Martin Witte, eds. *Global Energy Governance: The New Rules of the Game*. Berlin and Washington, DC: Global Public Policy Institute and Brookings Institution.

Daniel, Philip. 1995. Evaluating State Participation in Mineral Projects: Equity, Infrastructure and Taxation. In James Otto, ed. *Taxation of Mineral Enterprises*. London: Graham & Trotman.

Eifert, Benn, Alan Gelb, and Nils Bjorn Tallroth. 2003. The Political Economy of Fiscal Policy and Economic Management in Oil-exporting Countries. In Jeff M. Davis *et al.*, eds. *Fiscal Policy Formulation and Implementation in Oil Producing Countries*. Washington, DC: International Monetary Fund.

Extractive Industries Source Book, available at: www.eisourcebook.org.

Gillies, Alexandra. 2011. Reputational Concerns and the Emergence of Oil Sector Transparency as an International Norm. *International Studies Quarterly*, 54, 1: 103–126.

Goldthau, Andreas. 2010. Energy Diplomacy in Trade and Investment of Oil and Gas. In Andreas Goldthau and Jan Martin Witte, eds. *Global Energy Governance: The New Rules of the Game*. Berlin and Washington, DC: Global Public Policy Institute and Brookings Institution.

Gupta, Sanjeev, Benedict Clements, Kevin Fletcher, and Gabriela Inchauste. 2003. Issues in Domestic Petroleum Pricing in Oil-Producing Countries. In Jeff M. Davis *et al.*, eds. *Fiscal Policy Formulation and Implementation in Oil Producing Countries*. Washington, DC: International Monetary Fund.

Hartley, Peter, and Kenneth Medlock. 2008. A Model of the Operation and Development of a National Oil Company. *Energy Economics* 30, 5: 2459–2485d.

Heller, Patrick. 2012. Angola's Sonangol: Dexterous Right Hand of the State. In Victor *et al.* 2012.

Hults, David. 2012. Hybrid Governance: State Management of National Oil Companies. In Victor *et al.* 2012.

Humphreys, Macartan, Jeffrey Sachs, and Joseph Stiglitz, eds. 2007. *Escaping the Resource Curse*. New York: Columbia University Press.

IMF. 2007. *Guide on Resource Revenue Transparency*. Washington, DC: International Monetary Fund, available at www.imf.org/external/np/fad/trans/guide.htm.

Jaffe, Amy, and Roland Soligo. 2010. State-backed Financing in Oil and Gas Projects. In Andreas Goldthau and Jan Martin Witte, eds. *Global Energy Governance: The New Rules of the Game*. Berlin and Washington, DC: Global Public Policy Institute and Brookings Institution.

Karl, Terry Lynn. 1997. *The Paradox of Plenty: Oil Booms and Petrostates*. Berkeley, CA: University of California Press.

Karl, Terry Lynn. 2007. Ensuring Fairness: The Case for a Transparent Social Contract. In Humphreys *et al.* 2007.

McPherson, Charles. 2003. National Oil Companies: Evolution, Issues and Outlook. In Jeff M. Davis *et al.*, eds. *Fiscal Policy Formulation and Implementation in Oil Producing Countries*. Washington, DC: International Monetary Fund.

McPherson, Charles. 2005. Governance, Transparency, and Sustainable Development. In Jan Kalicki and David Goldwyn, eds. *Energy and Security: Toward a New Foreign Policy Strategy*. Washington, DC: Woodrow Wilson Center Press.

McPherson, Charles. 2010. State Participation in the Natural Resource Sectors. In Philip Daniel, Michael Keen, and Charles McPherson, eds. *The Taxation of Petroleum and Minerals: Principles, Problems and Practice*. New York: Routledge.

McPherson, Charles, and Stephen MacSearraigh. 2007. Corruption in the Petroleum Sector. In J. Edgardo Campos and Sanjay Pradhan, eds. *The Many Faces of Corruption: Tracking Vulnerabilities at the Sector Level*. Washington, DC: World Bank.

Natural Resources Charter, available at: www.naturalresourcescharter.org.

OECD. 2005. *OECD Guidelines on Corporate Governance of State-owned Enterprises*. Paris: OECD.

Soares de Oliveira, Ricardo. 2007. Business Success, Angola Style: Postcolonial Politics and the Rise and Rise of Sonangol. *Journal of Modern African Studies* 45, 4: 595–619.

Stevens, Paul. 2007. *Investing in the Middle East and North Africa: Institutions, Incentives and the National Oil Companies*. Report 40405-MNA. Washington, DC: World Bank.

Thurber, Mark C., David R. Hults, and Patrick Heller. 2011. Exporting the "Norwegian Model": The Effect of Administrative Design on Oil Sector Performance. *Energy Policy* 39, 9: 5366–5378.

Tordo, Silvana. 2011. *National Oil Companies and Value Creation*. Working Paper 218. Washington, DC: World Bank/ESMAP.

Victor, David, David Hults, and Mark Thurber, eds. 2012. *Oil and Governance: State-Owned Enterprises and World Energy Supply*. Cambridge: Cambridge University Press.

Victor, Nadejda. 2007. On Measuring the Performance of National Oil Companies. Working Paper 71, Palo Alto, CA: Stanford University Program on Energy and Sustainable Development.

Warshaw, Christopher. 2012. The Political Economy of Expropriation and Privatization in the Oil Sector. In Victor *et al.* 2012.

Yergin, Daniel, and Joseph Stanislaw. 1998. *The Commanding Heights*. New York: Simon & Schuster.

Part III   Global Energy and Security

The Handbook of Global Energy Policy, First Edition. Edited by Andreas Goldthau.
© 2013 John Wiley & Sons, Ltd. Published 2013 by John Wiley & Sons, Ltd.

Chapter 10

Global Resource Scramble and New Energy Frontiers

Øystein Noreng[1]

Introduction: Oil Schizophrenia

Throughout its commercial history, oil has been a theme of controversy, with fiery debates whether oil is scarce, plentiful, beneficial or evil. Statements of impending resource scarcity alternate with reports on environmental damages. In short, the world seems to have a schizophrenic relationship to its major source of energy: we use it and benefit from it, but we do not like it. There are concerns that both too little and too much oil is available, and that oil prices are both too high and too low. The blame is put on the oil suppliers – oil companies as well as exporting countries – and especially the member countries of OPEC, the Organization of Petroleum Exporting Countries.

Oil has been indispensable to the modern economy at least since 1900, not because of oil industry manipulation, but due to its physical properties (Muller 2008). Crude oil and oil products such as gasoline, diesel, and fuel oil are superior to competitors such as coal, batteries, and even hydrogen by the energy content in relation to weight and size, in addition to being easily transported and stored. Indeed, already in the late nineteenth century the oil-fueled combustion engine could beat batteries as the preferred power source for automobiles at a time when oil trade was subject to limited competition, with resulting high prices (Blair 1976).

In the twentieth century, oil has been essential for transportation and the exchange of goods, industrial specialization, and productivity gains that have permitted major improvements in real wages and living standards in large parts of the world (Ridley 2010). Correspondingly, disruptions of oil supplies and abrupt oil price rises have caused economic setbacks, social distress, and political discontent. During the twentieth century, oil has been a key factor in the outcomes of two world wars. The Allies' access to oil was essential to their victories. The military significance of oil is essentially unchallenged in the twenty-first century (Le Bideau 2010). Nuclear submarines and aircraft carriers do not diminish the need for oil for navies, armies, and air forces, and hence the military significance of oil.

The Handbook of Global Energy Policy, First Edition. Edited by Andreas Goldthau.
© 2013 John Wiley & Sons, Ltd. Published 2013 by John Wiley & Sons, Ltd.

Against this backdrop, security of oil supplies has become a major concern in many of the world's capitals. The actual meaning varies from secure volumes to secure prices. From the perspective of oil consumers, secure oil supplies should mean stable supplies at stable prices, or preferably, rising supplies at stable prices. Measured by turnover, oil remains the world's most important traded commodity. It is an essential input factor in the transportation sector. For consumers, the most costly oil is the volume that is needed, but not available. Referring to experienced or assumed economic setbacks due to oil shortfalls, the discussion of "Peak Oil" at times leads to apocalyptic visions of catastrophe because oil demand will rise faster than supplies (Roberts 2004). The bulk of literature on Peak Oil evades economic reasoning and neglects the importance of oil prices (Holland 2008). In some capitals, the temptation is to secure access to allegedly scarce oil by military means; in other capitals to use money to secure preferential access. Both approaches indicate distrust in the market. This is why debates on a global resource scramble have been vividly revised during the past years.

The present chapter aims at a critical review of some of the premises of the "Peak Oil" debate, emphasizing economic assumptions, political implications, and strategy options. The immediate focus is on our schizophrenic attitude to oil; it is essential but the world loves to hate it (Hofmeister 2011). The next issue is the "Peak Oil" controversy, why the theory builds on limited empirical experience and cannot have unqualified validity. The third issue is the erroneous economics of the "Peak Oil" theory, why demand and supply *cannot* develop independently of each other. The fourth issue is the perilous politics of "Peak Oil," why mistaken ideas about scarcity may provoke counterproductive military intervention. Finally, the conclusion aims at pointing out new perspectives for oil policies and strategies.

The "Peak Oil" Controversy

The "Peak Oil" debate is in many ways a controversy between geologists and economists. The former proclaim that the extraction of finite oil resources is heading toward an imminent shortage, and the latter assert that supply and demand of oil adapt dynamically to shifting oil prices so that oil shortages will not occur, provided prices correctly reflect the supply potential and the cost of substitutes (Mills 2008). From this perspective, an eventual shortfall of oil would imply that oil is underpriced in relation to the supply potential, substitutes, and its utility to consumers, meaning a market failure. Causes may be consumer subsidies or insufficient taxation. Correspondingly, an oil glut, as in a market with stagnant demand and rising supplies, would imply that oil is overpriced with regard to the resource base and consumer utility, meaning another market failure, but due to imperfect competition.

The "Peak Oil" theory is not monolithic. A discussion of absolute and irreversible decline of oil supplies and ensuing fuel scarcity is changing into a more differentiated debate of costly unconventional oil replacing cheap conventional oil (Meadows *et al.* 2010).

A common concern is that "Peak Oil" in the near future will lead to a shortage of key resources and input factors of our economy that will force adverse changes in living and social conditions. Some observers ascribe the financial crisis, unemployment, and the lack of growth prospects as the consequences of a nascent shortage of oil and higher energy prices (Rubin 2009). One argument is that factors such as deficit budgeting along with financial deregulation and the Iraq War have contributed to higher oil prices, regardless of the balance in the oil market (Reinhart and Rogoff 2011).

A widespread idea is that oil demand and oil supplies develop independently of each other. Therefore, a catastrophe has to occur when supplies do not cover demand. This is a common theme among many geologists, physicists, and journalists. They tend to disregard the potential for energy saving and substitution, pointing to the crucial fact that commercial energy is a prerequisite for a modern society to function. Consequently, commercial energy is essential to basic human needs, apart from bringing profits to producers of goods and services (Rosa *et al.* 1988).

The Sources of Controversy

Controversy about the oil scarcity began in recent times with the report *Limits to Growth*, commissioned by the Club of Rome, written by researchers at MIT (Meadows *et al.* 1972). Different scenarios came to the same result: the world would consume too much, environmental damage would increase dramatically, the supply of food and raw materials would decline. The use of energy was an important parameter and the message was that oil supplies would run out, probably in the early 1980s.

The model excluded factors that could alter the conclusion, such as efficiency improvements and technological development, and the crucial signaling and anticipatory role of prices. It ignored human ingenuity and adaptability. Critics have called the report an exercise in disinformation and confusion rather than an objective analysis with the aim of new insight (Smil 2003). In hindsight, it seems as if the aim was to notify a disaster as a fatality without regard to political, economic, or technological conditions.

The oil price increases of 1973/4, after a prolonged period of company controlled but flat to declining prices, led to greater awareness of oil resources. A larger study, sponsored by Shell among others, concluded that at the latest, in 2000, the supply of oil would not be able to meet demand (WAES 1977). In 1979, during the Iranian revolution, the CIA published a study arguing that oil demand imminently would surpass supplies, and that oil extraction would decline during the 1980s (CIA 1979).

In 1998 British geologist Colin Campbell and his French colleague Jean Laherrère in an article in *Scientific American* again set the "Peak Oil" debate in motion (Campbell and Laherrère 1998). Their arguments were briefly that an oilfield has a natural threefold production/recovery path: a fast rise, a top ("peak"), and a corresponding decline – the so-called "Hubbert's curve." This gives a logistic curve, conditioned by the physical properties of the reservoirs. It is ideally a bell-shaped curve where output decline mirrors the rise to the peak (Laherrère 2000). The key premise of the authors was that this pattern should also apply to other regions and the world. The argument was that the world's supply of oil is declining because extraction and consumption surpass new discoveries. They also argued that the members of OPEC exaggerated their reserve estimates in order to increase their quotas and to improve creditworthiness, and that discovery rates of oil reached a peak in 1964; subsequently, new discoveries did not offset the depletion of the older, larger fields, so that oil extraction would reach a peak and then decline. The peak of possible oil extraction appeared imminent; the decline would inexorably follow from physical conditions, regardless of technology or economics. "Peak Oil" theorists have been criticized for such a mechanistic approach to the problem, assuming that important parameters are static, and that economics and technology do not affect the production curve (Clarke 2007).

The counterarguments are briefly that the "Peak Oil" theory of an oilfield's natural extraction path refers to US experience with private ownership of resources, seeking rapid maximization of extraction and income, with no or weak government regulation. In sum, commercial considerations to cover immediate needs have driven US exploration for oil and the development of oilfields. These conditions do not apply without reservations outside the US and parts of Canada.

Oil is a heterogeneous product and there is no clear distinction between conventional crude oil and unconventional oil, such as shale oil or heavy oil (Smil 2010). As industry gains experience and matures, costs of unconventional oil are declining. Political conditions deny or restrict access to large parts of the world to the international oil industry. Canada, Iraq, Iran, Mexico, Norway, Russia, and Saudi Arabia have large prospective areas that have not been subject to exploration. This even applies to parts of the US, which prohibits drilling in many areas.

Oil Supplies are Dynamic

Technological development continuously lowers costs; a globalized oil industry disseminates innovation. New technology and lower costs make new resources available. Smaller and more difficult available prospects become economical. New sites with potentially large resources open up, as in the Arctic and in deep waters of the Atlantic off Angola and Brazil. Experience lowers costs for equivalent fields and enables new, smaller, and less accessible fields to replace mature fields. Although costs in the Arctic and in deep waters off Angola and Brazil are higher than in the Middle East, the North Sea or the Gulf of Mexico, they are well below current oil prices. In 1970, the North Sea was a marginal oil province with high geological, technical, and economic risk; today it is central.

The world's reserves of oil, conventional and unconventional, are considerably larger than what the public debate usually assumes (Mills 2008). Until the whole world has been the subject of intensive exploration, knowledge of the resource base will be incomplete and any conclusion about the end of oil is premature (Smil 2010). Moreover, experience shows that the volume estimate of reserves at the time of discovery of a new field in many cases represents but a fraction of the total volume extracted over the lifetime of the field, because knowledge and technology can enhance recovery rates, volumes, and lifetimes. Therefore, finding rates are not representative of the reserve growth that to a considerable extent takes place over time through technological development and investment in producing fields. US authorities estimate that in the Gulf of Mexico, the discovery of one barrel of oil will give a reserve growth of five barrels over 150 years (Laherrère 2000).

The conclusion is that no one knows the world's oil resources and how much the industry will recover with what technology, at what expense and at what time. Therefore, the theory of "Peak Oil" should not be accepted without reservations. Large areas of the world, not least in established oil provinces, among others Iraq, Iran, Libya, Russia, and Saudi Arabia, have been little or not at all explored. Any estimate of future oil production by Hubbert's or other models, needs assumptions on reserves and the producers' interests and strategies, but they are not constant. In the US, the oil reserve-to-production ratio has been about 10:1 for the past 30 years, but the country is still one of the world's leading oil producers.

To sum up, Hubbert's model and the theory of "Peak Oil" rest on six key assumptions:

- knowledge of the world's oil reserves is complete;
- reserve estimates are constant;
- extraction inevitably takes the shape of a fairly symmetric curve;

- technology is constant;
- oil prices do not matter;
- all oil producers have the same revenue/profit motive and goals.

None of these assumptions corresponds to reality. Knowledge of the world's oil reserves changes through exploration. Insight and technology cause a continuous update of reserves. The actual extraction can take many different shapes, depending on the interests and strategies of the resource owner, in addition to technology and economics. Technology is not constant, new equipment and new methods lower costs and open new opportunities. Oil prices do matter, especially for investment in marginal resources. Not all oil producers have the same objectives.

Referring to Hubbert's model, Colin Campbell in 1991 presented detailed predictions of annual output from major oilfields in most of the world's countries until 2050 (Campbell 1991). From 1980 to 2010, the world's total oil reserve estimate has increased by 75%, annual extraction by almost 40%, and the oil reserve-to-production ratio has increased from 30 to 40 years (Mohn 2010). A static approach is unsuited for understanding today's dynamic oil industry (Lynch 2002).

In the oil industry, important parameters are dynamic. Costs, reserves, prices, profits, competition, and regulations are constantly changing. No natural science dictates that extraction has to follow a symmetrical curve. Instead, a host of physical, technical, economic, and political factors determine oil extraction. The alternative to a quick peak is to keep extraction at a lower level but for a longer time, which may boost total recovery. Presently, governments are the owners of the bulk of the world's oil reserves; in many cases they do *not* maximize short-term volumes and income but instead seek the greatest ultimate long-term recovery.

Against this backdrop, it is important to emphasize uncertainty about resources and recovery, rather than the scarcity or abundance, not least since the definition of oil is unclear and the supply of substitutes for conventional oil is uncertain. Geologists often have a pessimistic view, emphasizing scarcity. The counterpart is a temptation among economists to overestimate the effects of incentives on energy demand and supplies, regardless of the physical resource base. So far, however, economists appear to have been more right than geologists (Mills 2008).

The Erroneous Economics of "Peak Oil"

Conventional economic theory on finite resources does not presume a peak, but takes for granted that prices will follow the cost of capital, but in the opposite direction, and that extraction over time will be more costly and growth rates decline, but not necessarily reach a peak (Dasgupta and Heal 1979). It does not consider technological change. The theory of "Peak Oil" moves ahead and introduces a peak, but without integrating geology with economics. The looming scarcity is presented as a fatality and volumes are assumed to be unaffected by prices. "Peak Oil" protagonists seem to think that physics only and not economics determines oil supplies.

Indeed, the common methodological error of the "Peak Oil" catastrophe warnings is to overlook the price mechanism. Whenever projections show a mismatch between supply and demand trends, a correction has to take place and the price is the trigger. This works both ways. A projected supply shortfall or excessive demand growth indicates that the price is too low to restore balance. This was the case in the early 1970s, as well as the early 2000s. A projected supply surplus or insufficient demand indicates that the price is too high for the market. This was the case in the 1980s, late 1990s, and might occur again.

In recent years, the volume of financial oil transactions has increased significantly, and it is an open question to what extent financial agents have caused oil price rises and volatility (Chevalier 2010). The issue is subject to research and results are not conclusive (Babusiaux *et al.* 2011). Many fundamental factors, such as the volume balance between supply and demand, production costs and consumer utility, technology, and government policies condition long-term crude oil price development, but also more sporadic factors such as interest rates, currency exchange rates, inflationary expectations, and financial agents who use oil positions as a storage of value condition short-term movements. Even if financial agents hardly influence the long-term oil price trend, they may well influence, amplify, and destabilize short-term movements. This may have been the case with the extreme oil price volatility in 2008/9. When financial markets perceive an oil undersupply, they rally to buy; when they perceive an oversupply, they rally to sell.

Prices Matter

The apocalyptic visions of "Peak Oil" disregard the salience of the market. Practically, oil consumption *cannot* exceed volumes available through extraction and inventory draw-down. Consequently, it is misleading to compare extrapolated trends for supplies and demand in order to demonstrate a gap. The oil price is the bridge between supplies and demand. In the oil market, the price mechanism can be brutal and slow, but it works in the longer run. If oil prices are too high for consumers unable or unwilling to pay, demand will subside and supplies diminish because unsold and unsaleable oil has no value for the suppliers. If, by contrast, oil prices are too low, incentives to explore for oil and develop oilfields will weaken and extraction will gradually shrink.

Indeed, low *short-term price elasticity* of both demand and supplies is typical of the oil market, meaning that oil price movements have limited immediate effect on volumes produced and consumed. Long lead times and high capital requirements for both supply and demand slow down adaptation and permit oil prices to stay well above supply costs, as long as consumers can pay. In the short term, incomes seem to be more important for oil demand than prices.

The prerequisite for oil demand is that consumers benefit more from using oil than from *not* using oil at available prices, so that they enjoy a consumer surplus, which varies with the price and users' income. Consumers use oil and other energy sources as an input factor to produce services, along with capital and labor, or time (Becker 1978). The use of energy, in this case oil, can save labor and time, for example by using a private car rather than public transport. Capital can replace oil by investing in new, more energy efficient equipment, in this case a more modern car.

Businesses operating for a profit have to consider a bottom line and adapt to higher energy prices by changing procedures and investing in equipment that is more efficient. Households ponder concerns about costly energy and the ensuing need to change habits and invest in more efficient equipment, and daily life convenience (Lutzenhiser 1993). Both businesses and households can use more energy efficient equipment to actually demand more energy, as the cost of energy as an input factor declines (Wirl 1997). This is one reason why household energy demand increased massively in the old industrial world in the twentieth century, a trend that in the twenty-first century is prevalent in the emerging economies. This also affects oil demand. Against this backdrop, energy and oil demand needs to be analyzed not only as an economic issue, but also as a behavioral and sociological issue. From this perspective, consumer utility, not only supply and demand, also conditions energy prices (Jesse and van der Linde 2008). For oil, the transportation sector is crucial (Beaudreau 1998). Economic growth raises incomes and consumers' time costs. Therefore, economic growth raises the cost of *not* using oil to save time by quick transportation and raises consumer tolerance for high oil product prices. The price elasticity of oil demand thus diminishes.

These factors make oil prices unstable and unpredictable. Consumers' increasing acceptance of high oil prices and willingness to pay, as long as their incomes are growing, provide incentives for the oil exporters not to flood the market and not compete for market shares. Indeed, the properties of oil demand can make it more profitable *not* to extract oil by withholding oil from the market rather than open the spigots, eventually by cooperation among major suppliers, as is the case with OPEC. Producing less can generate higher income. The low price elasticity of oil demand makes the risk of price decline and income disproportionate in relation to potential volume gains. For this reason, competition over market shares among oil producers is economically irrational.

A consistent flaw in the successive editions of the *World Energy Outlook* of the International Energy Agency (IEA) has been to show an undersupply of oil in relation to oil demand extrapolations, but not to mention the ability of higher prices to restore balance. The markets respond to higher prices by making new resource categories available, inciting technology development. Examples are pre-salt deepwater oil in Brazil and elsewhere. In the US, current high oil prices make the exploitation of shale gas and shale oil economical, but knowledge of the reserves is not new (Brodman 2012). Shale oil and gas are like conventional oil and gas for practical uses, but they are in different traps in the crust of the earth. Costs are high and pollution of air and water is a serious problem. At present oil prices, the procedure is economical, with important regional differences (IEA 2010). CO_2 emissions are much higher than those from conventional oil.

A Question of Market Power

Imperfect competition in the oil market enables a few large producers to affect the balance in a market whose price formation is highly sensitive even to anticipations of changes in the physical balance of supply and demand (Hofmeister 2011). Prices above extraction costs encourage the development of substitutes and alternatives.

Withholding volumes has nothing to do with "Peak Oil" but with market power and imperfect competition (Adelman and Watkins 2008). When oil demand surges, OPEC and other producers have the discretionary choice of meeting the surge by raising volumes and stabilizing the price, or by keeping volumes constant and letting prices rise, or a combination. Correspondingly, when oil demand recedes, the choice is cutting volumes to defend the price, or keeping volumes constant and letting the price decline, or a combination. At times, the key OPEC member countries disagree and act in different ways.

Consequently, rising oil prices do not necessarily lead to higher volumes to the oil market in the short term. Oil exporters with comfortable budget and current account surpluses have few incentives to invest in enhanced capacity. The domestic economy may have absorption limits; further use of oil revenues can be destabilizing or inflationary. Some oil exporters may sell oil according to income targets, meaning a negative price elasticity of oil supplies, as rising prices diminish the required volumes. Again, this has nothing to do with "Peak Oil" but with the fact that some major oil exporters consider keeping oil in the ground a worthwhile option compared to foreign investment or domestic use (Holland 2008).

Conventional economic theory on finite natural resources overlooks the possibility of producer cooperation and withholding supplies to defend a high price. Consequently, the theory misses a crucial characteristic of the oil market – the resource concentration and oligopoly centered in the Middle East (OPEC) and sporadic cooperation by other exporters. The Middle East constitutes the core of oil supplies because it has the largest reserves and the lowest costs, Saudi Arabia being the pivot. Even if oligopolistic markets – markets dominated by few large sellers – are generally unstable, OPEC has often been able to defend high prices for some time, by cutting supplies when demand falters or outside suppliers increase volumes. The outcome of all these factors is that price adjustment can be brutal rather than gradual, as witnessed by the oil price declines of 1986 and 1998 and the sudden oil price drop in 2008/9.

This market imperfection usually keeps oil prices far above marginal extraction cost, enabling the major producers to capture an economic resource rent. It means that on a trajectory of gradually rising costs, as prescribed by conventional theory, the market makes a leap. The counterpart to the economic resource rent is the entry of alternatives and substitutes in the market. High oil prices stimulate alternatives; they gradually become more cost efficient and more competitive (Nordhaus 1973). Over time, they will take market shares, put a ceiling on oil prices, and subsequently push oil prices down (Watkins 1992).

The historical precedent is that of the massive oil price increases of the 1970s, followed by a protracted period of declining real oil prices from the mid-1980s to about 1999, only interrupted by the first Gulf War in 1990/1. During this period, oil lost its position in stationary uses such as heating and power generation to coal, natural gas, and nuclear power. Unusually high oil prices for a decade triggered a wave of substitution as well as efficiency measures and investment in hitherto marginal oil provinces such as Alaska and the North Sea.

In the present situation, after several years of historically high real oil prices, the key issue is to what extent and how quickly the dominant position of conventional oil in the transportation sector will erode. Competition comes from unconventional oil, and hybrid and electric automobiles. In addition, OECD government policies mandate and promote the use of biofuels and other alternative energy through subsidies. They also put pressure through NGOs and international organizations, such as the World Bank, on emerging economies to move in that direction.

Technology Matters

High oil prices incite energy conservation and efficiency, as well as investment in conventional oil outside OPEC. They also enhance the competitiveness of natural gas and its conversion to liquid fuel. A final issue is to what extent rising oil demand in the emerging economies will make up for declining demand in the "old" industrialized world. To sum up, the threats to oil are galore. In 2012, the hazard of conflict in the Middle East keeps oil prices high, but judging by market fundamentals, the oil price risk is downward. Depending on the state of the world economy, a repetition of the post-1985 experience is possible.

In an imperfect market with prices well above extraction costs, conventional oil meets its competitors without using the full advantage of superior cost effectiveness. Oil prices at US$100 per barrel make competitors such as shale oil and heavy oil economical, as well as the conversion of natural gas to liquids (IEA 2010). Moreover, oil prices at this level encourage investment in conventional oil in hitherto marginal areas such as the

Arctic and the deep offshore Atlantic, again recalling the experience of the 1980s and 1990s in Alaska, the Gulf of Mexico, and the North Sea. As investment picks up, experience accumulates, and technology develops, costs will fall and competition will build up, adding to the downward price pressure. As prices fall, the economic resource rent from conventional oil will diminish, adding to pressure for cost cuts also in established oil provinces.

This outlook is the opposite of the prescription of conventional resource extraction theory and radically opposed to the "Peak Oil" theory that envisages steadily rising marginal costs and diminishing volumes. In an imperfect oil market, the price decline may wait for some time, but a belated adjustment may be more brutal. From this perspective, the issue of ultimately recoverable oil reserves is irrelevant. Oil in the ground that the resource owner cannot extract economically and sell at a profit has *no* value. This again undermines the theory of "Peak Oil." By contrast, technological change induced by competition and lower prices can lead to falling marginal costs that over time enhance the viability of marginal petroleum prospects. This was the case in the North Sea in the 1980s and 1990s and is being repeated around the world, first of all in the Gulf of Mexico. Indeed, depending upon key suppliers' income needs, lower prices may induce higher volumes, contrary to conventional economic resource theory as well as "Peak Oil" theory.

Policies Matter

Hubbert's model and the "Peak Oil" theory ignore resource policy and government intervention, and consequently fail to grasp the political framework and the economic realities of today's oil industry. Hardly any country gives an unqualified green light to petroleum exploration, development, and extraction. Consequently, the historical empirical reference for the "Peak Oil" theory, that is, the successive waves of exploration, development, and extraction that took place in the continental US from the 1930s to the 1960s, with little or no government interference, is not representative of the world.

Indeed, policies shape the framework and scope of action for the oil industry. Since governments can issue money and generally consider longer time horizons than private investors do, as resource owners they tend to stretch the lifetime of their reserves by limiting access and controlling volumes.

Alarmists point to Saudi Arabia as the epitome of "Peak Oil" because of the apparent decline of the giant Ghawar field (Simmons 2006). A close analysis of finds, development, and upgrades lends credibility to Saudi official estimates (Mills 2008). The issue in Saudi Arabia is to maintain export levels, as domestic oil consumption is surging due to low domestic prices and economic development (Alyousef and Stevens 2011). Large financial reserves provide freedom of choice in oil policy.

Iraq is one of the world's oldest oil producers and has the world's second largest proven oil reserves, in spite of limited exploration. Political instability, sanctions, wars, and foreign occupation have hampered exploration, development, and production. At the time of the US invasion in 2003, 98 oilfields were proven, but only 21 had ever been producing. Two giant fields, Rumaila and Kirkuk, made up 90% of the output. Iraq has a huge potential and might possibly increase present output several times, on the condition of political stability.

Iran also has an old oil industry. Exploration has been limited and sanctions have restricted access to and investment in modern technology. The bulk of Iran's oil production is from six giant fields, which are old and suffer from inadequate maintenance and investment, with a recovery rate of only 15–20%. Nevertheless, Iran has in recent years explored with success and the country has a potential for increasing oil output, dependent on access to capital and technology.

Venezuela has one of the world's oldest and largest oil industries, but with relatively higher extraction costs than the Middle East due to low-quality oil. It has huge deposits of heavy oil and oil sands, which are economical at present prices (IEA 2010). The country also has a largely unexplored offshore potential (Mills 2008).

Libya has Africa's largest proven oil reserves. The country has not been much explored, but has a huge potential of both oil and natural gas.

Norway has for decades practiced an oil policy with restrictive licensing, keeping much acreage off limits to the oil industry, with high taxes and a high degree of state participation, in addition to strict environmental safeguards. In spite of relatively little exploration, there are important new finds and the potential is considerable (Mills 2008). The middle part of the Norwegian Sea has, with little exploration, indications of oil and natural gas resources almost as large as those of the Gulf of Mexico. Technical challenges and costs are, however, substantially higher. The Barents Sea, the northernmost part, has promising geology with oil and natural gas finds in recent years. In Norway, the limitation is not "Peak Oil" but labor costs, taxes, environmental concerns, and restrictive licensing. The huge oil discovery in the southern part of the Norwegian continental shelf, a supposedly mature area, will raise Norway's oil output from a declining trend.

Russia has the world's second oldest oil industry and huge prospective areas, in addition to mature areas with a potential for upgrading and boosting output. The recovery since 2000 is due to higher oil prices as well as modern technology and management. Proven reserves for decades of extraction at present levels provide few incentives for accelerating exploration. "Peak Oil" is not the problem.

Mexico has large petroleum potential and major problems. The state monopoly Pemex does not get sufficient funding to maintain installations and make fresh investment under current fiscal arrangements with the central government. The constraint is politics, not "Peak Oil."

Brazil has seen a quick growth in oil production. The exploitation of pre-salt oil deposits in deep offshore waters represents a major technological breakthrough with worldwide ramifications. The matching geology of West Africa, which separated from South America millions of years ago, enhances the potential for making giant oil discoveries. Most of Africa has had little exploration and the petroleum potential is considerable.

Canada is an established, large producer of oil and natural gas, and an emerging producer of unconventional oil from oil sands. The potential is huge for conventional as well as for unconventional oil.

Exploitation of oil shale has long traditions in many countries. Costs are high and pollution of air and water is a serious problem. At present oil prices, the procedure is economical, with important regional differences. CO_2 emissions are much higher than those of conventional oil.

Synthetic oil from coal is an old procedure going back for about a century. It is costly, requiring heavy initial capital investment, high variable costs, access to water, and purification. Oil from tar sands has similar problems. At present oil prices, both processes are marginally economical. CO_2 emissions are even more than those of oil shale.

Liquid fuel from plants is a mature technology, but present agricultural technology does not permit the world to both grow its requirements of liquid fuel and feed its population (Smil 2010).

Electric cars and hybrids are subject to quickly developing technology. Consequently, oil products such as gasoline and diesel are meeting increasing competition in the transportation sector, although they are likely to dominate for decades (Bryce 2010).

The Perilous Politics of Peak Oil

When higher oil prices do not lead to rising supplies, as apparently has been the case with conventional oil in recent years and as was the case *initially* in the 1970s, the easy conclusion is that the world is running out of oil. As has been argued above, this is a simplistic notion that overlooks basic features of the oil market. The short-term price inelasticity of oil demand and supply and long lead times, imperfect competition due to the concentration of oil reserves relative to oil consumption, and the growing role of financial markets, slow down the response of the market. Nevertheless, the perception of "Peak Oil" has been stubbornly persistent. The controversy is not interest neutral. The idea of "Peak Oil" may serve business interests such as investors in competing energy sources and some oil companies in search of an excuse for high oil prices, as well as environmentalist movements, whereas a contrary perception may serve the interests of oil companies eager to expand, and their suppliers (Gorelick 2010). "Peak Oil" can serve as an argument to limit wages and consumption growth, whereas the notion of resource abundance can incite demands for pay rises.

A perception of oil scarcity can inspire a militaristic strategy to secure access to oil by force, eventually to deny access to others. The European imperial powers, Britain and France, in the twentieth century used military means to secure their oil interests (e.g., in Iraq and Algeria). The US–UK coup in Iran in 1953 was motivated by oil interests. The 2003 US-led invasion of Iraq was by some influential accounts largely motivated by oil (Greenspan 2008). Israel's drive for regime change in Iraq was also important (Cooley 2005). Likewise, oil was allegedly a motive for the British and French intervention in Libya in 2011, not only humanitarian concerns (Browser 2011). Oil interests are hardly alien from US arguments for a long-term NATO presence in Libya (Haass 2011).

War is a costly and ineffective way to secure oil supplies. The US occupation of Iraq caused chaos in its oil industry and declining output, which contributed to the subsequent oil price rise (Muttitt 2011). Likewise, the 2011 Libyan uprising caused damage to the oil industry, and the output loss contributed to rising oil prices. Oil importers' unsuccessful wars against oil exporters and their unintended consequences, not "Peak Oil," did cause the volume losses and ensuing oil price rises since 2003. Since the 1920s, in times of peace, there has never been a physical scarcity of oil in the world market; supply shortfalls have had their causes in political turmoil and acts of war. The 1970s' supply shortfalls in the US oil market had their roots more in ineffective regulation than outright shortfall.

If the purpose of the Iraq War had been to stabilize the oil market through incremental Iraqi supplies with a downward pressure on oil prices, the Iraq War appears a major failure. If by contrast, the aim had been to raise oil prices for the benefit of the independent upstream oil companies and the oil services firms that were a constituency for President Bush Jr. and Vice President Cheney, the Iraq War was a success. It was also beneficial to those with geostrategic interests who view high oil prices as an incentive to curtail US "dependence" on Middle Eastern oil. If the purpose had been to secure long-term US control of Iraq's oil, it appears a costly failure.

Nevertheless, control of Iraq in order to dominate the Middle East *politically*, in addition to securing control of Iraq's oil, appears as the most rational and most salient explanation for the US to go to war. From this perspective, the more assertive foreign policy of the George W. Bush administration caused the US to opt for military means to gain control of Iraq's oil; the driving force was not the need for oil (Bromley 2005).

Such an agenda could not have been expressed openly. This agenda included oil, but with a wider grasp than just to secure US oil supplies. Dominance of the Middle East would permit the US to control oil flows to third parties, especially China, seen as the new adversary (Learsy 2008). US strategy seems to be to control the world's major oil producing regions, regardless of its own needs, in order to be able to deny access to potential enemies. The establishment of a new US military command for Africa, AFRICOM, lends credence to this hypothesis. Against this backdrop, the Iraq War appears less of a blunder, in spite of the mismanagement of the US occupation, than as part of a larger mindful strategy to dominate the world increasingly by military means, as US economic predominance erodes. The record of US intervention in the region supports this view (Bromley 1991).

Perceptions of a tight oil market can also induce a mercantilist strategy to secure preferential positions for access to oil. China is the prime example (Ebel 2005). Through cooperation between its government, securing long-term finance at low interest rates, and its wholly or partly state-owned companies, China gains footholds in many oil exporting countries, in the Middle East as well as in Africa and South America. The strategy is to diversify supplies and risk, to secure preferential treatment by loans being reimbursed in oil and natural gas, and to promote interdependence through investment, credits, and market shares for Chinese industrial goods in the oil and gas exporting countries.

Through long-term comprehensive deals with preferential treatment, China secures a first call for oil from many sources. China therefore is less exposed to a physical supply risk, insofar as oil flows continue, but it is exposed to the price risk. Since 2000, China has become the major trade partner and investor for many oil exporters. China is the largest buyer of Saudi crude, the largest investor in Iran and Iraq, and the major trading partner for Brazil and Angola. The strategy is also to gradually conduct more trade, also oil trade, in Chinese currency, limiting the role of the US dollar. Money, trade, and investment thus secure China's oil interests without recourse to military force (Brautigam 2009).

New Perspectives for Oil Policies and Strategies

A decade of historically high oil prices has opened new frontiers for the energy industry. Because of resource abundance and lower costs, natural gas is set to be a winner. Natural gas is less polluting than coal, it is less costly and more reliable than wind and solar power, and it is available in large quantities from many sources. Its reserves are widely distributed, especially among high energy-consuming countries. Any breakthrough in gas-to-liquids technology lowering conversion costs would integrate oil and natural gas markets, causing oil prices to fall and gas prices to rise. At present oil prices, the procedure is economical (IEA 2010). CO_2 emissions are less than those of conventional oil.

The breakthrough for shale gas and shale oil in the US changes basic balances in the international energy markets. Only a few years ago, the outlook was for the US to become a major importer of liquefied natural gas, straining the world market. Instead, the technological progress in shale extraction is leading the US toward natural gas self-sufficiency and possibly net exports. For oil, the outlook is for higher US production and reduced imports. The immediate effect is a downward pressure on natural gas prices worldwide. Reduced oil imports to North America will make larger volumes available for the Asian markets at a time when oil production is surging in Africa and South America. The hurdle to lower oil prices is not resource scarcity, but the risk premium due to strife in the Middle East.

High oil prices stimulate investment in both conventional and unconventional oil, in natural gas, and in alternative energy sources. The lead times mean that the effect on the market is not immediately evident, but over time competition is building up, both within the oil market itself and to oil in the overall energy market, with new challenges to the position of oil in its prime market, the transportation sector. Some of the competitors to conventional oil have a substantial potential and their costs are likely to decline as investment picks up, technologies mature, and experience accumulates. Contrary to "Peak Oil" theory, the industry continuously makes new conventional oil discoveries, offshore as well as onshore, in new oil provinces such as Africa and Brazil, and in established oil provinces such as the Gulf of Mexico and the North Sea (Lynch 2002). Prospects for higher volumes and lower oil and natural gas prices weaken the competitiveness of nuclear power, coal, and renewable energy (Yergin 2011). This is particularly important in an economic recession.

For the oil industry, the outlook is more investment opportunities, in conventional and unconventional oil, in established as well as emerging oil provinces. The cutting edge is increasingly technology development and managerial competence that the oil and gas producers need.

The oil importers have to consider policy options against oil price assumptions. The historical record of forecasting oil prices is poor. The various crisis scenarios referring to "Peak Oil" have not materialized. "Peak Oil" serves as an argument for investing in renewable, "green" energy, supposedly good for the environment as well as a good business (Helm 2011). Experience teaches prudence.

In the late 1970s, with the perception of ever-rising oil prices, many countries opted for nuclear power, but the oil price decline in the mid-1980s and the Three Mile Island and Chernobyl accidents caused nuclear investment to decline. Referring to "Peak Oil," governments support solar and wind power, assuming that ever-rising oil and natural gas prices will make them competitive. The risk is that lower oil and natural gas prices will make them even less competitive, and that costs will force governments to cut support (Azelton and Teufel 2009).

The risk of not preparing for scarcer and costlier oil in the future is to confront energy supply and payment problems. The US is a case in point. In 2005, the US Department of Energy sponsored a report concluding that "Peak Oil" would hit the US brutally; it would be irreversible and have dramatic consequences for the economy (Hirsch 2005). A few years later, the outlook for the US is less dependence on oil imports and that "Peak Oil," if not called off, at least seems far away. In the meantime, oil prices have risen and provided incentives for technological breakthroughs as well as investment. The market has worked, although at a high social cost. US energy policy is to work with the market. Rhetoric about energy independence has been a smokescreen for political inaction (Bryce 2008).

The dilemma for the US is whether to continue to secure access to oil by force and a militaristic strategy or by more mercantilist means. In relation to the Middle East, which has a food deficit, the US has a favorable position as the world's major food exporter. The preference for the militaristic strategy is, however, embodied in the powerful position of the military-industrial complex in US politics. Even if the militaristic strategy has been a failure measured by costs and benefits, especially when taking high oil prices into account, it has been a success for bureaucratic and industrial interests that benefit from larger military budgets. Special interests dominate US politics, not least US foreign policy, and the pro-Israel lobby has weight (Friedman 2011). These interests need enemies to justify their budgets (Zenko and Cohen 2012). After the Soviet collapse, Islam seems to be a useful foe; China may be next in line (Alexander 2011). Whether the perceived threat is Soviet, Islamist or Chinese, it can justify a military presence in the Middle East. Oil is not the only driver in US policy toward the Middle East; Israel often seems more important (Mearsheimer and Walt 2007).

Even if the US remains the world's largest oil consumer, imports are falling due to growing domestic output. This is of critical importance to the world oil market. Oil use in the US is economically far less efficient than in Europe or Japan, indicating a large potential for energy conservation. It is, however, politically difficult to tax motor fuel, which would provide incentives for oil use efficiency and government revenues. A coalition of the automobile industry, oil, real estate, and construction interests has sufficient influence in Congress to prevent harsh measures to reduce such oil demand. The preference is for increased supply incentives that benefit private business interests. American consumers do not want to renounce large cars and large homes, shopping centers, or long driving distances.

Basing energy policy on the notion that oil (and natural gas) will be steadily scarcer and costlier carries a risk of overinvesting in energy conservation and alternative energy at high cost, at the expense of consumers and businesses, only to be challenged in the future by competitors with access to abundant and inexpensive energy. Energy efficiency and renewable energy do not come free of charge (Smil 2010). The European Union (EU) is a case in point with its plan to phase out almost all use of fossil fuels by 2050 (EU 2011). The policy is to detach the EU, as far as possible, from the world oil market. Implicitly, this is also a policy of limiting trade with Europe's "near abroad," the Middle East, North Africa, and Russia.

The emphasis is on clean technologies, together with efficiency. Economic constraints hardly get attention. Europe already has the world's most efficient use of energy, exceeding Japan, as compared to economic output; it has done the easy part of energy conservation. A more efficient energy use normally requires investment. At a time when most European governments are preoccupied with budget deficits, subsidies for new energy sources are easy prey. The energy roadmap of the European Commission does not refer to any cost-benefit analysis of investment costs or economic growth, industry structure, and employment. Implicitly, this is a vision of zero economic growth, stagnation like Japan, with a declining and ageing population that works less, consumes less, and pollutes less, but with higher unemployment and social tensions. Nevertheless, the vision is policy, detached from the real economy (Morris *et al.* 2011). EU energy policy is to work against the market, at a high risk. Persistently high oil and natural gas prices would ultimately make solar and wind power competitive, although at a high system cost for back-up (Newbery 2012). By contrast, a protracted period of declining world market oil and natural gas prices would raise the cost of supporting renewable energy, harming industrial competitiveness and employment.

The Chinese alternative is closer cooperation with the oil and natural gas exporters, aiming at balancing trade and securing supply diversification. Indeed, the oil suppliers have concerns about markets and revenues, complementary to China's concerns about security of supply. China is the major competitor for oil from the Middle East, Africa, and South America. The country currently imports well over half of its oil needs, a ratio that is increasing quickly. Total Asian oil import requirements may exceed the export capacity of the Middle East (Mitchell 2010). China largely procures oil within a mercantilist strategy based on cooperation between the government and fully or partly state-owned companies (Kong 2010). Tools are financing and trade to secure long-term contracts and preferential positions. The strategy is to diversify supplies and establish interdependence through Chinese investment, credit, and manufactured goods exports.

Since 2000, China has become the major trading partner and investor for many countries in Africa, the Middle East, and South America. China is the largest buyer of

Saudi crude, the major investor and trading partner for Iran, and the largest investor in Iraq's oil industry. It is also the major trading partner for Angola and Brazil. Capital, investment, and trade give China international influence and strong positions in key countries without using military power (Brautigam 2009). The long-term interest is to conduct a growing share of the trade, also oil, in the Chinese currency.

The agreement concluded in the autumn of 2004 between China and Iran in many ways heralds a historical shift in oil trading patterns, with the emergence of a new energy mercantilism, characterized by close links between governments and companies and the active use of state-owned enterprises to promote strategic national interests. The logic is for oil importing governments to provide their state-owned enterprises with cheap capital to promote long-term strategic thinking and risk-taking. The rationale is to offset energy dependence with more comprehensive economic relations, letting industrial exports pay for oil and natural gas imports and secure long-term supplies. The lesson is that oil with a secure supply has a value above the market price; eventually this also applies to natural gas.

China's oil imports grow as US oil imports recede. This may reduce tension and the conflict potential over access to oil. As the centre of import gravity in the oil market shifts from the North Atlantic to South and East Asia, important economic and political consequences follow. The flows of manufactured goods from China to compensate for the oil flows are shifting the patterns of world trade in general, not just oil and other raw materials. The flows of investment and payment are shifting the patterns of international finance, likely to be reinforced by the gradual internationalization and convertibility of the Chinese currency. Finally, the flows of oil, manufactured goods, and money are creating new patterns of political allegiances. China is already the leading economic power in the Middle East, with growing political influence. The drive by the US and the EU to impose harsher sanctions on Iran serves China's interests well.

Within this setting, the US appears to be heading for increasing marginalization in world oil trade. With oil imports receding, the US market will be less important to the major oil exporters, but the trade deficit will also diminish. The issue is to what extent higher self-sufficiency in oil and budgetary constraints will reduce the US military commitment in the Middle East, or by contrast lead to an even more unilateral pro-Israel stance, further unsettling the Middle East and the oil market (Parsi 2007). By uncritically supporting US policies on the Middle East, the EU chooses irrelevance in world affairs, neglecting distinctive European economic and political interests. At an earlier stage, the euro appeared as a rival to the dollar, also in oil trade; currently the Chinese yuan has that position (Subramanian 2011).

Note

1. The author is thankful to Sweden's Finance Ministry that commissioned and funded a study on "Peak Oil" – Theory and Objections.

References

Adelman, M. A., and G. C. Watkins. 2008. Reserve Prices and Mineral Resource Theory. *Energy Journal*, Special Issue to Acknowledge the Contribution of Campbell Watkins to Energy Economics: 1–16.

Alexander, Andrew. 2011. *America and the Imperialism of Ignorance*. London: Biteback.

Alyousef, Yousef, and Paul Stevens. 2011. The Cost of Domestic Energy Prices to Saudi Arabia. *Energy Policy* 39, 11: 6900–6905.

Azelton, Aaron M., and Andrew S. Teufel. 2009. *Fisher Investments on Energy*. New York: John Wiley & Sons, Inc.

Babusiaux, Denis, Axel Pierru, and Frédéric Lasserre. 2011. Examining the Role of Financial Investors and Speculators in the Oil Markets. *Journal of Alternative Investments* 14, 1: 61–75.

Beaudreau, Bernard C. 1998. *Energy and Organization*. London: Greenwood Press.

Becker, Gary P. 1978. *The Economic Approach to Human Behaviour*. Chicago, IL: University of Chicago Press.

Blair, John. 1976. *The Control of Oil*. New York: Panther Books.

Brautigam, Deborah. 2009. *The Dragon's Gift*. Oxford: Oxford University Press.

Brodman, John. 2012. *The U.S. Oil and Gas Boom*. Paris: IFRI.

Bromley, Simon. 1991. *American Hegemony and World Oil*. University Park, PA: Pennsylvania State University Press.

Bromley, Simon. 2005. The United States and the Control of World Oil. *Government and Opposition* 2: 225–255.

Browser, Derek. 2011. The War for Libya's Oil. *Petroleum Economist* (May): 12–19.

Bryce, Robert. 2008. *Gusher of Lies*. New York: Public Affairs.

Bryce, Robert. 2010. *Power Hungry*. New York: BBS Publications.

Campbell, Colin J. 1991. *The Golden Century of Oil 1950–2050*. London: Kluwer Academic.

Campbell, Colin J., and Jean Laherrère. 1998. The End of Cheap Oil. *Scientific American* (March): 78–83.

Chevalier, Jean-Marie. 2010. *Report of the Working Group on Oil Price Volatility*. Paris: Ministère de l'économie, de l'industrie et de l'emploi.

CIA. 1979. *The World Market in the Years Ahead*. Washington, DC: CIA, National Foreign Assessment Center.

Clarke, Duncan. 2007. *The Battle for Barrels*. London: Profile Books.

Cooley, John K. 2005. *Alliance Against Babylon*. London: Pluto Press.

Dasgupta, P. P., and G. M. Heal. 1979. *Economic Theory and Exhaustible Resources*. Cambridge: Cambridge University Press.

Ebel, Robert E. 2005. *China's Energy Future*. Washington, DC: CSIS Press.

EU. 2011. *Roadmap for Moving to a Competitive Low-Carbon Economy in 2050*. Brussels: European Commission.

Friedman, Thomas L. 2011. Newt, Mitt, Bibi and Vladimir. *New York Times*, December 12.

Gorelick, Steven M. 2010. *Oil Panic and the Global Crisis*. Oxford: John Wiley & Sons Ltd.

Greenspan, Alan. 2008. *The Age of Turbulence: Adventures in a New World*. New York: Penguin Books.

Haass, Richard. 2011. Libya Now Needs Boots on the Ground. *Financial Times*, August 22.

Helm, Dieter. 2011. The Peak Oil Brigade Is Leading Us into Bad Policymaking on Energy. *The Guardian*, October 18.

Hirsch, Robert L. 2005. The Inevitable Peaking of World Oil Production. *The Atlantic Council Bulletin* 16, 3.

Hofmeister, John. 2010. *Why We Hate the Oil Companies*. New York: Palgrave Macmillan.

Holland, Stephen P. 2008. Modeling Peak Oil. *Energy Journal* 29, 2: 61–79.

IEA. 2010. *World Energy Outlook 2010*. Paris: IEA/OECD.

Jesse, Jan-Hein, and Coby van der Linde. 2008. *Oil Turbulence in the Next Decade*. The Hague: Clingendael, Netherlands Institute for International Relations.

Kong, Bo. 2010. *China's International Petroleum Policy*. Westport, CT: Praeger.

Laherrère, Jean. 2000. Learn Strengths, Weaknesses to Understand Hubbert Curve. *Oil and Gas Journal*, April 17.

Learsy, Raymond J. 2008. *Over a Barrel*. New York: Encounter Books.

Le Bideau, Hervé. 2010. *Le pétrole, enjeu stratégique des guerres modernes*. Paris: Éditions l'esprit du livre.

Lutzenhiser, Loren. 1993. Social and Behavioural Aspects of Energy Use. *Annual Review of Energy and the Environment* 18: 247–289.

Lynch, Michael C. 2002. Forecasting Oil Supply: Theory and Practice. *The Quarterly Review of Economics and Finance* 42: 373–389.

Meadows, D. H., Dennis Meadows, Jørgen Randers, and William W. Behrens III. 1972. *The Limits to Growth*. New York: Universe Books.

Meadows, Donnella, Jørgen Randers, and Dennis Meadows. 2010. *Limits to Growth – The 30-Year Update*. London: Earthscan.

Mearsheimer, John J., and Stephen M. Walt. 2007. *The Israel Lobby and U.S. Foreign Policy*. New York: Farrar, Straus and Giroux.

Mills, Robin M. 2008. *The Myth of the Oil Crisis*. Westport, CT: Praeger.

Mitchell, John. 2010. *More for Asia*. London: Royal Institute of International Affairs.

Mohn, Klaus. 2010. *Elastic Oil*. Working Papers in Economics and Finance 2010/10. Stavanger: University of Stavanger.

Morris, Andrew P., William T. Bogart, Roger E. Meiners, and Andrew Dorchak. 2011. *The False Promise of Green Energy*. Washington, DC: Cato Institute.

Muller, Richard A. 2008. *Physics for Future Presidents*. New York: Norton.

Muttitt, Greg. 2011. *Fuel on the Fire*. London: The Bodley Head.

Newbery, David M. 2012. Contracting for Wind Generation. *Economics of Energy & Environmental Policy Issues* 1, 2.

Nordhaus, Wiliam D. 1973. *The Allocation of Energy Resources*. Washington, DC: Brookings Institution.

Parsi, Trita. 2007. *Treacherous Alliance*. New Haven, CT: Yale University Press.

Reinhart, Carmen M., and Kenneth P. Rogoff. 2011. *A Decade of Debt*. Washington, DC: Peterson Institute for International Economics.

Ridley, Matt. 2010. *The Rational Optimist*. New York: Harper Perennial.

Roberts, Paul. 2004. *The End of Oil*. Boston: Houghton Mifflin.

Rosa, Eugene E., Gary E. Machlis, and Kenneth M. Keating. 1988. Energy and Society. *American Review of Sociology* 14: 49–72.

Rubin, Jeff. 2009. *Why Your World Is About to Get a Whole Lot Smaller: Oil and the End of Globalization*. New York: Random House.

Simmons, Matthew. 2006. *Twilight in the Desert*. Oxford: John Wiley & Sons Ltd.

Smil, Vaclav. 2003. *Energy at the Crossroads*. Cambridge, MA: MIT Press.

Subramanian, Arvind. 2011. *Eclipse*. Washington, DC: Peterson Institute for International Economics.

WAES. 1977. *Workshop on Alternative Energy Strategies*. Cambridge, MA: MIT Press.

Watkins, G. C. 1992. The Hotelling Principle: Autobahn or Cul de Sac? *Energy Journal* 13, 1: 1–24.

Wirl, Franz. 1997. *The Economics of Conservation Programs*. Boston: Kluwer Academic.

Yergin, Daniel. 2011. *The Quest*. New York: Penguin Press.

Zenko, Micah, and Michael A. Cohen. 2012. Clear and Present Safety. *Foreign Affairs* 91, 2: 79–93.

Chapter 11

Cooperation and Conflict in Oil and Gas Markets

Dag Harald Claes[1]

Introduction

There are a number of factors influencing the developments of energy industries and markets. In the hydrocarbon age the physical availability of coal, oil, and gas resources is a fundamental condition for energy production and consumption. The choice of energy among individual energy consumers worldwide and the elasticity of their demand forms another condition for energy producers. However, to large extent, the development of the international oil and gas industries and their respective international markets has been formed through cooperation and conflict among key political and commercial actors. This is the topic of the present chapter. It is impossible to cover all aspects of all energy sectors. Today, the emerging renewable energy sectors gain importance and political attention on many levels. They will not be part of this chapter but are dealt with elsewhere in this volume. This chapter will focus only on the international oil and gas markets.

The chapter is split in two parts. The first one is historical and confined to past cooperation and conflicts in international energy. It centers on oil, reflecting that natural gas became an important part of the energy mix only more recently. The second part discusses three present aspects of cooperation and conflict: first, the new forms of cooperation among companies in the vertical markets structure; second, the role of international governmental organizations as instruments for political governance of the oil market; and third, a case of global cooperative behavior and regional conflictual behavior in the international gas markets. The chapter ends with a tentative attempt to look into the future of cooperative and conflictual energy relations. But first it is necessary to outline the perspective and key concepts guiding the discussion in this chapter.

Patterns of Conflict and Cooperation

Cooperation and conflicts in the oil industry appears in both the vertical and the horizontal market structure. The vertical structure is the stages of refinement of the product from raw materials to the final product sold to the end user. In the vertical structure

The Handbook of Global Energy Policy, First Edition. Edited by Andreas Goldthau.
© 2013 John Wiley & Sons, Ltd. Published 2013 by John Wiley & Sons, Ltd.

market power arises as actors are able to control other actors' outlets or access to the goods traded at the various stages of the product chain. In the oil industry we usually call exploration and production of crude oil the upstream segment of the market, and transportation, refining, marketing, and sales the downstream segment. The horizontal structure is the relationship between companies in the various stages of the vertical structure. In the horizontal structure market power arises as actors are able to form monopolies, oligopolies, or various forms of cartelization of the market. The horizontal structure forms the basis for the exercise of market power by coherent action by a group of actors that might lead to the realization of monopoly profit. One fundamental feature of the oil industry is that the upstream segment, exploration and production of crude oil, can take place only where the oil resources are located. Access to these resources is a premise for any related commercial activity.

Oil, as with natural resources in general, has a surplus value after all costs and normal returns have been accounted for, called the resource rent. The potential for capturing this rent makes the control over the upstream segments particularly important and lucrative. Thus, the access to natural resources is usually controlled by governments as owners of the resources. A special relationship emerges between companies that have capital and technological know-how but seek opportunities for foreign direct investment in the upstream segment of the oil industry and host governments that control the ownership of oil resources but seek revenues from extraction of the resources. Initially the companies have the upper hand as they can choose where to go, but once they have made a substantial amount of investments they are hostages in the hands of the government. The companies' investments are sunk costs and the government can impose additional conditions on the companies and increase taxation without risking the companies leaving the country. This phenomenon was named the obsolescing bargain by Raymond Vernon (1971: 46–59). The relative power of governments and companies in the oil industry and the value of the resource rent have varied over time. Thus, the relevance of the obsolescing bargain has also changed. Vivoda (2008) argues that its relevance has increased in recent decades.

Energy is an important input factor in most industries, a prerequisite for modern transportation and vital for almost all kinds of human activities and general welfare. This has motivated political interference in energy industries at the local, national, regional, and international levels. Prospects of shortages or shutdowns in the supply of oil have caused oil-consuming countries, in particular the US, to use foreign diplomacy, economic bargaining, and military force to regain security in oil supplies (Klare 2004: 26–74). Such cases of political interference gain high public attention. They also have implications for the analytical approach to the oil industry: "Oil, unlike other commodities, has a universality of significance which in the real world places limits on the application of economic rationale to the evolution of the industry" (Odell 1986: 38). Energy is a topic in need of "some analytical framework for relating the impact of states' actions on the markets for various sources of energy, with the impact of these markets on the policies and actions, and indeed the economic development and national security of the states" (Strange 1988: 191).

The political importance of energy supplies means that the conflict and cooperation takes place (1) among countries, like the producer cooperation among OPEC countries and the consumer cooperation in the IEA; (2) among companies, like the historic case of the Seven Sisters and today's various types of collaboration between companies both in the vertical and horizontal market structure; and (3) between governments and companies, like when the Church commission of the US Congress attacked the US oil companies for contributing to the oil price rise of the early 1970s. This pattern of relationship illustrates what Stopford and Strange (1991: 1) called the *triangular diplomacy*. The fact that political actors are increasingly involved in business activities and the fact that the behavior of companies has political implications leads to a dynamic interaction between these three different relationships. In the age of globalization and after the 2008 financial crisis this is all too obvious. The international oil industry has, however, always had a prominent political importance. In fact, one could say that oil became a strategic commodity before it gained its present broad commercial importance. As part of the naval arms race with Germany prior to World War I the British First Lord of the Admiralty, Winston Churchill, took the decision to switch from coal to oil-fired naval vessels. This had two major implications: first, the oil market became an international market, as the oil used in these vessels was located in the Middle East; second, it became highly politicized because of the immediate security importance of the oil resources for warfare. The next section provides a brief account of the history of conflict and cooperation step in the international oil market.

The History of Cooperation and Conflict in the International Oil Market

The strategic importance of access to the oil resources in the Middle East triggered a conflict among the victorious Allied Powers of World War I with direct implications for the private oil companies.

After the war Britain tried to make Mesopotamia a British mandate under the regulations of the League of Nations. One aim was to gain British control over oil resources in the region. However, according to the Anglo-French oil agreement signed in San Remo in April 1920, France got 25% of the oil from Mesopotamia. This upset the US government as it revoked the principle of the Paris peace negotiations that the war had been won by the Allied Powers fighting together and any economic benefits should be available to all Allied Powers (FRUS 1920: 649–659). The US also claimed that "the San Remo agreement discriminated against the rights of American nationals, that no rights in Iraq were vested in the Turkish Petroleum Co., and that no valid concessions could come into existence through the government of the people of the territory" (FTC 1952: 51–52). The British government argued that British nationals had "acquired rights." [2] The fact that the US had been an Allied Power gave it no right to trespass upon such rights.

After long negotiations, the US, Britain, and France reached a compromise in 1928. The American companies got one-fourth of the Iraq Petroleum Company (IPC, formerly the Turkish Petroleum Company) concession. The companies and authorities also agreed to the so-called self-denying clause of 1914 stating that all parties should work jointly – and only jointly – in the region (Yergin 1991: 204). The region included the Arabian Peninsula (except Kuwait), Iraq, and Turkey. This was the so-called Red Line Agreement (Yergin 1991: 203–206). In the areas inside the red line, the companies would pursue joint concessions. As soon as the US companies were included in the agreement, the open-door policy was abandoned and the door was shut to any new company (Anderson 1981: 19). Although the private international oil companies (IOCs) were important players, and to some extent pushed the governments into the Middle East, the governments of the US, Britain, and France took the leading role in defining the rules of the energy game in the area. Furthermore, the policy process was one of interstate bargaining. However, the companies where soon to gain full control over the international oil business through an elaborate but tacit form of cooperation.

By 1928 more than 50% of oil production outside the US was controlled by Exxon, Shell, and British Petroleum (BP). These companies took their cooperation one step further when they met secretly and worked out a market-sharing deal, the so-called "Achnacarry Agreement," also known as the "as-is" agreement (Yergin 1991: 260–265).[3] This was an agreement to keep the respective percentage market shares of sales in various markets and thus not challenge each other's positions. Another important point in the agreement was the "Gulf plus pricing system," according to which crude was to be priced as if produced in the Mexican Gulf regardless of actual origin. Later the companies also agreed to control production. In the following decades most of the other US companies joined the agreement. Various agreements covered operations in all countries except the US and the Soviet Union. By the end of the 1920s, the companies had set up agreements governing their interrelations in the whole production chain. After World War II the international oil market was totally dominated by seven companies, popularly known as the "Seven Sisters."[4] They were integrated companies in the sense that they controlled the entire production chain from exploration to sale of the refined products. As of 1953 these companies controlled 95.8% of the reserves, 90.2% of the production, 75.6% of the refining capacity, and 74.3% of the product sales outside the US and the communist bloc (Schneider 1983: 40). This created a stable structure.

Only a few firms were capable of the risky search for oil in remote and often harsh places. In each consuming country, refining and marketing was a small industry, protected by distance and government. Production was too risky without an assured outlet, known as "finding a home for the crude." Refining was too risky without an assured supply of crude. Hence in each country few sellers were confronted by few buyers, and neither side wished to be at the mercy of the other. The obvious solution was vertical integration (Adelman 1995: 44).

The integrated structure created high barriers to entry for other oil companies, also called the newcomers. The Sisters informally organized their operations in the Middle East through a consortium in which all the major companies were engaged in at least two countries (see Table 11.1).

In this way, the Sisters stood stronger against possible regulation by the producing countries, as none of them was totally dependent on the will of one government only. If the government of a producing country should put pressure on one of the Sisters in order to increase taxes or introduce less favorable conditions, this company could increase its operation in another country. The other Sisters would compensate the company that was under governmental pressure. This solidarity between the Sisters made it hard for governments of producing states to increase their control and revenues from oil production

Table 11.1 Ownership shares in Middle East production distributed by companies, 1972 (percent).

Company	Iran	Iraq	S. Arabia	Kuwait
Exxon	7	11.875	30	
Texaco	7		30	
SoCal	7		30	
Mobil	7	11.875	10	
Gulf	7			50
BP	40	23.75		50
Shell	14	23.75		
CFP	6	23.75		
Iricon	5			
Gulbenkian		5		

Source: MNC Hearings (1974: Part 5, 289).

within their own territories. The Sisters were able to recreate on the international scene some of the features of the logic behind the strategy of Standard Oil in the US market 70 years earlier, a position that had been broken by anti-trust legislation.[5] The Sisters' tacit cooperation in the Middle East would clearly have been illegal in the US (Odell 1986: 16). The position of the Sisters is an extreme case of how cooperation among multinational companies can define the rules of the game in a vital economic sector.

The Sisters' control over the international oil market lasted until 1971. There is a widespread belief that the decisive change in the oil market appeared in 1973 when the Arab oil producers initiated an embargo to the US in order to make the US change its support for Israel in the ongoing Arab–Israeli War. This is a misconception. The decisive change took place two years earlier in 1971 with the Tehran and Tripoli agreements between the IOCs and members of OPEC. Now the control over the oil market changed from the IOCs to the producing countries. A key factor behind this change was the increased unity among the producing countries and the emergence of new IOCs, the "newcomers," that did not adhere to the tacit agreement among the Sisters.

After the 1956 Suez crisis and the 1967 Arab–Israeli War, North African oil exploration intensified. North African production had an important advantage over that in the Persian Gulf in that the oil did not have to be transported through the conflict area around the Gulf and the Suez Canal. As Libyan oil contained less sulfur than most Gulf oil, it was cheaper to refine and could therefore be priced higher than the heavier crudes of the Gulf region. "Because of these advantages, Libya received the highest per barrel payments of any Arab government, but most observers still considered Libyan oil underpriced" (Schneider 1983: 140). Furthermore, Libya was not part of the Sisters' cooperation. Other independent US and European oil companies had almost 52% of the Libyan oil production. This, together with a new radical regime, meant that the road was open for a confrontation with the oil companies. In September 1969 a coup d'état took place in Libya and Muammar al-Qadhafi became leader. In January 1970 the new Libyan government started price negotiations with the companies individually, not as a bloc. By playing the independent Occidental and the other multinationals against each other, Libya managed to raise posted prices and the government take thereof.

After the Libyan affair, Iran and Venezuela increased their share of profits and a "game of leapfrog began" (Yergin 1991: 580). After some internal differences, the companies united in a common front and sought to negotiate with OPEC in two rounds of negotiations: one with the Gulf exporters and one with the Mediterranean exporters. It should be noted that the market structure had not changed; there was no scarcity caused by underlying changes in the relationship between supply and demand: "From early 1971 to nearly the end of 1972, prices increased despite continuing substantial excess supply" (Adelman 1995: 93).

In February 1971, the so-called Tehran agreement between the IOCs and the OPEC members exporting through the Persian Gulf was signed. In April a similar agreement was signed for the OPEC members exporting through the Mediterranean. The agreements covered tax and price increases, inflation compensation, and a fixing of such rates for future years. The effects of the agreements were a 21% price increase for Saudi Arabian crude (from $1.80 to $2.18) and an increase in revenue of 38.9%. What was more important, however, was the fact that the producer countries had now gained control over the price. By these agreements the distribution of market power in the international oil market changed dramatically.

The outcome of the Tehran and Tripoli negotiations was the result of a combination of lack of unity among the IOCs and a newfound unity among the OPEC members. One could argue that both factors were necessary, but neither of them alone was sufficient to explain the events of 1971. A unified OPEC strategy would not have succeeded unless the unity among companies had begun to crack. The lack of unity among the companies would have meant nothing unless the OPEC members had gained some ability to act in concord.

After 1971 the role of the IOCs in governing the international oil market diminished. A new group of companies emerged: the national oil companies (NOCs). The NOCs were initially created as instruments of government policy, whose aim was primarily to assert sovereign rights over national resources. In a context of resource nationalism, they were to give the state control over the pace of exploitation and the pricing of its finite resources. NOCs also ensured that the state received an equitable share of profits (Marcel 2006: 8). By the mid-1970s, therefore, most of the OPEC members had de facto nationalized their oil industry. In the 1970s the NOCs could be regarded as only part of the public policy of the producer governments and not as actors with independent aims and interests.

In the 1980s and 1990s the international oil market changed character once again. In particular the price fall of 1986 changed the market. With oil prices in the range of $15–25/barrel no super profit was to be gained either by companies or countries. Oil was to a large extent a commodity like any other commodity. The price increases over the last decade have again raised the stakes in the oil market. Most producing countries and some newly important consuming countries (especially China) have vitalized the role of their NOCs. Some analysts draw pessimistic conclusions based on the renewed market power of the NOCs:

> If an increasing proportion of global oil and gas resources are under the control of NOCs, it is reasonable to expect that an increasing majority of oil and gas developments will be driven with political objectives in mind. Relative to a commercial outcome, this will result in inefficiencies in the production of revenues, which can manifest through lower levels of production and higher prices than would otherwise occur. (Eller *et al.* 2007: 33–34)

However, along several dimensions the nature, behavior, and roles of NOCs are very different today from the situation in the 1970s. First, although the NOCs of the 1970s were almost totally oriented toward the domestic oil industry, several of today's NOCs are going abroad. There they face other NOCs on the opposite side of the negotiation table. Marcel (2006: 218) finds that Middle East NOCs distrust other NOCs and do not find them attractive partners, as they complement their own strengths and assets to a lesser extent than IOCs. Nevertheless various alliances between NOCs are emerging, something that did not happen in the old structure. One illustration is how "Sonatrach, Statoil, Petrobras and Saudi Aramco are active participants and organizers of the NOC Forum, which ... has brought CEOs of national oil companies [together] to share ideas on how to develop NOCs' core competences" (Marcel 2006: 220). Second, the oil industry has become more fragmented during the last three decades, as oil companies have outsourced several engineering and technological services. Thus NOCs can acquire technology from other sources than the IOCs (UNCTAD 2007: 118). Third, various forms of contractual relations between oil-producing governments and IOCs have emerged (Likosky 2009). Consequently, the IOCs have re-entered exploration and production in many of the countries that they were kicked out of in the 1970s. Fourth, the wide varieties of alliances between NOCs and IOCs have become the order of the day. This will potentially influence the operations and behavior of NOCs in a commercial direction. Marcel (2006: 209) observes "an increased blurring of differences between NOCs and IOCs."

Cooperation and Conflict in the Present Oil and Gas Markets

The blurring becomes even more prominent when we consider the vertical structure of the market. Here we find cooperation and conflict among companies acting as sellers and buyers in the various stages of the value chain of the oil industry.

Vertical Cooperation and Conflict Among Oil Companies

The OPEC countries dominated the upstream segment from the 1970s. However, the IOCs still controlled the downstream segment of the market, such as refineries, marketing, and product sales. From the early 1980s some of the major oil-exporting countries have made downstream investments in Western Europe and the US, gaining partial or total control over companies that can refine and distribute part of their crude oil exports in the main consuming countries, with the aim of securing outlets for sales and thereby more stable revenues in an increasingly volatile market. These exporters also bought tankers, harbor and storage facilities, and petrochemical plants in consuming countries. Among the largest downstream investors were Kuwait, Venezuela, Saudi Arabia, Libya, and Norway. By 1990 Kuwait and Venezuela had refining capacity (domestic and foreign) covering 90–100% of their production capacity; the corresponding figure for Saudi Arabia was about 50% (Finon 1991: 264).

In the upstream segment of the market there were similar problems. Between 1985 and 1987 the seven largest majors replaced only 40% of oil consumed in the US and 59% outside the US through new discoveries, extensions, and improved recovery rates in existing fields. When revisions of oil reserves and purchases are taken into account, the majors' replacement was still 11% short compared with production. The majors – primarily Exxon, Shell, and BP – purchased reserves from smaller companies that were either cutting back their oil activity or dropping out of the industry completely, but the majors were still "crude short," even if they were better off in this regard than some of their smaller competitors. The companies seemed to be preparing for a more competitive environment, in which increased size is perceived as necessary in order to take higher risks in upstream investments. For the OPEC countries the key problem was the ability, or rather lack of ability, to finance the necessary investments in their existing production facilities, on their own. As most OPEC countries produced close to capacity, increase in production capacity would imply new investments. The financial reserves of the OPEC countries were no longer what they were in the heyday of high oil prices. By 1990 virtually all OPEC countries were in need of more financing, more technology, and more organization (Finon 1991: 263). Many OPEC countries revised their policies and opened up for production-sharing agreements with foreign firms.

In the 1990s these developments created a new order based on a convergence of interests between the IOCs providing technology and financial resources for exploration and production and producing countries controlling the access to the resources. The large oil exporters searched for secure outlets for their crude oil in order to protect themselves against future market volatility. Downstream integration is an expression of a risk-averse attitude on their part, which is understandable given their experience in the oil market during the 1980s. Companies with financial difficulties need new investments, for example in their refineries, whereas crude-short companies are interested in arrangements improving their access to oil reserves. For the IOCs, joint ventures with NOCs from the OPEC countries provide protection against future scarcity, which was the companies' nightmare of the 1970s. This opened the way for the companies' return to the upstream market, and subsequently partly reversed the structural change of the 1970s.

International Governmental Organizations: OPEC and IEA

Cooperation among states depends in part on institutional arrangements. Any discussion of cooperation must take into account the role of institutions (Keohane 1989: 3). There are several ways in which international institutions can increase cooperation between states. First, institutions can provide the actors with necessary information about the other actors' preferences. Second, institutions can provide negotiation frameworks that reduce transaction costs and thus increase the effectiveness of interstate transactions. Such frameworks can also help coordinate actors' expectations by establishing conventions. Third, institutions can change the cost-benefit calculations of states, or the incompatibilities of their positions, by providing an arena for issue-linkages, mediating between states, and providing instruments for verification that actors are abiding with agreements or sanctions for those who are not. Fourth, institutions can "frame" how decision-makers view their collective options, as well as contribute to the creation of collective identities.

By far the most important intergovernmental cooperation in the oil market over the last 40 years has been the cooperation among the OPEC members. In the 1970s OPEC as an institution played a minor role, as the oil producers where able to increase the price of oil and increase sales volumes at the same time. As a producer you don't need any common rules or regulations to do that. However, following the oil price rise after the Iranian revolution and the outbreak of the Iran–Iraq war, the demand was weakened. The price increases during the 1970s made possible increased energy efficiency among consumers and the development of new oil provinces, like the North Sea and Alaska. This development made it necessary for OPEC in 1982 to start behaving like an operative cartel, adjusting production levels in order to sustain prices. There were wide differences between the OPEC members over their willingness to adhere to the individual production quotas. In general, Saudi Arabia took on the role as swing producer, leaving room for the smaller members to exploit the big one. However, the weakening of the market continued even though Saudi Arabia had cut its production from 8.5 mbd in early 1982 to about 2.5 mbd in the summer of 1985. This caused Saudi Arabia to change its market strategy in 1985 in order to regain market share. Subsequently, the Saudi strategy became more of a coercive hegemonic power. By changing its strategy from trying to sustain prices by cutting production, Saudi Arabia tried to regain its market share by flooding the market with oil and thus instigated a price war with other producers. Toward the other OPEC members the aim was to force them to cooperate by cutting their production. Toward the new oil producers outside OPEC the aim was to force them out of business, in particular as they generally had higher production costs than the OPEC countries (Claes 2001: 281–297).

The consuming countries tried to counter the market power of OPEC by creating the International Energy Agency (IEA) in 1974. The core aim of the IEA was to handle future oil supply disruptions using an emergency oil crisis management system, originally triggered by a 7% reduction in daily oil supplies, but in 1979 a more flexible system of crisis cooperation was adopted, and this was used again in the Gulf War in 1991 and following Hurricane Katrina in 2005. The IEA has become a vital institution for providing information on international energy and its agenda-setting role has increased in recent years. However, as a market-governing institution it is safe to conclude that the IEA "has limited authority in rule creation and enforcement" (Kohl 2010: 198), although the organization might contribute to coordinated consumer behavior by other means, such as information and statements regarding the market situation and proposals for joint action by the member states. The OPEC cartel had a similar market control to the Sisters, and the benefit of a dominant position in the upstream sector was transferred from the big private oil companies to the treasure chest of oil-producing countries.

Already in the late 1970s some initiatives were taken in order to create a dialogue between oil producers and consuming countries. In 1991 a gathering of ministers from both oil-producing and oil-consuming countries met in Paris. Such meetings have continued every two years and grown into a semi-formal organization called the International Energy Forum, which today has a permanent secretariat in Riyadh (Lesage *et al.* 2010: 61–63). On the political level it is fair to say that the cooperation among oil producers has been highly important over the last 40 years. The cooperation among oil consumers has been less important, but not totally unimportant. There has also been some attempt to establish cooperation between producing and consuming countries. These attempts have not yet had significant impact on the international oil market.

Cooperation and Conflict in International Gas Trade

Gas trade differs from oil trade in various ways. Most importantly in relation to the topic of this chapter is the fact that there is no genuine global gas market. There are three regional gas markets in the world, the North American, the European, and the Asian market. Thirty percent of world total gas consumption is traded across national borders. Almost 70% of this is traded through pipelines (see Table 11.2). Pipeline gas trade physically connects the seller and buyer, reducing the room for spot or short-term trade on competitive markets, unless the infrastructure is developed in order to facilitate gas-to-gas competition in various gas hubs. This is the case with the domestic US gas market, and is emerging in the European market (see below). Nevertheless, such gas-to-gas competition will be limited to the areas covered by the pipeline network.

The only way to create gas-to-gas competition between the three regional gas markets is by liquefied natural gas (LNG). Total global traded LNG constitutes less than 10% of world gas consumption and about 30% of total traded gas. Almost half of this LNG trade (46.34%) is Japanese and South Korean imports. Thus, LNG is not an important factor in most parts of the various gas markets. A substantially increased role for LNG is a prerequisite for the future evolution of a global gas market (de Jong *et al.* 2010: 221).

In 2001 some of the countries with large gas reserves created the Gas Exporting Countries Forum (GECF). The 11 member states account for 42% of gas production and 70% of world gas reserves.[6] The Forum is not at all an operative cartel like OPEC, but signals awareness among gas producers of the potential for coordinated behavior in a situation where gas trade is globalized.

Table 11.2   Gas consumption and trade, 2010.

	BCM	Share of Consumption	Share of Imports
Total consumption	3193,30		
Total trade	975,22	30,54	
Total pipeline imports	677,59	21,22	69,48
Total LNG imports	297,63	9,32	30,52

Source: *BP Statistical Review of World Energy*, June 2011.

As indicated above there is no genuine global gas market that resembles the situation in the oil market. A number of factors influencing the future development of such a global gas market are presently changing. However, the different factors point in different directions. The reduced costs of LNG relative to pipelines increase the competitiveness of LNG in the established regional markets. On the other hand, the technological revolution making shale gas commercial, at least in the North American market, has made the US self-sufficient in gas supplies. This reduces the globalization of gas trade. Finally, the economic downturn in the US and Europe reduces prospects of a substantial increase in gas demand, moving the globalization of gas trade further into the future. If the aim is to create a global gas market, the main challenge seems to be how to increase the amount of traded LNG at the same time as making investments in LNG infrastructure (LNG tankers and export and import terminals) commercially viable. For such a development to create a genuine global market, different suppliers of LNG must meet competitive hubs in LNG-receiving terminals. Such a competitive marketplace would normally put pressure on prices, making LNG investments even more risky.

In the present situation it is hard to see a genuine global gas market emerging in the foreseeable future. Some LNG trade will take place between the three regional markets, but the volumes will most likely be insufficient to facilitate price harmonization and true interregional competitive behavior among consumers. The need for global producer cooperation is simply not here yet. Compared to the changing structure of the global oil market described above this suggest that in this respect the gas and oil market are becoming structurally even more different than they were before.

The most prominent displays of the cooperative and conflictual aspects of gas trade are found in the European market. There are of course important cooperative aspects in the energy policy of the EU. However, in this section I will confine the discussion to the conflict between the EU and its main gas supplier – Russia.

Russia exports about 50% of its gas to Europe, with currently no opportunity to diversify its gas export. EU countries, on the other hand, import about the same percent of their total gas consumption from Russia, thus making both parties dependent on each other. However, even though Russia is dependent on the income from gas exports to the European market, the European gas consumers seem relatively more dependent on Russian gas supplies, mainly because gas is an important commodity, which citizens rely upon in order to fulfill basic needs (heating, cooking, and so forth), and because gas in the short run is almost impossible to substitute, due to the fact that machinery operating on gas cannot use any other source of energy to function. A possible shut-down in gas supply from Russia to Europe will be extremely costly for the latter. When Russia temporarily suspended its flow of gas to Ukraine in January 2009, it became apparent how vulnerable certain European states are to disruption of their gas supply (see below). Structurally this relationship is highly asymmetric. In addition a shut-down in gas supply from Russia would hurt certain European countries harder than others, as the Russian share of gas imports varies across the European countries (Figure 11.1). This creates a competitive situation between the countries, and thus undermines the European Commission's ambition to create a common approach to external energy relations. The individual trade structure increases the probability that some EU members that are dependent on imports of natural gas are vulnerable to political pressure from Russia. While some countries are worried about dependency on Russian gas, others are not. On the one hand, the Baltic States and Poland are extremely worried about the new Nord Stream pipeline going directly from Russia to Germany, as it could isolate them from western Europe. Their role as transit countries would disappear, which in turn

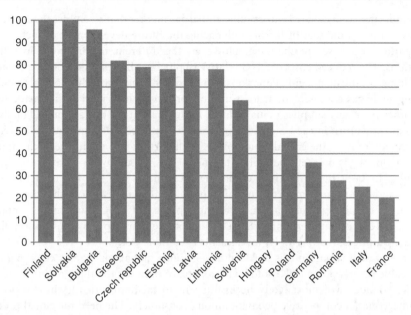

Figure 11.1 European countries' dependency on Russian gas supplies, 2006 (percentage of total gas imports).
Source: Eurostat 2007.

could leave them vulnerable to political coercion from Russia. Thus, the Russian threat of a shut-down in gas supply to eastern Europe could be a more realistic foreign policy instrument, as the new pipeline would ensure that western Europe would not be affected. On the other hand, Russia has no share of Belgian, Italian, and UK gas imports. In the case of France, imports from Russia constitute 23% of total gas imports, but only 3.6% of total energy consumption. These countries are thus not worried about importing gas from Russia.

The increase in Soviet gas export to Europe during the 1980s triggered high-level political concerns both in Europe and in Washington (Jentleson 1986). After the end of the Cold War such concerns eased, but the general goal of diversification of energy imports remains an important aim of all European governments. The eastern European countries are in a special situation since their infrastructure in electricity and gas was developed when they were part of the Soviet-dominated Comecon. The energy relationship among former allies became particularly conflictual between Russia and Ukraine. During the 2000s both the price of gas imports to Ukraine and the Ukrainian fee for transit of gas to the rest of Europe became highly disputed both on the company and political level. As most of Russian gas exports to Europe were then transferred in pipelines across Ukraine this conflict also infected the relationship between all European gas-consuming countries and Russia.

On January 1, 2009, Russian exports to Ukraine were halted completely. The gas intended for transfer to other European customers was maintained. However, the next day the pressure in the pipelines to Hungary, Romania, and Poland dropped. After two weeks the dispute was settled, but both the high level of dependency and the dubious reliability of Russian gas were fully demonstrated. Furthermore, the Europeans are worried that Russia can use their dependency on Russian gas for political purposes, and interpreted the Russo-Ukrainian gas disputes as an example of such politicization of the European energy trade. Almost the opposite perspective can be applied to the understanding of the Russian policy. During the Cold War the Soviet Union used cheap energy supplies as a political instrument in order to maintain the political loyalty of the East European communist states. After the end of the Cold War Gazprom started behaving more commercially and demanded a market price for Ukrainian gas consumers, and also no longer accepted delayed payment. To put it bluntly: the more commercially Gazprom behaves toward Ukraine the more Russian behavior is interpreted as political.

The Future of Cooperation and Conflict in Global Energy Markets

In this chapter a long historical view has been applied. A hundred years have passed since Churchill initiated the politicization and internationalization of oil. For most of these years the oil business has been controlled by a small group of actors through close cooperation among themselves: first the victorious Allied Powers of World War I, next the Seven Sisters, then the member states of OPEC. Today, the set of influential actors in the oil industry has increased, and the structure has become more complex. It is unlikely that this "network" structure will imply a shutout of the traditional IOCs. However, their role is changing, and will continue to do so in partnerships with NOCs of both producing and consuming countries. In most cases producing governments are also likely to welcome the IOCs as partners in the development of their energy resources. After a period of transition such a structure could possibly be as stable as both the era of the Sisters and the OPEC era. But it will not give a single group of actors the kind of market power that characterized the previous periods. At face value this increased complexity implies weaker governance and increased instability, which potentially can result in increased conflict among both companies and countries. To avoid this "policy-makers need to adapt and strengthen the institutional architecture of international oil and gas relations" (Goldthau and Witte 2009: 390).

Since oil trade has become globalized and the interdependence among the actors has increased, political institutions should also be renewed. This can lead to a more cooperative climate. As pointed out by Goldthau and Witte (2010: 355) the "energy world of the future is unlikely to be a world of producers versus consumers, or old consumers versus new ones. The main reason for this is that all actors in the energy domain have shared interests." They suggest the need to establish institutions for correcting market failures, for lowering transaction costs, and for setting standards and rules (Goldthau and Witte 2010: 344–350). To what extent commercial actors and policy-makers have a similar perspective on the energy future of the world remains to be seen.

It is unlikely that the role of hydrocarbons will be the same in 2112 as it is today. Most likely the next 100 years will include a global energy transformation. The transition can happen sooner if concerns regarding environmental degradation and climate change lead to implementation of energy policies in key countries (potentially globally) promoting renewable energy sources at the expense of hydrocarbons. The transition can happen later with increased recovery rates of existing oil and gas reserves, increased ability to make new discoveries in new territories, and development of new technology able to extract unconventional oil and gas resources. What is certain is that this transition will create conflicts among companies and countries and its success rests on the ability of all actors to find cooperative arrangements. Both in the case of climate change and the depletion of hydrocarbon energy resources, the companies and countries all over the world are truly in the same boat.

Notes

1. This chapter has been written as part of the Geopolitics in the High North programme (www.geopoliticsnorth.org), funded by the Research Council of Norway.
2. The term "acquired rights" referred to the rights held by the Turkish Petroleum Company and the rights promised to that company by the Ottoman Grand Vizier, as evidenced by his letter of June 28, 1914, to the British and German ambassadors (FTC 1952: 51–52).
3. A draft version by BP is included in Bamberg (1994: 528–534). It is assumed to correspond to the final text.
4. The name "Seven Sisters" was phrased by Enrico Mattei, the director of ENI, the Italian national oil company, and later used by Anthony Sampson as the title of his book about the big oil companies (Sampson 1975), and includes Exxon, Texaco, SoCal, Mobil, Gulf, BP, and Shell.
5. In 1890 the US Congress passed the Sherman Antitrust Act with the aim of restricting monopolistic and anti-competitive market behavior of large corporations. In 1906 the US administration prosecuted Standard Oil for violation of the Sherman Act. In 1911 the US Supreme Court rejected the company's final appeal and Standard Oil was broken up into 34 different companies (Yergin 1991: 106–110).
6. Interview with GECF Secretary General L. Bokhanovskiy, *Gulf Times*, July 5, 2010.

References

Adelman, Morris A. 1995. *The Genie out of the Bottle – World Oil since 1970*. Cambridge, MA: MIT Press.

Anderson, Irvine H. 1981. *Aramco, the United States and Saudi Arabia: A Study of the Dynamics of Foreign Oil Policy, 1933–1950*. Princeton, NJ: Princeton University Press.

Bamberg, James H. 1994. *The History of the British Petroleum Company*, vol. 2: *The Anglo-Iranian Years, 1928–1954*. Cambridge: Cambridge University Press.

Claes, Dag Harald. 2001. *The Politics of Oil-Producer Cooperation*. Boulder, CO: Westview Press.

de Jong, Dick, Coby van der Linde and Tom Smeenk. 2010. The Evolving Role of LNG in the Gas Market. In Andreas Goldthau and Jan Martin Witte, eds. *Global Energy Governance – The New Rules of the Game*. Washington, DC: Brookings Institution.

Eller, Stacy L., Peter Hartley, and Kenneth B. Medlock III. 2007. *Empirical Evidence on the Operational Efficiency of National Oil Companies*. Houston, TX: James A. Baker Institute for Public Policy, Rice University.

Finon, Dominique. 1991. The Prospects for a New International Petroleum Order. *Energy Studies Review* 3, 3: 260–276.

FRUS 1920. *Papers Relating to the Foreign Relations of the United States, 1920*. Washington, DC: US Department of State.

FTC. 1952. *The International Petroleum Cartel*. Staff Report. Washington, DC: US Federal Trade Commission.

Goldthau, Andreas, and Jan Martin Witte. 2009. Back to the Future or Forward to the Past? Strengthening Markets and Rules for Effective Global Energy Governance. *International Affairs* 85, 2: 373–390.

Goldthau, Andreas, and Jan Martin Witte, eds. 2010. *Global Energy Governance – The New Rules of the Game*. Washington, DC: Brookings Institution.

Jentleson, Bruce. 1986. *Pipeline Politics: The Complex Political Economy of East–West Energy Trade*. Ithaca, NY: Cornell University Press.

Keohane, Robert O. 1989. *International Institutions and State Power: Essays in International Relations Theory*. Boulder, CO: Westview Press.

Klare, Michael T. 2004. *Blood and Oil – The Dangers and Consequences of America's Growing Dependency on Imported Petroleum*. New York: Metropolitan Books.

Kohl, Wilfrid L. 2010. Consumer Countries' Energy Cooperation: The International Energy Agency and the Global Energy Order. In Andreas Goldthau and Jan Martin Witte, eds. *Global Energy Governance – The New Rules of the Game*. Washington, DC: Brookings Institution.

Lesage, Dries, Thijs Van de Graaf, and Kirsten Westphal. 2010. *Global Energy Governance in a Multipolar World*. Farnham: Ashgate Publishing.

Likosky, Michael. 2009. Contracting and Regulatory Issues in the Oil and Gas and Metallic Minerals Industries. *Transnational Corporations* 18, 1: 1–40.

Marcel, Valérie. 2006. *Oil Titans – National Oil Companies in the Middle East*. London: Royal Institute of International Affairs.

MNC Hearings. 1974. *Multinational Corporations and the United States' Foreign Policy*. Hearings before the Subcommittee on Multinational Corporations of the Committee on Foreign Relations, United States Senate, Ninety-third Congress. Washington, DC: Government Printing House.

Odell, Peter R. 1986. *Oil and World Power*. Harmondsworth: Penguin Books.

Sampson, Anthony. 1975. *The Seven Sisters: The Great Oil Companies and the World They Made*. London: Hodder & Stoughton.

Schneider, Steven A. 1983. *The Oil Price Revolution*. Baltimore, MD: Johns Hopkins University Press.

Stopford, John, and Susan Strange. 1991. *Rival States, Rival Firms – Competition for World Market Shares*. Cambridge: Cambridge University Press.

Strange, Susan. 1988. *States and Markets*. London: Pinter.

UNCTAD. 2007. *World Investment Report. Transnational Corporations, Extractive Industries, and Development*. New York and Geneva: The United Nations.

Vernon, Raymond. 1971. *Sovereignty at Bay*. New York: Basic Books.

Vivoda, Vlado. 2008. *The Return of the Obsolescing Bargain and the Decline of Big Oil – A Study of Bargaining in the Contemporary Oil Industry*. Saarbrücken: VDM Verlag Dr Müller.

Yergin, Daniel. 1991. *The Prize – The Epic Quest for Oil, Money and Power*. New York: Simon & Schuster.

Chapter 12

The "Gs" and the Future of Energy Governance in a Multipolar World

Charles Ebinger and Govinda Avasarala

Introduction

Amid dramatic global geopolitical, economic, and financial change, interest at the multilateral level in energy and environment policy has diminished. Despite relative uncertainty in the global energy sector, including volatility in international oil markets surrounding unrest in the Middle East, the uncertainty of the future of the nuclear power industry following the accident at the nuclear power plant in Fukushima in Japan, and the urgent need for international action on climate change, the various multilateral fora – the G8 (and G8+5), G20, and G77 – have all tended to prioritize more prominent issues, such as providing responses to the global financial collapse of 2008 and the ensuing global economic downturn in 2009. Some even argue that such multilateral venues for negotiations have been turned into "crisis committees," responding to the most pressing issues of international concern at a specific moment.[1] As a result, international energy- or environment-related policy prescriptions deriving from such meetings have been limited in scope.

Looking forward, it is unlikely that multilateral negotiations will yield many energy policy prescriptions or mandate any energy or environmental policy reforms to national practices. Much of this is owing to the shifting balance in geopolitical relations. The industrialized economies of the US, the European Union, and Japan, have been affected significantly by the most recent economic recession and a sudden economic recovery is not only unlikely but also may take longer than expected.[2] By comparison, emerging market nations largely have weathered the economic recession and, as a result, are rewarded with more geopolitical clout in multilateral negotiations. This complicates decision-making. Smaller multilateral fora, such as the G8 (or the less formal G8+5) have now relinquished much of their issue platforms to larger, more inclusive gatherings, such as the G20.[3] This transition means that multilateral negotiations are now subject to the local politics in emerging market nations, which often are not aligned with the local politics and priorities of industrialized nations. In some cases this either encourages, or is encouraged by, the

The Handbook of Global Energy Policy, First Edition. Edited by Andreas Goldthau.
© 2013 John Wiley & Sons, Ltd. Published 2013 by John Wiley & Sons, Ltd.

politics of industrialized economies as they jockey to maintain competitiveness with their emerging market counterparts.

The apparent schism between emerging and industrialized economies foreshadows a limited but still productive future for policy-making within multilateral settings. While this may not result in sweeping agreements, it will result in significant multilateral research, financing, and cooperation, all of which will help ease some of the uncertainty and volatility in international energy markets.

After analyzing the history of energy and environmental policy making within multilateral fora, focusing specifically on the three major "Gs" (G8, G8+5, and G20),[4] this chapter highlights the changing energy landscape and discusses some of the looming difficulties for policy-makers looking to encourage global energy governance and international energy or environmental negotiations.

History of the Gs and International Energy Policy

G8

In November 1975, the leaders of the governments of France, Germany, Italy, Japan, the UK, and the US met in Rambouillet, France, on the initiative of French President, Valéry Giscard d'Estaing. Among the various economic issues that the heads of government discussed, resolving the "serious energy problems" in the wake of the oil price shock of 1973–1974 was a prominent item on the agenda.[5] In the final declaration, the parties agreed that "world economic growth is clearly linked to the increasing availability of energy sources" and that they would "spare no effort in order to ensure more balanced conditions and a harmonious and steady development in the world market."[6] Canada joined the group in 1976, informally making it the G7. Following the second oil price shock of 1979, the group's meetings took on greater importance. At the Tokyo Summit of 1979, which coincided with a meeting of the Organization for Petroleum Exporting Countries (OPEC) oil ministers, G7 leaders agreed that "the most urgent tasks [were] to reduce oil consumption and to hasten the development of other energy sources."[7]

However, there was little substantive depth to the resolutions put forth in Tokyo by the seven leaders. This is because, despite the surface unity, there was no agreement among the participants on whether governments should regulate spot oil markets. While France favored regulating the market, Germany and the UK were adamantly opposed to any market intervention. In addition, France publicly assailed the US for doing nothing to conserve oil, since US oil imports had risen 27% since the 1973–1974 Organization of Arab Petroleum Exporting Countries oil embargo. In contrast, the European Economic Community (EEC) had cut oil imports by 5%. While the French critique of US policy had some resonance, in reality in 1978 US oil imports were 7% below 1977 levels and had dropped another 3% during the first half of 1979. Further, energy consumption per unit of GDP had declined each year since 1973.[8]

At the Tokyo Summit, the participants agreed to cut oil consumption but could not agree on how to do it. The US and Japan advocated strict targets for import cuts in 1979 and 1980 while the Europeans wanted a freeze on oil imports until 1985. The leaders also disagreed on a range of other issues, including the share of oil imports allocated to each country or region and the need to coordinate oil consumption policies with other nations that would be impacted by the high price of oil.[9] G7 leaders reached little consensus, despite a rash of events in 1979 that illustrated the importance of a secure supply of oil to global economic growth and demonstrated the developed world's dependence on the

Persian Gulf for the secure supply. These events included the sabotage of a major export terminal in Nigeria that cut oil production from the West African nation; the abrogation by Iran of its triangular gas contract with the Soviet Union and the EEC; and Nigeria's nationalization of British Petroleum's assets. As 1979 ended and 1980 saw further rises in oil prices there was real concern about the stability of the international financial system. The mood at the time was best summed up in a farewell address by US Energy Secretary James Schlesinger who in August 1979 warned that "the energy future is bleak and likely to get bleaker in the decade ahead."[10]

In the short term, Secretary Schlesinger was right. In the months following Tokyo the world was thrown into convulsions by the Soviet Union's invasion of Afghanistan, the seizure of the American hostages in Iran, the attempted takeover of the Grand Mosque in Mecca, the assault on Mohammed's tomb in Medina, riots among the Shiite oilfield workers in the eastern province of Saudi Arabia, a serious deterioration in Iraqi-Iranian relations and an ongoing stalemate in Israeli-Palestinian peace negotiations. As a result, high oil prices continued into 1980, eliciting more urgent rhetoric from the member nations of the International Energy Agency (IEA) and the participants at the 1980 G7 Summit in Venice. "The economic issues that have dominated our thoughts are the price and supply of energy," agreed the participants in the final statement, adding that "unless we can deal with the problems of energy, we cannot cope with other problems."[11] This concern about energy centered on the fact that by the time of the Summit many countries had been forced to draw on their reserve oil inventories because of high energy prices and the radical changes in the commercial terms of access to oil imposed by Saudi Arabia and other OPEC countries. The Summit leaders paid special attention to the effect high energy prices were having on the economies of the less developed countries, urging the World Bank, the oil exporting nations, and industrial nations to give high priority to spurring oil exploration, development, and production of both conventional and renewable energy sources. While lacking specificity, the Summit leaders agreed to reduce oil's share in their economies from 53 to 40% and develop non-oil energy resources equivalent to 15–20 million barrels per day by 1990. To meet these goals, the Summit leaders called for a "large" increase in coal use, "enhanced " use of nuclear power, and in the longer run the development of synthetic fuels, solar power, and other renewable resources.

The singular focus on energy that was evident at the Summits in Rambouillet, Tokyo, and Venice, in addition to other Summits in the 1970s, has not been replicated in recent years. Despite a continued increase in oil prices in the early 1980s following the outbreak of the Iran–Iraq war in September 1980, by 1982 oil prices had fallen substantially and remained low for much of the rest of the 1980s and 1990s (other than a brief spike following the first Gulf War). As a result, during this period the G7/G8 (Russia officially joined the group in 1997) rarely discussed energy.[12]

In 1997, at the 5th Conference of Parties of the United Nations Framework Convention for Climate Change, the Kyoto Protocol was created, binding signatory nations to caps on carbon dioxide emissions. The Protocol also marked the pivot in the energy dialogue at future G8 meetings – if energy was even discussed in depth. Deliberations on abating the impacts of oil supply and price shocks gave way to negotiations over climate change, global warming, and the importance of renewable energy development.

In 2000, at the G8 Summit in Okinawa, Japan, leaders created the Renewable Energy Task Force to "prepare concrete recommendations for consideration at our next Summit regarding sound ways to better encourage the use of renewables in developing countries."[13] The final report on the creation of the Renewable Energy Task Force was delivered in July 2001 at the G8 Summit in Genoa, Italy. It recommended that G8 nations reduce technology costs of renewable energy generation by expanding market access, create a stronger marketplace for renewable energy development, mobilize financing, and encourage market-based mechanisms to address the economic competitiveness of renewable energy projects.[14] However, the report received scant attention at the 2001 Summit. (The only reference to the report in the Summit's final communiqué was an acknowledgment of the efforts of the Task Force participants.)[15] Some of the stalemate over climate change and renewable energy goals was attributed to the new US President, George W. Bush, who earlier that year had pledged not to sign the Kyoto Protocol. Further, some commentators argue that the Bush Administration saw no use in the task force's work as it was a "Clintonian exercise of no interest to this administration."[16]

In 2002, energy and climate was pushed even further off the G8 agenda as the September 11 attacks and international security concerns dominated the Canada meeting. The discussion of energy issues was similarly absent at the 2003 Summit in France other than passing references to commitments to use cleaner, more efficient energy, and reassertion of goals to "promote the safe and secure use of civil nuclear energy technology."[17]

It was not until the Gleneagles Summit in 2005 that energy regained a prominent place on the G8's agenda. In the several years prior to the Summit, oil markets had tightened considerably as OPEC limited production, a second war in Iraq was under way, and the major exporters of Venezuela and Nigeria experienced internal political and social strife that curbed production. The rising price of oil (it reached a new record high of $60/barrel in August 2005) as well as the priority of climate change to the Summit host, UK Prime Minister Tony Blair, mandated that energy would be a central topic on the agenda. The result was the Gleneagles Plan of Action on Climate Change, Clean Energy and Sustainable Development, which contained 63 non-binding energy or climate change agreements.[18]

The remainder of the decade saw the promulgation of a number of initiatives relating to energy and the environment. The 2006 Summit, hosted by Russia, established the St Petersburg Plan of Action on Global Energy Security and the Global Energy Security Principles. While the supporters of the G8 process laud the Plan of Action for providing a high-level international energy policy, some analysts highlight that much of the Plan lacked tangible goals and targets for its conclusions. However, the Global Energy Security Principles represented some incremental progress, as the IEA found in its report to the G8 Summit in Japan in 2008.[19] At this Summit, leaders agreed to adopt at that year's United Nations Framework Convention on Climate Change the goal of reducing global emissions by 50% by 2050. (Two issues are notable about this agreement. First, it did not specify a base year against which emissions reductions were targeted. Second, the communiqué of the Major Economies Forum, which joined some of the meetings in Hokkaido and includes non-G8 major energy consumers, included no such agreement on emissions targets.) In 2009, at the Summit in L'Aquila, Italy, the leaders agreed that the increase in global temperatures should not exceed 2 °C and that industrialized countries should reduce greenhouse gas emissions by 80% by 2050, from 1990 (or a more recent year) levels.

Finally, the G8 has made efforts to expand its representation in some energy areas. The G8+5 emerged as a forum in 2005 as part of the Gleneagles Plan of Action on Climate Change, Clean Energy, and Sustainable Development. The addition of the G5 – Brazil, China, India, Mexico, and South Africa – represented a significant step as it was the recognition that action on climate change required dialogue between the industrialized powers and new, emerging economies. In 2007, as part of the Heiligendamm Process that came out of that year's Summit in Heiligendamm, Germany, the G5 were integrated further into the G8 process. But while other emerging economies, such as Indonesia and South Korea, had been included in G8+5 negotiations on climate change and clean energy, other major developing economies such as Saudi Arabia and Turkey were refused entry.[20]

G20

Although it is typically viewed as a forum for resolving financial issues, the G20 is more representative of the changing energy order than its G8 and G8+5 counterparts. Unlike the G8, the G20 includes many of the major energy exporters, consumers, and emerging consumers, such as Saudi Arabia, India, China, and Turkey.

Founded during the throes of the Asian financial crisis in September 1999, the G20 was initially tasked with reinstating financial and economic stability to a world in the midst of a tectonic shift in the economic landscape. That the financial collapse in Asia had such dire ramifications for the global financial system indicated that the world was more economically interconnected than ever before, and that the economic importance of the traditional hegemons (the G7 nations) had eroded substantially.[21] By the early 2000s, the G20 was holding regular meetings looking at a range of financial issues from financial crisis prevention and resolution to development financing.

Given the G20 meetings' early focus on finance, energy was rarely a discussion topic. This changed during the rise in oil and commodity prices between 2005 and 2008, during which it became evident that the G20 was transitioning from a finance-focused organization to one looking to tackle obstacles to short and long-term economic growth.[22] By 2006, commodity prices became part of the G20 dialogue, with member nations looking for ways to smooth the volatile nature of commodity price fluctuations.[23] This was followed up by a similar declaration at the 2007 G20 Summit in South Africa.

By the time the 2008 G20 Summit occurred in Washington, DC, oil prices had already peaked at over $147/barrel and were in free fall. On Friday, November 14, the day before the G20 meetings began, the price of crude oil was roughly $57.[24] Partly as a result of this steep decline in the price of oil, but more likely because of the depth of the economic and financial crisis, energy was a negligible part of the discussion, with the only energy or environmental resolution being: "we remain committed to addressing other critical challenges such as energy security and climate change . . ."[25] While the leaders' statement at the London G20 Summit in 2009 also contained only one reference to energy and environmental policy, it was more substantial, reaffirming the leaders' commitment "to reach agreement at the U.N. Climate Change conference in Copenhagen" that December.[26]

Five months after the London Summit, the G20 leaders reconvened in Pittsburgh in the US to "transition from crisis to recovery." Energy and climate change were two important pillars of the G20's comprehensive plan for international economic recovery. Indeed, many of the resolutions involved delegating responsibility and action to other international organizations including the IEA, the International Energy Forum (IEF), and OPEC. The resolutions sought to address several issues including oil market transparency and data collection, regulation of commodities trading, energy efficiency, fossil fuel subsidies, and clean and renewable energy development. The Pittsburgh Summit sustained the commitment to "reach agreement in Copenhagen through the UNFCCC negotiation."[27]

Since the 2009 Pittsburgh Summit, energy has been a marginal component of G20 Summits, mostly surfacing as a topic with respect to leaders' continued support for climate change negotiations or their interest in improved oil market transparency and better regulated commodity financial markets. But as the G20 has usurped the G8's position as the main forum for international economic governance issues, most of the remainder of this paper will analyze where the G20 has been successful and what the future holds for the G20 and energy governance.

How Has the G20 Done?

Regarding energy, the G20 has focused on six core issues: oil market stability and relatively affordable oil prices, international climate change negotiations, international energy data transparency, commodity financial markets, clean energy and energy efficiency promotion, and fossil fuel subsidies.

While the G20's actual success on many of these issues has been limited, there are some areas for optimism. Not surprisingly, the successful efforts have emerged in areas where there is greater consensus among G20 members. For instance, G20 members are more likely to agree on the need for a more predictable and affordable oil market given that a number of key members are heavily dependent on imported oil. The US imports over 40% of its oil needs, South Korea and Japan are almost exclusively reliant on imported oil for domestic consumption, and by 2035 the IEA estimates that China and India will import 85% and 91% of their oil needs, respectively.[28] Although two G20 members, Russia and Saudi Arabia, are the world's largest producers of crude oil and therefore benefit from high oil prices, oil prices and price volatility have a negative impact on their economies. Too high an oil price leads to demand destruction as in 2008, which can be harmful to national economies dependent on oil revenues. As a result, the G20 has made valuable progress on encouraging oil market transparency and data collection, and in limiting oil market price volatility.

However, in other areas, the characteristics of energy consumption are not as conveniently uniform across G20 members. For instance, coal consumption in emerging Asia continues to increase, while some Western industrialized nations attempt to (slowly) phase coal out of their electricity mix. Similarly, policies on nuclear power vary from G20 member to member (Germany is phasing out nuclear power while China has 27 reactors currently under construction) as do policies on carbon emissions and energy subsidies. Together, these differences create obstacles to developing a framework for international energy governance. In this regard, the obstacle to international consensus is more often local as opposed to geopolitical. Given the importance of energy policy to the citizens, and therefore voters, in each member nation, consensus on such issues is normally derailed by domestic politics. There is a substantial amount of literature that analyzes the political dynamics of the "new" energy consumers and how internal politics will impact international energy trade.[29]

In these areas, where energy supply and demand characteristics diverge from country to country and, therefore, where the priorities of voters diverge from country to country, the G20 has been less successful. Its ability to shepherd the leading energy consumers and carbon emitters to a global consensus on climate change has been dismal. On encouraging the reduction in energy subsidies the G20 has been similarly impotent, in spite of the valuable research from the IEA highlighting how subsidies are leading to rising global demand for fossil fuels.[30]

An Accord on Climate Change

Many analysts correctly argue that although international climate negotiations cannot be labeled a success, in recent years there has been incremental progress. Countries that disagree on the contours of a broad agreement have nevertheless been able to agree on smaller programs, including the institutionalization of a Green Climate Fund and other financing measures. However, the prospect for a comprehensive agreement on emissions reductions, the ultimate goal of the UNFCCC, remains bleak. Despite a vague agreement on the part of the major emitters to commit to cutting carbon emissions and a forward-looking promise for another "legal instrument or agreed outcome" to be implemented by 2020, there is still disagreement between emerging and developed economies.[31] Emerging economies such as China and India remain adamant that the burden for curbing carbon emissions should not be shared equally between developed and developing countries. The difficulty of domestically enacting the terms of an international agreement is an additional obstacle to international consensus. In this, the US has more in common with the emerging economies than industrialized ones, as climate change legislation looks unlikely given its polarized political system.

The point of dispute is not a divergence of opinion within the UNFCCC between G20 and non-G20 members. Much of the standstill is due to conflicting interests between G20 nations that (as will be discussed later) have differing views of the appropriate balance of economic development and responsibility for mitigating climate change. As a result, G20 meetings have done little to give momentum to climate change negotiations beyond general reassurances that G20 members are committed to pursuing an international agreement.

Reducing Energy Subsidies

Another issue that the G20 has raised is tackling the energy subsidies that are prevalent around the world, but particularly in developing economies. In its 2010 World Energy Outlook the IEA included a thorough assessment, at the request of the G20, of existing energy subsidies around the world and the implications of subsidy rationalization. According to the IEA, in 2009 consumption subsidies worldwide totaled $312 billion.[32] Four of the five biggest subsidizers of fossil fuel consumption were G20 nations: Saudi Arabia, Russia, India, and China.[33]

Many emerging G20 economies, including South Africa, China, India, Indonesia, and Mexico, have made initial steps toward a more market-based pricing structure for energy products. However, despite the IEA's and the G20's efforts this progress has been marginal. In many cases, early initiatives to implement market pricing for fuels has been met by fierce public opposition, forcing governments to cut back on their proposals. Even if reforms are implemented, enforcement is not guaranteed. In June 2010, India agreed to allow the price of gasoline to float at market rates; however, Indian oil marketing companies still ask for the approval of the Ministry of Petroleum and Natural Gas before raising prices and in some cases they have been asked to forego requested price increases.[34]

The inability of the G20 to successfully encourage the move toward a rational energy pricing structure, which would benefit all parties, illustrates a growing divide within the group that is scarcely addressed but is the result of a shift in the global energy landscape that has occurred in recent years.

The Changing Global Energy Landscape

Concurrent to the G20's rise as a central economic forum for the world's leading policy-makers were three major trends that would help transform the global energy landscape:

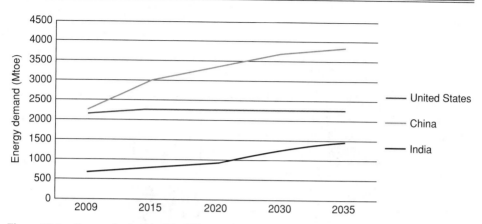

Figure 12.1 Energy demand in US, China, and India, 2009–2035.
Source: Based on International Energy Agency data.

the emergence of new energy consumers, the stagnation, if not absolute peak, in Western energy demand, and the "revolution" in US hydrocarbon production. The confluence of these trends will have a number of impacts on fundamentals of world energy supply and demand as well as on the ability of international groupings such as the G8 or G20 to reach substantive progress.

The changes in the global energy landscape are astounding. According to the IEA, non-OECD countries are projected to account for 93% of global energy demand growth between 2008 and 2035. Asia's energy demand is expected to double over the same period, with China and India alone accounting for roughly one third and one fifth, respectively, of the total increase in global demand. In contrast, the OECD's share of global energy consumption, which stood at nearly 70% in 1965, is set to fall below 40% by 2035 (see Figure 12.1).[35] Net declines in oil consumption in the OECD, coupled with increasing domestic production in the US and other Western Hemisphere countries, are accelerating the shift away from a world characterized by westbound oil and gas to one in which Asia becomes the major demand center for Middle East energy exports.

The consequences of this tectonic shift in traditional energy producer–consumer relations are profound. They include a potential change in the US presence in the Middle East, a change in the definition of "energy security," and a reexamination of the role of governments and markets in energy consumption. The role of the "Gs" and other international fora will have to adapt to the dramatic changes occurring in the global energy market in order to have relevance.

The Potential Changing Role of the US in the Middle East[36]

Three concurrent trends are limiting the US reliance on energy imports from the Middle East: flat or declining oil consumption, increasing domestic production, and increased production in the Western Hemisphere.

As a result of increasing vehicle efficiency and modest macroeconomic growth prospects, US oil consumption is set to remain level and possibly decline in the coming years. According to the US Energy Information Administration, oil demand (including natural gas liquids and biofuels) was nearly 21 million barrels per day in 2007 but has since dropped to less than 19 million barrels per day in 2011.[37] The IEA forecasts that US

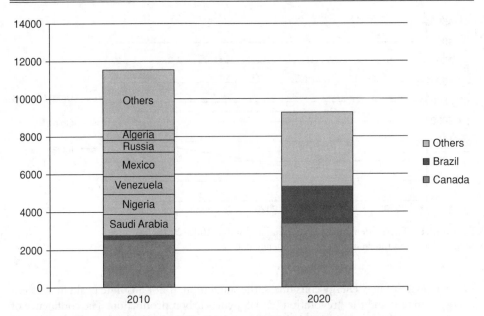

Figure 12.2 Current and potential US oil imports, 2010 and 2020.
Source: Based on data from US Energy Information Administration, National Energy Board (Canada), and Petrobras.

oil demand has peaked and will decline by nearly 1% per year to 2035.[38] This demand moderation coincides with a rapid increase in US hydrocarbon production. The hydraulic fracturing process that has yielded dramatic increases in natural gas production is helping boost domestic production of shale oil, with some estimates that US oil production could increase by more than 200,000 barrels per day annually until 2020. US oil production is currently at historically high levels, and the country's import dependence – the proportion of oil it gets from overseas – has dropped below 50% for the first time in more than a decade.

Finally, the resurgence in US production is being paralleled by increasing production from near neighbors, most notably Canada and Brazil (see Figure 12.2). As a result of Canada's oil sands and Brazil's pre-salt oil reserves, the two could export as much as 5 million barrels of oil per day. Together, these three factors (flat oil demand, increasing domestic production, and increasing production in the Western Hemisphere) will likely limit the amount of oil imported from the Middle East to the US, which, historically, has been a major rationale in the US military and strategic presence in the region.

The consequences of decreased US oil dependency on the US presence in the Middle East are unclear. Even if the US is self-sufficient in energy production, the world's largest oil consumer will still be exposed to a global market and will have an interest in securing supply routes and preventing disruptions. The US also has substantial major non-energy security interests in the region that are likely to justify a continued presence both militarily and diplomatically.

However, in a period of strained military budgets, fatigue among the US public with engagement in the Middle East, and a focus on Asia as the center of strategic priority, there is likely to be pressure on the US to curtail either its presence in the region or to share the burden of providing security. Whether China and others act to complement or compete with the US as security guarantors for Middle East oil and gas will have wide-ranging strategic implications for the region and for global energy markets.

New Definitions of "Energy Security"

During the post-World War II period of economic development in the US and Western Europe, "energy security" was defined as the ability to secure adequate amounts of energy – principally oil – at prices that enabled and facilitated an increasingly affluent, mobile, consumer society. With the eastward shift in consumption patterns, this definition is changing. To the governments of China, India, and other countries in emerging Asia, energy security is underpinned by a focus on "energy access" – ensuring at least basic energy services for all citizens. For these countries, energy is a matter of survival, and is an essential component of economic development and political stability. The new rationale for energy security is already shaping relations in the multilateral arena where major developing economies see emissions caps as a limitation on their right to pursue industrialization and economic growth, and in geopolitical negotiations, where Western-led restrictions on oil imports from Iran are facing opposition from China and India.

For Middle East producers, energy security is defined primarily as security of demand. However, the nature of shifting consumption from West to East holds risks. In the past suppliers were able to diversify exports to a basket of consumers from both industrialized and developing economies. The reduction in demand from the West means that the major producers from the region are going to be increasingly dependent on emerging, and often more volatile, economies for oil and gas exports. Economic uncertainty in China and India, for example, could have more dramatic consequences for OPEC exports than in the past. And while the economies of both China and India are growing at rates that exceed any other major economy in the world, there are signs of potential economic turmoil in both countries. The Chinese government is concerned about a property bubble and the country is facing a potential demographic crisis. India's gridlocked political landscape has made much-needed economic and energy reforms impossible, which has cooled the interest of many foreign investors.

Reexamination of the Respective Roles of the State and the Market in Shaping Energy Flows and Consumption Patterns

The changing definitions of "energy security" are likely to result in a new set of actors in the realm of international energy trade and investment that will change the way energy is consumed. When the primary centers of energy demand were the US and Western Europe, energy consumption was largely market-driven, and energy was viewed as a commodity for efficient consumption. In developing countries, many of which subsidize fuel to promote economic development and stability, energy is viewed as a vital good that must be secured and supplied by the state at any cost (see Figure 12.3).

One major ramification of this trend is a reduced acceptance of the primacy of market mechanisms. Already the largest growing energy consumers, China and India, are skeptical of "free markets" and believe they need to be managed or moderated to some extent. Both engage in differing types of subsidy regimes, both explicit and implicit, to ensure that the domestic population has affordable access to energy. Such rejection of pure market principles is likely to have wide-ranging implications for international energy trade. For producer nations, the emergence of Asia as major energy consumer will mean interacting with a new set of counterparties. As government-backed companies,

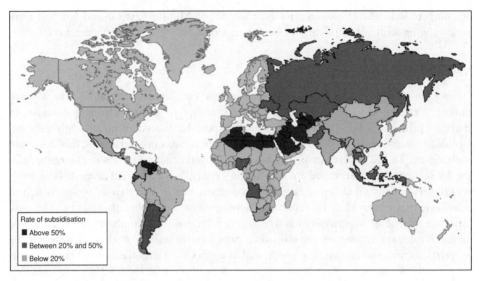

Figure 12.3   Rate of subsidization of fossil-fuel consumption, 2010.
Source: IEA (2010).

motivated partially by political and geopolitical considerations, compete directly with
private-sector companies, new business models, relationships, and potential sources of
conflict are likely to emerge.

The Rise of the Middle East as Energy Consumer

The Middle East can no longer be viewed as just an energy consumer. By 2020, the
Middle East will account for roughly one-third of the world's oil demand growth, as it
continues to fuel economic and population growth (see Figure 12.4).

The Arab Spring will make curbing this demand more difficult. In order to minimize the
risks of public discontent in their own countries, several major oil producers have implemented generous increases in social disbursements, subsidies, and other state-provided
benefits. This largesse, seen by many as an insurance policy against unrest, has placed
increasing pressure on public finances. This in turn has raised the price at which states
need to sell oil in the global market in order for their fiscal budgets to balance. According
to the International Monetary Fund, Saudi Arabia, the UAE, Iraq, and Iran all need a
price between $80 and $100 per barrel to break even in their domestic fiscal budgets.[39]

This new reality has manifested itself in statements from Saudi Arabia's oil ministry
that the Kingdom, the world's swing producer and *de facto* OPEC leader, is looking to
stabilize the global price around $100 per barrel.[40] In the longer term, the response of
oil and gas producers to regional instability is not sustainable: increasing subsidies and
social spending will simply extend the benefits that many citizens, particularly among
the younger generation, already take for granted. The communication technology-driven
demands for political reform among this group, which is less conscious of the "rentier"
state dynamics than the previous generation, are likely to continue irrespective of the level
of social spending. Indeed, to the extent that increased social spending entrenches the
status quo and prevents more fundamental reform of the economic and political system,
it may serve as an amplifier of domestic challenges.

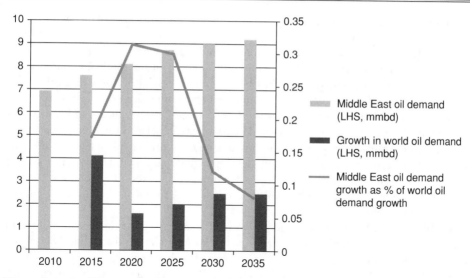

Figure 12.4   Middle East growth in oil demand relative to world oil demand growth.
Source: Based on International Energy Agency data.

What Do These Changes Mean for the Future of the "Gs" and Energy Policy?

In most cases, the aforementioned changes to the global energy sector will make it more difficult for the "Gs" to reach international consensus. The geographical flattening of energy consumption means that domestic politics in emerging economies will have a more important role in shaping future global energy trends in production and consumption. This also means that industrialized economies will have a lesser role, relative to before, in shaping international energy trade.

The radical change to the international energy landscape has not been kind to those hoping for sustained progress in international energy governance. The rise of new energy consumers driven primarily by economic development goals, coupled with dramatic increases in oil and gas production in the US, has fractured the traditional, convenient balance between industrialized energy importers and developing energy producers. The G8 and G20 as energy governance institutions have been directly affected by this transition. The G8 is no longer as relevant, and the G20 is far more disjointed. However, owing to its importance to international finance markets and the global economy, the G20 will remain a critical forum for national leaders to convene. For energy to be a successful part of this dialogue, rising above simply an acknowledgment of opposing views on sensitive energy policy issues, G20 members need to focus on three areas of improvement.

First, the industrialized nations in the G20 must recognize the shift in global energy consumption and the priorities of the new major energy consumers if the G20 is to overcome the schisms that exist today. The dominant energy consumers in the industrialized world relied more heavily on market signals to drive resource allocation and investment. The rise of emerging economies has shifted this preference for free-market fundamentals to more government control in the energy sector, particularly when it comes to rent allocation and fuel subsidies. While the traditional energy consumers of the OECD viewed "energy security" merely as reliable access to affordable energy resources, today's consumers in non-OECD nations view "energy security" as a more fundamental right to economic development and poverty alleviation. Global energy governance will have to adapt to these new realities.

Second, the G20 should be more targeted and realistic in its goals when tackling energy issues. For instance, while it is unwise to expect an international consensus on capping carbon emissions, the G20 has been more successful in improving oil market transparency and data collection. A sustained effort to minimize oil price volatility and maximize oil market liquidity is in the interest of all G20 oil producers and consumers. Regarding climate change, the G20 should build on its early efforts to develop and deploy carbon capture and sequestration programs around the world. While it has embarked on some early investments and technology development programs in this sphere, targeted focus on this issue would help accelerate the commercialization of a technology that is integral to a world that faces the dual (and conflicting) problem of a changing climate and energy poverty.

Finally, the G20 needs to develop a plan for the emergence of the next tier of major energy consumers. A number of countries, including Vietnam, Pakistan, and Nigeria, do not currently belong to the G20 but are projected to become large energy consumers within the next decade. How these countries will be integrated into the international energy governance framework will be critical. In many cases, the next tier of major energy consumers is likely to be less predictable than today's, which tend to have better developed domestic energy institutions.

With its economic clout, the G20 will remain a critical economic and financial institution for decades to come. However, for it to have a productive hand in guiding the future of energy governance in an increasingly multipolar world, it will need to recognize the tectonic shift in the global energy landscape, understand the implications of this change, and adapt to accommodate the new and emerging major energy consumers.

Notes

1. Subacchi and Pickford (2011).
2. The International Monetary Fund (IMF 2011a) estimates that in 2012 and 2013, the economies of the US, Japan, and the Eurozone will grow by an average annual rate of 2.16%, 2.17%, and 1.75%, respectively.
3. The final Leaders' Statement at the 2009 G20 Summit in Pittsburgh designated the G20 to be the premier forum for international economic cooperation (http://www.g20.utoronto.ca/2009/2009communique0925.html).
4. For an thorough background summary of the history of energy governance within the G8 and G20, see Van de Graaf and Westphal (2011).
5. Declaration of Rambouillet, November 17, 1975. (http://www.g8.utoronto.ca/summit/1975rambouillet/communique.html).
6. Ibid.
7. Tokyo Summit: Declaration, June 29, 1979. (http://www.mofa.go.jp/policy/economy/summit/2000/past_summit/05/e05_a.html).
8. Ebinger *et al.* (1982: 19).
9. Ibid.
10. Ibid., 20.
11. Declaration of Venice, June 23, 1980. (http://www.g8.utoronto.ca/summit/1980venice/communique/index.html).
12. Van de Graaf and Westphal (2011).
13. G8 Communiqué Okinawa 2000, July 23, 2000. (http://www.mofa.go.jp/policy/economy/summit/2000/pdfs/communique.pdf).

14. G8 Renewable Energy Task Force: Final Report, July 2001. (http://www.climate.org/PDF/g8_ren_energy.pdf).
15. G8 Communiqué Genoa 2001, July 22, 2001. (http://www.g8.utoronto.ca/summit/2001genoa/finalcommunique.html).
16. Florini (2009: 163). Also see Disarray in Genoa. *The Economist*, July 22, 2001.
17. G8 Communiqué Genoa 2003, June 3, 2003. (http://www.g8.utoronto.ca/summit/2003evian/communique_en.html).
18. Van de Graaf and Westphal (2011).
19. IEA (2008).
20. Kirton (2010) (http://www.g20.utoronto.ca/biblio/kirton-g20-g8-g5.pdf).
21. It is important to note that prior to the formalization of the G20 as an economic forum, nations experimented with the G22 and G33 for discussing financial stability and economic stabilization. It was deemed, however, that these fora were too ad hoc and inclusive to stimulate candid, productive dialogue and solutions (G20 2008: 16).
22. Ibid.
23. Ibid., 41.
24. According to the US Energy Information Administration.
25. Declaration of the Summit on Financial Markets and the World Economy, Washington, DC, November 15, 2008 (http://www.g20.utoronto.ca/2008/2008declaration1115.html).
26. London Summit, Leaders' Statement, April 2, 2009 (http://www.g20.utoronto.ca/2009/2009communique0402.pdf).
27. G20 Leaders Statement, Pittsburgh, September 24–25, 2009 (http://www.g20.utoronto.ca/2009/2009communique0925.html).
28. Information on the US oil import dependence is from the US Energy Information Administration. Information on Japan, South Korea, India, and China is from the IEA (2011).
29. See Goldthau and Witte (2010). Also see the September 2011 issue of *Global Policy*, an international academic journal, which published a special issue on global energy governance. Some articles in this issue (Bo Kong; Navroz Dubash; Antonia La Viña *et al.*) analyzed the internal politics of energy in emerging economies such as China, India, and The Philippines, and the impact internal politics have on international energy governance.
30. IEA (2010) included a thorough special report on the amount and implications of fossil fuel subsidies around the world.
31. UN (2012).
32. IEA (2010: 587).
33. Ibid., 579.
34. Oil Cost May Not Raise Fuel Prices Today." *The Hindu*, March 31, 2012.
35. Based on calculations of data from IEA (2011).
36. Some of this material has been adapted from conference proceedings from the Brookings Doha Energy Forum in Doha, Qatar, February 19–20, 2012.
37. According to the Energy Information Administration, US Product Supplied of Crude Oil and Petroleum Products.(http://www.eia.gov/dnav/pet/hist/LeafHandler.ashx?n=PET&s=MTTUPUS2&f=A).
38. IEA (2011: 107).
39. IMF (2011b).
40. Saudi Arabia Targets $100 Crude Price. *Financial Times*, January 16, 2012.

References

Ebinger, Charles, Wayne Berman, Richard Kessler, and Eugenie Maechling. 1982. *The Critical Link: Energy and National Security*. Washington, DC: Center for Strategic and International Studies.

Florini, Ann. 2009. Global Governance and Energy. In Carlos Pascual and Jonathan Elkind, eds. *Energy Security: Economics, Politics, Strategies, and Implications.* Washington, DC: Brookings Institution, pp. 148–183.

Goldthau, Andreas and Jan Martin Witte. 2010. *Global Energy Governance.* Washington, DC: Brookings Institution.

IEA. 2008. *National Reports on Global Energy Security Principles and St. Petersburg Plan of Action.* Paris: IEA.

IEA. 2010. *World Energy Outlook 2010.* Paris: IEA.

IEA. 2011. *World Energy Outlook 2011.* Paris: IEA.

IMF. 2011a. *World Economic Outlook Database.* Washington, DC: IMF.

IMF. 2011b. *World Economic and Financial Surveys: Regional Economic Outlook, Middle East and Central Asia.* Washington, DC: IMF.

Kirton, John. 2010. The G20, the G8, the G5, and the Role of the Ascending Powers. Paper presented to "Ascending Powers and the International System," a seminar in Mexico City, December 13–14 (http://www.g20.utoronto.ca/biblio/kirton-g20-g8-g5.pdf).

Subacchi, Paolo and Stephen Pickford. 2011. *Legitimacy vs. Effectiveness for the G20: A Dynamic Approach to Global Economic Governance.* Briefing Paper. London: Chatham House.

UN. 2012. *Report of the Conference of Parties on Its Seventeenth Session, Held in Durban from 29 November to 11 December 2011.* United Nations Framework Convention on Climate Change, March 15, FCCC/CP/2011/9/Add.1.

Van de Graaf, Thijs and Kirsten Westphal. 2011. The G8 and G20 as Global Steering Committees for Energy: Opportunities and Constraints. *Global Policy* 2 (September): 19–30.

Chapter 13

Nuclear Energy and Non-Proliferation

Yury Yudin

Introduction

From the outset of the nuclear age, the challenge has been to facilitate the peaceful use of nuclear energy while inhibiting the proliferation of nuclear weapons. But as Robert Oppenheimer, who is often called the "father of the atomic bomb" for his role in the building of the first nuclear weapons, observed, the close technical parallelism and inter-relation of the peaceful and the military applications of atomic energy make countering nuclear proliferation an especially difficult task. Both civilian and military applications of nuclear energy depend essentially on the same key ingredient: fissile materials. Peaceful and military applications of nuclear energy cannot be clearly separated because there are no clear technological barriers between producing fissile materials for peaceful or military applications. So-called "sensitive" nuclear technologies – uranium enrichment and plutonium separation – can provide enriched uranium or separated plutonium either for generating electricity or for explosive applications. In the absence of technological barriers, additional international institutional mechanisms for controlling access to sensitive materials, facilities or technologies may help separate the "Siamese twins" of atoms for peace and atoms for war.

This chapter starts by describing the current status and future prospects of nuclear energy worldwide and the challenges it would face from the nuclear non-proliferation standpoint. It then discusses some technical aspects of sensitive technologies used for the production of fissile materials, as it is impossible to grasp the problem of nuclear non-proliferation without understanding its core element: what are fissile materials and what proliferation-relevant technologies are used to produce them. After addressing the main elements of the existing international nuclear non-proliferation regime, the chapter discusses possible future multilateral approaches to mitigate nuclear proliferation risks stemming from further dissemination of sensitive nuclear technologies.

The Handbook of Global Energy Policy, First Edition. Edited by Andreas Goldthau.
© 2013 John Wiley & Sons, Ltd. Published 2013 by John Wiley & Sons, Ltd.

Nuclear Energy: Past, Current, and Future

Today nuclear power is an established part of the world's energy mix and the fourth largest source of electricity (after coal, natural gas, and hydro) providing about 12.4% (2.63 trillion kWh) of the world's electricity production and 5.3% of the commercial primary energy production in 2010 (IAEA 2011b: 19). For comparison, in 1995 more than 16% of electricity generated worldwide was derived from nuclear power (2.19 trillion kWh). While in absolute terms the production of nuclear electricity increased, this growth of about 20% over 15 years is rather modest when compared to the 60% rise in global electricity production for the same period.

Nuclear power generation grew rather rapidly from the early 1960s until the mid-1980s. There have been two major peaks in the connection of nuclear power reactors to the electrical grid: in 1974 with 26 reactor start-ups and in 1984 and 1985 with 33 grid connections each year (IAEA 2011b: 82). However, the escalating start-up costs coupled with growing public opposition after the Three Mile Island accident of 1978 and the Chernobyl accident of 1986 have led to a global slowdown in nuclear power plant construction. As of December 2011, 435 nuclear power reactors were in operation globally with a total net installed electrical generating capacity of 368 GW(e). Sixty-five plants are under construction in 14 countries and Taiwan, to have a net capacity of 63 GW(e) (IAEA 2011d).

Since about 2001 there has been much talk that nuclear energy is poised for a global expansion, or "renaissance." What factors are driving this change? The first factor is energy security. Limited supply, uneven distribution, and the volatile prices of fossil fuels have created a need to pursue alternative cost-competitive energy sources including nuclear power. The second factor is the increased awareness of the dangers of anthropogenic climate change from the emission of carbon dioxide and other greenhouse gases. Over the last decade, the nuclear industry has been successful at presenting the environmental case for nuclear power as a clean, "emission-free" energy source. Although nuclear power is not absolutely "emission-free," it indeed produces fewer greenhouse emissions than power plants that burn fossil fuels. The third factor is prestige. Civilian nuclear energy has often been identified as a symbol of modernity and economic progress, despite the fact that almost all nuclear power technologies used today were developed in the 1950s and 1960s. Building nuclear reactors and fuel-cycle facilities could be perceived by some states as a way to "catch up" with more developed states and to increase their rank, role, and prestige at the international level. Recently some 60 states have turned to the International Atomic Energy Agency (IAEA) for guidance as they consider whether to introduce nuclear power to their energy mix, and some of them are actively developing the appropriate technical, regulatory, and safety infrastructure. IAEA Director General Yukiya Amano stated in February 2011 that "We expect between 10 to 25 new countries to bring their first nuclear power plant online by 2030" (IAEA 2011c).

The severe nuclear accident at the Japanese Fukushima power plant in March 2011 certainly dealt a blow to the nuclear power industry. Initially, it seemed that the Fukushima accident might harshly constrain future growth of civilian nuclear power. Soon after Fukushima a number of nations with nuclear power reactors, including Belgium, Germany, and Switzerland, announced that they would gradually withdraw from nuclear energy, while some potential newcomers indicated deferment of their plans to go nuclear. However, after a period of reconsideration, it is clear that nuclear power is far from dead. A recently updated report by the World Nuclear Association (2011) says that nuclear power is still under serious consideration in over 45 countries that do not have it, mostly in

Asia, the Middle East, and Africa. This global trend, when countries with well-developed nuclear power programs are slowing or shutting them, while states that have only a few nuclear reactors or do not have them at all are pushing to build more, is not new. But the post-Fukushima developments have made it more prominent.

This would lead to even more of a focus on nuclear nonproliferation. First, countries planning nuclear reactors are also making choices about their fuel supply and fuel-cycle options. An anticipated expansion of nuclear power could potentially result in the further dissemination of sensitive nuclear technologies. More and more states could acquire the capability to produce materials directly usable for, or easily converted to, nuclear weapon capability. Second, many of the potential newcomers to nuclear power are located in politically turbulent regions with historical animosities and the lack of trust among states. Among the key factors that could drive the desire of states in these regions to obtain sensitive nuclear know-how could be the intention to diversify security capacities by maintaining a viable option for the relatively rapid acquisition of nuclear weapons, based on an indigenous technical capacity to produce them, in case a political decision to develop actual nuclear deterrence is made. The anticipated deployment of new nuclear technologies, such as laser uranium enrichment, fast breeder reactors, and pyroprocessing, may also lead to an increase in the potential proliferation risks. These new technologies can produce fissile materials that are directly usable in nuclear weapons and they will present new challenges for the IAEA to safeguard and verify these technologies.

Fissile Materials and Dual-Use Nature of Nuclear Technology

Nuclear energy is released in nuclear reactions, processes that affect atomic nuclei, the dense cores of atoms. Ordinary chemical reactions, such as the burning of fossil fuels, by contrast, involve only the orbital electrons of atoms without affecting their nuclei. Nuclear fission is a nuclear reaction in which the nucleus of an atom splits into lighter nuclei. In the heaviest nuclei the fission process may occur spontaneously, but it may also be induced by irradiating nuclei with a variety of subatomic particles, such as neutrons, protons, deuterons, or alpha particles. Some nuclides are capable of undergoing fission after capturing neutrons of all energies, including slow low-energy neutrons. Those nuclides are referred to as "fissile material" to distinguish them from nuclides that are capable of fission only by sufficiently fast and energetic neutrons. The nuclear fission process produces radioactive products, several free neutrons, and releases significant amounts of energy. Under proper conditions, these product neutrons can react with additional nuclei and thus give rise to a chain reaction in which a large number of nuclei undergo fission and an enormous amount of energy is released. If the energy release from nuclear fission chain reaction is controlled and sustained over an extended period, as in a nuclear reactor, it can be used to generate electrical power for society's benefit. If uncontrolled, as in a nuclear weapon, the energy release occurs in fractions of a second and can produce tremendous explosive force.

Common fissile materials include uranium-233, uranium-235, and plutonium-239. Uranium-235 is the only fissile nuclide that occurs naturally while all others are manmade and have been present on Earth for only about the last 60 years. The uranium-235 isotope accounts for merely 0.71% of natural uranium, while 99.28% is uranium-238. Although the isotopes of an element have very similar chemical properties, their nuclear properties may be very different, as is the case with the uranium isotopes. The 0.71% uranium-235 is not sufficient to produce a self-sustaining fission chain reaction in nuclear weapons or light water reactors, the most prevalent commercial power reactors, which constitute more than 80% of the world's power reactors. For these applications, uranium must be "enriched", that is, the abundance of the fissile uranium-235 isotope must be increased. According to the classification used by the IAEA, low enriched uranium (LEU) is uranium containing less than 20% of the isotope U-235, while high enriched uranium (HEU) is uranium containing 20% or more of the isotope U-235 (IAEA 2002: 31–32). Uranium is considered weapon-grade when it contains 90% or more of U-235 (Albright *et al.* 1993). However, theoretically, a nuclear explosive device could be constructed using enriched uranium with a U-235 fraction ranging from as low as 20% or even less, even if such a device would be more difficult to design and fabricate. The IAEA considers any HEU as direct-use material, which means that it "can be used for the manufacture of nuclear explosive devices without transmutation or further enrichment" (IAEA 2002: 33).

Light water reactors use uranium enriched to 3–5% uranium-235. Uranium enrichment is not needed for some types of commercial nuclear reactors, such as the Canadian-designed heavy water CANDU reactor, which uses natural uranium fuel. Special technologies are used to increase or "enrich" the proportion of fissile uranium atoms. The enrichment processes that have been demonstrated industrially or in the laboratory are: gaseous diffusion; thermal diffusion; gas centrifuge separation; atomic vapor laser isotope separation (AVLIS); molecular laser isotope separation (MLIS); separation of isotopes by laser excitation (SILEX); aerodynamic isotope separation; electromagnetic isotope separation; plasma separation; and chemical separation. Of these only two – the gaseous diffusion process and the gas centrifuge process – are currently operating on an industrial scale, with centrifuge enrichment technology gradually replacing the outmoded and non-competitive gaseous diffusion process. Two other uranium enrichment processes – electromagnetic isotope separation and aerodynamic isotope separation – have been used in the past to produce fissile uranium for nuclear weapons. In June 2009, General Electric submitted to the US Nuclear Regulatory Commission a license application to construct the first commercial-scale laser enrichment plant using the SILEX process.

There is the obvious proliferation risk attendant to uranium enrichment. There is no technological barrier between the production of LEU for nuclear reactors and HEU for weapons. Weapon-grade material can be produced using the same enrichment equipment that otherwise is used to produce LEU for civilian power generation. The risk is compounded by the fact that centrifuge enrichment facilities – and probably future laser enrichment facilities – are difficult to detect using satellite imagery or any other method of observation because their physical buildings are inconspicuous, they consume modest amounts of electricity (about 30 to 40 times less per SWU[1] than gaseous diffusion facilities), and the atmospheric emissions from these facilities are nearly non-existent. Currently, there are 14 states with military or civil uranium enrichment capabilities: Argentina, Brazil, China, the Democratic People's Republic of Korea (DPRK), France, Germany, India, Iran, Japan, the Netherlands, Pakistan, Russia, the UK, and the US.

Uranium enrichment is one of two developed methods to produce fissile material for nuclear explosive applications. The other is the production of plutonium. The uranium-238 isotope, which comprises most of the fuel mass in a nuclear reactor, is converted into heavier isotopes through nuclear reactions. This is how plutonium-239 and other isotopes of plutonium, including plutonium-240, plutonium-241, and plutonium-242, are formed in the process of nuclear reactor operation. The presence of various plutonium isotopes determines the material's grade. Weapon-grade plutonium, that is, the plutonium most suitable for the use in nuclear weapons, is defined as being predominantly plutonium-239, typically no less than 93% plutonium-239 (Albright *et al.* 1993). Plutonium containing larger quantities of other plutonium isotopes is usually referred to as "reactor-grade" plutonium. Certain practical obstacles such as greater heat emission, greater neutron emission, and higher exposure to radiation, make nuclear explosive devices made from reactor-grade plutonium more difficult to design, fabricate, and handle as compared to weapons-grade plutonium. However, Carson Mark, who headed the theoretical division of the Los Alamos National Laboratory for 27 years, emphasized that "the difficulties of developing an effective design of [a nuclear weapon of] the most straightforward type are not appreciably greater with reactor-grade plutonium than those that have to be met for the use of weapons-grade plutonium."[2] In fact, the IAEA classifies almost any isotopic composition of plutonium as direct-use material.[3] This effectively defines any plutonium discharged from commercial nuclear reactors as direct-use material. The isotopic composition of that plutonium, and thus its potential for use in a weapon, depends mainly on how the reactor is operated. The lower the fuel "burn-up" – that is, the shorter the duration of irradiation of nuclear fuel in the reactor – the higher is the percentage content of plutonium-239 and the more preferable this plutonium would be for weapons use. Commercial light water reactors can also be used for the production of weapons-grade plutonium, even if they are less suitable for this than reactors specifically designed for weapons programs.

Fuel discharged from a nuclear reactor after irradiation still contains most of the uranium-238 that was present in the fuel when charged, appreciable concentrations of fissile nuclides (uranium-235, isotopes of plutonium), other artificial elements (called "actinides") produced by nuclear reactions, and large amounts of radioactive fission products. Reprocessing technologies can be used for the chemical separation of plutonium and uranium from the radioactive fission products and actinides. Separated plutonium can be used to fabricate fuel – so-called mixed oxide uranium/plutonium fuel (MOX) – for different types of nuclear power reactors. However, it is also a key fissile component in nuclear weapons. As with uranium enrichment, there is no technological barrier between separating plutonium for peaceful purposes or for military purposes.

Separation of plutonium from spent nuclear fuel is less technologically sophisticated than enrichment of uranium and could be mastered more easily, but spent fuel reprocessing facilities are usually large and produce distinct signatures, such as emissions of certain radionuclides, so these facilities are easier to detect. Spent fuel reprocessing also involves serious environmental and safety risks as reprocessing facilities handle large amounts of highly radioactive and dangerous materials. Today, there are nine states with military or civil spent fuel reprocessing capabilities: China, the DPRK, France, Israel, India, Japan, Pakistan, Russia, and the UK. Several countries, for example Belgium, Germany, Italy, and the US, have had reprocessing capabilities in the past, but have had them shut down or reoriented to other missions, such as treating radioactive wastes.

Thus, obvious proliferation risks derive directly from the "dual-use" nature of certain nuclear technologies: uranium enrichment, reprocessing of spent fuel, and handling of plutonium, including manufacture of MOX fuel. The IAEA (1979) classified them as "sensitive technological areas.". These technologies are required for nuclear power generation, but at the same time they can provide states, and even non-state actors, with weapon-usable materials. The interchangeability and interdependency of military and civilian nuclear production cycles is illustrated by Figure 13.1. Historically, nearly all principal technologies of the uranium fuel cycle were developed in dedicated nuclear weapon programs by the US, the Soviet Union, and the UK before civilian nuclear energy existed. Only after those technologies and the fissile materials they produced had been used to make the atomic bomb did politicians, scientists, and engineers start to ponder seriously their peaceful applications. Today, more than 60 years later, over much of

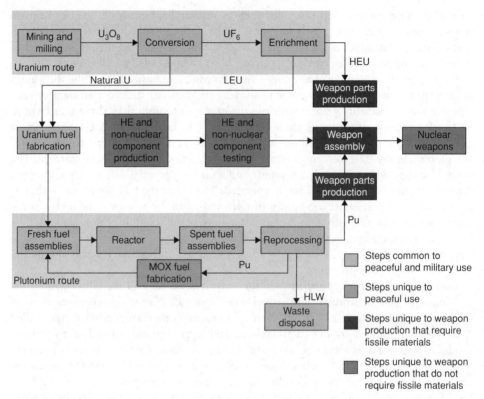

Figure 13.1   Overlap of nuclear energy and nuclear weapons production cycles. HLW = high level waste.
Source: Yudin (2010).

their course military and civilian nuclear production cycles still use identical, or nearly identical, materials, technology, and equipment.

At the dawn of the nuclear age, it was believed that some nuclear technologies, in particular uranium enrichment, were forbiddingly complex and costly to be mastered by the overwhelming majority of would-be proliferators. However, over the past decades the world witnessed a steadily accelerating dissemination of nuclear knowledge and technology. This, together with the rapid growth in dual-use technology applications and globalized trade, has made the sensitive nuclear technologies more accessible to less advanced countries. In recent decades, states with a relatively limited scientific and industrial base such as the DPRK, Iran, and Pakistan have proven capable of mastering sensitive nuclear technologies. The nuclear programs of those states have highlighted a worrisome trend. While those states have not established commercially competitive enrichment or reprocessing programs because of economic or technological constraints (their programs have been based on out-of-date technologies that cannot compete with established suppliers of nuclear services), they have been able to acquire technologies that provide, or can provide, them with materials that are directly usable in a nuclear weapon or a nuclear explosive device. There is considerable overlap between the basic scientific disciplines required for a civil nuclear program and a weapons-related nuclear program. Such disciplines include physics, chemistry, mathematical and computing science, nuclear engineering, chemical engineering, mechanical engineering, and others. Civil nuclear programs would certainly add to the state's dual-use nuclear expertise and experience through training the personnel, getting assistance from other states and the IAEA, and setting up nuclear research centers furnished with research reactors, hot cells, and other equipment.

It is worth noting that designing and manufacturing a nuclear explosive device and designing, building, and operating a commercial nuclear reactor or a fuel-cycle facility require different scientific knowledge and technical expertise. While these differences could present certain barriers for a would-be proliferator, they should not be overestimated, especially when a proliferator's goal is to design and build relatively unsophisticated nuclear weapons. In 1964, scientists at the Lawrence Livermore National Laboratory, California launched an exercise called the Nth Country Experiment to find out if "nuclear innocents" could design a working nuclear device. They hired two young physicists, who held no security clearance and had no experience with nuclear weapons, and charged them with designing a nuclear explosive device with a militarily significant yield without access to any classified information. The two scientists were supposed to represent an imaginary country with fewer resources than an industrialized nation, but which met certain minimum requirements, such as having "a good university library, some competent machinists to shape plutonium or uranium, and an explosives team" (Stober 2003: 58). Seven months after they began, the physicists realized that it was so easy to make a gun-type weapon with high enriched uranium (HEU) that it did not present any real challenge and thus would not help to build their reputations. The two scientists picked a plutonium implosion design specifically because it would present a greater challenge. By April 1967, they had designed a working plutonium device that was "run through the computers and brains of the bomb designers" (Stober 2003: 61), which confirmed that it would function. Designing a nuclear explosive device would be easier today than it was in the 1960s because the information revolution has facilitated the dissemination of knowledge, and computing tools have become much more powerful and affordable. Thus, the greatest technological barrier to the manufacture of a nuclear explosive device is the acquisition of fissile material, mainly in the form of HEU or separated plutonium.

The Global Nuclear Non-proliferation Regime

The international nuclear non-proliferation regime is a complex system of multilateral and bilateral agreements, arrangements, and mechanisms intended to prevent the spread of nuclear weapons and ultimately to achieve a world without nuclear weapons. At the same time, the regime is intended to provide a framework to enable states to foster the peaceful uses of nuclear energy. The Treaty on the Non-Proliferation of Nuclear Weapons (NPT) is a cornerstone of the regime and is commonly described as having three main "pillars": non-proliferation, peaceful use of nuclear energy, and disarmament. The treaty is supported by nuclear-weapon-free zones. In accordance with the NPT provisions, the international community has entrusted the IAEA with carrying out independent inspections of all nuclear material and facilities subject to safeguards agreements in order to verify compliance of states with their non-proliferation commitments. Through establishing and administering safeguards, the IAEA seeks to verify that states do not use nuclear material or equipment to acquire nuclear weapons. Two major nuclear export control regimes, the Zangger Committee and the Nuclear Suppliers Group (NSG), stipulate guidelines on the kind of sensitive materials and technologies that can be transferred across borders.

The Nuclear Non-Proliferation Treaty

The NPT is a founding document of multilateral non-proliferation endeavors and the cornerstone of the international nuclear non-proliferation regime. When opened for signature in 1968 after three years of negotiation in the then Eighteen-Nation Disarmament Committee (the predecessor of the current Conference on Disarmament), the treaty was intended primarily to stop the spread of nuclear weapons beyond the five states that had "manufactured and exploded a nuclear weapon or other nuclear explosive device prior to 1 January 1967", these being France, China, the Soviet Union (whose obligations and rights are now assumed by Russia), the UK, and the US. No international arms control treaty is as widely adhered to as the NPT. Today, treaty membership stands at 189 states. Only four states remain outside of the NPT: India, Israel, and Pakistan, which have not signed it, and the DPRK, which announced withdrawal from the treaty in 2003, although the validity of the withdrawal is debated. The NPT has been remarkably successful in limiting, albeit not entirely preventing, the further spread of nuclear weapons. This success can be judged by what might have happened had the treaty not existed: the spread of nuclear weapons would probably have been uncontrollable. During the 1950s and 1960s many experts predicted a rapid increase in the number of states possessing nuclear weapons. At one point or another, Argentina, Brazil, Egypt, Germany, Iraq, Japan, Libya, Poland, Romania, South Africa, South Korea, Sweden, and Switzerland had nuclear weapons-related programs. However, only four states have developed nuclear weapons since 1970; all of them are non-NPT states.

The NPT legitimizes, at least temporarily, the nuclear arsenals of the five nuclear-weapon states and sets different rights and obligations for the two groups of its states parties, nuclear weapon "haves" and "have-nots." The treaty represents a complex compromise, with certain weaknesses, ambiguities, and contradictions inherent to any compromise, reached after extensive debates between the nuclear-weapon states and the non-nuclear-weapon states. The NPT rest on three fundamental issues, or "pillars":

- Pillar 1: *non-proliferation*. Each nuclear-weapon state party undertakes not to transfer "nuclear weapons or other nuclear explosive devices" or control over them and "not in any way to assist, encourage, or induce" a non-nuclear-weapon state party to acquire nuclear weapons (Article I). Non-nuclear-weapon states parties agree not to receive, manufacture or acquire nuclear weapons or to "seek or receive any assistance in the manufacture of nuclear weapons or other nuclear explosive devices" (Article II). Non-nuclear-weapon states parties also agree to accept safeguards by the IAEA to verify that they are not diverting "nuclear energy from peaceful uses to nuclear weapons or other nuclear explosive devices" (Article III);
- Pillar 2: *peaceful use of nuclear energy*. The NPT recognizes "the inalienable right of all the Parties to the Treaty to develop research, production and use of nuclear energy for peaceful purposes without discrimination," in conformity with their non-proliferation obligations (Article IV.1). As a result, all states parties undertake "to facilitate, and have the right to participate in, the fullest possible exchange of equipment, materials and scientific and technological information for the peaceful uses of nuclear energy" (Article IV.2); and
- Pillar 3: *disarmament*. "Each of the Parties to the Treaty undertakes to pursue negotiations in good faith on effective measures relating to cessation of the nuclear arms race at an early date and to nuclear disarmament" (Article VI).

The NPT's non-proliferation Articles I and II are relatively unambiguous. Non-nuclear-weapon states are not to acquire nuclear weapons in any way and nuclear-weapon states are not to assist them in acquiring nuclear weapons in any way whatsoever. Where the interpretation of the articles becomes clouded is in the word "manufacture." Does the NPT commitment not to manufacture nuclear weapons incorporate a prohibition on all, or some, related activities, such as research and development applicable to nuclear weapons design, production of weapon-usable materials, and fabrication of components? Or is it applicable only to the final assembly of a nuclear explosive device? The exact meaning and scope of "manufacture" remains undefined.

The delegations participating in the NPT's drafting were aware that prohibiting only the final stage of the "manufacture" of nuclear weapons could be insufficient. When in 1965 the Soviet Union publicly presented its first draft of a non-proliferation treaty, article I of the draft obliged nuclear-weapon states not to assist non-nuclear-weapon states in the manufacture, in preparation for the manufacture or in the testing of such weapons and *not to transmit to them any kind of manufacturing, research or other information or documentation which can be employed for purposes of the manufacture or use of nuclear weapons* (UN 1965). In February 1966 Swedish ambassador Alva Myrdal argued before the Eighteen-Nation Disarmament Committee:

> We could, of course, all agree that it is important to block the road to nuclear-weapon development as early as possible. But we must be aware that what we are facing is a long ladder with many rungs, and the practical question is: on which of these is it reasonable and feasible to introduce the international blocking? [...] To prohibit just the final act of "manufacture" would seem to come late in these long chains of decisions. (UN 1966: 11–12)

However NPT negotiators were unable to come to an agreement on what constituted the term "manufacture" and where on this "long ladder with many rungs" such international blocking should be introduced, because many states opposed the clause calling for a prohibition of transfer of dual-use knowledge and technology, fearing that such prohibition would have negative repercussions on their ability to master nuclear technology. By 1968, these states had successfully lobbied for full access to nuclear knowledge and technology that could be considered peaceful.

This NPT's entitlement, referred to as an "inalienable right," has sparked many controversies. Some governments have interpreted Article IV as implying a "sovereign" right to nuclear activities, in other words an absolute and unconditional right. They view this article as giving non-nuclear-weapon states a right to engage in development of any nuclear technology short of actual weaponization, provided that a state promised to use nuclear energy exclusively for peaceful purposes and concluded a comprehensive safeguards agreement with the IAEA. Their opponents argue that the acquisition of certain nuclear technologies – preeminently uranium enrichment and spent fuel reprocessing – would bring states a long way toward nuclear weapons by shrinking the time needed to manufacture them even if those states do not directly violate the IAEA safeguards. They also argue that the treaty can be interpreted as "not recognizing the 'inalienable right' of signatories to those nuclear materials, technologies, and activities that the IAEA cannot effectively safeguard" (Zarate 2008: 226).

Although the claim that enrichment and reprocessing facilities are difficult to effectively safeguard against diversion of nuclear material is basically correct, nothing in the text of the NPT can be unequivocally interpreted as prohibiting the development and operation of domestic enrichment and reprocessing facilities. Moreover, the history of

NPT negotiations shows that negotiating parties had no intention to impose such a prohibition. William Foster, the US representative in the Eighteen-Nation Committee on Disarmament, stressed that "a non-proliferation treaty which prohibits the manufacture or acquisition of nuclear explosives would not restrict the dissemination of application technology in any fashion" (UN 1967: 9). In an *aide-mémoire* of November 17, 1967 the Swiss Government interpreted Articles I and II as not covering "exploitation of uranium deposits, enrichment of uranium, extraction of plutonium from nuclear fuels, or manufacture of fuel elements or heavy water, when these processes are carried out for civil purposes" (US Arms Control and Disarmament Agency 1969: 81).

The inalienable right of NPT states parties to use nuclear energy for peaceful purposes has been reaffirmed on numerous occasions. The final document of the 2010 Review Conference of the Parties to the Treaty on the Non-Proliferation of Nuclear Weapons stated:

> The Conference reaffirms that nothing in the Treaty shall be interpreted as affecting the inalienable right of all the parties to the Treaty to develop research, production and use of nuclear energy for peaceful purposes without discrimination and in conformity with articles I, II, III and IV of the Treaty. The Conference recognizes that this right constitutes one of the fundamental objectives of the Treaty. [. . .] In this connection, the Conference confirms that each country's choices and decisions in the field of peaceful uses of nuclear energy should be respected without jeopardizing its policies or international cooperation agreements and arrangements for peaceful uses of nuclear energy and its fuel cycle policies. (UN 2010: para. 31)

By requiring respect for the NPT states parties' fuel-cycle choices and policies, the language of the final document further reaffirms their right to engage in dual-use nuclear activities, including uranium enrichment and spent fuel reprocessing.

The ambiguity around the legal meaning and technical modalities of the inalienable right to the peaceful application of nuclear energy creates the tension between the non-proliferation and the peaceful use of nuclear energy pillars of the NPT. While the primary purpose of the NPT is to halt the further spread of nuclear weapons, the treaty allows states parties to engage in diverse nuclear activities short of actual weaponization. This leads to a growing number of what former IAEA Director General Mohamed ElBaradei referred to as "virtual nuclear weapons states" (IAEA 2006). Such states have developed the capability to produce plutonium or HEU and possess the knowledge and non-nuclear materials and components needed to manufacture nuclear weapons. But they stop short of assembling nuclear weapons. They would therefore remain technically compliant with the NPT while maintaining the latent capability for the rapid acquisition of nuclear weapons. The spread of the latent capability for the rapid acquisition of nuclear weapons under one NPT pillar, which the NPT actually does little to restrict, comes into collision with the commitment to forsake such weapons under another NPT pillar, which would weaken the non-proliferation regime as a whole.

Nuclear-Weapon-Free Zones

Among the most important of the additional multilateral legal instruments of the international non-proliferation regime are nuclear-weapon-free zone (NWFZ) treaties. The concept of NWFZs was devised to prevent the emergence of new nuclear-weapon states. Article VII of the NPT affirms the right of countries to establish specified zones free of nuclear weapons. The General Assembly of the United Nations reaffirmed that right in 1975 and outlined the criteria for such zones. A NWFZ is any zone, recognized by the UN General Assembly, which a group of states, in the free exercise of their sovereignty, has established by virtue of a treaty or convention, in which states parties commit themselves not to develop, manufacture, test, or deploy nuclear weapons in a given area. A NWFZ has to have a system of verification and control to guarantee compliance with the obligations (UN 1975).

Five NWFZs exist today, with four of them spanning the entire Southern Hemisphere: Latin America and the Caribbean (the 1967 Treaty of Tlatelolco); the South Pacific (the 1985 Treaty of Rarotonga); Southeast Asia (the 1995 Treaty of Bangkok); Africa (the 1996 Treaty of Pelindaba); and Central Asia (the 2006 Treaty of Semipalatinsk). In addition, in 2000 the UN General Assembly recognized Mongolia's status as a one-state NWFZ. The 1959 Antarctic Treaty also forbids the deployment and testing of nuclear weapons in the Antarctic. The 1967 Outer Space Treaty prohibits the stationing of nuclear weapons in outer space, on the moon or on other celestial bodies, while the 1971 Seabed Treaty prohibits the placement of nuclear weapons on the seabed.

In contrast to the NPT, NWFZ treaties do not permit the "stationing" of nuclear weapons on the territories of states parties or "nuclear sharing" arrangements among nuclear-weapon and non-nuclear-weapon states. The most recent NWFZ treaties, for example the Treaty of Pelindaba and the Treaty of Semipalatinsk, incorporate more stringent and less ambiguous non-proliferation provisions as compared to the NPT. These treaties obligate their states parties "not to conduct research on, develop, manufacture, stockpile or otherwise acquire, possess or have control over any nuclear weapon or other nuclear explosive device by any means anywhere." All NWFZs seek to promote international cooperation in the peaceful uses of nuclear energy. Similarly to provisions of Article III of the NPT, each state party of the NWFZ treaties is required to conclude a comprehensive safeguards agreement with the IAEA, but the Treaty of Semipalatinsk goes further by requiring its states parties to adopt the Additional Protocol to their safeguards agreements, which provides for expanded verification.

IAEA Safeguards

The principal organizational embodiment of the international nuclear non-proliferation regime is the IAEA, which was established in 1957 as the world's "Atoms for Peace" organization within the UN family to promote safe, secure, and peaceful use of nuclear technologies. As of November 2011, the IAEA had 152 member states. The IAEA is endowed by its statute with the authority to establish, administer, and apply safeguards to verify that safeguarded nuclear material and facilities are not used for military purposes. Today, the IAEA safeguards system remains the only mechanism expected to provide credible assurance to the international community that nuclear material and other specified items are not diverted from peaceful uses.

The IAEA safeguards system has undergone substantial changes since its inception. Throughout the 1960s, all safeguards agreements were voluntary arrangements initiated by a state or group of states that covered only the nuclear material, facilities, equipment, and non-nuclear materials specified in an agreement. Since 1968, such voluntary safeguards agreements have been based on the provisions in document INFCIRC/66/Rev.2. The IAEA currently implements voluntary safeguards agreements in three non-NPT states: Israel, India, and Pakistan.

Pursuant to their obligation under the NPT, non-nuclear-weapon states agree to accept safeguards for the exclusive purpose of verification of the fulfillment of their non-proliferation commitments. In 1972 the IAEA approved INFCIRC/153(Corr.), a so-called "comprehensive safeguards agreement" that enacted safeguards on all of a state's source or special fissionable material in all peaceful nuclear activities within its territory, under its jurisdiction, or carried out under its control anywhere. In 2010, 167 states had comprehensive safeguards agreements in force with the IAEA (IAEA 2010: 79). Although these agreements are comprehensive, they are not complete in the sense that the safeguards are actually applied to declared nuclear material and facilities. Under comprehensive safeguards agreements, the agency has the authority to verify the absence of undeclared nuclear activities, but the tools available to it to do so, under such agreements, are limited.

Prompted by the discovery of clandestine nuclear weapons-related activities in Iraq and the DPRK in the early 1990s, the IAEA approved the Model Additional Protocol (INFCIRC/540) in May 1997. Additional protocols are not stand-alone documents. They, by definition, are additional to existing safeguards agreements. Additional protocols equip the IAEA with additional legal authority, more information, greater rights of access, and enhanced technical tools to verify the correctness and completeness of states' declarations under comprehensive safeguards agreements. However, states are not obliged to sign or bring into force additional protocol agreements with the IAEA, so today those agreements remain voluntary. As of December 31, 2010 only 99 of 184 non-nuclear-weapon states have brought additional protocols into force (IAEA 2010: 79). Some states with significant nuclear activities (for example Argentina, Brazil, Egypt, Iran, Syria, and Venezuela) have refused either to sign or enforce an additional protocol for various national reasons.

Despite all the measures that have been taken since the early 1990s to strengthen the IAEA safeguards, the extent to which the agency can actually effectively safeguard nuclear materials and nuclear facilities – especially so-called bulk-handling facilities where "nuclear material is held, processed or used in bulk form" (IAEA 2002: 42), such as plants for conversion, enrichment, fuel fabrication, and spent fuel reprocessing – remains debated.[4] In such facilities different forms of nuclear materials are present (gases, solutions, powders, pellets) and some forms are converted into others. Moreover, the ingoing and outgoing quantities of nuclear materials, such as plutonium in reprocessing and uranium in enrichment plants, can be measured only with an unavoidable and rather high degree of uncertainty. During bulk-handling facility operations, a significant amount of material inevitably becomes stuck inside processing equipment, piping, and filters. All these factors make the application of "materials accountancy as a safeguards measure of fundamental importance" (IAEA 1972: para. 29) rather difficult. A non-zero difference between the book inventory and the physical inventory of nuclear facilities can be expected for bulk-handling facilities even in the absence of diversion and material accountancy must use statistical methods to distinguish possible diversion from measurement uncertainty.

According to the document INFCIRC/153, a technical objective of IAEA safeguards is specified as the timely detection of the diversion of significant quantities of nuclear material from peaceful nuclear activities to the manufacture of nuclear weapons or of other nuclear explosive devices or for purposes unknown and deterrence of such diversion by the risk of early detection (IAEA 1972: para. 28). A significant quantity is defined as "the approximate amount of nuclear material for which the possibility of manufacturing a nuclear explosive device cannot be excluded" (IAEA 2002: 23). For HEU a significant quantity is judged to be 25 kg of uranium-235, and for plutonium 8 kg. Miller (1990) demonstrated that for large bulk-handling facilities, such as the Rokkasho Reprocessing Plant in Japan with an annual throughput of 800t of heavy metal, the minimum loss of nuclear material that can be expected to be detected by material accountancy would be 236 kg of plutonium, or about 30 significant quantities, if physical material balance inventories were performed on an annual basis.

The IAEA does not rely exclusively on material accountancy. The agency routinely uses other techniques such as containment and surveillance, environmental sampling, design information verification, unannounced inspections, and so on. But as Pierre Goldschmidt, former IAEA Deputy Director General for Safeguards and Verification, admitted "there are still problems inherent in ensuring that, in 'bulk facilities', even small amounts of nuclear material – a few kilograms among tons – are not diverted without timely warning" (Goldschmidt 2008: 295). As for timely warning, it would be preferable to detect a diversion well before the perpetrator could assemble a weapon from the diverted material, in order to give the international community enough time to organize and coordinate some form of response. Thus, detection time for a given nuclear material should be shorter than the conversion time for that material. But the estimated time required to convert some forms of nuclear material into components of a nuclear explosive device is extremely short: 7–10 days for metallic plutonium and HEU; one to three weeks for pure non-irradiated compounds of these materials such as oxides or nitrates, or for mixtures; and one to three months for plutonium or HEU in irradiated fuel (IAEA 2002: 22). This does not allow time for much political maneuvering.

Other factors limit the effectiveness of international safeguards. The NPT does not have a built-in response mechanism for non-compliance. According to the IAEA statute, the agency's Board of Governors is to call upon the violator to remedy non-compliance with IAEA safeguards and should report non-compliance to the UN Security Council and General Assembly. The IAEA statute does not stipulate any specific deadlines for this reporting, so the Board could dwell on the decision to report. The UN Security Council may impose specific penalties, such as diplomatic and economic sanctions, but the violator could still choose to remain non-compliant. Moreover, any of the Security Council's five permanent members could use its veto power to delay or dilute sanctions.

Export Controls

Article III.2 of the NPT imposes the conditions of supply of nuclear items. Each state party to the Treaty undertakes not to provide: (a) source or special fissionable material, or (b) equipment or material specially designed or prepared for the processing, use or production of special fissionable material, to any non-nuclear-weapon state for peaceful purposes, unless the source or special fissionable material shall be subject to the safeguards required by this Article.

In 1971, the Zangger Committee was formed by a group of interested states to specify exactly what material and equipment are relevant under that clause. Three years later, the committee drafted a "trigger list" of applicable items that was published in IAEA document INFCIRC/209. The NPT states parties agree that any transfer of any of the listed materials to a non-nuclear-weapon state for peaceful purposes would "trigger" three conditions of supply: a non-explosive use assurance, an IAEA safeguards requirement on the transferred materials, and a retransfer provision to apply the same conditions if the materials are to be exported by the importing country. Today the Zangger Committee has 38 members, which participate in the regular revision of this trigger list. The list embraces nuclear materials, including plutonium and HEU, as well as nuclear facilities.

The Zangger Committee considers itself to be an informal organization, specifying that its export controls are not legally binding on the states involved. Rather, each state makes unilateral declarations to each other and to the IAEA to abide by such controls. The restrictions on export of sensitive nuclear materials and technologies are enforced by states through their national export control laws.

A more formalized export control regime called the Nuclear Suppliers Group (NSG) was created following the 1974 Indian nuclear test that revealed the ease with which nuclear materials and equipment allegedly acquired for peaceful purposes could be diverted to military use. As of December 2011, the NSG comprised 46 nuclear supplier states, which have voluntarily agreed to coordinate their export controls governing transfers of civilian nuclear and nuclear-related material, equipment, and technology to non-nuclear-weapon states. The NSG has established and regularly revises two sets of guidelines: guidelines for nuclear transfers (INFCIRC/254/Part 1), and guidelines for transfers of nuclear-related dual-use equipment, material, and technology (INFCIRC/254/Part 2). The NSG guidelines are more comprehensive than the Zangger Committee's list. To be eligible for importing items included in the Part 1 list from an NSG member, a state must have a comprehensive IAEA safeguards agreement in force. The NSG guidelines also include extra export conditions: physical protection measures, particular caution in the transfer of sensitive facilities, and a strengthening of the retransfer provisions of the Zangger Committee.[5]

In June 2011 the NSG issued a new set of guidelines governing the transfer of sensitive nuclear technologies. The NSG members have agreed on a "criteria-based" approach to such transfers. The transfer of enrichment and reprocessing facilities, equipment, and technology can be authorized only if a recipient is an NPT member state in a good standing with the IAEA and has brought into force a comprehensive safeguards agreement and an additional protocol based on the Model Additional Protocol or, pending this, "is implementing appropriate safeguards agreements in cooperation with the IAEA, including a regional accounting and control arrangement for nuclear materials, as approved by the IAEA Board of Governors" (IAEA 2011a). The latter clause effectively gives an exception to Argentina and Brazil, NSG members that refuse to bring into force an additional protocol.

At the NPT Review Conferences many states parties have expressed their support for effective and transparent export controls. But some non-nuclear-weapon states have criticized the discriminatory nature of export controls and claimed that they harm their economic development and the inalienable right to the peaceful use of nuclear energy. Another valid question pertains to the effectiveness of export controls, particularly given the nuclear weapon programs of the DPRK, India, Iraq, Israel, and Pakistan. The emergence of nuclear black markets, such as the operation set up by Pakistan's A. Q. Khan, has called into question the effectiveness and adequacy of export control regimes. In 2004, then IAEA Director General Mohamed ElBaradei stressed:

> The present system of nuclear export controls is clearly deficient. The system relies on informal arrangements that are not only non-binding, but also limited in membership, and many countries with growing industrial capacity are not included. Moreover, at present there is no linkage between the export control system and the verification system. (ElBaradei 2004)

Moreover, the geopolitical and economic interests of supplier states may be out of sync with non-proliferation concerns and powerful states could use their political and economic weight to pressure others, even through multinational bodies such as the NSG.

Under pressure from states such as France, Russia, the UK, and the US, for example, other members of the group agreed to give a waiver to India in 2008, thus not requiring India to have comprehensive IAEA safeguards as a precondition for NSG members to export nuclear material and fuel for use in Indian civilian nuclear facilities.

Even the most effective export control regime, which can impede transfer of sensitive nuclear technologies from established suppliers or their illicit acquisition from the "black market," cannot prevent indigenous development of these technologies.

Need for New Institutional Mechanisms

The absence of technological barriers between the production of fissile materials for peace and production for war makes it impossible to use only technical measures to compensate for the limitations of the existing nuclear non-proliferation regime. What is needed are additional international institutional mechanisms involving various political, economic, and diplomatic strategies for controlling access to sensitive materials, facilities or technologies. Since 2003 there have been renewed calls for the utilization of multilateral approaches to the nuclear fuel cycle in order to strengthen the non-proliferation regime and provide states with secure and equitable access to the benefits of peaceful nuclear energy. Multilateralization of the nuclear fuel cycle, in a broad sense, means taking certain aspects of nuclear energy (those considered to be intrinsically dangerous) out of national hands and placing them in multinational or international hands.

This idea is not new. Interest in institutional arrangements for the nuclear fuel cycle dates back to the dawn of the nuclear age. The first effort to define a policy on the international control of atomic energy, the Acheson–Lilienthal Report, dates back to 1946. In the 1970s and 1980s several feasibility studies on multilateral approaches to the nuclear fuel cycle were undertaken. Although these studies generally drew favorable conclusions regarding the technical and economic viability of multilateral approaches, no further pursuit of those approaches followed, not least due to the general lack of political will and the disinclination of some states to renounce sovereign control over nuclear technology. As their advocates argue, multilateral fuel cycles could provide a better way to reduce proliferation risks as compared to purely national fuel cycles. Even if multilateralization of the nuclear fuel cycle cannot change the dual-use nature of nuclear technology, it could change the way in which nuclear technology is managed. Without nationally controlled enrichment or reprocessing facilities, no state would have direct access to weapon-usable fissile materials, or be able to quickly and secretly manufacture nuclear weapons.

If properly arranged, multilaterally owned and operated fuel cycles would present the following advantages:

- they would be important confidence-building measures by providing assurance to the partners and the international community that the most sensitive nuclear technologies are less vulnerable to misuse;
- they would ensure a greater degree of peer scrutiny from participants, providing less opportunity for diversion or theft of nuclear material;
- they would facilitate the application of IAEA safeguards by guaranteeing higher standards of transparency and cooperation;
- they would complicate a "break-out scenario" because to seize the facility, the host state would have to expel other partners, which would constitute a considerable political barrier to such an action;

- they would make the nuclear intentions of states more apparent;
- they could provide cost-effectiveness and economies of scale; and
- they would provide any state that wishes to make use of this option with some vested interest in the major elements of fuel-cycle services.

Comprehensive multilateralization of sensitive components of the nuclear fuel cycle, primarily uranium enrichment and spent fuel reprocessing, would require a new binding international norm, because it goes beyond the NPT legal framework in that it would amount to a change in the scope of Article IV of the NPT. Such a norm is not entirely impossible, but would likely only be agreed upon in the context of broad negotiations, and for many states could probably only be acceptable through the application of universal principles, thus bringing all states to the same level of obligation without exception. Given the deep divisions that today exist between groups of states on issues pertaining to nuclear non-proliferation, disarmament, and the peaceful use of nuclear energy, this would be not an easy task to achieve.

In 2005–2007, in response to Mohamed ElBaradei's call to revisit multilateral approaches, governments, nuclear industry, and non-governmental organizations put forward a dozen proposals regarding multilateral approaches to the nuclear fuel cycle and assurances of nuclear fuel supply. These proposals vary considerably in their vision, scope, goals, implementation timelines, and degree of elaboration.[6] Many of these proposals are similar to those made in the 1970s and 1980s: additional guarantees from nuclear suppliers, fuel banks, and multilateral fuel-cycle facilities were discussed then as now. But today we see greater progress in the direction of the practical implementation of multilateral fuel-cycle arrangements. Four proposals have been pursued: the Russian International Uranium Enrichment Center, the Russian guaranteed LEU reserve, the IAEA LEU bank, and the UK nuclear fuel assurance proposal.

Despite recent success in the implementation of a few proposals for multilateral mechanisms, the fate of multilateral approaches to the nuclear fuel cycle is far from clear. Supplier states do not have a coherent policy on the issue and do not show any interest in discussing the conversion of their national fuel-cycle facilities to multilateral operations. Many non-supplier states resist multilateral approaches because of fears of possible infringement on their NPT right to peaceful nuclear research and development. They also have voiced fears that a "cartel" of suppliers might be formed under the guise of multilateralization. The general lack of political will and the disinclination of states to renounce sovereign control over sensitive nuclear technology are among the strongest barriers to multilateralization today as they were 30 years ago. To overcome this unwillingness and engender political will toward multilateralization, it would be necessary to build a broad coalition of states, which would include both supplier and non-supplier states, with a politically attractive and economically sound strategy toward multilateralization of the nuclear fuel cycle. The focus should be changed from assertions of sovereign "rights" to the common interests of non-proliferation, energy security, economic development, and strengthened international cooperation.

Assurances of fuel supply in the form of additional guarantees from suppliers and LEU banks are likely to have only limited impact on states' decisions to develop sensitive nuclear technologies. Nevertheless, fuel assurance arrangements should be further promoted because they still could be attractive for some states that may be concerned about security of supply. If successful, such arrangements could encourage further developments on the issue of multilateralization of the nuclear fuel cycle. As developing an international norm to have all sensitive fuel-cycle facilities internationalized appears unlikely in the current political climate, a gradual approach to multilateralization should be taken. Serious consideration should be given to placing existing and future enrichment and reprocessing facilities under some form of multilateral control. Multilateral enrichment and reprocessing facilities could help to create an equitable environment for any state to participate in joint ownership, management, operation, decision-making, profit-sharing, and in other activities not necessarily involving direct access to sensitive nuclear technologies. Being an equity partner in an advanced multilateral technology enterprise could be a meaningful attraction in its own right to many states that are interested in nuclear energy to boost economies, reduce reliance on increasingly costly fossil fuels, and reduce pollution.

Conclusions

Two features of nuclear energy underscore the principal weaknesses of the international non-proliferation regime:

- the dual-use nature of nuclear technology. Peaceful and military applications of nuclear energy cannot be clearly separated. There is a wide grey zone of "sensitive" nuclear technologies – uranium enrichment, plutonium separation, the manufacture of plutonium and mixed uranium/plutonium fuel – that can provide fissile material either for generating electricity or for explosive applications; and
- the predominantly national management and control of nuclear activities, which can provide states with readily available sources of weapon-usable materials, namely high enriched uranium and separated plutonium.

The ambiguities in the NPT's provisions allow its states parties to engage in development of national sensitive nuclear technologies, thus stimulating the spread of virtual nuclear weapon capabilities, which is a major non-proliferation concern. Technological thresholds to proliferation, and in particular to acquisition of uranium enrichment technology, have decreased significantly in recent decades and may decrease even further. Incentives for acquisition of national fuel-cycle capabilities remain high, showing a continuing interest in energy independence, economic development, higher national prestige, and/or nuclear hedging.

The IAEA safeguards system and export controls, while certainly important and which should be further strengthened, suffer from certain limitations in their ability to verify sensitive nuclear technologies and prevent their further dissemination. New political approaches and multilateral institutional approaches are needed for the future, or the ghost of nuclear proliferation will continue to go hand in hand with the provision of nuclear energy, affecting the decision of states about their energy options and complicating the international security situation.

Multinational institutional mechanisms and arrangements, that have been widely discussed recently, show a certain promise of compensating for the limitations of the existing nuclear non-proliferation regime by helping curb the spread of national sensitive nuclear technologies and ultimately taking proliferation-sensitive aspects of nuclear energy out of national hands. They would be able to provide all states with non-discriminatory and equitable access to the benefits of peaceful applications of nuclear technology and simultaneously strengthen all three pillars of the NPT.

Notes

1. The work of isotope separation is measured in "separative work units" (SWUs). The SWU is a complex unit which is a function of the amount of uranium processed and the degree to which it is enriched (i.e., the extent of increase in the concentration of the U-235 isotope relative to the remainder) and the level of depletion of the remainder.
2. Mark (1993). The theoretical division at Los Alamos played a major role in US nuclear weapon design, and Carson Mark was intimately involved in the design of both fission weapons and thermonuclear weapons.
3. IAEA (2002: 33). The IAEA does not include plutonium containing more than 80% plutonium-238, as this isotope is very difficult to use for explosives due to its high heat and radiation emission. Because of its properties, plutonium-238 is used for radioisotope thermoelectric generators and radioisotope heater units. Plutonium-238 accounts for no more than a few percent of plutonium in spent fuel of commercial nuclear reactors.
4. For a more detailed discussion of limitations pertaining to safeguarding bulk-handling facilities, see Yudin (2010).
5. IAEA document INFCIRC/539/Rev.3 provides a useful summary history of the two nuclear export control regimes.
6. For details and discussion of the proposals, see Yudin (2009; 2011).

References

Albright, David, Frans Berkhout, and William Walker. 1993. *World Inventory of Plutonium and Highly Enriched Uranium 1992*. Oxford: Oxford University Press.
ElBaradei, Mohamed. 2004. Nuclear Non-Proliferation: Global Security in a Rapidly Changing World. Speech at the Carnegie International Non-Proliferation Conference, Washington, DC, 21 June. www.iaea.org/newscenter/statements/2004/ebsp2004n004.html, accessed December 19, 2011.
Goldschmidt, Pierre. 2008. The Nuclear Non-Proliferation Regime: Avoiding the Void. In Henry D. Sokolski, ed. *Falling Behind: International Scrutiny of the Peaceful Atom*. Carlisle, PA: Strategic Studies Institute, US Army War College, pp. 293–310.
IAEA. 1972. *The Structure and Content of Agreements between the Agency and States Required in Connection with the Treaty on the Non-Proliferation of Nuclear Weapons*. Document INF-CIRC/153 (Corrected). Vienna: IAEA.
IAEA. 1979. *The Revised Guiding Principles and General Operating Rules to Govern the Provision of Technical Assistance by the Agency*. Document INFCIRC/267. Vienna: IAEA.
IAEA. 2002. *IAEA Safeguards Glossary*. Vienna: IAEA.
IAEA. 2006. Addressing Verification Challenges by Director General Dr. Mohamed ElBaradei. Symposium on International Safeguards, Vienna, October 16. www.iaea.org/NewsCenter/Statements/2006/ebsp2006n018.html, accessed December 17, 2011.
IAEA. 2010. *Annual Report 2010*. Vienna: IAEA.
IAEA. 2011a. *Guidelines for Nuclear Transfers*. Document INFCIRC/254/Rev.10/Part 1. Vienna: IAEA.
IAEA. 2011b. *Nuclear Power Reactors in the World 2011 Edition*. Vienna: IAEA.
IAEA. 2011c. Powering Development: IAEA Helps Countries on the Path to Nuclear Power. http://www.iaea.org/newscenter/news/2011/powerdevelopment.html, accessed November 27, 2011.
IAEA. 2011d. Power Reactor Information System. http://www.iaea.org/programmes/a2/, accessed December 20, 2011.
Mark, J. Carson. 1993. Explosive Properties of Reactor-Grade Plutonium. *Science and Global Security* 4: 111–128.

Miller, Marvin M. 1990. Are IAEA Safeguards on Plutonium Bulk-Handling Facilities Effective? www.nci.org/k-m/mmsgrds.htm, accessed December 19, 2011.

Stober, Dan. 2003. No Experience Necessary. *Bulletin of the Atomic Scientists* 59, 2: 56–63.

UN. 1965. Letter Dated 24 September 1965 from the Minister for Foreign Affairs of the Union of Soviet Socialist Republics Addressed to the President of the General Assembly. UN document A/5976, art. I, para. 2 (emphasis added).

UN. 1966. *Final Verbatim Record of the Two Hundred and Forty-Third Meeting.* Conference of the Eighteen-Nation Committee on Disarmament. UN document ENDC/PV.243.

UN. 1967. *Final Verbatim Record of the Three Hundred and Third Meeting.* Conference of the Eighteen-Nation Committee on Disarmament. UN document ENDC/PV.303.

UN. 1975. Resolution 3472 (XXX). Comprehensive Study of the Question of Nuclear-Weapon-Free Zones in All Its Aspects. UN General Assembly, December 11.

UN. 2010. *Final Document of the 2010 Review Conference of the Parties to the Treaty on the Non-Proliferation of Nuclear Weapons.* UN document NPT/CONF.2010/50 (Vol. I).

US Arms Control and Disarmament Agency. 1969. *International Negotiations on the Treaty on the Nonproliferation of Nuclear Weapons.* Washington, DC: US Arms Control and Disarmament Agency.

World Nuclear Association. 2011. Emerging Nuclear Energy Countries. http://www.world-nuclear.org/info/inf102.html, accessed December 19, 2011.

Yudin, Yury. 2009. *Multilateralization of the Nuclear Fuel Cycle: Assessing the Existing Proposals.* Geneva: UNIDIR.

Yudin, Yury. 2010. *Multilateralization of the Nuclear Fuel Cycle: Helping to Fulfil the NPT Grand Bargain.* Geneva: UNIDIR.

Yudin, Yury. 2011. *Multilateralization of the Nuclear Fuel Cycle: A Long Road Ahead.* Geneva: UNIDIR.

Zarate, Robert. 2008. The NPT, IAEA Safeguards and Peaceful Nuclear Energy: An "Inalienable Right," But Precisely to What? In Henry D. Sokolski, ed. *Falling Behind: International Scrutiny of the Peaceful Atom.* Carlisle, PA: Strategic Studies Institute, US Army War College, pp. 221–290.

Part IV Global Energy and Development

The Handbook of Global Energy Policy, First Edition. Edited by Andreas Goldthau.
© 2013 John Wiley & Sons, Ltd. Published 2013 by John Wiley & Sons, Ltd.

Chapter 14

Energy Access and Development

Subhes C. Bhattacharyya[1]

Introduction

Looking at the developments in global energy situation in the past and pondering over the future energy prospects, it becomes apparent that the extreme form of disparity existing in the production, consumption, and access of energy characterizes the unsustainable nature of the development. The critical role played by energy in achieving sustainable development has been well recognized and the access to energy has been identified as a major challenge. There is also a consensus that without affordable, reliable, and clean energy services to the population, the sustainability objective cannot be achieved. Yet, billions of people are without access to such vital services in the world in the twenty-first century and many may continue to be so even in the future. To raise awareness about the problem and take concrete actions toward universal energy access, the United Nations has declared 2012 the International Year of Sustainable Energy for All. As a contribution toward this theme, this chapter has three objectives: to provide a review of the energy access problem and its link with development; to critically examine the existing the policies being used toward energy access; and to suggest sustainable alternatives.

The term "energy access" is used to mean access to modern and clean energies by the population of a country.[2] Generally the term is used to reflect the condition of people in developing countries who rely on traditional energies to satisfy their energy needs. It focuses on whether consumers have physical access to the supply of energy, and access to markets for equipment. This is different from the concept of "energy poverty" used in an economic sense, which focuses on the share of income spent by a household on energy services: when the share exceeds a certain threshold level, the household is considered to be suffering from energy poverty or fuel poverty (Bhattacharyya 2011: 507). In this chapter, our focus is on energy access and not on energy poverty.

Energy access is predominantly a rural problem of poor developing countries, although the urban poor can also face such challenges. Although our attention in this chapter will be more on the rural poor, the urban poor will not be ignored either. The chapter is organized as follows: first an overview of the energy access problem is discussed. This is followed

The Handbook of Global Energy Policy, First Edition. Edited by Andreas Goldthau.
© 2013 John Wiley & Sons, Ltd. Published 2013 by John Wiley & Sons, Ltd.

by a review of indicators used to measure and capture energy access and development links. Various approaches used to enhance energy access are reviewed thereafter. Finally, the challenges and ways out are presented, followed by some concluding remarks.

Overview of the Energy Access Problem

Energy demand in poor households normally arises from two major end-uses: lighting and cooking (including preparation of hot water).[3] Cooking energy demand is predominant in most cases and often accounts for about 90% of the energy demand by the poor. Such a high share of cooking energy demand arises partly from its low energy efficiency and partly due to limited scope for other end-uses. As electricity is considered the appropriate form of energy for lighting, it is customary to associate the access to clean lighting to the level of electrification of a country. Access to clean cooking energies on the other hand can take different paths and therefore the access-related information is generally presented for electricity and for cooking energies separately.[4] We maintain this approach in this chapter.

This section is organized in three sub-sections: first, a review of the current electrification status is presented, followed by a discussion on the cooking energy status. The final sub-section indicates the future outlook.

Status of Electrification in Various Regions

The regional picture of electrification is presented in Table 14.1.[5] In 2009, more than 1.4 billion people (about 22% of the global population) did not have access to electricity. Two regions stand out in this picture: South Asia, with 614 million (42% of the global population without access), comes first while Sub-Saharan Africa comes second with 587 million (40%). After these two regions, East Asia has 195 million without access to electricity (about 13% of those without access).

A closer look at the data shows that about 70% of those lacking access to electricity reside in just 12 countries while the remaining 30% is dispersed in all other countries (see Figure 14.1). The rural population in most of these countries is lacking access, although in a few countries the urban population also lacks access. While the total number of people without access to electricity is high in South Asian countries, Sub-Saharan Africa

Table 14.1   Level of electrification in various regions, 2009.

Region	Population Without Electricity (millions)	Electrification Rate (%)		
		Overall	Urban	Rural
North Africa	2	98.9	99.6	98.2
Sub-Saharan Africa	587	28.5	57.5	11.9
China and East Asia	195	90.2	96.2	85.5
South Asia	614	60.2	88.4	48.4
Middle East	21	89.1	98.5	70.6
All Developing Countries	1453	72.0	90.0	58.4
Transition economies + OECD	3	99.8	100.0	99.5
Global total	1456	78.2	93.4	63.2

Source: IEA (2010).

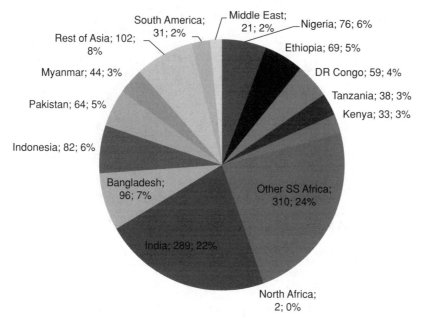

Figure 14.1 Major concentration of population without access to electricity, 2009 (in millions followed by global percentage).
Source: IEA (2011).

fares worse in terms of rate of electricity access. In fact, out of 10 least electrified countries in the world, nine are from Sub-Saharan Africa and Myanmar is the only country from Asia.

It can also be noted that most of these countries (Ailawadi and Bhattacharyya 2006) have low per capita GDP compared with the world average. Except Indonesia, all named countries in Figure 14.1 have national average per capita GDP less than 10% of the world average, low per capita primary energy consumption ranging from 8% to 42% of the world average, and very low per capita electricity consumption – the national average per capita electricity consumption in these countries ranges between 1% and 15% of the world average.

Interestingly, the most populous country in the world, China, has achieved a very impressive record of providing electricity with only 8 million (or about 0.6% of its population) lacking the facility. Similarly, a number of Southeast Asian countries such as Thailand, Malaysia, Vietnam, and the Philippines have made impressive progress on the electrification front. In South America, Brazil, the most populous country of the region, has achieved an impressive record of about 2% of its population without access to electricity, most of whom are located in the Amazon region. This suggests that there exists a great potential for cross-learning from successful and failed cases around the world.

Status of Cooking Energy Access

IEA (2011) provided some details about biomass use in the developing countries and estimated that about 2.7 billion people use biomass for cooking and heating purposes in these countries (see Figure 14.2).

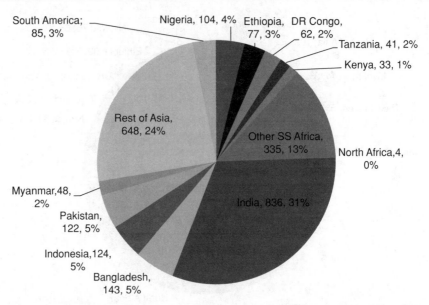

Figure 14.2 Distribution of lack of cooking energy access in the world, 2009 (in millions followed by global percentage).
Source: IEA (2011).

At a disaggregated level, more than 80% of the people lacking access to clean energies live in rural areas (see Table 14.2). Asia, with more than 72% of those lacking access, has the largest share followed by Sub-Saharan Africa, but in contrast to electricity access, where both regions share similar sizes of population without access, here the picture is quite different. The size of urban population lacking access to clean energies in both the regions is very similar but the rural population lacking access to clean cooking energies in Asia is 3.5 times more than that of Sub-Saharan Africa. India has the single largest concentration of people lacking clean cooking energy access in the world.

Although a range of fuels is used for cooking, according to the UNDP-WHO (2009) study, about 2.6 billion people rely on traditional energies and 400 million rely on coal. There is significant regional variation in terms of fuel use (see Figure 14.3), but the rural population is generally more reliant on solid cooking fuels, including traditional energies. Moreover, the use of improved cooking stoves is limited to only 30% of those relying on solid fuels, but the dependence on traditional stoves is predominant in the least developed countries (LDCs) and Sub-Saharan Africa.

Table 14.2 Reliance on biomass for cooking energy needs, 2009 (millions).

Region	% of Total Population	Total	Rural	Urban
Sub-Saharan Africa	78	653	476	177
Total Africa	65	657	480	177
India	72	836	749	87
China	32	423	377	46
Rest of Asia	63	731	554	177
Latin America	19	85	61	24
Total	51	2662	2221	441

Source: IEA (2011).

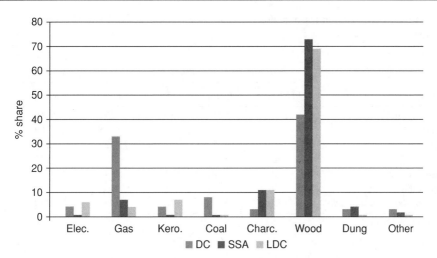

Figure 14.3   Share of different cooking fuels in developing countries, 2007 (by developing countries, Sub-Saharan African countries, and least developed countries).
Source: UNDP-WHO (2009).

Clearly, such a heavy reliance on traditional energies imposes economic cost on the society. Use of solid fuels is attributed to health effects such as child pneumonia, chronic obstructive pulmonary disease, and lung cancer. About two million premature deaths occur that is directly attributable to solid fuel use in developing countries (UNDP-WHO 2009). The regional distribution of these premature deaths follows the biomass use patterns and South Asia and Sub-Saharan Africa suffer the maximum loss in this respect. Disease-adjusted life years (DALY), which captures the years of life lost as a result of premature death and the years lived with a disease, is a broader indicator that is also used to show the health damage. Approximately 40 million DALYs are lost per year due to solid fuel use in cooking (UNDP-WHO 2009). As women and children are more exposed to such conditions, they are more vulnerable.

The problem of cooking energy access is much more widespread than that of electricity access. The predominance of heating and cooking energy needs by the poor demands greater attention to this widely neglected issue.

It should be noted that the Human Development Index (HDI) generally improves with higher levels of electricity access and access to cooking energy. Similarly, the life expectancy and mean schooling years are also positively correlated to electricity access. However, it is interesting to note that a large number of countries have scored between 0.3 and 0.4 HDI even with a very low level of energy access (both electricity and cooking energy) and that an average life expectancy of 50 years can be achieved even without access to clean energies. This perhaps shows that certain components of HDI do not depend on energy access or can be achieved without energy access. Despite this, an improvement in energy access level is likely to lead to a better living condition.

Future Outlook

Forecasts by IEA (2011) suggest that almost one billion people will still lack access to electricity in the 2030 horizon while 2.7 billion people will not have access to clean cooking energies. Although the forecast assumes a significant level of investment

Table 14.3   Expected number of people without electricity and cooking energy access, 2030 (millions).

Region	Without Electricity Access			Without Cooking Energy Access		
	Urban	Rural	%	Urban	Rural	%
Sub-Saharan Africa	107	538	49	270	638	67
India	9	145	10	59	719	53
China	0	0		25	236	19
Rest of Asia	40	181	16	114	576	52
Latin America	2	8	2	17	57	14
Middle East	0	5	2			
Total of developing world	*157*	*879*	*16*	*485*	*2230*	*43*

Source: IEA (2011).

($13 billion per year on average), increases in the population in developing countries of South Asia and Sub-Saharan Africa will mean that electricity access will remain a problem. According to IEA (2011) 356 million in South Asia and 645 million in Sub-Saharan Africa will still live without electricity access (see Table 14.3).

In terms access to clean cooking energies, the situation will be even worse. IEA (2011) suggests that 485 million urban population and 2.2 billion rural population will still continue with traditional energy even by 2030 unless more specific interventions are made (see Table 14.3). The size of population relying on traditional energies in Sub-Saharan Africa will increase to 900 million by 2030 and one-third of the global population without clean cooking energy will reside here. This represents a deterioration of the condition compared to the present situation. The situation in Asia improves marginally but still about two-thirds of the population without clean cooking energy will be found here by 2030. India will continue to remain the country with the highest concentration of population without clean cooking energy. Clearly, the future does not appear to be very promising and serious thoughts in terms of policy analysis and implementations will be required to address these issues.

However, governments of most of the countries are now aware of the problem and a large number of countries have set energy access targets. According to UNDP-WHO (2009), almost half of all the developing countries have set targets for electricity access, with Sub-Saharan countries emerging as leaders in setting targets. In the case of clean cooking energies, only a few countries have set targets, but Sub-Saharan countries are appearing to be more proactive here as well.

Examining Current Approaches to Enhance Energy Access

A number of approaches have commonly been used by countries to enhance energy access but the emphasis was mostly on electrification, with cooking energy access receiving limited and often sporadic efforts. This section provides a critical review of existing approaches, thus forming the basis for policy implications in the subsequent section. The approaches to enhance electricity access are presented first, followed by that for cooking energies.

A Brief Review of Electrification Experience from Around the World

There exists a well-developed body of literature on rural electrification and electricity access.[6] Although electricity can be supplied by extending the central grid system and through decentralized systems (or off-grid systems), particularly in remote areas where extending the grid becomes prohibitively costly, the general trend all over the world is to rely on grid extensions where possible and to use off-grid solutions either as a pre-electrification process or as a stop-gap measure until grid extension materializes. This bias is evident in all countries where such electrification has been successful, including South Africa, Brazil, Thailand, Vietnam, and the Philippines. While off-grid options have received favor and support of international organizations and donor agencies, and a few technologies such as solar home systems have emerged as leaders in this sector, there has been a relatively limited penetration of this option globally. High cost, limited application, and poor performance of the technologies as well as an image of the "inferior or temporary" nature of such options have hindered their development.

As indicated earlier, a significant progress has been noticed in terms of electrification in East Asia and Southeast Asia. China has achieved universal electrification despite its billion plus population. Countries in Southeast Asia (including Thailand, Malaysia, the Philippines, and Vietnam) have also recorded rapid progress in a limited period of time and have provided successful examples of electrification. The main concentration of people lacking electricity access is now located in South Asia and Sub-Saharan Africa. While some South Asian countries like Sri Lanka have made a significant progress and others including India are striving to improve the situation through various efforts, universal electrification in South Asia and Sub-Saharan Africa is unlikely to materialize in the near future.

A selected set of country cases is presented briefly below. More detail is available in Palit and Chaurey (2011), Bhattacharyya (2010), and Bhattacharyya and Ohiare (2011).

China   China, the most populous country in the world, presents a very impressive case of successful electricity access provision. With an almost 100% electrification rate, China stands out in the developing world. Unlike other developing countries that followed a top-down approach to electrification, China has relied on a bottom-up approach, where local administration and participation was responsible for the local solution. The approach allowed flexibility and was anchored in self-reliance. The success of electrification can be attributed to the pragmatic approach which allowed local administrative responsibility for the projects while retaining the overall planning at the central level.

The successful electrification of more than 900 million people represents the most important success story in the world. The economic reform began in rural areas in 1979, and by 1997 almost 96% of the rural households had access to electricity. According to recent information, only two million households in the country lack access to electricity and there are plans to electrify them with off-grid solutions. Rural electrification relied on three modes of delivery: local grid-based, central grid-based and a hybrid system of local and centralized grids. Local grids played an important role in areas with large hydro potential where county water bureaus or small hydropower companies are responsible for electricity supply. However, the dominant mode of supply remains the extension of central grid.

China relies on three levels of management: central, provincial, and local. Management through the decentralized local governments was a main driving force behind the success of rural electrification in China. Funds for rural electrification flowed from central and local governments, and even local residents participated in providing funds. The decentralized, local management of rural electrification initiatives and the emphasis on rural development through agricultural activities, town and village enterprises, and poverty reduction programs contributed to the success story.

China also adopted a phased development approach where local grids at the village or community level were established initially, followed by an upgrading of the system to link to the regional or national network. This also proved to be a pragmatic approach, as the expansion and upgrading of the system at a later date proved less challenging due to the better financial and economic standing of the country.

While most other countries have taken up electrification as a social policy objective of the government, China recognized that rural electrification and rural energy supply is closely linked to rural economic development. Its focus on agricultural development in the planned economy phase and on township and village enterprises in the reform era clearly highlights this recognition. Through sustained rural economic activities, China was able to reduce rural poverty rapidly and improve the living conditions of its population.

In addition, through its emphasis on local resources, China allowed selection of locally relevant energy sources and as a consequence allowed technological diversities to coexist. Although the main emphasis was on small hydropower and coal initially, there was never a "single solution fits all" approach. Technological flexibility has also allowed local resource utilization and avoided highest cost options for difficult locations. The sense of local ownership has also ensured the success of projects in remote areas.

Further, strong state support and the ability to engage the local communities in the creation of local infrastructure have surely contributed to the success. Moreover, the pricing system ensured almost full cost recovery, which in turn allowed future sustainability of the system. In fact, China has avoided the trap of high electricity subsidy noticed in many South Asian countries.

Simultaneously, programs have hardly attempted to integrate local resources to ensure long-term sustainability of activities. In addition, the quality of service tends to be much poorer in the rural locations compared to the urban areas. This in turn reduces the social benefits of electrification to some extent.

India   Although the government has been making conscious efforts on rural electrification, a significant share of the rural population still lacks access to electricity. According to IEA (2011), 25% of the country's population lacks electricity access. In fact, household electrification did not receive adequate attention in the initial years, when energizing the irrigation pumps in rural areas was the priority. The central government made attempts to enhance electricity access through a number of initiatives, which either focused on rural electrification or were linked to rural development. But multiplicity of programs meant that inadequate funding was available for each program and their coordination and management was weak. Unlike China which did not try to distort the price signal in rural areas, India relied on heavily subsidized supplies (including offering free electricity) for rural areas, which imposed financial burdens on the electric utilities. Consequently, utilities were not enthusiastic about rural electrification. In the 1990s when the focus was on electricity sector reform, rural electrification and electricity access was put on the back burner, but the issue of rural electrification has re-emerged as a main political issue since 2001. A new program was launched in 2005 (Rajiv Gandhi Grameen Vidyutikaran Yojana or Rajiv Gandhi Rural Electrification Scheme) to electrify rural households by 2012. The central government bears 90% of the infrastructure costs and there is provision for energy subsidies to the poor as well. Although significant progress has been made through this initiative, it is unlikely that universal electrification will be achieved by 2012.

The Philippines   This Southeast Asian archipelago of more than 7000 islands is considered to be a success with an overall electrification rate of above 90%: 77% of households have access to electricity at the moment and the country plans to achieve 90% household electrification by 2017. The government initiatives for rural electrification in the Philippines started in 1960 when the Electrification Agency was set up. Significant progress was made after the National Electrification Administration was set up in 1969 and the rural electrification cooperatives were promoted. The progress in electrification slowed down in the 1990s, when the market reform received priority. In 2003, the government launched the Expanded Rural Electrification Program to achieve 100% electrification by 2008 (extended to 2010) and 90% household electrification by 2017. The program focuses on a combination of approaches including extension of distribution network, setting up of micro/mini-grids, and the use of off-grid systems. The program has allowed participation of non-government and non-utility agencies in electricity provision and resource generation by involving qualified third parties (QTP). Where a cooperative or a franchisee finds it unviable to provide electricity, the Missionary Electrification project is undertaken, which receives a continuous flow of subsidy from a fund created by levying a universal charge, set by the electricity regulator, on electricity users. The country has relied on both state support and market-based mechanisms to enhance electricity access.

Vietnam   Vietnam provides another example where rapid progress has been made in terms of rural electrification. From a mere 1.2 million population with electricity access in 1976 the country managed to provide electricity to 82 million people by 2009. About two million households living in remote areas lack access to electricity grids, and there off-grid electrification methods are being used. Rapid progress in electrification was made since the mid-1990s when the electricity generating capacities and transmission networks became available, and when Electricity of Vietnam (EVN) was established to ensure integrated development of the electricity supply industry. Vietnam also relied on grid extension as the main mode of electrification. The state played an important role in the entire electrification process – policy-making, strategy development, and delivery. Vietnam followed a logical approach in building the capacity and infrastructure first and then expansion of the system to rural areas. It also prioritized the process by putting emphasis on productive use of energy, which helped create demand for electricity. The creation of EVN and its effective support in promoting rural electrification contributed to the success of the program as well.

In all cases, the electrification process has relied heavily on government subsidies for infrastructure development and in many cases for system operation. The case for subsidizing the required infrastructure to provide the supply relies on the public good argument: that the limited demand will not justify private investment and the market will remain unserved if profitability alone is considered. But the social benefit of enhancing the access can be far greater than the private cost, due to reduced environmental damages and better living conditions, thereby making state intervention to remedy the market failure justified. The support for energy use (or the operational cost) on the other hand invokes the idea of minimum energy needs for sustaining livelihood. If a section of the population is unable to procure the minimum level of energy needs, they could be supported on social equity grounds. Lifeline rates for electricity and subsidized electricity rates are common in such cases. Some countries have also relied on cross-subsidies – for example, the consumers of the Provincial Electricity Authority in Thailand benefit from a cross-subsidized power purchase rate for rural supply. However, such subsidies lead to the issue of their long-term viability and sustainability as subsidized supply imposes financial burdens on the supplier and the state, and poorly targeted subsidies accrue to the rich, thereby creating disincentives for efficient energy use (see Bhattacharyya 2010 for details).

Electrification Strategies

Countries have relied on different strategies for electrification: for example, in China three levels of management have been used – central government, provincial government, and local committees. In other Southeast Asian countries, a strong state utility with a dedicated focus on rural electricity supply (e.g., the Provincial Electricity Authority in Thailand, and EVN in Vietnam) has generally played an important role. In South America, private distribution licensees have been effectively used to provide electricity within their area of service. At the same time, a number of African countries with a dedicated rural electrification agency have failed to improve electricity access due to managerial and funding constraints.

Unfortunately, the electrification programs in most countries have paid limited attention to sustainability issues and rural development issues. Electrification programs have hardly been linked to rural development programs, whereby productive activities are created to generate better income opportunities. Consequently, rural electrification has resulted in limited direct income-generating activities or led to rural development. Yet, unless the poor have income-generating opportunities that provide them with a source of regular monetary income, they are unlikely to allocate a share of it to purchase a marketable energy good or the required appliances to use such energies (Bhattacharyya 2006). This remains a major issue in electrification.

In respect of off-grid options, diesel generators (both portable and stationary ones) are widely used as a conventional solution to supply electricity for a few hours per day. However, price fluctuations in the international market and environmental pollution make this source an unsustainable solution. Out of the renewable energy alternatives, hydropower, biomass, and solar energy form the most common options. The solar home systems have emerged as the lead technology in off-grid supply. While the off-grid option is donor-driven and grant-funded in general, markets have started to develop in some countries. For example, Kenya has a thriving market for solar home systems. Similar is the case in Bangladesh and Sri Lanka. This often takes the form of fee-for-service, leasing or outright sales, and in some countries consumer financing through micro-finance has been attempted.

Where off-grid solutions are used, they appear to cater to limited needs of the consumers for lighting and some entertainment through radio/TV connections. However, very limited efforts have been found where these solutions have promoted productive use of energy for income generation. Similarly, very limited effort has so far gone into hybrid off-grid solutions to provide a reliable and affordable solution. While some private partners are participating in some off-grid supply activities, it is the general experience that donor assistance or state support has been the catalyst for off-grid solutions. Better results have been achieved where the entire program is well coordinated with adequate support services and clear assignment of responsibilities. The development of a local supply chain has also played a major role in the successful delivery of the systems.

The above experience suggests that countries have relied on different strategies to record different levels of success, but the electrification process has heavily depended on government subsidies. But state involvement alone is not a guarantee of success.

Can Rural Electrification Resolve the Energy Access Problem?

For any commercial energy to successfully penetrate the energy demand of the poor, the following economic factors have to be satisfied (Bhattacharyya 2006):

- The energy should be suitable and perhaps versatile for satisfying the needs;
- It should have a competitive advantage that would place no or little demand for money transactions (in other words, the low cost supplies) in the present circumstances, and/or
- the use of modern energy should result in supply of adequate money flows to the poor so that they become willing to spend some part of the money on purchasing commercial energies.

Applying these factors to electricity provides insights about its potential for resolving the energy access issue. Electricity normally accounts for less than 10% of poor households' energy needs as it is rarely used for cooking purposes. This in turn implies that demand for lighting cannot alone resolve the problem of energy access in rural areas, as other fuels would be used by the poor to meet cooking demand. It appears that policy-makers tend to ignore or forget this simple truth, maybe because of better prestige and visibility of electrification projects (and hence for better political mileage).

In addition, electricity is unlikely to be competitive when compared with traditional energies, unless it is supplied at highly subsidized rates. However, the viability of a subsidized supply is questionable. Moreover, as rural areas in many countries suffer from poor power quality, reliance on alternative forms of energy cannot be avoided in practical terms. Moreover, electricity for cooking entails significant initial investment when compared with traditional energy use, which the cash-strapped poor households are unlikely to be able to afford. Thus, electricity has less chance of succeeding in the cost competition with other fuels.

Experience of Providing Clean Cooking Energy Access in Developing Countries

As solid fuels dominate the cooking energy sector in the developing world, a number of initiatives have been taken by various countries to reduce the negative effects of solid fuel reliance by promoting clean energies. As in the case of electricity, both conventional energies and renewable energy options have been tried on the supply side. In addition, interventions to improve the efficiency of solid energy use through improved technology have also been common (demand-side intervention). This section provides a brief review of the global experience.

Promotion of Modern Fuels

One of the major strategies followed by many countries was the promotion of modern energy carriers such as LPG and kerosene for cooking to displace solid fuel use for cooking and heating (kerosene was promoted for lighting as well as cooking). The main policy intervention here was to subsidize the supply to ensure that consumers change their consumption behavior in favor of the modern fuel. However, in many cases little attention was given to the fact that the appliance required to use the fuel can be costly as well and unless a consumer is able to pay for the appliance, the subsidized fuel does not benefit her. Consequently, the benefits accrued to the wealthier section of the population and did not really help the poor. Further, the subsidized supply also found alternative uses: for example, kerosene found application in motor fuel adulteration and LPG was diverted for the use of personal transport vehicles. Moreover, for import-dependent countries, the subsidized supply becomes a financial burden when prices in the international market harden, and governments operating under severe budget constrains faced difficulties in maintaining such subsidies in the long run. India offers an example where the benefits of such non-targeted subsidies have imposed huge financial burdens on successive governments.

Also, the supply chain issues did not receive adequate attention, implying that even when consumers decided to use the modern fuel, its regular availability was not ensured. The problem arose due to limited demand of such energies from poorer households, weak organizational arrangement of often state-owned suppliers for rural supply, and poor financial viability due to high transaction costs. Innovations have taken place as well: smaller bottle sizes for LPG have emerged that make transportation easier and reduce the initial payment as well as the recurring costs; to prevent adulteration, dyed products for different market segments (controlled and decontrolled markets) have been attempted; support for initial investment in appliance purchase and connections was provided in some cases. Although IEA (2010) suggests that the transition of 1.2 billion people to LPG will increase the oil demand by merely 0.9 million barrels per day, yet the long-term sustainability of this approach is doubtful due to price volatility in the international market and the increasing foreign exchange and subsidy requirements for preferential support. While LPG can still play a role, it is unlikely to emerge as a global rural solution.

Promotion of Local Resources

Faced with the challenge of avoiding energy security issues by promoting import-dependent fuels, countries have also tried to harness local resources for cooking and heating. Promotion of biogas is the most prominent example of this category, although other examples, such as charcoal, exist. The Chinese example is worth mentioning here. Biomass plays an important role in rural energy supply in China but instead of traditional burning, bio-gasification has been promoted widely in the 1980s, making China the world leader in biogas production. About 26.5 million biogas plants are currently being used in the country (Chen *et al.* 2010) and the country has been successful in promoting this form of energy. The use of biogas has replaced coal and reduced GHG emissions (Chen *et al.* 2010). Countries in South Asia (such as India, Sri Lanka, and Nepal) also use biogas to a lesser extent, and the use of biogas is also increasing elsewhere, such as in Vietnam, Brazil, and Tanzania.

China used pilot projects to gain vital information before implementing it on a large scale. This small-scale experimentation has allowed program adjustments and helped the country to direct resources where necessary. In addition, the emphasis on training and capacity building, standardization, and dissemination has helped in spreading the knowledge widely across the country. The development of a cadre of skilled technicians and project staff and the performance improvement through feedback loops were also essential factors. The existence of a large manufacturing base has also allowed the country to take advantage of the technology. Continued growth in demand and consequent exploitation of scale and scope economies have resulted in lower supply costs, making supply more affordable. Moreover, local supply also reduced external dependence and project completion time. But above all, the buying-in of local participants through sustained awareness campaigns has also played a role.

Technology Intervention

The demand-side intervention essentially followed from the realization that solid fuel use perhaps will continue and may be difficult to change in the near term. Better use of such resources through improved technology therefore remains a valid strategy option. Many initiatives have been supported over the past 25 years through international cooperation and funding agency support (see Ekouevi and Tuntivate 2011 for the World Bank initiatives in this area). According to UNDP-WHO (2009), about 65% of those using improved cooking stoves live in China and another 20% in other Asia-Pacific countries. But Sub-Saharan Africa, where 80% of the solid fuel using population lives, accounts for only 4% of the improved cooking-stove using population. This implies that the region with most needs has not benefited much from this intervention.

As Foell *et al.* (2011) argue, despite the gravity of the problem, the global attention on clean cooking and heating energies has been relatively low compared to that for electrification. However, as argued above, substantial environmental and economic benefits cannot be achieved without ensuring clean cooking energies. Moreover, most of the interventions so far have been assisted by international organizations or donor agencies and have at best undertaken pilot studies. Although they are useful, large-scale replication and up-scaling of the programs are not possible unless the national governments become active partners in the transformation process and engage local populations. This in other words means that sporadic or ad hoc international support and non-governmental organization involvement alone will not resolve the problem.

Policy Imperatives and Options

Given the magnitude of the problem and the undesirability of such a situation to continue in the future, clearly the global challenge is to find a way out within a reasonable time. Various scenarios for improving the situation have been analyzed by international organizations like IEA and UNDP-WHO (see IEA 2009, 2010, 2011, and UNDP-WHO 2009). For example, UNDP-WHO (2009) estimated that in order to achieve the national targets for electrification or to achieve electrification compatible with UN Millennium Development Goals (MDGs), 1.2 billion people have to be electrified by 2015 in the developing world. Similarly, to achieve the MDG-compatible goals for providing cooking energy access will require providing two billion people with clean cooking energies by 2015. Foell *et al.* (2011) indicate that to meet the UN Millennium Project recommendation of halving the number of those relying on traditional energies by 2015 will mean that 880,000 people per day have to be provided with clean energy access. This clearly shows the magnitude of the problem. Given that most of the supply provisions have to be made in the least developed countries of Asia and Africa, the issue becomes more challenging, especially when half of these countries have not set targets for clean energy provision.

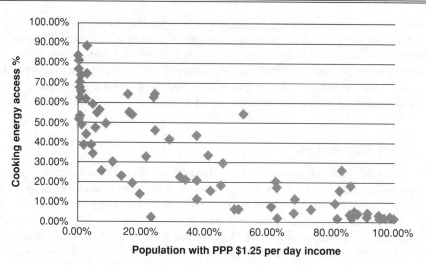

Figure 14.4   Cooking energy access versus population below poverty line.
Source: HDI database, http://hdr.undp.org/en/statistics/hdi/.

Need for Developing Strategies and Proper Planning

The main challenge then relates to developing strategies and planning for clean energy supply in developing countries, especially in South Asia and Sub-Saharan Africa. The problem is aggravated due to the large share of population below the income level of PPP $1.25 per day (purchasing power parity, see Figure 14.4) in most of these countries and the weak economic strength of the countries themselves. Poor governance, limited economic diversities, and weak human capital as well as industrial base in many countries would also imply that even if strategies are developed, their implementation can be difficult and time-consuming. Given that most ground-level activities have to take place in rural areas, international support for pilot projects is not sufficient and without a strong local capacity, any significant progress is unlikely.

Need for Adequate Financing

A related challenge is the volume of investment required to improve the situation. IEA (2011) reports that to provide electricity for all by 2030, an additional investment of $641 billion (constant 2010 dollars) will be required, implying on average an annual investment of $30 billion during 2011–2015 rising to $55 billion during 2026–2030. Similarly, for cooking energy access, another $74 billion (constant 2010 dollars) will be required to ensure clean cooking energy for all. Clearly, these are significant investments by the standard of low-income countries, although they are minuscule in terms of global energy sector investments, which run into trillions of dollars.

Financing such investments for socially relevant projects will have to rely on state finances or international support. It is unlikely that private investors with profit motives will play an important role, although some private investment especially in large countries cannot be overruled, especially when the state provides guarantees for adequate cost recovery of private investments. Despite rising international awareness about the energy access issue, according to IEA (2011) only over $4 billion of investment in 2009 flowed from bilateral and multilateral support for energy access in the developing world. Considering the fact that international support is often tied to specific programs or technologies and for technical assistance which in turn pays for expatriate specialists, the real contribution of such aid remains questionable and surely inadequate to resolve the problem. Moreover, the aid flow may decline in the future due to worsening economic conditions in the developed economies. Further, financially constrained developing economies may find it difficult to allocate adequate funds for energy access improvements and the affordable funding for such social causes may not materialize. As the least developed countries need a large share of the investment, mobilizing it will remain a major concern.

Balancing Local and International Contexts

Further, while the effect of increased clean energy supply to meet the energy access targets on global energy demand and supply will be minimal, the same cannot be said about the developing countries themselves. Where the clean energy requires higher imports, the macro-economic impacts in terms of trade balance and balance of payments can be important. Weak industrial capability is likely to impose higher material and appliance import burdens as well. Consequently, the issue of long-term viability of the adopted solutions can be a challenge, particularly due to supply chain management issues, affordability concerns, and vulnerability of the economies due to price fluctuations in the international markets.

Finally, the solutions have to adapt to the local conditions of each country and therefore searching for one-size-fits-all solutions does not resolve the problem. While most studies just try to find replicable generic solutions, attention is not always paid to the local aspect. Generic solutions would require local experimentation, a process which cannot be underestimated.

Alternative Options

Sustainable energy access solutions require a holistic approach that places the issue of energy access in its proper context of income generation, monetization of income, and provision of affordable supply. Such solutions cannot rely on subsidized supply of clean energies alone. Similarly, piecemeal solutions that address only a part of the problem will not help either. What is required in the long term is to ensure adequate supply of monetary resources to households to sustain a lifestyle that relies on clean energies and other monetized inputs. Thus the energy access issue joins here the problem of ensuring economic development, which in turn calls for an integrated approach combining various development efforts at a decentralized level as opposed to treating electrification or energy supply issues in an isolated manner.

Similarly, the solutions have to be essentially local solutions but can be coordinated centrally, where the policy objective is to promote innovative solutions rather than to prescribe templates. Such a flexible approach, which has been successfully deployed in China, allows each decentralized unit to take ownership of the solutions, while the state provides direction and support in the form of selective but judicious use of market interventions to make energy supply affordable while ensuring financial viability of the efforts. Implementation of such a policy would require development of a common framework that can be adapted to each situation, creation of an organizational set up to carry out the policy, building organizational capacity, adequate funding arrangements, and above all a complete review and perhaps an overhaul of the mode of functioning of the government, existing organizations, and economic activities to facilitate such a decentralized mode of functioning.

While a long-term approach as indicated above aims for permanent cure of the problem, the short-term issue cannot be ignored either. Thus a phased approach is more appropriate, where problem containment and immediate relief can be aimed at in the near term but a durable solution is targeted in the long term. Local mini-grids or solar home systems can be such short-term options for electricity while improved cooking stoves can help reduce externalities arising from cooking energy use. As the users are more interested in the services they receive, a bundled approach to the service at least initially may be more financially viable or business friendly.

Conclusion

This chapter has presented an overview of the energy access challenge by considering the access to electricity and cooking energies. The chapter has indicated the current and future status and presented the commonly used indicators of access, including the energy development index suggested by the IEA. The review of experience with energy access promotion suggested that state support has played an important role and that electricity access has received a disproportionate attention, although the challenge is more pronounced in the case of clean cooking energy access. The emphasis on productive use of energy has been quite limited and the link between energy access and economic development has not received adequate attention at the project/program level. The chapter suggested a bottom-up approach to resolve the problem, although such a policy is inherently difficult to implement. Further research is required in this area to identify practical, viable solutions.

Notes

1. The work reported here was funded by an EPSRC/DfID research grant (EP/G063826/1) from the Research Councils UK Energy Programme. The Energy Programme is a RCUK cross-council initiative led by EPSRC and contributed to by ESRC, NERC, BBSRC, and STFC. The author gratefully acknowledges the support.
2. There is no universal definition of the term but the IEA (2011) gives the following definition: "a household having reliable and affordable access to clean cooking facilities, a first connection to electricity and then an increasing level of electricity consumption over time to reach the regional average." However, the definition focuses on the household alone and its reference to increasing level of energy consumption adds an unsustainable dimension. We have therefore avoided such a formal definition in this chapter.
3. In some climatic conditions space heating may also be an important source of energy demand.
4. It is important to mention that the quality of data on this subject is relatively poor, although major efforts are being made by the international organizations to improve the situation. The data quality is affected, among others, by the distributed and dispersed nature of the population being considered, lack of any administrative arrangements for systematic records on traditional energies, definitional issues, limited availability of comprehensive surveys, and poor communication and infrastructure facilities. This aspect needs to be kept in mind in respect of any analysis on the subject.
5. See also UNDP-WHO (2009) for a detailed review of energy access.
6. This section relies on the research carried out under the Off-grid Access Systems for South Asia (OASYS South Asia) Project funded by Research Councils UK. A number of Working Papers (e.g., WP1, WP2, WP4, and WP10) have been used in writing this material. See www.oasyssouthasia.info/ for further details. See also Palit and Chaurey (2011), Barnes (2011), and Cook (2011).

References

Ailawadi, V. S., and S. C. Bhattacharyya. 2006. Access to Energy Services by the Poor in India: Current Situation and Need for Alternative Strategies. *Natural Resources Forum* 30, 1: 2–14.

Barnes, D. F. 2011. Effective Solutions for Rural Electrification in Developing Countries: Lessons from Successful Programs. *Current Opinion in Environmental Sustainability* 3, 4: 260–264.

Bhattacharyya, S. C. 2006. Energy Access Problem of the Poor in India: Is Rural Electrification a Remedy? *Energy Policy* 34, 18: 3387–3397.

Bhattacharyya, S. C. 2010. *Off-Grid Electrification Experience Outside South Asia: Status and Best Practice*. Working Paper 2. OASYS South Asia Project, see www.oasyssouthasia.info.

Bhattacharyya, S. C. 2011. *Energy Economics: Concepts, Issues, Markets and Governance*. London: Springer.

Bhattacharyya, S. C., and S. Ohiare. 2011. *The Chinese Electricity Access Model for Rural Electrification: Approach, Experience and Lessons for Others*. Working Paper 10. OASYS South Asia Project, see www.oasyssouthasia.info.

Chen, Y., G. Yang, S. Sweeney, and Y. Feng. 2010. Household Biogas Use in China: Opportunities and Constraints. *Renewable and Sustainable Energy Reviews* 14: 545–549.

Cook, P. 2011. Infrastructure, Rural Electrification and Development. *Energy for Sustainable Development* 15, 3: 304–313.

Ekouevi, K., and V. Tuntivate. 2011. *Household Energy Access for Cooking and Heating: Lessons Learned and the Way Forward*. Energy and Mining Sector Discussion Paper 23. Washington, DC: World Bank.

Foell, W., S. Pachauri, D. Spreng, and H. Zerriffi. 2011. Household Cooking Fuels and Technologies in Developing Economies. *Energy Policy* 39, 12: 7487–7498.

IEA. 2009. *World Energy Outlook 2009*. Paris: International Energy Agency.

IEA. 2010. *World Energy Outlook 2010*. Paris: International Energy Agency.

IEA. 2011. *Energy for All: Financing Access for the Poor*. Special early excerpt of World Energy Outlook 2011. Paris: International Energy Agency.

Palit, D., and A. Chaurey. 2011. Off-grid Rural Electrification Experiences from South Asia: Status and Best Practices. *Energy for Sustainable Development* 15, 3: 266–276.

UNDP-WHO. 2009. *The Energy Access Situation in Developing Countries: A Review Focusing on the Least-Developed Countries and Sub-Saharan Africa*. New York: United Nations Development Programme.

Chapter 15

Resource Governance

Andrew Bauer and Juan Carlos Quiroz

Introduction

Oil, natural gas, and mineral resources do not only constitute a source of energy and raw materials; they are also a principal source of income for resource-rich countries. From 2006–2010, oil and gas constituted over half of government revenues or over 60% of export earnings in 27 producing countries. Over the same period, minerals, including coal, constituted over 30% of government revenues or over 40% of export earnings in an additional 10 countries (see Figure 15.1 for an illustration of resource-dependent countries). What's more, undiscovered resources and uncollected revenues offer many countries the possibility of becoming resource-rich, especially in Africa, Asia, and South America (Collier 2010). These vast resources should provide an opportunity for growth and development. In reality, this potential often remains unrealized.

Oil-producing countries are less democratic and more secretive than those without oil. Most have not grown as quickly as they should have given their natural resource wealth (Ross 2012). The instability, uncertainty and, in some cases, violent conflict stemming from frustrations associated with resource mismanagement, has implied higher costs and price volatility for energy consumers around the world. As such, poor resource governance threatens global energy supplies and has stymied national development in many regions, particularly in Africa, Central Asia, and the Middle East.

This chapter will examine the steps necessary for transforming extractive resource wealth into well-being and stability. While corporate responsibilities toward the societies where they operate are an essential component for good resource governance, the focus will be on public policy and the interaction between non-state actors – companies, international institutions, civil society, and citizens – and the state. The following section will define resource governance and briefly diagnose the reasons why countries rich in natural resources tend to perform below potential. The next section will provide some basic principles for managing oil, gas, and mineral resources for the public good using the *Natural*

The Handbook of Global Energy Policy, First Edition. Edited by Andreas Goldthau.
© 2013 John Wiley & Sons, Ltd. Published 2013 by John Wiley & Sons, Ltd.

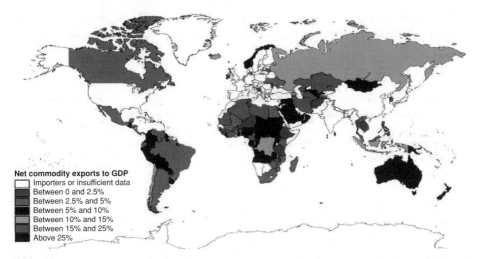

Net commodity exports to GDP
- Importers or insufficient data
- Between 0 and 2.5%
- Between 2.5% and 5%
- Between 5% and 10%
- Between 10% and 15%
- Between 15% and 25%
- Above 25%

Figure 15.1   Resource-dependent countries.
Source: IMF (2012).

Resource Value Chain as a framework. The third section will describe specific policies that resource-rich countries can implement to take full advantage of their resources. The chapter concludes by offering recommendations on how to improve resource governance globally.

Resource Governance

Good resource governance is the effective, accountable, and transparent management of oil, gas, and mineral resources. This definition implies the enactment of rules to promote the use of natural resources to improve public welfare and as well as strengthening public institutions, like the justice system and oversight bodies, to enforce these rules. In most cases, it also requires political will to transform subsoil assets into tangible benefits for citizens.

Managing oil, gas, and mineral resources for the public good has never been easy. As far back as the fifth century BC, industrial expansion of iron demand and production – a result of the invention of the double-acting bellow – created a class of iron industrialists in China that threatened national stability and the power of the Han dynasty (Lynch 2002). Intensive production drove labor and capital from the Chinese economy's traditional strength, agriculture, to the mining sector. Emperor Wu's response to the perceived misbalance between private and public benefits was to nationalize the iron industry by 117 BC.

Similarly, in 483 BC, a major silver vein strike at the Athenian mines of Laurium provoked a fierce debate between Themistocles and Aristides over the management of resource revenues. Foreshadowing twenty-first-century debates in Iraq and Mongolia, Aristides argued that mining revenues should be parceled out to Athenian citizens as per tradition. Themistocles, in turn, made the novel suggestion to invest these large revenues for the public good, in this case to build a fleet of ships. The investment paid off two years later when the Persians attacked Athens and were repulsed by those same ships (Plutarch).

The Chinese Imperial Order banning mining activity in AD 1078 is the first well-documented regulatory response to poor resource governance. A mining commission at the time wrote:

> Nature has provided us with excellent deposits. These deposits were capable of producing much profit to the people. The officials, thinking that there was very much money in mining business, wished to take it for themselves, so that in every mining district corrupt practices grew up amongst them, to the very great injury of the people. For this reason the rich refuse to devote their capital to mining and mining enterprises are gradually ruined. [. . .] It will thus be seen that Chinese mining affairs are exceedingly badly managed. (Collins 1918)

Thankfully, our understanding of the causes of poor social outcomes from resource extraction has evolved over the past centuries. The observation that countries rich in natural resources tend to perform below potential has given rise to a considerable body of literature seeking an explanation. Although the empirical evidence is not uncontested, the most compelling explanations of the negative relation range from purely economic arguments to political failures.

First, governments are often at a disadvantage relative to companies in negotiating contracts. Oil and mining companies often know more about the value of the resource, the geology, and the terms of international contracts, putting them in strong bargaining positions relative to governments (Humphreys et al. 2007). In many cases, they also have better access to economic and legal expertise, not to mention strong industry associations working hard to tout the benefits of favorable fiscal terms. This can result in countries collecting a negligible share of resource revenues. In 2008, for example, the Democratic Republic of the Congo (DRC) collected only $92 million in mineral taxes and tariffs on estimated mineral exports of $2 billion. And Cameroon collects approximately 12 cents on the dollar for its oil compared to Norway's 78 cents. A fiscal regime that fails to distribute enough revenue to the host country can fail to effectively compensate the state and communities for the value of its depleting resources, and can foster citizen dissatisfaction and national instability. A lack of revenue can also starve the state of necessary funding for domestic investment.

A second set of explanations centers on the role that large capital inflows have on the macro-economy. In the late 1970s, economists began documenting the rise of a "booming sector" and decline of other export sectors in a number of resource-rich countries, namely Australia, the Netherlands, the UK, Norway, and OPEC countries. Nicknamed the "Dutch Disease" after the perceived negative effects of North Sea natural gas discoveries in the 1970s, the phenomenon was explained as an interaction between two effects, a spending effect and a resource pull effect. In the first place, large capital inflows cause an appreciation of the real exchange rate, making non-oil or mineral exports less competitive and leading to a decline of these sectors. Second, the appreciated exchange rate provokes consumers to shift expenditures from domestic to foreign goods, leading to a further shift of domestic resources away from export sectors to the natural resource sector and "non-tradeables" (Corden and Neary 1982). There is significant evidence of continuous Dutch Disease impacts in Iran, Nigeria, Russia, Trinidad and Tobago, and Venezuela over the past 30 years with several other countries experiencing milder forms of the "disease" for shorter periods of time (Darvas 2012; Ismail 2010).

A third set of explanations arises from the effects of volatile and unpredictable government and private sector revenues stemming from variable prices and production volumes. In countries dependent on resource revenues, large and unpredictable shocks to public revenues have three effects. First, they make long-term planning difficult, especially for multi-year capital expenditures that require guaranteed financing over many years. This often results in "pro-cyclical" fiscal policy, underspending when revenues are low and overspending in good times. These boom-bust cycles can have an impact on the types of expenditures chosen by governments. Large revenue windfalls lend themselves to spending on expensive legacy projects such as airports, hospitals, and monuments rather than slow scaling-up of education or health programs. In countless cases, governments have engaged in wasteful spending when revenues have been high and undertaken painful cuts when revenues have declined. Boom-bust cycles can also lead to significant inefficiencies where bureaucracies find it difficult to adjust to a large scaling-up of expenditure, leading to poorly conceived, designed, and executed projects (Ramey and Ramey 1995). That said, succumbing to boom-bust cycles is not inevitable. While some governments have exacerbated the negative effects of oil revenue volatility with pro-cyclical "money-in, money-out" fiscal policy (e.g., Iran; Venezuela), others, like Norway and Saudi Arabia, have tried to counteract these effects through "counter-cyclical" fiscal policy, as illustrated in Figure 15.2. The positive consequences of these actions is not negligible.

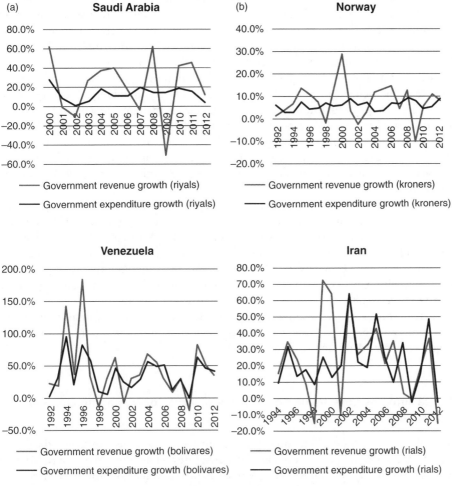

Figure 15.2 Counter-cyclical and pro-cyclical fiscal policy in four oil-dependent countries.
Source: Data from the IMF World Economic Outook database.

Pallage and Robe (2003) estimated that eliminating output volatility in Sub-saharan Africa would increase annual GDP growth by 1% indefinitely. Second, private businesses are affected by revenue volatility, expanding production and overborrowing when government expenditures are high, and suffering when the government cuts spending sharply and demand drops. In this way government expenditure volatility can lead to bankruptcies in the wider economy, as we saw in Kazakhstan post-crisis and are starting to see in Mongolia today (Esanov and Kuralbayeva 2011). Finally, large windfalls can encourage overborrowing as resource-rich governments and businesses are considered more creditworthy when projected revenues are high. In the 1970s, for example, the governments of Mexico, Nigeria, and Venezuela borrowed heavily against their oil revenues, provoking debt crises in the 1980s when oil prices, and hence revenues, declined (Humphreys *et al.* 2007).

A fourth set focuses on rent-seeking activities associated with oil, gas, and mineral extraction and their effects on institutional development. Large single-point sources of revenue are relatively easily captured by powerful elites. As such, elites in natural resource-rich countries are not only less likely to invest in productive enterprises like job-creating manufacturing industries, but may even fight over control or the right to allocate these resources, purposefully dismantling societal checks, a process nicknamed "rent-seizing" (Auty 2007; Ross 2001). Some have also argued that elite focus on rent-seeking promotes corruption and is damaging to institutional development, thereby engendering weak states, low levels of public service provision, and low growth (Arezki and Brückner 2009; Bulte *et al.* 2005; Isham *et al.* 2005; Karl 1997).

Lastly, oil and mineral revenues may lead directly to less government willingness to improve public welfare since citizens demand less public accountability. According to this view, a dependence on natural resource revenues means that governments are less reliant on broad-based taxation. As a result, the linkages between government and citizens are weak and citizens demand less accountability from their governments. In turn, politicians and officials are less interested in responding to the public interest, provide less services, and focus less on diversified and broad-based growth (Brautigam *et al.* 2008; Humphreys *et al.* 2007).

Despite the lack of consensus about the causality between extractive resources and poor economic performance, there is a general agreement that this so-called "'resource curse" is not inevitable. For instance, according to the IMF (2007) it is "prudent and transparent management practices" that have allowed a range of countries including Botswana, Canada, Chile, Norway, and the United Arab Emirates to benefit from resource wealth. Given the strong link between institutional development and benefiting from resource wealth, the logical prescription is to improve the quality of institutions that manage the extractive sector and public finances. The next two sections outline these institutions and the policies that can be implemented to promote the transformation of natural resource wealth into well-being.

The Natural Resource Value Chain: A Framework for Managing Oil, Gas, and Mineral Resources

Drawing on Collier's (2007) treatment of the topic, organizations working in the field of resource governance, like the Revenue Watch Institute and the World Bank, have adopted various forms of the Natural Resource Value Chain as a framework for thinking through the steps needed to turn natural subsoil assets into broad-based sustainable development.

While the details of the value chain differ from organization to organization, in general it follows five steps.

Step 1 – Deciding to Extract: This stage covers the process of discovery and the decision to engage in extractive activities, of which the licensing of exploration rights is a principal activity. Decisions must be taken on the size of plots to be licensed, the financing of exploration, and who gets the right to prospect. Once a discovery is made, the decision to extract should be based on a rigorous assessment of the local and national costs and benefits of drilling a hole in the ground or digging a pit and extracting the resource, including a full analysis of the social, economic, and environmental impacts. All too often, this step is skipped or public authorities do not have the willingness or capacity to actively oversee the process or make their own independent assessments.

Step 2 – Getting a Good Deal: Once a petroleum or mineral deposit has been discovered, a decision must be made on how to share the rents between extractive companies, governments, and communities. In most cases, rent-sharing is determined via tax regimes, contracts, and legal jurisdiction between national and local governments. What constitutes a "fair" or "good" deal is up for considerable debate. However, as previously mentioned, it is clear that in many countries rents are not shared equitably, leading to unstable fiscal regimes and, as we have seen in the Niger Delta (Nigeria) and Mindanao (Philippines), violent conflict from a lack of perceived equity.

Step 3 – Revenue Assessment and Collection: Often, governments do not collect either what was due or what was expected. Reasons for undercollection range from tax avoidance via transfer pricing and exploitation of archaic tax incentives to poor administrative systems and controls, lack of government oversight, and corruption. Communities are often similarly disappointed, having been promised transformative jobs, revenues, and economic development only to find that extractive activities are not major job producers, they come with significant environmental impacts, and large benefits do not accrue to local populations. Full disclosure of contracts, production volumes, payments, costs, and profits – along with improved public sector capacity to monitor production and contracts, oversee tax collection, and enforce rules – would enable revenue collection and help manage public expectations about impacts of extraction.

Step 4 – Managing Large, Finite, and Volatile Revenues: Even if all revenues due are collected, these revenues may still be mismanaged or wasted. First, governments may not have the capacity to increase public investment as quickly as revenues rise, leading to wasteful spending. This is referred to as a lack of "absorptive capacity." Second, volatile revenues lead to pro-cyclical fiscal policy, as outlined in the preceding section. Third, since revenues are finite, once they run out the government may have to cut public spending drastically, leading to a severe contraction of the economy, as Yemen is currently discovering. This step therefore requires the government to make revenue projections, think through how much revenue to save and spend, and then decide how to allocate revenues between the national government, local governments, special funds, state-owned oil or mining companies, communities, and individuals.

Step 5 – Investing in Sustainable Development: Resource-rich developing countries that collect a large share of the rents are wealthy and have access to capital yet are investment poor. In those that are authoritarian or controlled by oil or mining elites, spending decisions are generally less susceptible to public pressure. Many also lack the capacity to spend effectively. Development plans and expenditure systems must be tailored to these unique environments. Countries must strengthen capital investment processes such as procurement systems and project assessments, make budgets transparent, introduce accountability mechanisms that promote spending for the national interest, and focus on diversifying the economy away from the resource sector.

Principles of transparency and accountability straddle each of these five steps.[1] As the IMF (2007) and the Natural Resource Charter (2010) make clear, citizens can only feel confident about the integrity of the public decision-making processes if they are informed. Transparency aligns public expectations with government objectives, builds public trust, and reduces internal conflict by creating a consensus around the role of extractives.

Public disclosure requirements can also improve the quality of data the government gathers and maintains, thereby making the jobs of ministries and regulatory agencies easier. This can improve the efficiency and effectiveness of government policies. In a remarkable example, the 2005 Extractive Industries Transparency Initiative (EITI) report on Nigeria – which was designed to shed light on previously secretive oil sector payments – identified $4.7 billion in unpaid bills from the national oil company, the Nigerian National Petroleum Corporation (NNPC), to the government, in addition to over $560 million in unpaid taxes and other payments from private oil companies.

As well, transparency can foster more favorable access to domestic and international capital markets by strengthening credibility and investor confidence, as well as public understanding of government policies and choices. It can also help to highlight potential risks, resulting in an earlier and smoother fiscal policy response to changing economic conditions and thereby reducing the incidence and severity of crises.

Finally, transparency can improve public accountability. A well-informed public with the capacity to act can engage in a constructive discussion around policy formulation and government oversight of resource governance. Through public scrutiny, officials can be deterred from acting unethically and held accountable for abuses of power for private gain. Accountability, in turn, is critical to ensuring the sustainability of fiscal terms, revenue management systems, and budget decisions because it encourages adherence to rules and principles of efficient economic policy-making and effective management of public resources on timelines beyond officials' own tenure in power.

Policies for Improved Resource Governance

Understanding the principles of good resource governance is not enough to guarantee good resource governance. Public policy must reflect the complex and unique political realities of resource-rich countries. This section will elaborate on some of these policies and will offer suggestions on how to maximize the public benefits of extraction in real world contexts. It will begin with the assumption that a country has made the decision to extract. Given the relatively generic nature of Step 5 – Investing in Sustainable Development, we focus on Steps 2 through 4 of the Natural Resource Value Chain.

Getting a Good Deal

The extractive sector is different from other types of industries due to the central role that economic rents play in it. Discovery is a geological lottery as much for countries as for companies; some countries are endowed with very large reserves while others have almost none. However, once a discovery is made, the cost of extraction is generally a fraction of the value of the resource, creating potential for large profits and economic rents.[2] Countries must therefore decide how to share that rent between different levels of government, special interest groups, and extractive companies. This balance is sometimes difficult to achieve in practice. If rent distribution is overly favorable to companies, citizens and governments can demand a larger share, leading to unstable contracts or even violence. On the other hand, if companies do not receive an adequate share, investment may dry up or production may stop.

Further complicating matters, oil, gas, and mineral exploration and production is fraught with uncertainty. The risk of drilling a dry oil well can be as high as 9-in-10 in underexplored petroleum basins and mineral exploration typically only follows after screening hundreds of mineral occurrences. Production requires large capital investments with long time-horizons. Return on investment is uncertain given the extreme volatility in oil and mineral prices. The risks of negative return, renegotiated contracts, and outright expropriation by the state after large investments have been made provide a justification for a larger than bare minimum share of rents accruing to companies.

That said, companies are at a distinct advantage in negotiating many contracts (see Box 15.1 for common types of fiscal arrangements between governments and oil, gas, and mineral companies). In general, they know more about the geology, costs of production, and international contractual standards than their negotiating partner, the government. They also often have superior legal expertise and strong lobbying powers. As a result, a significant number of contracts in both developed and developing countries strongly favor companies. In Alberta, where political and social risks are limited, the province has captured only 47% of economic rents on the oil sands over a decade, a $121 billion windfall for companies from 1999 to 2008 (Campanella 2010). The Indonesian government takes 45% of the net revenue of copper production; in contrast, government take for petroleum in Angola is 89% (McPherson 2010).[3]

Box 15.1 Types of relationships between governments and extractive companies.

In general, there are three types of fiscal arrangements employed in the oil, gas, and mineral industries.

- *Concession agreements*, which grant ownership over the resource to a company, including the rights to explore, develop, produce, and market the resource, in exchange for taxes and royalties. These are common in Canada and the US.
- *Production sharing contracts* (PSCs), which typically allocate oil, gas, or minerals to companies as reimbursements on production costs, and then split the remaining "profit oil" or gas between the operating group of companies and the government. The government maintains ownership of the resource and either sells its portion on its own, or takes cash payment from the operating companies in lieu of physical delivery of the commodity. PSCs are most common in Africa, the former Soviet Union, and parts of the Middle East (e.g., Bahrain).
- *Service contracts*, where the government owns the resource and the company is paid a fee for producing in exchange for oil, minerals, or cash. Service contracts are common in Latin America and parts of the Middle East (e.g., Iraq).

Strong domestic resource mobilization and use is the only path to genuine sustainable development and an end to aid dependency for many countries. Oil, gas, and mining very likely represent the single most significant untapped source for financing development in resource-rich low- and middle-income countries, especially in Africa. In 2008, aid flows to Sub-Saharan Africa reached $36 billion per year. Natural resource rents, by contrast, stood at $240 billion (Revenue Watch 2011). How then to ensure that countries get a good deal?

One way is to promote royalty and tax regimes that are sufficient to compensate the country for the value of the asset being depleted. The right set of fiscal terms enables a government to strike a balance between attracting the best investors and getting a good deal for the country (see Box 15.2 for key tax instruments for extractives).

Fiscal terms influence not just expected national revenue under various resource price and production scenarios, but also the enforcement of extractive agreements. The success or failure of a legal system to provide benefits for the country depends on the state's ability to manage its commitments and ensure that all parties are adhering to the rules. Thus, an analysis of any system must consider how effectively it empowers the government to enforce the terms that capture benefit for the state. Good terms will enable governments to minimize the risk of corruption, non-compliance, and use of loopholes.

Among these "loopholes" is transfer pricing. An integrated international company may use sales among various subsidiaries as a means to reduce its fiscal obligations within a particular country. A sale of mineral or petroleum output from one subsidiary to another at a price under the fair market value may serve to reduce the revenue the company reports to the government and thus limit the royalty or tax payments it owes. Similarly, by purchasing a good or service from a related company at an inflated price, a company can raise its reported costs, thereby increasing deductions and decreasing income tax liabilities. In order to limit transfer pricing abuse, a government should put in place a firm policy for the valuation of transactions between related parties, linking the prices utilized for revenue-collection assessments to objective market values wherever possible.

As well, interest payments on loans are often deductible for income-tax purposes. Integrated international companies sometimes finance subsidiaries in extractive-rich countries with extremely high levels of debt in the form of related-party loans, which means that interest payments made from the subsidiary to its parent company are deducted, limiting the subsidiary's tax liability. Governments can combat this problem by capping the level of debt that an extractive subsidiary can take on in relation to its total capitalization, or by mandating that interest payments made on debt exceeding a certain debt-to-equity ratio will not be deductible for tax purposes.

Companies that have multiple activities within one country sometimes use losses incurred in one project (say, exploration expenses from a new mine that has not yet begun production) to offset profits earned in another project, thereby reducing overall tax payments. Governments can overcome this situation through "ring-fencing," the separate taxation of activities on a project-by-project basis, which facilitates the government collecting tax revenue on a project each year that it earns a profit.

Many tax systems also allow a taxpayer to deduct losses generated in one year from income earned in a subsequent year. Such a system takes into account the heavy up-front costs necessary to get a project off the ground. But in an effort to prevent unfettered carry-forwards from overwhelmingly reducing long-term revenue generation, some governments have placed limits on them, restricting either the period of time that a loss can be kept on the books or the amount of income in any given year that can be offset by past losses.

Box 15.2 Key tax instruments for extractives.

Governments use a wide range of tax instruments in their arrangements with extractive companies. Among the most common are:

- *Bonus:* A one-time payment made upon the finalization of a contract, the launch of activities on a project, or the achievement of certain goals laid out in the law or contracts. Sizes vary, ranging from tens of thousands to even hundreds of millions of dollars for a few large petroleum projects.
- *Royalties:* Payments made to the government to compensate it for the right to extract (and purchase) a non-renewable natural resource. Most royalties are either ad valorem (based on a percentage of the value of output, e.g., 5% of the value of the minerals produced) or per unit (based on a fixed amount, e.g., $10 per ton).
- *Income Tax:* In some cases, oil, gas, and mining companies are subject to the general corporate income tax rate prevailing for all businesses in a country; in other cases, there is a special regime for these extractive sectors. Because petroleum and mining projects require heavy capital and operational investments, rules on how the tax system handles costs and deductions – the deductibility of interest payments, the depreciation of physical assets, the ability to count losses from one tax year to offset profits in a future tax year, etc. – play a major role in determining how governments and companies benefit.
- *Windfall Taxes:* Some countries have set up special tax instruments designed to give the government a greater share of project surpluses, through additional tax payments, when prices or profits exceed the levels necessary to attract investment.
- *Government Equity:* In some cases, petroleum and mining projects are set up as locally incorporated entities for which shares are divided between a private company and a state-owned company or another public body. Holding these equity stakes can give the state access to a portion of dividend payments.
- *Other Taxes and Fees:* Additional sources of fiscal revenues for the state include withholding tax on dividends and payments made overseas, excise taxes, customs duties, and land rental fees.

Finally, petroleum and mineral contracts often have clauses that establish that the law that exists on the day that the contract is signed will govern the agreement, and that subsequent legal changes will not have any effect on the contract. These "stabilization clauses" offer investors some assurance that they will not be subjected by legislative action to a drastically different fiscal regime than the one on which they based their decision to invest. But in order to protect the interests of citizens, preserve state sovereignty, and remain flexible to changing economic and political circumstances, stabilization clauses should be narrowly drafted and limited to major revenue streams such as royalties, taxes, duties, and major fees. Stabilization clauses should not freeze environmental, labor, or other similar rules.

In each country the best fiscal terms are characterized by variations in economic priorities, administrative capacities, mineral/petroleum endowments, and levels of political risk. However, there are certain considerations that governments should include in the design of any fiscal regime.

First, fiscal regimes should be clearly established by easily accessible laws and regulations. Minimizing parties' discretion to alter fiscal terms in individual contracts facilitates contract enforcement and the application of a coherent sector-wide fiscal strategy, and reduces the risk of corruption in negotiations.

Second, fiscal regimes should contain progressive elements that give the government an increasing share of revenues as profitability increases in order to improve stability of contracts and respond to citizen demands when commodity prices rise. This can be achieved using a variety of instruments, including progressive income taxes, windfall profits taxes, and variable-rate royalties.

Third, all contracts should be made public. When contracts are publicly available, government officials have an incentive to stop negotiating bad deals and citizens can better understand the complex nature of their country's agreements with industry (Maples and Rosenblum 2009).

Improved fiscal terms and closing loopholes can have huge impacts on government take. Guinea's new mining code, for example, could bring in as much as $3 billion a year in extra revenues just from iron ore. This would triple Guinea's national budget. However, fiscal terms are only one aspect of the rights and obligations between companies and governments. Other key terms include those related to the environment, local economic development, extractive work programs, community rights, dispute processes and rights to information. In order to evaluate whether or not the country is getting a "good" or "fair" deal it is necessary to weigh all of these terms together.

Revenue Collection

In any given year, average government take for oil, gas, or minerals can range from 90 cents on the dollar in countries like Azerbaijan, Iraq, and Norway to nearly zero cents in countries like the DRC. In some cases, low government share may be a result of unbalanced fiscal terms or generous tax incentives. But weak enforcement of contracts or revenue "loss" can also be contributing factors.

There are many reasons why taxes are not always fully collected. First, poor control, management, and accounting practices in tax administrations are common in developing countries. Among the challenges are complex filing and payment regimes, too many agencies responsible for collection, poorly qualified staff, and poor information management systems (Calder 2010). Second, many countries do not adequately monitor the volume and quality of oil or minerals being extracted and exported, making valuation of extraction near impossible. Informal estimates in the Philippines put underdeclaration as high as 70% of mineral value in some cases. Third, even if governments are aware of the volume and quality of resources being produced, the price used in revenue calculations can be manipulated, also leading to undervaluation. Fourth, multinational resource companies sometimes engage in aggressive tax planning, misstating prices in order to shift apparent sources of profit to the jurisdiction which provides the most advantageous tax outcome (Mullins 2010). Fifth, there are instances of outright theft by company or government officials. In one well-documented case, AmLib, a mining company, was found to have withheld $100,000 from the Liberian government. An investigation revealed that an AmLib employee had been falsifying receipts to the government and pocketing the payments.

It is to address some of these problems that the Extractive Industries Transparency Initiative (EITI) was launched in 2002. Based on voluntary principles of disclosure and compliance with validation criteria, the initiative was designed with a simple premise in mind: If companies disclose their payments to governments and governments disclose payments received from companies, these figures can be verified against each other to ensure that all revenues make their way into the public coffers. Furthermore, if these payments are transparent, citizens are better able to hold their governments to account for the revenues generated from extractive resources. As of December 2012, 16 countries have been declared "EITI compliant" and 21 are "candidate countries," of which one has been suspended.

While the EITI has faced difficulties, namely regarding inconsistent data quality and coverage, it has nonetheless proven invaluable at drawing citizen and international attention to resource governance where it is most needed (Gillies 2011). The impacts of oversight and awareness-raising are difficult to quantify; after all, how can researchers prove that an official chose to be less corrupt because of a transparency initiative? That said, several EITI reports have had clear and tangible benefits. A Nigerian EITI report, for example, uncovered $800 million in discrepancies between what companies said they paid and what the government received, an amount exceeding the individual budgets of the ministries of health and education. Of the $800 million, $560 million proved to be a shortfall in company payments (EITI 2012).

More recently, the US Congress passed the Cardin-Lugar Energy Security Through Transparency (ESTT) Amendment as part of the 2010 Dodd-Frank Act. The provision, sponsored by a Republican and a Democrat, requires extractive sector companies listed on US stock exchanges to disclose payments made to governments on a country-by-country and project-by-project basis, and to make this information available through an easy-to-use online database. While the goals of the legislation were to improve energy security by reducing the amount of revenue collected by violent groups or corrupt elements in unstable regimes, and to add stability to markets by providing information to investors on political and social risks of oil and mining projects, it also acts as a mandatory rule that complements EITI's voluntary standards. Since the US passed the provision, the European Parliament and European Council have published legislative proposals requiring similar rules. Norway has also committed to adopt these rules.

Though the EITI and Cardin-Lugar rules improve transparency of payments, much still needs to be done to ensure full tax collection. Mandatory disclosure of payments should be expanded to all major jurisdictions, including Canada and Australia where the majority of mining companies are listed and headquartered. The scope of transparency initiatives needs to be expanded to include full contract transparency and disclosure of production, cost, and profit figures, and government capacity to monitor contracts and collect taxes must be improved. This will take time but can be supported through partnerships with international financial institutions and development partners. Finally, tax treaties can be updated and international tax transparency standards can be improved to reduce opportunities for aggressive tax planning and scope for hiding profits in tax havens.

Revenue Management

The governments of Azerbaijan and Norway both get a good deal on their oil, making approximately 80–90 cents for every dollar of revenues net of costs. They are also both large producers and have similar production paths. According to their respective EITI reports and the BP statistical review, Azerbaijan collected $19 billion in 2010 and has 7 billion barrels remaining while Norway collected over $44 billion in 2009 with 6.7 billion remaining barrels. Yet their management of oil revenues has been very different. Norway has invested heavily in health, education, infrastructure, and the oil sector, went from one of Europe's poorest countries per capita to one of its richest in four decades, and has saved $553 billion of oil revenue in its Government Pension Fund as of end 2011. In contrast, over 130 years after industrial oil production started on what is now Azerbaijani territory (Azerbaijan gained independence in 1991), 29% of the Azerbaijani rural population still does not have easy access to clean water, the government spends only 11% of its national budget on education, and it has saved only $33 billion in its State Oil Fund as of October 2012 (World Bank statistics).

Norway and Azerbaijan exemplify two extremes of a public policy choice: to consume for the benefit of the present generation or invest or save for the benefit of future generations. Following a large discovery, the temptation to consume today can sometimes be overwhelming, both for direct political gain and to respond to public expectations of immediate transformational benefits. Moreover, once money has started to flow, it is often difficult to shut the faucet. However consumption can also be well justified. Over 1.5 billion people in resource-rich countries live off less than $2 per day. These acute needs must be addressed today.

Many countries have employed special administrative rules and mechanisms to help rebalance the extractive revenue savings-investment-consumption ratio while also reducing budget volatility. The most common are formal or informal *fiscal rules* limiting public spending. For example, Botswana has enacted a rule limiting public expenditure to 40% of GDP while Australia and Peru have restricted their public debt to GDP ratios. However the most common set of fiscal rules in resource-rich countries are resource-based "revenue rules," ceilings on the amount of oil or mining revenue that is allowed to enter the budget. Ghana, Mexico, Nigeria, and Timor-Leste have each enacted a legally binding revenue rule. For example, Ghana's rule says that a maximum of 70% of a seven-year average of oil revenues shall be allocated to the national budget. Of the remaining amount, a minimum 30% is deposited into a Heritage Fund for future generations and a maximum of 70% is deposited into a stabilization fund to mitigate unanticipated revenue shortfalls. Figure 15.3 illustrates how this kind of fiscal rule works, creating fiscal surpluses to be saved for future generations and "smoothing" public expenditures to mitigate the negative effects of revenue volatility on the national budget.

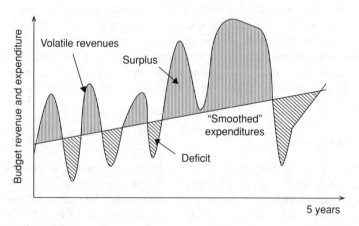

Figure 15.3   Effect of an expenditure or revenue rule.

Table 15.1 Selected natural resource funds.

Country or Province	Current Name	Objective	Founded	Assets (billion USD, 2011)
Abu Dhabi (UAE)	Abu Dhabi Investment Authority	Savings	1976	627
Norway	Government Pension Fund Global	Savings	1990	520
Kuwait	Kuwait General Reserve Fund	Multiple	1953	296 combined
	Reserve Fund for Future Generations	Savings	1976	
Kazakhstan	National Fund	Stabilization	2000	39
Chile	Pension Reserve Fund	Savings	2006	4
	Economic and Social Stabilization Fund	Stabilization	2007	13
Alberta (Canada)	Alberta Heritage Savings Trust Fund	Savings	1976	15
Timor-Leste	Petroleum Fund	Savings	2005	9
	Infrastructure and Human Capacity Development Funds	Development	2011	1
Botswana	Pula Fund	Savings	1996	7
Mexico	Oil Revenues Stabilization Fund	Stabilization	2000	6

Source: Bauer (2012).

Among fiscal rules, revenue rules in particular require governments to decide what to do in instances when revenues are greater than the rule allows. In other words, they must decide where to save revenue surpluses. In cases where they are not held by the central bank as reserves or in private accounts of corrupt officials, they are saved in natural resource funds (NRFs), a subset of sovereign wealth funds (SWFs). These funds have proliferated in resource-rich countries; in 2009, there were 48 active NRFs in 34 countries, representing more than $2.03 trillion in assets (see Table 15.1 for selected natural resource funds).

NRFs have been established with several objectives in mind. First, they save resource revenues, accumulating interest for the benefit of future generations and for use in the event of crises (e.g., Abu Dhabi Investment Authority; Trinidad and Tobago Heritage Fund). Second, some provide a source of funds to be drawn upon in cases of revenue shortfall in the budget. These "stabilization funds" absorb revenue volatility, their balance sheets oscillating with the cyclical rise and fall of oil and mineral prices and production (e.g., Mexico's three Oil Income Stabilization Funds and Chile's Economic and Social Stabilization Fund). Third, they can help "sterilize" capital inflows, essentially redirecting capital inflows to foreign investments to mitigate Dutch Disease effects. The Kazakh, Kuwaiti, and Saudi NRFs may have helped sterilize capital inflows from 2001 to 2007 when oil prices increased by about 250% but none exhibited large-scale real exchange rate appreciation despite being on fixed exchange rates. Fourth, provided their financial statements are published, they can improve the transparency and accountability of resource revenues by collecting all resource revenues in one place and streamlining saving-spending decisions, as in Ghana and Timor-Leste. Fifth, they can encourage investment by earmarking resource revenues for development projects or capital expenditures (e.g., Timor-Leste Infrastructure and Human Capacity Development Funds).

Many funds have multiple objectives. For example, the Wyoming Permanent Mineral Trust Fund was established to limit government spending and provide funding for the US State in cases of crisis, though in practice it also acts as a stabilization fund.

The deposit, withdrawal, investment, and governance rules that regulate NRFs often determine how effective they are at helping countries overcome overspending, expenditure volatility, Dutch Disease, and rent-seizing. While a one-size-fits-all approach to these rules is not appropriate, there are conditions under which NRFs are more likely to function in the public interest.

First, NRF objectives should be clearly stated by the government and should be functions of country-specific revenue management challenges such as excessive expenditure volatility, excessive recurrent spending, or lack of capital investment.

Second, the rules governing which revenue streams (e.g., royalties; government equity; excise taxes; corporate taxes; fees) are deposited into the fund and the conditions for withdrawal should be a reflection of the fund's objectives. For example, if a fund is given a stabilization objective, it may be appropriate to include only volatile streams, as in the case of the Wyoming Permanent Mineral Trust Fund. On the other hand, *all* revenue streams ought to be deposited into a fund designed to save for future generations. Similarly, it may be appropriate to prevent withdrawal from a savings fund until production has ceased, as in the case of the Ghana Heritage Fund, while a stabilization fund requires rules that help absorb budget volatility (see Figure 15.4 for two natural resource fund models with example deposit and withdrawal rules). That said, the vast majority of NRFs do not have clear or legislated rules. Iran is a case in point. Based on an analysis of data from the Central Bank of Iran and the IMF, Heuty (2012) estimates that the government withdrew over $150 billion from the Oil Stabilization Fund between 2006 and 2011 without clear economic justification. The Azerbaijani and Kuwaiti funds similarly lack withdrawal rules, with similar results.

Third, some countries have generated additional fiscal space or greater fiscal flexibility by manipulating the price and revenue assumptions necessary for the calculation of permissible withdrawals. In order to adhere to the spirit of the rules, it is essential that the formula for projecting resource revenues not be subject to manipulation. Price assumptions and revenue projections should follow strict criteria. Including an objective price assumption in legislation or regulation, such as the World Energy Outlook's intermediate-scenario oil price forecast, may be helpful in this regard.

Fourth, NRFs should have clear investment rules consistent with objectives and capacity to invest. Since they must be able to finance quarterly or annual shortfall in government revenue, stabilization fund assets must be sufficiently liquid and low-risk. Conversely, it may be appropriate to invest in higher-risk or alternative assets to generate a higher return if the fund has a savings objective. Of course, the fund's risk profile should also reflect the capacity to manage investment or to oversee the investment managers. The Libyan Investment Authority (LIA) offers a cautionary tale, having paid high management fees and received poor advice as a result of inadequate safeguards, rules or transparency under the Gadhafi regime, resulting in over $1 billion in losses. Institutions that invest public money should be held to a higher standard of transparency, accountability, and risk mitigation.

Fifth, governments may find that using resource funds as collateral on debt, especially if the fund is large, can provide access to finance on good terms. However the value of improved borrowing terms must be balanced against the risk of squandering the public's savings. Furthermore, if savings are large, there is less justification for borrowing externally since the government can draw on these resources. The choice to encumber all

One fund approach

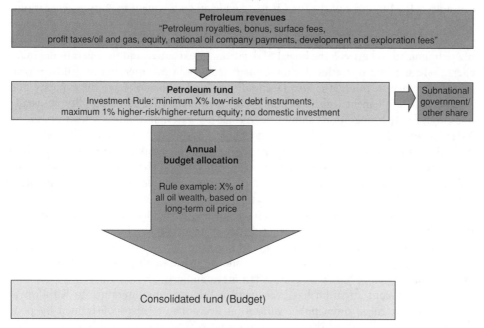

Two fund approach

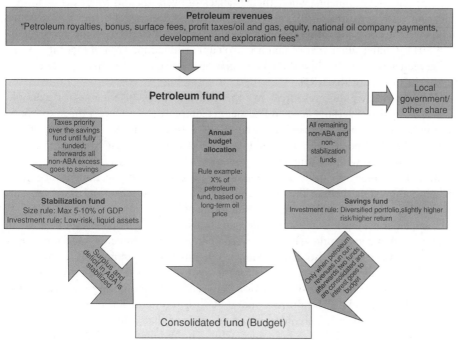

Figure 15.4 One and two fund approaches to managing resource revenues.

or part of resource revenues should, again, be a reflection of the fund's objectives. For example, if the objective of the fund is to provide an endowment for future generations, over-risking the fund may not serve this objective.

Sixth, faithful implementation of fiscal rules requires clear responsibilities and lines of communication between the board of directors or executive and the operational managers. Safeguarding operational management from political interference can help promote adherence to rules. It is also essential that fund managers have the capacity and incentives to implement legislated policies. However internal accountability is usually not enough. NRF operations should be verified by external auditors and overseen by independent bodies with the capacity and technical knowledge to analyze NRF behavior. Formalized monitoring of activities by parliament and the establishment of a specialized oversight committee consisting of government officials and civil society, like the Ghana Public Interest and Accountability Committee (PIAC), may be necessary for promoting adherence to the rules.

Finally, in order for oversight actors (civil society, parliamentarians, media) to fulfill their roles, there must also be a strong degree of transparency in all NRF operations. Information on annual activities, investment portfolio, flow-of-funds, governance structure, board of directors, and managers should be made publicly available in a timely manner and reporting should meet international standards.

While significant attention has been paid in the academic literature to withdrawal and investment rules, often public allocation decisions are also constrained by poor deposit rules, specifically as they pertain to transfers from state-owned resource companies (SOCs). National oil companies (NOCs) can be a particular drain on public resources in low-capacity environments, where there is a lack of oversight and where dividend, cost calculation, and reinvestment rules are unclear or poorly designed. Often, NOC revenues remain with the company rather than making their way to the NRF or budget. For example, in 2011, Ghana's first year of major petroleum production, 47% of the state's share of oil revenues was retained by the Ghana National Petroleum Company, leaving very little for the budget or the NRFs.

NOCs can drain public resources in other ways as well. In Nigeria, for example, "cash calls" by the Nigerian National Petroleum Company (NNPC) cost the taxpayer $7 billion in 2010, and petrol subsidies by the NNPC cost the state $11 billion in 2008/9. In Mexico, PEMEX posted a $30 billion loss in 2009, covered by the Mexican taxpayer. The risks associated with creating and empowering a SOC deserve a full chapter. Still, in general, these challenges can be overcome by aligning SOC and government incentives, improving and strengthening the rules governing SOCs, and ensuring that SOCs are accountable to the government, regulators, and the public.

The saving-consumption-investment decision is also a function of the resource revenue allocation system between the national government, local governments, special interests, and citizens. Revenues are distributed to local governments or communities for four main reasons: To compensate producing areas for the costs of production (e.g., environmental damage); equalize the benefits of extraction between richer and poorer regions; prevent or reduce conflict; and decentralize accountability for revenues. The challenge countries face is how to design a widely accepted stable system that achieves these objectives. In fact, very few countries have been successful. The Indonesian and Peruvian oil and mineral revenue-sharing regimes have arguably increased mismanagement and waste. The Nigerian and Mongolian regimes have increased inequality between producing and non-producing regions. It is unclear whether revenue sharing has mitigated or exacerbated violent conflict in the DRC, Iraq, and Nigeria. On the other hand, as Canada, Mexico, and the UAE have shown, a widely accepted formula-based, predictable, and transparent regime can be conducive to improving development outcomes and can help countries meet their revenue-sharing objectives.

Recently, researchers at the Center for Global Development (CGD) and the World Bank have advocated for a specific type of revenue allocation as a means of improving accountability of resource revenues and spending efficiency: direct transfers of oil and mineral revenues to citizens. Under the CGD proposal (Moss 2011), "a government would transfer some or all of the revenue from natural resource extraction to citizens in a universal, transparent, and regular payment. Having put this money in the hands of its citizens, the state would treat it like normal income and tax it accordingly – thus forcing the state to collect taxes and fueling public demand for the government to be transparent and accountable in its management of natural resource revenues and in the delivery of public services." Already, Alaska, Bolivia, Iran, and Mongolia have adopted these types of payments and Iraqi officials are currently considering doing the same. Iran's "oil-to-cash" transfer program started awarding about $40 a month to over 70 million citizens in 2010, costing about $45 billion a year (Heuty 2012).

Since unconditional lump-sum transfers would benefit the poor relative to the rich, there is justification for these direct transfers in countries with high levels of poverty. However income taxes are difficult to introduce in many less developed resource-rich countries and direct distribution of cash may starve the public sector of much needed financing for investment. Governments may also not develop the capacity to deliver on infrastructure and human capital projects. After all, governments, like businesses, learn by doing. In summary, managing resource revenues is not simply a matter of collecting revenues and depositing them into the national budget. The flow of funds is often muddled by the institutions involved, generating opportunities for waste, mismanagement, and rent-seeking (see Figure 15.5). These systems can also make it more difficult to address macroeconomic challenges commonly found in resource-dependent countries, namely budget volatility, a tendency to overconsume, and Dutch Disease. Key institutions should be governed by predictable and appropriate rules, their activities should be transparent and they should be subject to rigorous oversight by independent, competent and credible bodies. While these steps do not guarantee that resource revenues will be transformed into useful investments, the probability that governments will make the right choices increases with each.

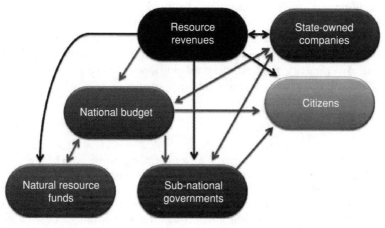

Figure 15.5 Typical resource revenue management system.

Conclusion

Oil, gas, and mineral resources are different from other sources of income. They are non-renewable, therefore only worth extracting if the country is adequately compensated for asset depletion. They are location-specific with high upfront costs and production timelines, providing leverage to government in contract negotiations, notwithstanding the information asymmetries between companies and the government. They generate substantial economic rents. Revenues generated from these resources are highly volatile. And, in many countries, they are large, with the potential to transform economies and societies. As such, these resources should be managed differently.

For example, contracts and company financial information should be made public. Fiscal terms should reflect practical limits to full tax collection, like transfer pricing or poor domestic administrative capacity. The management of resource revenues should be rules-based and transparent, with strong fiscal rules that promote appropriate levels of public savings, consumption, and investment, and "smooth" public expenditures. Fiscal rules should be enforced and compliance should be subject to oversight, for example by auditors, legislatures, the media, and civil society groups. Public institutions, like natural resource funds and state-owned companies, should also be administered largely by rules and be similarly subject to oversight. In general, resource revenues should be spent in the best public interest and governments should be accountable to the public for their use.

Energy consumers have a strategic interest in promoting good resource governance. In many producing countries, extractive revenues have fed corruption, wasteful spending, violent conflict, and instability rather than productive investment. This has exacerbated poverty, increased attacks on extraction sites, and empowered unstable regimes, creating significant uncertainty in world resource supplies. Improved governance fosters trust with local communities, reducing investment risk and creating more stable operating environments for companies. In turn, improved reliability of commodity supplies promotes greater energy security.

As such, the international community has an incentive to support good governance initiatives in large producing countries and in the home markets of extractive companies. For example, many developing countries may need to enhance domestic capacity to negotiate contracts, analyze fiscal regimes, collect taxes, manage resource revenues, and spend efficiently on projects that support growth and economic diversification. While capacity building and development initiatives should be driven by domestic imperatives, development partners and international financial institutions can help finance and support these activities. The entire international community should also support and adhere to best international standards of resource management, namely mandatory disclosure of payments to governments, publication of all contracts, elimination of tax havens, and improving accountability of companies in home jurisdictions.

Notes

1. *Transparency* is defined as clarity of government agencies' and officials' roles and responsibilities, public availability and access to information, open budget preparation, execution and reporting, and assurances of integrity, including high data quality and adequate oversight (IMF 2007). *Public accountability* is defined as the obligation of public officials to explain and justify their conduct and make decisions based on a concept of public service.

2. Economic rents are the returns to investment over and above the minimum required to invest.

3. "Government take" is the sum of royalties, taxes, and the value of "profit oil." See Box 15.1 for explanation of "profit oil."

References

Arezki, Rabah, and Markus Brückner. 2009. Oil Rents, Corruption, and State Stability: Evidence from Panel Data Regressions. IMF Working Paper 09/267. Washington, DC: International Monetary Fund.

Auty, Richard. 2007. Patterns of Rent-Extraction and Deployment in Developing Countries: Implications for Governance, Economic Policy and Performance. In G. Mavrotas and A. Shorrocks, eds. *Advancing Development: Core Themes in Global Economics*. London: Palgrave Macmillan, pp. 555–577.

Bauer, Andrew. 2012. *Managing and Spending Revenues Well*. New York: RWI. http://www.reve nuewatch.org/publications/managing-and-spending-resource-revenues-well, accessed November 28, 2012.

Brautigam, Deborah, Odd-Helge Fjeldstad, and Mick Moore. 2008. *Taxation and State Building in Developing Countries: Capacity and Consent*. Cambridge: Cambridge University Press.

Bulte, Erwin, Richard Damania, and Robert Deacon. 2005. Resource Intensity, Institutions and Development. *World Development* 33 7: 1029–1044.

Calder, Jack. 2010. Resource Tax Administration: Functions, Procedures and Institutions. In P. Daniel, M. Keen, and C. McPherson, eds. *The Taxation of Petroleum and Minerals: Principles, Problems and Practice*. London: Routledge.

Campanella, David. 2010. *Misplaced Generosity: Extraordinary Profits in Alberta's Oil and Gas Industry*. Edmonton: Parkland Institute.

Collier, Paul. 2007. *The Bottom Billion*. Oxford: Oxford University Press.

Collier, Paul. 2010. *The Plundered Planet*. Oxford: Oxford University Press.

Collins, William F. 1918. *Mineral Enterprise in China*. London: W. Heinemann.

Corden, Max, and Peter Neary. 1982. Booming Sector and De-Industrialization in a Small Open Economy. *Economic Journal* 92 (368): 825–848.

Darvas, Zsolt. 2012. *Real Effective Exchange Rates for 178 Countries: A New Database*. Working Paper 2012/06. Brussels: Bruegel.

EITI. 2012. *Nigeria EITI: Making Transparency Count, Uncovering Billions*. Oslo: EITI. http://eiti.org/document/case-study-nigeria.

Esanov, Akram, and Karlygash Kuralbayeva. 2011. Kazakhstan: Public Saving and Private Spending. In P. Collier and A. Venables, eds. *Plundered Nations? Successes and Failures in Natural Resource Extraction*. London: Palgrave Macmillan.

Gillies, Alexandra. 2011. *What Do the Numbers Say? Analyzing Report Data*. New York: RWI. http://data.revenuewatch.org/eiti/, accessed November 28, 2012.

Heuty, Antoine. 2012. *Iran's Oil and Gas Management*. New York: RWI. http://www.reve nuewatch.org/publications/iran%E2%80%99s-oil-and-gas-management, accessed November 28, 2012.

Humphreys, Macartan, Jeffrey Sachs, and Joseph Stiglitz. 2007. Introduction: What is the Problem with Natural Resource Wealth? In M. Humphreys, J. Sachs, and J. Stiglitz, eds. *Escaping the Resource Curse*. New York: Columbia University Press.

IMF. 2007. *Guide on Resource Revenue Transparency*. Washington, DC: International Monetary Fund.

IMF. 2012. *World Economic Outlook: Growth Resuming, Dangers Remain*. Washington, DC: International Monetary Fund.

Isham, Jonathan, Michael Woolcock, Lant Pritchett, and Gwen Busby. 2005. The Varieties of Resource Experience: Natural Resource Export Structures and the Political Economy of Economic Growth. *The World Bank Economic Review* 19, 2: 141–174.

Ismail, Kareem. 2010. *The Structural Manifestation of the "Dutch Disease": The Case of Oil Exporting Countries.* IMF Working Paper 10/103. Washington, DC: International Monetary Fund.

Karl, Terry Lynn. 1997. *The Paradox of Plenty: Oil Booms and Petro-States.* Berkeley, CA: University of California Press.

Lynch, Martin. 2002. *Mining in World History.* London: Reaktion Books.

Maples, Susan, and Peter Rosenblum. 2009. *Contracts Confidential.* New York: RWI.

McPherson, Charles. 2010. State Participation in the Natural Resource Sectors: Evolution, Issues and Outlook. In P. Daniel, M. Keen, and C. McPherson, eds. *The Taxation of Petroleum and Minerals: Principles, Problems and Practice.* London: Routledge.

Moss, Todd. 2011. *Oil to Cash: Fighting the Resource Curse Through Cash Transfers.* Washington, DC: Center for Global Development. http://www.cgdev.org/content/publications/detail/1424714/, accessed November 28, 2012.

Mullins, Peter. 2010. International Tax Issues for the Resources Sector. In P. Daniel, M. Keen, and C. McPherson, eds. *The Taxation of Petroleum and Minerals: Principles, Problems and Practice.* London: Routledge.

Natural Resource Charter. 2010. http://naturalresourcecharter.org.

Pallage, Stephane, and Michel A. Robe, 2003. On the Welfare Cost of Economic Fluctuations in Developing Countries. *International Economic Review*, 44, 2: 677–698.

Plutarch. *Life of Themistocles.*

Ramey, Garey and Valerie A. Ramey. 1995. Cross-Country Evidence on the Link between Volatility and Growth. *American Economic Review*, 85, 5: 1138–1151.

Revenue Watch. 2011. International Development Committee. Written evidence submitted by the Revenue Watch Institute. UK Parliament. http://www.publications.parliament.uk/pa/cm201213/cmselect/cmintdev/130/130vw22.htm, accessed December 4, 2012.

Ross, Michael. 2001. *Timber Booms and Institutional Breakdown in Southeast Asia.* Ann Arbor: University of Michigan Press.

Ross, Michael. 2012. *The Oil Curse.* Princeton, NJ: Princeton University Press.

Chapter 16

The "Food Versus Fuel" Nexus

Robert Bailey

Introduction

The use of biofuels (liquid fuels derived from biomass: see Box 16.1) is growing. Production more than quintupled between 2000 and 2010, from just over 9 Mtoe to more than 52 Mtoe, accounting for 3% of worldwide transport energy (BP 2011). This has not happened without controversy however. In particular, the large programs of the US and European Union (EU) have driven up food prices and increased food price volatility, with serious consequences for food-insecure populations elsewhere in the world.

This chapter considers the interlinkages between food security and energy policy – the so-called food versus fuel nexus. It focuses specifically on biofuels for transport as these are by far the largest source of energy-based demand for foodstuffs. The chapter begins by examining the different rationales for biofuel policies and the social and environmental risks these policies entail. It then explores the links between biofuels, food prices, and development before considering the options available to policy-makers.

Box 16.1   Definitions

Biofuels are fuels derived from organic materials (biomass). Though this could for example include wood, generally (and in this chapter) the term biofuel is taken to refer to liquid fuels used in transport. There are two basic types of biofuel for transport: ethanol and biodiesel. Ethanol is derived from sugar or starch crops such as sugarcane, corn (maize), and wheat by a process of fermentation. It can be blended with gasoline (petrol), or with small engine modifications used unblended. Biodiesel is primarily derived from vegetable oils such as palm, sunflower, rapeseed or canola via a process of transesterification. It can be blended with diesel fuel, though again can be used pure in modified engines.

The Handbook of Global Energy Policy, First Edition. Edited by Andreas Goldthau.
© 2013 John Wiley & Sons, Ltd. Published 2013 by John Wiley & Sons, Ltd.

Biofuel Policies: Rationales and Risks

There are three principal policy rationales for biofuel support: energy security, climate change, and agricultural development.

Energy Security: Advanced economies are dependent on liquid fuels for industrial uses and transport. Meanwhile demand from emerging economies is rapidly increasing: all of the demand growth for liquid fuels over the coming decades is expected to occur outside of industrialized countries (BP 2011). No meaningful alternative to liquid fuels is foreseen in the medium term, with new technologies such as electric vehicles expected to make only marginal contributions in the coming decades. This leaves countries facing increasing competition for oil: a finite resource of vital economic importance produced predominantly in unstable regions. Biofuels offer a potential alternative.

Climate Change: Displacing oil from the energy mix also offers environmental benefits if it is done using a lower carbon alternative. In theory, biofuels offer significant greenhouse gas (GHG) savings compared to petroleum products, because the carbon they release on combustion is the same carbon that was removed from the atmosphere as the feedstock was grown: the two must net to zero. However, in practice biofuels are not carbon neutral; there are other sources of emissions associated with their production – from farm machinery, chemical fertilizers, and industrial processing. Still, after incorporating these emissions, many biofuels may offer attractive levels of GHG savings (Figure 16.1), with one important caveat: that feedstock production does not trigger emissions from land-use changes such as deforestation.

Agricultural Development: Whilst developed countries justify biofuel support on the grounds of energy and climate security, for many developing countries the principal attraction is agricultural development. It is estimated that as much as 80% of the world's hungry live in rural areas. Investing in agriculture is seen as core to poverty alleviation, hunger reduction, and for kick-starting broader-based economic development. Developing countries hope to benefit from the biofuel boom, as they have a comparative advantage in feedstock production due to low labor costs, cheap land, and the potential to grow productive feedstocks such as sugarcane or palm oil.

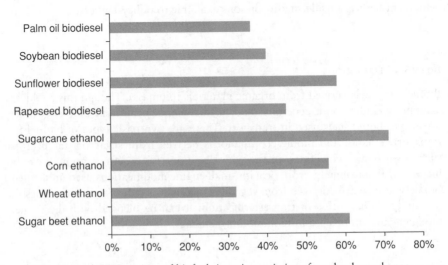

Figure 16.1 Typical GHG savings of biofuels ignoring emissions from land-use change. Source: European Commission.

Because biofuels are substitutable with petroleum products, the economics of production depend not only on the cost of the feedstock but also on the oil price. When the oil price is high, biofuels are more competitive; when the feedstock price is high, they are less competitive. Whether or not biofuel production is economically viable at any point in time therefore depends on the relative prices of oil and feedstock. Often this balance is unfavorable, with the result that public resources are required to sustain production. Governments employ a number of different policy instruments in this regard, most commonly:

- Mandates specifying an absolute usage requirement. For example, the US Renewable Fuels Standard requires the blending of 36 billion gallons of renewable fuels in 2022.
- Mandates specifying a proportion of biofuel to be blended with fossil fuel, for example the UK's Renewable Transport Fuel Obligation specifies that biofuels must constitute 5% (by volume) of road transport fuel by 2013.
- Subsidies and tax breaks to help make biofuels more competitive relative to petroleum products, for example until 2012 blenders in the US received a $0.45 tax credit for each gallon of ethanol blended with gasoline.
- Import tariffs to help shelter domestic producers from more competitive foreign producers, for example the EU maintains an import tariff on ethanol that limits imports from Brazil.

All carry economic costs. Mandates force energy consumers to buy biofuels when they are more expensive than petroleum products. Import tariffs limit the same consumers' ability to access cheaper imported biofuels. Subsidies and tax breaks represent direct transfers from taxpayers to industry. Unsurprisingly, many of these have come under pressure as governments have adopted greater fiscal discipline following the 2008 financial crisis (Jung *et al.* 2010). Nevertheless, governments show little sign of scaling back their ambitions. The International Energy Agency (IEA) forecasts that annual support for biofuels will expand from $20 billion in 2009 to $45 billion by 2020 (IEA 2010). By 2030, BP predicts that global production will have almost quadrupled (BP 2011). In particular, governments remain committed to increasing biofuel production through the expansion of mandates. There are now over 50 countries with biofuel mandates or targets in place or in implementation (IEA 2011).

Despite an increasing number of countries pursuing biofuel programs, globally production is highly concentrated. Ethanol production is now led by the US (Figure 16.2), where production and consumption have expanded rapidly in response to domestic support measures, followed by Brazil, for some time previously the world's largest producer. Biodiesel production (Figure 16.3) is currently about a quarter of ethanol output. The biggest producer is the EU, where production is increasing to meet the target of the 2009 Renewable Energy Directive, that all member states derive at least 10% of their transport energy from renewable sources by 2020. There is scope for electric vehicles to contribute, but in practice the overwhelming share will come from biodiesel and (to a lesser extent) ethanol.

The US and EU between them account for about 80% of government support worldwide for biofuels (IEA 2010). Their recent expansion has come at an economic cost to taxpayers and motorists, but has been justified by governments on energy security and environmental grounds. As biofuel production continues to expand, and particularly as more governments seek to pursue similar policies themselves, it is important to consider how successful US and EU policies have been.

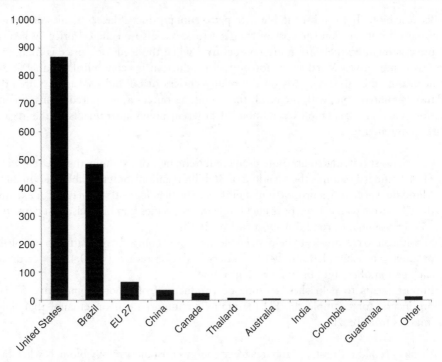

Figure 16.2   Global ethanol production, 2010 (thousand barrels per day).
Source: US Energy Information Administration.

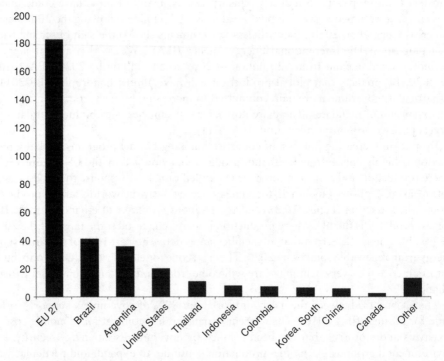

Figure 16.3   Global biodiesel production, 2010 (thousand barrels per day).
Source: US Energy Information Administration.

In terms of energy security, current biofuel technologies offer only marginal gains. Biomass has a much lower energy density than oil, so large quantities of food are needed to displace relatively small amounts of gasoline or diesel. In 2010 the US, the world's largest producer of corn, diverted about 40% of its harvest to ethanol production in order to displace less than 10% of gasoline consumption. In 2020, despite further expansion, ethanol is expected to displace 8.4% of US gasoline consumption (OECD-FAO 2011). Globally, biofuels contribute about 3% of transport energy, but use significant amounts of food production to do so: in recent years biofuels accounted for 11% of coarse grains and vegetable oil use and 21% of sugarcane use (OECD-FAO 2011).

As for climate change, biofuels are at best an expensive means of reducing emissions. Research by the Global Subsidies Initiative estimates that for the EU, the subsidy cost per tonne of CO_2-equivalent avoided is between €669 and €1422 for grain-based ethanol, compared to an average European market price of €16.25 per tonne of CO_2-equivalent over the period analyzed (Jung *et al.* 2010).

More problematically, emissions savings are only achieved under a special set of conditions: namely that feedstock production does not trigger significant emissions from land-use change. When new land is brought into production, vegetation is cleared and soils are turned over. If this land is of high carbon stock, such as rainforest or peatlands, then the emissions can be huge. One estimate found that converting Indonesian rainforest to oil palm (a biodiesel feedstock) would generate emissions so large they would take 420 years' of replacing fossil diesel with the resultant biodiesel to pay back (Fargione *et al.* 2008).

Assuming adequate auditing and verification, emissions from land-use change can be controlled by placing restrictions on the types of land used to grow biofuels – an approach being pursued by the EU. However land-use change may also occur indirectly, as biofuels lead to increased aggregate demand for agricultural commodities and remote expansion of the agricultural frontier. Indirect land-use change (ILUC) is transmitted via agricultural markets, from one commodity to another, and from one geographic region to another, making it impossible to manage with any confidence.

Consider the US, where rapid growth in ethanol demand has elevated corn prices. In response, farmers have expanded corn production and limited soybean, resulting in upward pressure on soybean prices. This signals soybean farmers in South America to increase production, potentially driving agriculture and ranching further into rainforest. The tightening of corn available for feed and food also leads to substitution, of wheat for corn in feed, and rice for corn and wheat as food, pushing up prices and potentially triggering land-use change in diverse regions as farmers respond. The size of any such effects depends on how commodity and livestock markets adjust, the extent to which increased demand is met by yield increase as opposed to land-expansion, and price transmission within and between countries. However the potential effects are huge. A controversial paper published in 2008 modeled direct and indirect land-use change from US ethanol, estimating it would take 167 years of ethanol consumption to pay back the resulting emissions (Searchinger *et al.* 2008). A more recent paper modeled the global land-use change that would arise if all biofuel mandates and targets currently announced are implemented by 2020. It predicted a net increase in emissions until 2043, after which point emissions from land-use change would be paid back (Timilsina and Mevel 2011).

ILUC remains an area of great controversy, contested by the biofuels industry and some academics as a significant source of emissions (e.g., Kim and Dale 2011; Oladosu *et al.* 2011). Its potential to flip biofuels from climate cure to disease makes it a particular headache for the US and the EU, which justify biofuel support as a means to tackle climate change. It remains unclear how policy-makers will deal with the issue because ILUC occurs away from the plantation, making it impossible to manage. And because emissions are potentially enough to negate biofuels as a climate policy option, and the calculations are extremely technical and sensitive to modeling assumptions and methodologies, the debate becomes highly politicized.

In the US, where regulations state biofuels must offer 20% GHG savings relative to fossil fuels, a 2009 study by the Environmental Protection Agency (EPA) incorporating estimates of ILUC emissions found that US corn ethanol would not qualify. The ethanol and corn lobbies mobilized immediately, and a bipartisan group of Senators warned the EPA against premature regulation. The following year, the EPA revised corn ethanol's GHG savings to 21% (IPC 2011).

In the EU, biofuels must offer at least 35% GHG savings, rising to 50% by 2017 and 60% by 2018 for new plant. But these targets also look challenging once ILUC is taken into account. Modeling for the European Commission calls into question the efficacy of the EU's target, indicating that without any action to reduce ILUC, there are unlikely to be significant (if any) net emissions reductions from European biofuels (Laborde 2011; Malins 2011). Biodiesel was found to be particularly damaging because it depends upon vegetable oils, especially rapeseed oil, increasing demand for which may indirectly drive oil palm expansion into tropical peatland rainforest in Indonesia and Malaysia. Again, the European biodiesel industry has lobbied intensively, and any remedial action appears unlikely before 2017 (IPC 2011).

Despite the rhetoric about climate change and energy security, the original attraction of biofuels for both the US and the EU was as a new source of transfers to agriculture. The US began subsidizing biofuels in 1978 as a mode of farm support, whilst the principal motivation for the EU's first biofuel target in 2003 was its potential to support farmers. In both cases, the hope was that by increasing farm gate prices, direct payments through farm programs might be reduced. But as with other agricultural subsidies, the result has been to create dependencies and sectoral interests that organized to defend and increase support. The biofuels lobby includes crop farmers, landowners, seed and input companies, agribusiness, and energy companies.

The alignment of these powerful interests behind biofuels has driven the expansion of policies in the US and EU. Biofuels now use almost 40% of the US corn harvest and two-thirds of EU vegetable oil production. This is good for farmers, but bad for food consumers. Among the most vocal critics of biofuel policies have been those industries experiencing higher variable costs, such as the livestock industry, which has seen the cost of feed increase, and food and beverage companies which have been hit by higher edible oil, sugar, and cereal prices.

Among people, higher international commodity prices hit the poor hardest. This is because poor households spend a higher proportion of their incomes on food, up to 75% in the poorest countries. Furthermore, they consume less processed and packaged foods, where the commodity price accounts for a small share of the overall production cost. Unprocessed staples such as corn, wheat, and rice make up a greater share of expenditures. When the price of corn doubles it may not be noticed by someone eating cornflakes in a rich country, but for a poor family that relies on corn meal as its staple, the impact could be disastrous.

In sum, biofuels offer only marginal improvements in energy security, and due to the high levels of public support they receive, they are at best an expensive way of reducing emissions from transport. But by increasing demand for agricultural crops, biofuel policies create significant social and environmental risks in the form of increased emissions from land-use change and higher food prices.

Biofuels and Food Prices

After decades of stagnation, international markets have entered a period of high and volatile prices beginning with the global food price crisis of 2007/8. Between 2005 and 2008, corn prices almost tripled, with similarly extreme movements in wheat, rice, and oilseed prices. Though prices receded following the inevitable supply response and a collapse in demand following the financial crisis, they settled at a higher level than before. They then began to rise again, reaching new records in 2011 (see Figure 16.4). Numerous analyses have identified biofuels as a causal factor.

Biofuels raise food prices in the long run by diverting agricultural commodities from food and feed into fuel production. This may happen directly, for example US ethanol producers competing with food consumers and the livestock industry for corn. Or it may happen indirectly via substitution effects, for example European food manufacturers buying palm oil to replace rapeseed oil diverted to biodiesel. Biofuels may also compete with food production for land, meaning that even biofuels made from non-food feedstocks may still push up food prices if they displace food production from existing land. Some (e.g., Baffes and Haniotis 2010) have argued that because biofuels account for only a very small share of global cropland (1.5% for grains and oilseeds), they cannot be a significant driver of price rises. This is misleading. During the period 2008 to 2010, biofuels accounted for 11% of coarse grain and vegetable oil use (OECD-FAO 2011). These shares are significant, and growing.

Biofuel policies also exacerbate food price volatility. By rendering agricultural commodities substitutable with petroleum products, biofuel policies facilitate price transmission from energy to food markets: a high oil price incentivizes consumers to switch to biofuels as an alternative, increasing demand for the feedstock and pushing up food prices (OECD-FAO 2011). Meanwhile mandates create inelastic demand for agricultural commodities, making prices more volatile (OECD-FAO 2011). Abbott *et al.* (2011)

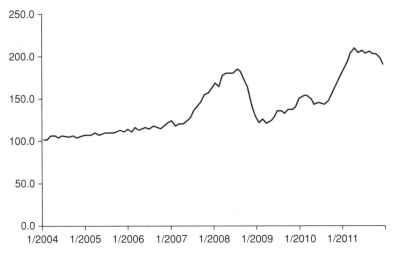

Figure 16.4   FAO Food Price Index, 2004–2011.
Source: FAO.

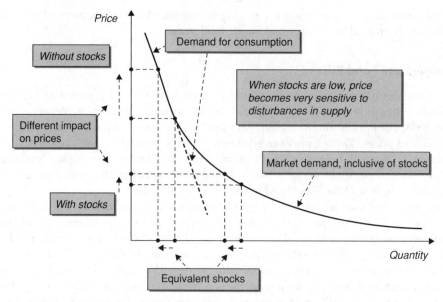

Figure 16.5 Low stocks increase volatility.
Source: Wright (2009). Used by permission.

further argue that US ethanol mandates limit the flexibility of production to move between crops, leading to inelasticity on the supply side.

Biofuels also increase volatility by driving down stock-to-use ratios. Price spikes in post-1945 cereal markets have always occurred when the ratio of stocks to use is low (e.g., Wright 2009). Because agricultural commodities are storable, market participants can build stocks when prices are low and release them when prices are high to make a profit. The impact of this behavior is to increase the elasticity of demand and supply and dampen price movements. But when stocks are depleted, market participants are unable to make releases in response to high prices, demand becomes inelastic, and prices become highly sensitive to supply or demand shocks (see Figure 16.5.) Stock-to-use ratios were at historical lows before the 2007/8 price spike, with increasing biofuel use having been a major factor (Tangermann 2011). Prices were therefore highly sensitive to supply shocks from weather events and the demand shock from biofuels, estimated to have accounted for 60% of the demand increase for cereals and oilseeds in the run-up to the price spike (OECD 2008). Following the 2010/11 price spike, stock-to-use ratios remain close to critical thresholds, and will struggle to recover over the next decade (OECD-FAO 2011).

Other factors have also increased food prices and volatility. Weather played a role in both the 2007/8 and the 2010/11 price spikes by disrupting harvests (Trostle *et al.* 2011). High oil prices increase food prices by raising transport and production costs, as well as via biofuels (Baffes and Haniotis 2010). Some commentators have argued that financial activity in derivative markets has contributed to price spikes (e.g., De Schutter 2010), however this remains fiercely contested. Although some statistical analyses have indicated a possible relationship (e.g., Robles *et al.* 2009), crucially a causal link between derivative prices and agricultural spot prices has not been established (Irwin and Sanders 2010).

Growing demand from emerging economies has certainly contributed to structural price rises and is likely to have been an indirect driver of volatility in so far as it contributed to long-run excess demand and declining stock levels. Finally, there is no doubt that the unilateral imposition of export controls by governments has been a significant factor in both the 2007/8 and 2010/11 price spikes. The former saw over 30 governments impose export controls on their agricultural sectors, not only choking off supply and forcing prices higher, but also panicking others into doing the same. Controversially, a paper published by the World Bank argued that these responses would not have occurred in the absence of biofuels pushing up prices and running down stock-to-use ratios in the first instance (Mitchell 2008).

The response of the biofuel lobby to mounting evidence of biofuels' detrimental impacts on food security has been to complain of a campaign of misinformation orchestrated by groups threatened by biofuels, from the food and beverage, livestock, and oil industries. They point instead to other factors, in particular speculation and growing demand from China, arguing by extension that biofuels cannot be to blame.

They are wrong on both counts. The case linking biofuels to high and volatile food prices is not made by industry lobbies, but by credible and impartial analysts. In 2011, 10 international organizations including the World Bank, International Monetary Fund (IMF), Food and Agriculture Organization (FAO), World Food Program (WFP), Organization for Economic Cooperation and Development (OECD), and World Trade Organization (WTO) presented a report to the G20 on food price volatility, identifying biofuels as a causal factor and recommending the dismantling of policy support (FAO *et al.* 2011). A similar report from UN High Level Panel of Experts on Food Security and Nutrition came to exactly the same conclusion and recommendation (HLPE 2011). This is not to say that biofuel policies are the only factor behind high and volatile food prices. They are clearly one of many, as indeed both reports recognized. Nor is it to say that they are the biggest factor. But they are arguably the most egregious. This is because they are politically created and can be politically dismantled, but instead have been continued and expanded in the face of mounting evidence regarding their detrimental impacts on food security.

Biofuels and Development

As shown above, the biofuel boom spearheaded by the US and EU has had a serious impact on the food security of poor food consumers by driving prices higher and contributing to price spikes such as the 2007/8 food price crisis. During this episode, protests broke out in over 60 countries as poor consumers took to the streets in desperation. WFP, hamstrung by rising prices and a fixed budget, was unable to secure the supplies it needed to respond. By 2009, the number of hungry people in the world was estimated to have passed 1 billion for the first time in history, and the number of people living in extreme poverty was estimated by the World Bank to have increased by over 100 million (Ivanic and Martin 2008).

Higher prices are of course good for poor food producers. Given that most poor people still live in rural areas, it might be supposed that biofuels indirectly contribute to development by increasing the prices poor farmers receive. Where poor farmers are able to access functioning markets, this will happen. But the majority of poor rural households are in fact net food consumers and therefore lose out. Analysis of the 2011 food price spike by the World Bank estimated that 24 million people were lifted out of poverty as a result of higher prices, but 68 million fell into poverty (Ivanic *et al.* 2011). This picture extends to the international level, where most poor countries are net food importers: of the 81 poorest countries currently qualifying for grant-based financing from the World

Bank, the FAO categorizes 66 as Low Income Food Deficit countries. These countries face deterioration in the balance of payments, inflation, and strain on public finances when food prices rise.

As we have seen, biofuel policies not only increase international prices in the long run, but make them more volatile, so prices move up *and down* over time around this higher level. This presents a further set of challenges to development. Because volatility is around a higher price level, price spikes are more extreme, exacerbating the problems for consumers described above. In addition, the uncertainty that volatility creates means problems for producers, who will be hit when prices fall. Farmers who expand production in response to high prices are locked in to an investment that may not pay off if prices fall before harvest. In particular, small farmers who lack significant savings or access to credit or hedging may be put out of business by such price swings. At the macro-level in producer countries, downswings will have immediate balance of payments impacts, whilst volatility more broadly can result in poor resource allocations, reduced investment, and lower growth (FAO *et al.* 2011).

Whilst the impact of US and EU biofuel production on international food prices has, on balance, been detrimental to food security and development, many poor countries are seeking to increase their own consumption or become exporters. As already noted, one reason for this is agricultural development. A second is oil prices. Many poor countries are dependent on imports for their oil needs, and a future of high oil prices makes domestically produced biofuels appear attractive. Brazil and Malawi have both succeeded in reducing their oil dependency through the development of national biofuel programs. One of the principal reasons why this has not been replicated elsewhere has been cost: poor countries lack the budgetary resources to subsidize production. However the fiscal burden of biofuels will decline in the future if oil prices rise.

National production of biofuel feedstocks, either for export or domestic use, raises important considerations for governments in countries where food insecurity is high. The main issues are likely to arise from the following (FAO 2010):

- Food displacement, if food crops, or land previously used to grow food, is diverted to biofuel production.
- Soil degradation, as biofuel feedstocks tend to be resource intensive, potentially affecting long-term soil quality and land productivity.
- Displacement of farmers, where investors acquire or lease tracts of land for biofuel production that were previously used by poor communities, undermining their livelihoods and ability to grow or purchase food.
- Water scarcity, as biofuel feedstocks tend to require significant water use and irrigation, potentially placing them into competition for water with food production.
- Incomes and employment, where biofuel production leads to new jobs and higher incomes for local people, improving their food security (though if the production leads to higher food prices, the impact may be ambiguous, and negative for the broader population in the absence of wider economic spillovers).

As global demand for biofuels grows, many developing countries are experiencing significant investment from companies wishing to grow biofuels for export. One recent study estimated that globally biofuels account for 40% of land acquired in recent years, and 66% in Africa (Anseeuw *et al.* 2012). The study found that many deals are not leading to development or job creation however, and that investors are often avoiding tax, and local communities are frequently losing access to land and water, or being displaced without adequate compensation.

Of course, not all biofuel projects are bad for poor people. The Brazilian national biodiesel program has raised and diversified incomes for thousands of small family farms, whilst small projects designed to increase access to energy among marginalized rural populations have had notable successes in Brazil and Tanzania among others (Bailey 2008). But these examples are exceptional in that they were designed specifically to address the needs of poor people. In developing countries where governance and institutions are weak, the rapid and large-scale expansion of biofuel production, whether for domestic use or export, is more likely to undermine rather than enhance rural livelihoods.

To summarize, in developing countries, biofuel projects may have positive and negative impacts on the food security, energy security, and incomes of different groups. This will depend on how projects are managed, how food availability and prices are affected, and how any benefits in terms of job creation, energy access, and economic spillovers are shared. Projects that are well managed and create significant livelihood opportunities may lead to benefits for rural communities. Generally speaking, urban populations might be expected to lose out in the short run if food production is displaced, though in the longer run if biofuel production stimulates broader agricultural development or reduces energy prices then this may not hold. In countries with already high levels of poverty and hunger, where government capacity and institutions are weak, or where budgetary resources are low, attempts to develop biofuels at scale present significant downside risks.

At the global level, growing biofuel production will increasingly compete with food production, which faces a daunting challenge. By 2050, it is estimated that demand for food will have increased by 70%, and 100% in some developing countries (FAO 2011). What is more, cereal yield growth is slowing, and arable land is increasingly scarce, with one authoritative review concluding "we should work on the assumption that there is little new land" (Foresight 2011). Water is also becoming a scarce resource. Agriculture accounts for 70% of global freshwater use, but this share will come under increasing pressure from industrial and municipal uses as countries develop and urbanize. Finally climate change will exacerbate land and water scarcity through processes such as desertification, sea level rise, and glacial melt, and will threaten food production directly as rising temperatures increase the drag on yields and extreme weather raises the incidence of harvest losses. In this context, given the highly contestable contribution biofuels make to reducing GHG emissions and enhancing energy security, the continued expansion of biofuel production appears mistaken.

Policy Options

There are various options available to policy-makers wishing to limit the negative consequences of biofuels for food security.

Reform of National Policies

Given the contestable contribution that biofuels make to climate and energy security, and the fiscal challenges most developed countries face, there is a strong argument for eliminating public support measures. Removing import tariffs would be consistent with the objectives of reducing GHG emissions and enhancing energy security. In the case of the EU for example, it would facilitate imports of Brazilian sugarcane ethanol offering improved GHG savings, and diversify the biofuel mix, so enhancing energy security. It would also be good for food security: shifting feedstock from wheat to sugarcane would reduce upward pressure on cereal prices.

Removing subsidies and tax breaks can help reduce food price rises in times of high oil prices, when market forces may push demand beyond mandated levels (Babcock 2010). Otherwise, whilst a mandate remains in place, the removal of fiscal support simply transfers the economic burden from the taxpayer to the consumer. Removing mandates is more politically challenging, as it would be fiercely contested by biofuel and farm lobbies.

A more feasible option would be to make mandates flexible, such that they are temporarily reduced or abandoned at times of acute food scarcity or high prices (de Gorter and Just 2010). Whilst not as threatening as a permanent dismantling of mandates, this approach is still deeply unattractive to biofuel producers, as it raises the prospect of a sudden curtailment in demand should the trigger be activated, leaving plant and workers idle until prices subside. Agreeing the precise details of the trigger and how it would be activated, though straightforward in principle, would prove difficult in practice given the intense interest it would receive from biofuel lobbies.

Importantly, whilst the removal of subsidies and mandates would do much to reduce biofuel consumption, it may now do less than one might suppose. The generous and sustained public support afforded biofuels has resulted in considerable plant and capacity that no longer needs mandates and subsidies to operate. All that is needed now is a favorable differential between oil and feedstock prices. One analysis of the US ethanol program found that between 2006 and 2009, ethanol was responsible for 36% of corn price rises; crucially only 8% was attributable to support measures whilst the remaining 28% was due to market forces (Babcock and Fabiosa 2011).

A way to manage this residual risk could be through the use of option contracts (Wright 2011). In this approach governments would purchase call options from biofuel producers. The contracts would specify a trigger, which when activated, would obligate the producer to release feedstock back into food chains, introducing flexibility into biofuel production, even when market forces have driven biofuel production beyond mandated levels. This approach should also be more acceptable to biofuel producers, which would enter into contracts freely and receive payment for doing so. There are still potential difficulties of course. Ultimately, the effectiveness of the approach is constrained by the extent of biofuel producers' participation. By auctioning contracts, governments could in theory achieve satisfactory participation through an appropriate option price, though should the price be too high, taxpayers may baulk at having to buy expensive contracts from already heavily subsidized industries.

Standards

Attempts to develop sustainability standards for biofuels, where they consider food security, typically require producers to *consider* the impacts of biofuel production on local or national food security, see for example GBEP (2011) or RSB (2011). This approach fails with regard to the US and EU, where the most serious impacts of biofuel production on food security occur in other countries. These schemes are also only of limited use in developing country contexts, as they fail to place restrictions or limits on what producers may do.

Might it therefore be possible to develop food security standards with teeth? Extending existing EU standards preventing the use of land of high biodiversity value and high carbon stock could be one approach. Methodologies already exist to categorize arable land according to its quality, so it is perfectly feasible to set a standard preventing the use of land of high food value. Another approach might restrict the use of feedstocks that impact most on prices of key foodstuffs, such as cereals and oilseeds. Inevitably, US and European industries would resist such standards, as they would penalize their preferred feedstocks such as corn, wheat, and rapeseed oil. Nor of course would they accept standards restricting their use of prime agricultural land. Policy-makers also argue that the use of binding standards to ensure food security would rapidly run into trouble at the WTO, which does not allow governments to differentiate between like products on the basis of how they are produced. That said, the EU has not been challenged on its use of environmental standards.

Alternative Feedstocks and Second Generation Technologies

Biofuels have had a negative effect on global food security because they divert food crops into energy production. One argument follows that by using non-food feedstocks, such as jatropha or switch grass, competition with food can be avoided. Unfortunately this argument only holds if these non-food feedstocks do not compete with food production for land, water, and other inputs. In the event that non-food feedstocks displace food production, or prevent the expansion of food production, then the result will still be upward pressure on food prices.

Much is pinned on so-called second generation technologies: biofuels manufactured using different production pathways allowing the use of alternative feedstocks, or improving efficiency so that less feedstock is required to produce a given quantity of biofuel. As we have seen, feedstocks that compete with food for inputs exert upward pressure on food prices, no matter whether the production pathway is first or second generation. For this reason, one of the most promising technologies may be biodiesel from algae, as the feedstock can be grown in water.

Unfortunately, second generation technology has failed to make the contribution to supply hoped for, and this looks unlikely to change in the foreseeable future (Cheng and Timilsina 2010). Nevertheless, if these technologies do eventually become commercially viable and biofuel yields are significantly increased, land demands to fulfill mandates would decline and some of the pressure on food prices alleviated. However, Wright (2011) points out that two less happy scenarios might follow instead. In one, mandates grow due to lobbying from special interests arguing that greater efficiency should mean bigger mandates. In another, biofuels become competitive with oil on a permanent basis. In both, the pressure on land and food prices remains or increases.

Non-Biofuel Policies

The relatively small contribution that biofuels make to transport energy could easily be met through improvements in vehicle efficiency with readily available technologies. Such efficiency enhancements would be more economical, would generate greater GHG savings, and would have no negative impacts on food security (Bailey 2008). In the longer term electric vehicles should offer an alternative to the internal combustion engine, drastically reducing the need for liquid fuels in transport. This would not only render biofuels obsolete beyond aviation and shipping, but massively reduce oil dependency.

Given the challenges faced by agriculture in meeting future demand for food, and the added strain biofuels represent, increasing efficiency in the food system is critical. There are two major opportunities. The first is to increase agricultural productivity.

Yields for major crops in poor countries are often a fraction of those in industrialized countries, where farmers enjoy greater public support as well as greater access to infrastructure, land, credit, insurance, and technologies such as fertilizers, seeds, machinery, and irrigation. Closing the yield gap by extending poor farmers' access to resources and technologies is key to ensuring future food security, but it requires significant efforts from governments and donors to correct market failures and provide public goods such as infrastructure, weather data, and research.

The second opportunity is to reduce losses. Estimates suggest that, worldwide, approximately a third of all food is lost before it is consumed (Foresight 2011). In developed countries this occurs at the consumption end of the value chain, as waste from shops, households, and restaurants. In developing countries, food is lost at the production end of the value chain due to poor farm management practices, poor infrastructure and a lack of access to appropriate technologies. Policies to address waste and post-harvest losses could have a significant impact on food availability.

Social Protection and Emergency Relief

Without significant action, biofuels will continue to contribute to high and volatile food prices and increasingly compete with food production in the developing world. Therefore policies to protect the most vulnerable people are crucial. As well as using options contracts to divert food from biofuel production to hungry communities in times of crisis, governments and international organizations can also develop decentralized systems of emergency stocks.

Social protection schemes in developing countries have been shown to reduce hunger and vulnerability when effectively administered and tailored to the needs of the poor. But such schemes are expensive and likely to increase the fiscal burden on the poorest states just as the food import bill is rising. Though donors can help plug the gap, it is also worth considering levies on biofuel producers to support poor countries seeking to provide such assistance.

Global Measures

In general, the options above pertain to national policy-making, however clear opportunities for action at the global level follow from them. In particular:

- International coordination of mandate flexing – for national efforts to reduce mandates during price spikes to be fully effective, an agreement to coordinate action is needed to militate against domestic lobbying and maximize impact.
- Agreement and institutionalization of *adequate* global sustainability standards – a globally accepted standard that adequately protects the food security of vulnerable populations within and outside producer countries is needed. To be effective, this must be binding and enforceable by producer and consumer governments.
- A global levy on biofuel producers to fund safety nets in poor countries – producer governments could coordinate to tax their biofuel sectors in order to fund an international mechanism to reinsure the safety net liabilities of poor, food-insecure countries. This would effectively act as a form of compensation to vulnerable communities, with the greatest burden falling on the biggest biofuel producers.
- Hedging WFP with call options – building on Wright (2011), producer governments could collectively agree to purchase call options from their biofuel sectors that trigger

the diversion of food to WFP during times of crisis, hedging WFP's exposure to price spikes and underwriting its capacity to respond to future food price crises. This system would effectively transform national biofuel sectors into a virtual global emergency food reserve.

- New global commitment to boost agricultural efficiency – increasing the productivity of agriculture in poor countries and reducing post-harvest losses will help agriculture cope with increasing competition from biofuels, as well as reducing poverty in rural areas. It is high on the agenda of national governments, the G8 and G20, and international organizations such as the World Bank, IFPRI, FAO, and IFAD. Despite this, there is considerable scope to increase the resources – financial, human, and technological – dedicated to the challenge, necessitating new commitments and partnerships from developed and developing country governments alike.

Conclusion

Biofuel production is already having a serious impact on food security and development. At the international level, the large programs of the US and EU have driven up food prices, reduced stocks, and increased price volatility. In developing countries, if properly managed and targeted at the needs of poor people, biofuel production may offer some development opportunities, however the downside risks are significant and the experience to date not encouraging.

The policy justifications for biofuels, that they reduce GHG emissions and enhance energy security, are highly contestable. At best, they offer an inefficient means of doing so. At worst, they crowd out more promising alternatives such as electric vehicles, and *increase* GHG emissions by driving the expansion of agriculture into land of high carbon stock.

For these reasons, the interests of taxpayers and fuel consumers in developed countries, and vulnerable populations in developing countries, would be best served by dismantling all biofuel support, including mandates, subsidies, tax breaks, and import tariffs. This would reduce biofuel production, and the burdens imposed on taxpayers and energy consumers. It would also ease upward pressure on food prices, and reduce food price volatility, reducing the burden on poor food consumers. Some biofuel production would likely still occur during periods of high oil prices, but the negative consequences would be greatly lessened.

Unfortunately this looks unlikely. Biofuel policies are best understood as a means to effect transfers from taxpayers and consumers to farm and biofuel lobbies. These interests mobilize effectively to resist policy reforms, making it difficult for governments to act in the wider interest. This makes international coordination of policy responses particularly important. Not only will coordination increase the overall impact of action, but it will mitigate special pleading from domestic lobbies by ensuring national biofuel sectors are treated equally.

In general, the most politically feasible responses are those least threatening to special interests. Therefore the use of call options to divert feedstocks from biofuel production into the food or humanitarian systems at times of crisis is among the most promising opportunities. At the global level, this could be achieved through governments purchasing options that specify WFP as the recipient. Policies to increase efficiency in the food system and to extend the provision of social protection and safety nets to vulnerable populations should also be prioritized.

References

Abbott, P., C. Hurt, and W. Tyner. 2011. *What's Driving Food Prices in 2011?* Oak Brook, IL: Farm Foundation.

Anseeuw, W., L. Alden Wily, L. Cotula, and M. Taylor. 2012. *Land Rights and the Rush for Land: Findings of the Global Commercial Pressures on Land Research Project*. Rome: International Land Coalition.

Babcock, B. 2010. *Impact on Ethanol, Corn, and Livestock from Imminent U.S. Ethanol Policy Decisions*. CARD Policy Brief 10-PB 3. Ames, IA: Iowa State University.

Babcock, B., and J. Fabiosa. 2011. *The Impact of Ethanol and Ethanol Subsidies on Corn Prices: Revisiting History*. CARD Policy Brief 11-PB 5. Ames, IA: Iowa State University.

Baffes, J., and T. Haniotis. 2010. *Placing the 2006/8 Commodity Price Boom in Perspective*. Policy Research Working Paper 5371. Washington, DC: World Bank.

Bailey, R. 2008. *Another Inconvenient Truth: How Biofuel Policies are Deepening Poverty and Accelerating Climate Change*. Oxfam International Briefing Paper 114. Oxford: Oxfam International.

BP. 2011. *BP Statistical Review of World Energy June 2011*. London: BP.

Cheng, J., and G. Timilsina. 2010. *Advanced Biofuel Technologies: Status and Barriers*. Policy Research Working Paper 5411. Washington, DC: World Bank.

de Gorter, H., and D. Just. 2010. The Social Costs and Benefits of Biofuels: The Intersection of Environmental, Energy and Agricultural Policy." *Applied Economic Perspectives and Policy* 32, 1: 4–32.

De Schutter, O. 2010. *Food Commodities Speculation and Food Price Crises. Regulation to Reduce the Risks of Price Volatility. Briefing note by the Special Rapporteur on the Right to Food*. New York: United Nations.

FAO, IFAD, IMF, *et al*. 2011. *Price Volatility in Food and Agricultural Markets: Policy Responses*. Rome: Food and Agriculture Organization of the United Nations.

FAO. 2010. *Bioenergy and Food Security: The BEFS Analytical Framework*. Rome: Food and Agriculture Organization of the United Nations.

FAO. 2011. *The State of the World's Land and Water Resources for Food and Agriculture (SOLAW) – Managing Systems at Risk*. Rome: Food and Agriculture Organization of the United Nations / London: Earthscan.

Fargione, J., J. Hill, D. Tilman, *et al*. 2008. Land Clearing and the Biofuel Carbon Debt. *Science* 319 (5867): 1235–1238.

Foresight. 2011. *The Future of Food and Farming, Final Project Report*. London: Foresight Project, Government Office for Science.

GBEP. 2011. *GBEP Report to the G8 Deauville Summit 2011*. Rome: Food and Agriculture Organization of the United Nations.

HLPE. 2011. *Price Volatility and Food Security. A Report by the High Level Panel of Experts on Food Security and Nutrition of the Committee on World Food Security*. Rome: Food and Agriculture Organization of the United Nations.

IEA. 2010. *2010 World Energy Outlook*. Paris: International Energy Agency.

IEA. 2011. *Technology Roadmap: Biofuels for Transport*. Paris: International Energy Agency.

IPC. 2011. *Biofuel Policies in the US and EU*. IPC Policy Focus. Washington, DC: International Food & Agriculture Trade Policy Council.

Irwin, S., and D. Sanders. 2010. *The Impact of Index and Swap Funds on Commodity Futures Markets: Preliminary Results*. OECD Food, Agriculture and Fisheries Working Papers 27. Paris: Organization for Economic Cooperation and Development.

Ivanic, M., and W. Martin. 2008. *Implications of Higher Global Food Prices for Poverty in Low-Income Countries*. Policy Research Working Paper 4594. Washington, DC: World Bank.

Ivanic, M., W. Martin, and H. Zaman. 2012. *Estimating the Short-Run Poverty Impacts of the 2010–11 Surge in Food Prices*. Policy Research Working Paper 5633. Washington, DC: World Bank.

Jung, A., P. Dörrenberg, A. Rauch, and M. Thöne. 2010. *Biofuels – At What Cost? Government Support for Ethanol and Biodiesel in the European Union – 2010 Update*. Geneva: International Institute for Sustainable Development.

Kim, S., and B. Dale. 2011. Indirect Land Use Change for Biofuels: Testing Predictions and Improving Analytical Methodologies. *Biomass and Bioenergy* 35, 7: 3235–3240.

Laborde, D. 2011. *Assessing the Land Use Change Consequences of European Biofuel Policies*. Washington, DC: International Food Policy Research Institute.

Malins, C. 2011. *Indirect Land Use Change in Europe – Considering the Policy Options*. Washington, DC: The International Council on Clean Transportation.

Mitchell, D. 2008. *A Note on Rising Food Prices*. Policy Research Working Paper 4682. Washington, DC: World Bank.

OECD. 2008. Rising Food Prices: Causes and Consequences. OECD paper prepared for the DAC High Level Meeting, May 20–21, 2008.

OECD-FAO. 2011. *OECD-FAO Agricultural Outlook 2011–2020*. Paris: OECD / Rome: Food and Agriculture Organization of the United Nations.

Oladosu, G., K. Kline, R. Uria-Martinez, and L. Eaton. 2011. Sources of Corn for Ethanol Production in the United States: A Decomposition Analysis of the Empirical Data. *Biofuels, Bioproducts & Biorefining* 5: 640–653.

Robles, M., M. Torero, and J. von Braun. 2009. *When Speculation Matters*. IFPRI Issue Brief 57. Washington, DC: International Food Policy Research Institute.

RSB. 2011. *RSB Principles and Criteria Version 2.0*. Lausanne: Roundtable on Sustainable Biofuels.

Searchinger, T., R. Heimlich, R. A. Houghton, *et al.* 2008. Use of US Croplands for Biofuels Increases Greenhouse Gases Through Emissions from Land-Use Change. *Science* 319 (5867): 1238–1240.

Tangermann, S. 2011. *Policy Solutions to Agricultural Market Volatility*. Geneva: International Centre for Trade and Sustainable Development.

Timilsina, G., and S. Mevel. 2011. *Biofuels and Climate Change Mitigation: A CGE Analysis Incorporating Land-Use Change*. Policy Research Working Paper 5672. Washington, DC: World Bank.

Trostle, R., D. Marti, S. Rosen, and P. Westcott. 2011. *Why Have Food Commodity Prices Risen Again?* Washington, DC: United States Department of Agriculture.

Wright, B. 2009. *International Grains Reserves and Other Instruments to Address Volatility in Grain Markets*. Policy Research Working Paper 5028. Washington, DC: World Bank.

Wright, B. 2011. Addressing the Biofuels Problem: Food Security Options for Agricultural Feedstocks. In Adam Prakash, ed. *Safeguarding Food Security in Volatile Global Markets*. Rome: Food and Agriculture Organization of the United Nations, pp. 481–492.

Chapter 17

Energy Efficiency: Technology, Behavior, and Development

Joyashree Roy,[1] Shyamasree Dasgupta, and Debalina Chakravarty

Introduction

In traditional economic growth theory, technology and continuous technological progress are seen as drivers of long-term growth. Yet what matters equally is the behavior of the potential adopter. "Users" in a social context, through adoption of a technology, give life to the inventors' creativity expressed through design of a piece of hardware/equipment commonly called "technology." Both technology and the behavior of its potential adopter matter. The technology adoption rate by producers and by final consumers, in many cases, determines the rate of productivity growth. Along with access to technology, user behavior matters with no less importance in determination of the outcome of technology diffusion and its scale of adoption. Examples of how the first diffusion happened of fountain pens, typewriters, sewing machines, candles, light bulbs, and so on, have given rise to many novels, stories, and movies to narrate the process of changing social practices. Besides professional literature there are several sources – oral tradition, chap book, ephemera, old literature, newspapers, films – which narrate how various "inventors" reached out to "users" to explain the benefits of a piece of a new technology and create a viable development regime through a positive bandwagon effect. In recent times, the superfast proliferation of mobile phones (in India 880 millions in the 15 years 1995–2011)[2] also shows a similar "planned strategic technology diffusion" and subsequent transformation in the communication service. The path dependency of technological progress is also discussed in the literature (Allen 1983; Barro and Sala-i-Martin 1992; Solow 1956). Appropriate artifacts, infrastructure, and institutions provide options to actors to make the right technology choice.

However, the point to notice is that what is "strategically marketed" to users is not the technology per se but the service it provides. How it enhances quality of life by providing access, how it is cost (money/time/resource) saving, how it gives a social status, competitive edge over peers, environmental and low carbon benefits, social equity, etc. become important in the process of diffusion. Thinking holistically about socio-economic systems is key to capturing the complex dynamic between energy

The Handbook of Global Energy Policy, First Edition. Edited by Andreas Goldthau.
© 2013 John Wiley & Sons, Ltd. Published 2013 by John Wiley & Sons, Ltd.

technology and energy use. A system transition (coherent change) in social practice (Grin 2011) can start through small/incremental steps (Clark 2002) or can be radical through system reconfiguration. The system reconfiguration encompasses technology invention and innovation (new technologies, and resulting commercially available products and processes), their selection and adoption by intermediate and final consumers (integration in user practice), and the broader process of social embedding (e.g., regulation, markets, infrastructures, and cultural symbols) (Geels and Schot 2011). It is a complex interplay of social, economic, ecological, technological, and institutional developments (Rotmans *et al.* 2000). There are multiple actors in any societal context who act and react through their decisions to create a sequence of events and the resultant "process" (Abbott 1992), which finally determines the outcome in favor or against a new technology and its role in transition from the incumbent socio-technical regime. Policy is key to influencing technology development and transfer, correct energy pricing, and also lowering energy intensity. India provides ample evidence, particularly in the development context, and this is why this chapter draws heavily on the Indian experience.

The following sections of this chapter show first that energy efficient technology is no single invention and thus needs a very well planned global strategy for economy-wide technology deployment in various social contexts, then deal with various facts on energy efficient technology adoption with examples from Indian industries – at what rate this has happened historically, what role energy price has played in the process of adoption, why there is need to break away from historical trend, and how the impact can be assessed through an index of energy intensity. The third section shows, with examples, that technology deployment policy, unless supplemented by pricing policy and incentive designs, can end up with a lower achievement level due to the take-back effect arising out of operational behavior of technology users. The chapter concludes with some remarks highlighting the need for national and global efforts.

Energy Efficiency: Need to Focus on Socio-Technical System Transition

There is consensus in the literature that there are now enough technological options (IPCC 2007; Stern 2007) awaiting deployment to make production and consumption processes energy efficient. UNIDO (2011) estimates show 25–30% technical potential awaiting adoption by industries. There is high untapped technical potential in home appliances, especially in space conditioning (Roy *et al.* 2011). In lighting appliances a 70–90% efficiency gain can be achieved in many social contexts. There has been low penetration of energy efficient technologies so far in the agricultural sector (Roy 2007). The energy efficiency gap, reflected through wide variation in energy intensity numbers across regions (UNIDO 2011), also shows the potential of technology deployment.

Energy efficient technology is no single isolated invention which can be seen through a traditional technology innovation pathway, where marketability of an invention/idea can bring the desired outcome. Energy efficiency is like a "genre" of invention, a category into which a wide variety of multiple inventions (Figure 17.1) with the common goal of natural/energy resource saving, and more recently to achieve environmental goals and other macro benefits (Kanbur and Squire 1999; Sathaye *et al.* 2005; Worrell *et al.* 2003), could be clubbed. The traditional innovation theory focused on direct benefits observable through market variables. The current goal of diffusion of energy efficient technology to achieve many indirect environmental and other benefits is far more complex, because the goal is not of direct incremental or radical changes but is of system transition where

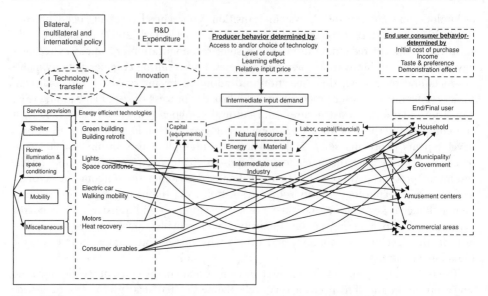

Figure 17.1   Energy efficiency: "technology–user" complex interconnectedness.

behavioral responses need to go beyond direct market benefits across various sectors and actors simultaneously.

Achieving energy use efficiency regime shift means system-wide actions and interactions that can reconfigure the current low efficiency energy use behavior system to a higher efficiency energy use system. This needs strategic management of the system transition at the global scale. In this interconnected network of actors and institutions and technology a holistic vision through multidisciplinary approaches is necessary. This complex management is not going to be achieved by taking a technology-centric or price-centric view alone. Case studies (see below, and Ghosh and Roy 2011) show that information about technology, financial incentives to overcome first-cost barrier, proactive policy announcements, target setting through command and control, subsidies for energy audit, are only few of the steps necessary, as they cannot guarantee the outcome given uncertainty in technology adopters' behavior. One of the main challenges comes from the lack of a market or institutions to create economic incentives for the multiple benefits being used to promote energy efficiency. A market price for carbon, or environmental waste-sink service pricing (Ethridge 1973), might be important additional triggers which can help faster energy efficient technology adoption. But economic behavior is not the only behavioral variable that matters. There are social, political, and cultural behaviors, all of which matter (Roy and Pal 2009) and make the issue complex. Understanding these intermediate and end use behavioral barriers would be useful for deciding on future policy options in making a system transition through sustained energy efficient technology penetration.

New technologies enabling production of an increased level of output with the same level of input through enhanced efficiency of inputs is an extremely important driver of development in a resource constrained world. This technology change can happen either through an invention–innovation pathway or though a technology-transfer pathway. Strategic actions can direct both innovation and transfer pathways (Baumol 1994; Clark 2002). Technological progress and adoption is in fact a "network phenomenon" (Figure 17.1) and is not determined by the action of any particular firm or person.

However, given the lifetime of any technology and network externalities, path dependence can emerge: the dependence of successive developments on prior events and a tendency for particular system to become "locked in" beyond a certain point (Allen 1983; Arthur 1988; David 1975; Katz and Shapiro 1985; Tomson 1988). So technology choice at any point of time is an important decision with long-term lock-in consequences.

Methodologically it has always been challenging to quantify the drivers of technological progress and its contribution to the development process of any economy. The standard growth accounting approach, pioneered by Solow (1956) and further developed by Denison (1972, 1979, 1985) and others, is employed to study long-run trends in energy use and its relationship to other economic variables (Roy et al. 1999). Total Factor Productivity Growth (TFPG) tries to express numerically the contribution (without looking into the source or driver) of the aggregate inputs' productivity change with technological progress over time to output growth (for a survey see Berndt and Watkins 1981; Dasgupta 2010; Goldar and Kumari 2003; Pradhan and Barik 1998; Roy 1992; Roy et al. 1999; Sarkar and Roy 1995).

Energy Efficient Technology Adoption: Cases from Indian Industries

Technology for Enhancing Input Productivity in Production Process

Growth accounting exercise and estimates of TFPG reveals that during 1973–1993 in the Indian manufacturing sector an average 5% of growth in output was contributed by the growth in aggregate productivity of inputs (Roy et al. 1999). This implies that 5% of the output growth could happen without any increase in quantity of inputs but due to technological progress. An extension of this growth accounting for the period beyond 1993 (Table 17.1) exhibits a higher contribution of productivity growth (14.4%) in the output growth. Table 17.1 compares the annual growth rates of output of energy intensive industries with the input and productivity growth for the period 1994–2008. It shows that over this period the output of aggregate manufacturing grew at an average annual rate of 9.7%. Average annual rates of growth of the selected energy intensive industries within the manufacturing sector, however, varied from 9.7% (in cement) to 5.1% (in pulp and paper). The contribution of TFPG remained positive for the selected industries, except for the chemical industry. So, historically, these energy intensive manufacturing processes experienced increasing input efficiency gain, the highest being in iron and steel (41%), fertilizer (34%), textiles (31%), and cement (27%), for the period 1994–2008. But in output growth, on average, the contribution of input growth remained unambiguously much higher than the growth in input efficiency. Although the contribution

Table 17.1 Growth accounting for energy intensive sectors in India, 1994–2008 (in %).

Industry	Output	Capital	Labor	Material	Energy	Total Input	Productivity
Aggregate Manufacturing	9.7	0.9	0.2	7.0	0.2	8.4	1.4
Cement	9.7	2.2	0.1	3.3	7.4	7.1	2.6
Chemical	7.1	1.2	0.1	5.8	0.1	7.2	−0.1
Fertilizer and Pesticide	8.7	0.5	0.1	5.0	0.1	5.7	3.0
Iron and Steel	6.9	1.3	0.1	1.7	1.1	4.1	2.8
Pulp and Paper	5.1	1.0	0.1	3.4	0.2	4.7	0.4
Textile	5.9	0.8	−0.1	3.4	0.0	4.1	1.8

Source: For details of methodology see Roy et al. (1999), Dasgupta (2010).

Table 17.2   Bias in technological progress of Indian manufacturing industries, 1973–2008.

Input	Aggregate Manufacturing	Cement	Chemical	Fertilizer and Pesticide	Iron and Steel	Pulp and Paper	Textile
Material	Using*	Saving*	Using	Using*	Saving	Using*	Using*
Labor	Saving*	Saving*	Saving*	Saving*	Saving*	Saving*	Saving*
Energy	Saving	Using*	Using	Using	Saving	Saving	Using
Capital	Saving	Using	Using	Saving*	Using*	Saving*	Using*

*Statistically significant at 5% level.

of input growth decreased compared to the output growth of aggregate manufacturing, it was still found to be as high as 85.6% during 1994–2008 (compared to 95% in 1973–1993).

The point to notice is that the contribution of productivity growth in output growth in manufacturing varied over time and across industries. If the goal is to make steadily increasing technological progress supplementary to autonomous historical trends, strategic intervention is needed. This is even more necessary to avoid the lock-in effect if technological progress is effected more via technology transfer. Technology transfer is not as easy as it implies, and involves much more than the transboundary movement of the design of a machine and its physical installation. Even if there are no patents, mastering a new technology calls for coordination between several groups of actors including engineers, planners, workers, investors, and users. Command over a piece of equipment and its functions comes through learning by doing (Nelson and Wright 1994).

It is evident that Indian industries are experiencing a positive trend in autonomous technological progress. From the perspective of energy use, however, the major question remains: How energy saving has this autonomous technological advancement (as reflected in productivity growth) been so far? The parameter estimates from econometric methods of cost/production function show that among energy intensive manufacturing sectors, other than iron and steel, pulp and paper, technological progress has not always been consistently energy saving in nature (bias in technological progress in Table 17.2). Rather the technological change has been *energy using* for quite long (Roy *et al.* 1999). This implies that the share of energy cost increased as a proportion of total cost along with autonomous technological advancement. It is however interesting to note that aggregate manufacturing is showing a long-run average energy saving technological trend (1973–2008), while for the earlier period (1973–1993) it exhibited an energy using technological trend (Roy *et al.* 1999). This is interesting because it captures a behavioral change among the producers, who are paying greater attention to the reduction of the energy cost share of the total cost of production.

The reason behind most of the energy intensive manufacturing sectors exhibiting energy using technological change may be technology lock-in, which is not tested and reported here. But one other reason that emerged from analytical assessment of econometric analysis is the relative cost share of energy inputs vis-à-vis other inputs in total production (Dasgupta 2010; Roy 1992; Roy *et al.* 1999). Estimates show that for none of the selected manufacturing industries did energy costs on average remain more than 20% during 1994–2008 (except 34% in cement). If business as usual (BAU) technological progress has been energy using historically, then over time, despite technological progress, energy use will increase in industries (Roy *et al.* 1999). So there is need for strategic policy decisions for technology development, deployment, and transfer.

Table 17.3 Inter-factor substitutability of inputs, 1973–2008.
(C Complementary, S Substitute.)

Inputs	Aggregate Manufacturing	Cement	Chemical	Fertilizer	Iron and Steel	Pulp and Paper	Textile
K-L	S	C	C	C	C	S	S
K-M	S	C	S	S	C	S	S
K-E	C	S	S	S	C	S	S
L-M	S	S	S	S	S	S	S
L-E	S	S	S	S	S	S	S
M-E	S	C	S	S	S	S	S

Technologies in various industries are evolving to substitute energy inputs either by capital (K-E relation in Table 17.3) or by materials/labor (M-E and L-E relations in Table 17.3) except for iron and steel and cement. In the cement industry material use and energy use are moving together, and the same is true for capital and energy in the iron and steel sector. But the remaining sectors are showing technological changes that are finding substitutes for energy. In strategic management these features can be accommodated.

How Do Price Induced Changes Matter in Input Use Behavior?

Following the oil shock of the 1970s a large body of econometric work on producer behavior tried to answer how behavioral responses matter in the face of changing energy prices. Christensen *et al.* (1971), Hogan and Jorgenson (1991), and Jorgenson and Fraumeni (1981) developed and applied methods to analyze relations between substitution effects induced by changes in relative factor prices, and pure "productivity" trends, on a sector specific basis over long periods. They have demonstrated that combining a finer level of analysis with a form of "endogeneity" in the modeling of technological change can reveal patterns that are not readily detected by more traditional methods. Studies in the context of manufacturing industries in Canada during 1957–1976 (Berndt and Watkins 1981) and in the US during 1948–1979 (Hogan and Jorgenson 1991; Jorgenson and Fraumeni 1981) demonstrated that long term growth came in the form of incremental changes over time and energy prices played an important role in changing behavior. For developing countries similar results hold.

Empirical estimates (Roy *et al.* 2006) of negative own price elasticity, especially for energy inputs, have far-reaching positive implications as far as energy consumption and resultant CO_2 emissions are concerned. Although technical parameter estimates (b_{ee} in Table 17.4) indicate that with rising energy price the cost share of energy increases, behavioral estimates (price elasticities E_{ee} in Table 17.4) indicate that industries do take decisions to reduce energy consumption (Roy *et al.* 1999, and Table 17.4). In balance there is reduction in energy use. For example, during 1973–1993 (Roy *et al.* 1999), for aggregate manufacturing with a 1% increase in energy price the cost share of energy inputs can go up by 0.052 (reflected by b_{ee} in Table 17.4) in the absence of behavioral response, but later leads to a 0.2% decline in reality due to own price elasticity (E_{ee} in Table 17.4). Energy price elasticities varied between –0.02 for total industry to as high as –0.57 for the cement sector. Using similar methodology for the period extended to more recent years until 2008, our estimates suggest that behavioral responses can reduce energy use at a higher rate with price changes (with E_{ee} magnitudes higher in recent years

Table 17.4 Own price elasticity of inputs in Indian manufacturing industries.

	Aggregate Manufacturing	Aluminium	Cement	Chemical	Fertilizer	Iron and Steel	Pulp and Paper	Textile	Glass	Total Industry	Source
E_{ee} (1973–1993)	−0.198	−0.389	−0.568	–	−0.132	−0.382	−0.238	–	−0.040	−0.018	Roy et al. 1999
b_{ee} (1973–1993)	0.052	0.091	0.040	–	0.101	0.066	0.096	–	0.085	0.076	
E_{ee} (1973–2008)	−1.34	–	−0.54	−0.85	−0.71	−0.68	−0.34	−0.68	–	–	Authors' estimation
b_{ee} (1973–2008)	−0.031	–	0.04	−0.007	0.021	0.025	0.081	0.02	–	–	

in Table 17.4) and also show that technological progress is gradually being associated with the potential for lowering energy cost as a proportion of total cost. However, this behavioral response is not true for all energy intensive industries.

The price elasticities also reflect the behavior of average productivity of the various factors. For example, during 1973–1993 the own price elasticity of –0.24 for energy in the paper industry implies that a 1% increase in the price of energy would increase energy productivity by 0.24%. Now given that energy and capital are substitutable (Table 17.3) to each other, an increase in the price of energy would on the one hand improve energy productivity (reduce energy intensity) because of the negative own price elasticity, but would on the other hand additionally reduce capital productivity and hence increase capital intensity. An increase in energy prices would have varying impacts across industries so far as average factor productivity is concerned, and may lead to a less than proportionate change in energy use. Coupled with the findings of energy using bias and slow incremental technological change, this suggests that price-based policies alone in the Indian economy may have limited impact, and may result in economic costs through output reduction in the longer run. The case for strategic management of technological progress through technology deployment gets even stronger with these results of fluctuating technological progress with energy using bias and material using bias, high material cost share, and less than proportionate price responsive behavior. In fact the sustainability of the production subsystem has so far remained contingent upon availability of natural resources (mainly in terms of energy and material in case of manufacturing industries), and hence in coming years the level of efficiency with which the industries would be able to use these natural resources is going to be an important indicator of sustainability.

One of the important drivers to enhance the input efficiency is definitely availability, diffusion, and adoption of new technology. In India, the potential of energy efficient technology adoption is very high as there are significant interplant variations in energy use per unit of output produced within an industry (Goldar 2010). The Perform Achieve and Trade scheme adopted under the National Action Plan on Climate Change (NAPCC 2008, para. 4.2), energy intensity targeting, and sale of energy saving certificates by the over-achiever to the under-achiever, is indeed a strategic technological progress management strategy with an energy saving bias. But unless supplemented by a well-designed energy certificate price policy, the outcome cannot be anticipated given the behavioral responsiveness of industries (Dasgupta et al. 2011; Roy 2010).

Energy Efficient Regime Dynamics: Need to Break Away from Historical Trends

The energy efficiency regime began to change globally in the 1970s. One indicator of energy efficiency improvement is energy intensity decline. Energy intensity declines have been experienced most rapidly in developed countries during the late 1970s and early 1980s, a period of unstable oil prices (Roy 2007). Over the 1970–1990 period, energy intensity (or energy use per unit of GDP) declined by 29% in the industrialized countries while it rose by 30% in the developing countries (Clark 2002). The early movers in the development pathway – the OECD countries – experienced declining energy intensity (energy efficiency gain) in double digits from the late 1970s and early 1980s (Roy 2007). Studies by Boyd et al. (1988), Farla et al. (1997), Golove and Schipper (1996, 1997), Greening and Greene (1998), Greening and Khrusch (1996), Howarth and Andersson (1993), Reitler et al. (1987), Schipper et al. (1992), Schipper et al. (1998), Schipper and

Grubb (2000), Sun (1999), Torvanger (1991), and Worrell *et al.* (1997), show that falling aggregate energy intensity is caused by high energy prices, autonomous trends in technology progress, particularly in the manufacturing sector, and strategically targeted energy efficiency programs. Intensity reduction occurred during periods of stable energy prices due to uptake of new process technologies. Later studies (Liaskas *et al.* 2000) recorded 40% decline during 1973–1993 in energy intensity for the industrial sector. Most of this fall was the result of reductions in volume of fuel requirements for individual industry branches, particularly cement, iron and steel, chemicals, and paper. Only a minor amount of the savings occurred because of product mix change, fuel switch, or movement of heavy industries to other countries with less expensive energy or raw materials. However, in OECD countries the energy intensity of manufacturing has declined more slowly than it did before 1990. For latecomers in the development process – the developing countries – the intensity trend started declining almost a decade later in the late 1980s and early 1990s. This might be an indication of technology transfer and deployment. Ang and Zhang (1999), Sheinbaum and Rodriguez (1997), Sheerin (1992), Dasgupta and Roy (2001, 2002) have shown for Taiwan, China, Mexico, Thailand, and India that in the manufacturing sector (and that too in energy intensive sectors) substantial improvements in energy intensity similar to OECD countries have been achieved.

However, since 2000 the vision to change energy efficiency shifted from oil price-led adjustments to strategic deployment of available technology for environmental and especially climate change mitigation benefits. Globally, between 1990 and 2008 industrial energy intensity has fallen by 1.7% per year, but now the target has become 3.4% per year until 2030 (UNIDO 2011). So the goal has shifted from incremental change to radical change. The need for a system transition is also embedded in climate change discourse now. NAPCC (2008) shows how India is becoming proactive in strategically managing this transition. UNIDO (2011) states how energy efficiency can change through system configuration changes by changing consumer preferences through training and awareness building, international cooperation for technology transfer, etc.

Although the energy intensity of the Indian manufacturing sector has declined over time, the historical trend has not touched the world best for the whole sector. For example in the case of iron and steel industries in India during 1985–2008 the rate of average annual decline of specific energy consumption (SEC) remained 1.6%. To catch up with the world average prevailing at least in 2008, by 2030 it needs to decline by at least 2.5% per annum (Figure 17.2).

The process used in steelmaking has significant implications toward energy use and carbon emission. The blast furnace route still dominates the production process although its share is falling (Figure 17.3). Kim and Worrell (2002) benchmarked and estimated the potential for energy efficiency of steel production to the best practice performance in five countries with over 50% of world steel production. The study found that the energy efficiency induced CO_2 emission reductions potential could be as high as 40% in India (compared to 15% in Japan and also 40% in China and the US). This calls for an urgent need to induce or stimulate technology adoption at a much faster pace. The target of 3% annual declining trend can attain the target by 2020 and a 5% annual reduction can achieve target world average SEC by 2015/16. But how to enhance the energy intensity reduction rate is the major challenge. Another example is the case of cement industries in India. Although India is one of the most efficient cement producing countries, using about 3.2 gigajoules of energy per tonne (GJ/t) of cement, compared to 3.0 GJ/t cement for the most energy efficient country (Japan) and a world average of 3.6 GJ/t (Saxena 2010), the technology adoption took decades to reach the world best level (Figure 17.4).

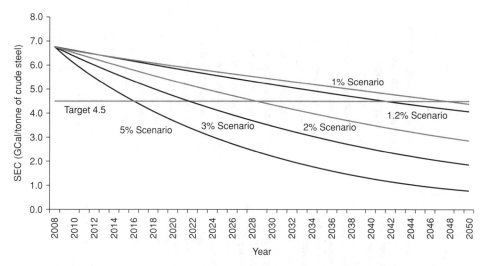

Figure 17.2 SEC reduction scenarios for iron and steel industries in India.
Source: Dasgupta (2010).

In the 1960s energy efficient dry cement plant technology contributed only 1% to total production and by 2008 the share had gone up to 97% (Saxena 2010), so that it took almost 50 years to transform the cement manufacturing process. Significant upgrading of the technology used in cement manufacturing happened in the early 1990s (Figure 17.5). Use of energy efficient dry process plants went up, mini-plants came into operation, and there was a significant change in the entire process of production. So, the transformation path consisted of incremental change since the 1960s as well as some radical changes in the 1990s. The post-1990 era, however, is characterized by a direct penetration of the importance of energy efficiency and climate change mitigation. The cement industry also witnessed steady decline in energy intensity (Figure 17.4).

The energy efficiency gap is there in the chemical industry as well. India's average energy use per tonne of ammonia was 37.5 GJ/t in 2007, compared to 28 GJ/t for the best available gas-based technology. The presence of high technology adoption potential

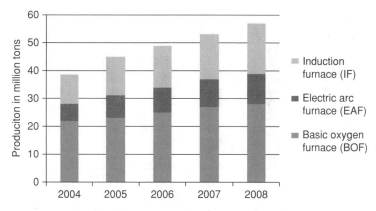

Figure 17.3 Transition profile of iron and steel industries in India.
Source: Dasgupta (2010).

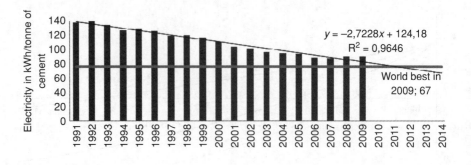

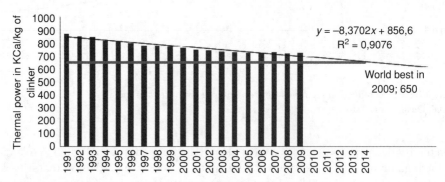

Figure 17.4 Falling energy intensity of cement industry in India.
Source: Data from www.iea.org/work/2010/india_bee/saxena.pdf.

in India is also evident given the fact that there are huge inter-plant variations in existing technology and resulting energy intensity within industries. For example, in the aluminum industry the SEC among the plants varied from 0.18 to 6.4, in textiles 0.01 to 6.3, in paper 0.22 to 1.6, and in iron and steel 0.05 to 1.91 (Ministry of Power, Government of India 2012). Thus a breakaway from incremental change through historical autonomous trend of technology adoption will be needed to achieve the target reduction in energy

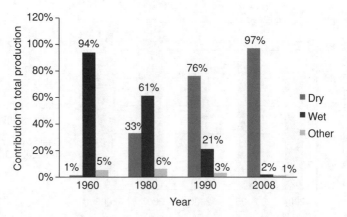

Figure 17.5 Process technology transition profile of Indian cement industries.
Source: Data from www.iea.org/work/2010/india_bee/saxena.pdf.

intensity faster. Now national pledges (INCCA 2010) and commitments are boosting the enhancement of energy efficiency gains not only in industry but in all economic sectors.

Targeting Energy Intensity Reduction

Decomposition of indicators of energy use drivers based on basic Kaya type identity (Kaya 1990; Kaya and Yokobori 1993) has been done using a number of methodologies (Albrecht *et al.* 2000; Duro and Padilla 2006; Kawase *et al.* 2006; Sheerin 1992; Sun 1999; Wang *et al.* 2005) to understand the contribution of technological and behavioral drivers in energy use. Energy intensity in the industrial sector, especially in the energy intensive manufacturing sector, has decreased over the past decades. Using methodologies in literature the world energy decomposition analysis shows that during 1980–2010, energy intensity reduction has helped in neutralizing energy use growth emerging out of population and per capita demand growth. While activity growth without technological and behavioral change could have led to 3.5% increase in world energy use, energy intensity could reduce it by 2.4% and structural change could reduce it by 0.13% (Figure 17.6). For the Indian manufacturing sector these numbers are 3.8%, 2%, and 0.81% (Figure 17.7). This implies that although the activity growth will continue to increase energy use intensity change, technological progress and behavioral response can pull it down. What is needed for the goal of bending the energy use growth curve is a stronger downward pull from intensity and structural change to counteract the upward push in energy use through activity. This of course necessitates a strategic technology innovation and deployment plan.

Following a close methodology UNIDO (2011) showed that during the period 1995–2008 average industrial energy intensity declined in all income-based groups of countries. What is interesting is the fact that in developed countries, being more advanced in the industrial development pathway, with a high share of skill and technology intensive output, the structural effect dominated the trend of decline there by showing economic activity moving away from energy intensive manufacturing processes. But in developing economies the trend was dominated more by the reduction in energy intensity from technological change.

Autonomous energy efficiency improvement (AEEI: Babiker *et al.* 2001; Manne and Richels 1990, 1992; Sanstad *et al.* 2006) which reflects the rate of change in energy

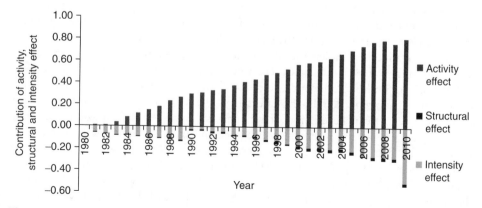

Figure 17.6 Decomposition of increase in total energy use in the world.
Source: World Bank.

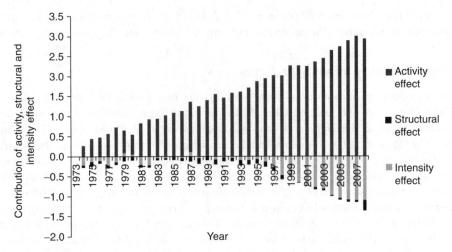

Figure 17.7   Decomposition of increase in energy use in Indian industries.
Source: *Annual Survey of Industries in India* (various vols.).

intensity (the energy-to-GDP ratio) holding energy prices constant, helps to understand the rate of penetration of technological change over time (Ausubel 1995). But BAU estimates (Roy 2007) show they are not enough. If we add to this the behavioral response through observed (through price elasticity values) or effective price change (rebound effect), then also BAU does not guarantee that. If not strategically managed, how the outcome can diverge from desired goals is demonstrated through the case study examples in following section.

Energy Efficient Technology Adoption Feedback: Cases from Indian Energy End Users

Can Rebound Take Back Benefits of Energy Efficiency Gain?

Technology improvement is supposed to help in promoting a society where it is possible to maintain the current or an improved standard of living with less energy consumed (Weizsacker *et al.* 1998). But rebound effect may outweigh the energy gains from technological improvement such that instead of decrease in energy consumption, it might increase under certain conditions (Roy 2000). Rebound effect was initially defined narrowly as direct increase in demand for an energy service whose supply has increased as a result of improvement in technical efficiency in the use of energy service (Khazzoom 1980, 1987, 1989; Khazzoom and Miller 1982). Over time the theoretical significance, nature, and magnitude of rebound effect have evolved. It is shown how Life Cycle Assessment (LCA) can be different if rebound is considered (Girod 2009). Rebound effect implies an increase in energy service demand with a corresponding decrease in effective price due to improvement in energy efficiency (Khazzoom 1980; Binswanger 2001; Brookes 1990; Greene 1992). Rebound effect can be explained through simple examples. For instance when a 75 W incandescent bulb is replaced by a 15 W compact fluorescent lamp, the consumer can enjoy the same level of illumination while saving 60 W of electricity. This implies an energy saving of 80%. In that case, many consumers, realizing the fact that the energy service now costs less and they are effectively saving some money, become less concerned about switching it off. Indeed they may leave it on all night for increased individual satisfaction arising out of the illuminated ambience. An effective fall in relative prices of energy services leads to an income effect, other things being equal. The income effect is likely to lead to further increase in energy service demand because the same budget constraint now allows for more purchasing power than before. The income effect would primarily increase the demand for energy service because beyond a certain level of income energy service or good becomes an inferior good (Lovins 1988). Studies by Jungbluth *et al.* (2007: mentioned in Girod 2009) show that in Switzerland, despite efforts to promote efficiency, Green House Gas (GHG) emissions remained more or less constant from 1990 until 2004, attributable to the "Rebound Effect." This explains how technological improvements evoke behavioral responses and as a result why it is not always the case that 1% increase in energy efficiency results in 1% decrease in energy demand (Khazzoom 1980) and sometimes even causes an increase in resource use (Roy 2000). Literature identifies three types of rebound effect:

- Direct Rebound: Improved energy efficiency for a particular energy service will decrease the effective price of that service and lead to an increase in consumption of energy. This will offset the expected reduction in energy consumption provided by the efficiency improvement (Greening *et al.* 2000; Herring and Roy 2007; Sanne 2000).
- Indirect Rebound: A lower effective price of energy services will lead to changes in demand for other goods and services (Greening *et al.* 2000; Herring and Roy 2007; Sanne 2000).
- Economy-wide Rebound Effects: A fall in the real price of energy services will reduce the price of intermediate and final goods throughout the economy, leading to a series of price and quantity adjustments with energy intensive goods. Energy efficiency improvements may also increase economic growth, which may itself increase energy consumption (Greening *et al.* 2000; Herring and Roy 2007; Sanne 2000; Saunders 1992).

In case of rebound analysis, the basic mechanisms are widely accepted but their magnitude and importance are disputed. Quantification of rebound effect is generally not easy (Girod 2009; Greening *et al.* 2000; Sorell and Dimitropoulos 2008). Some analysts argue that rebound effects are of minor importance for most energy services (Schipper and Grubb 2000), while others argue that the economy-wide effects can be sufficiently important to completely offset the energy savings from improved energy efficiency (Brookes 1990; Saunders 1992). Empirically the magnitude of rebound in household appliances varies from 0% to 50% for a 100% (Greening *et al.* 2000) or even 200% (Roy 2000) increase in energy efficiency where the size of the responses depends on consumer awareness during consumption of the services.

In the presence of rebound effect the energy efficiency improvement may be below the full technical potential. Hence, an effective energy efficiency policy through technological improvement needs to be carefully designed with appropriate information on rebound. Otherwise a significant portion of the energy or carbon savings will be lost due to the rebound effect, and the credibility of energy efficiency policies will vary.

Table 17.5   Ranges of rebound effect.

Rebound Effect	Implication for Energy Efficiency Improvement
0%	The energy efficiency improvement is fully realized.
0–100%	The energy efficiency improvement is partially realized and partially offset by increased demand for energy as relative price of energy falls.
100%	The whole energy efficiency improvement is just offset by increased demand for energy as relative price of energy falls.
>100% (Backfire Effect)	The energy efficiency improvement is outweighed by increased demand for energy as relative price of energy falls.

Source: Turner (2009), Anson and Turner (2009).

Rebound Matters: Two Case Studies from India

Two bulk consumers are studied below and the results contrasted to provide an idea how market-led behavioral responses for efficient illumination and space conditioning give rise to diverse outcomes. To estimate rebound effect, we need intensive and exhaustive information on home appliances (number of appliances, wattage consumption of each appliance, hours in use, method of usage, etc.) for two time points, the pre-action year (base year) showing the BAU scenario and the post-action year showing the scenario after the energy efficiency policy is implemented. To compute the magnitude of direct rebound effect, we use methodology suggested by Roy (2000) and Berkhout *et al.* (2000) for two major end use activities: illumination through artificial lighting, and living and workspace ambient temperature control service or air-conditioning for space cooling.

A private urban health service centre (case study I) took an initiative under the guidance of a professional consultant to make its lighting and space cooling "technically" more energy efficient, and we followed in detail the implementation and outcome. A high rebound effect can be seen in case study I through an increase in the operational load of efficient equipment (Figure 17.8 and Figure 17.9). Although inefficient lighting equipment is reduced, operational load increased due to increase in usage hours of efficient equipment, as the reduction in electricity costs provided additional financial space. Comparing the operational load in 2006 and 2010 we can calculate the actual savings in kWh. In lighting it is found that there is approximately 488% rebound effect. In space cooling the rebound is found to be 80%.

Case study II is of a bulk consumer with office cum residential space. Here the consumer took and monitored a strategic decision to change technical and behavioral responses. In illumination services, a comprehensive plan of converting the appliances to

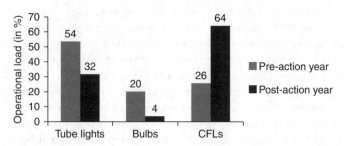

Figure 17.8   Illumination for case study I: proportion of operating load in lighting for pre- and post-action year.

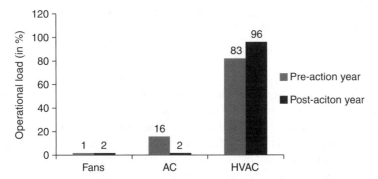

Figure 17.9 Space cooling for case study I: proportion of operating load in pre- and post-action year.

energy efficient lamps was carried out. Rebound in illumination service was estimated to be −0.028 to −0.005%. Negative rebound effect implies that the actual savings are *greater* than the expected savings (Turner 2009). Compared to case study I, we see here that operational load in the post-action period is lower (Figure 17.10). If we go by these two case studies we observe that rebound effect estimate can be a very important guiding parameter to understand how much of technical potential is realized through behavioral response or net gain through technical energy efficiency-based policy measures. The ranges vary widely as observed in these two case studies.

It is clear that in case study I a technically guided consumer ended up with higher operational load due to rebound effect in both sectors. Due to the rebound effect, loss is of 60.53% technical potential for lighting and 46.95% for space cooling. In case study II, for both technical and behavioral reasons, there is no positive rebound, rather there exists "superconservation." Thus we can conclude that the full economic and environmental

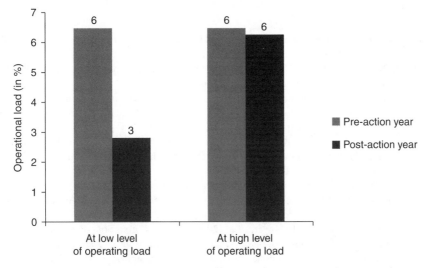

Figure 17.10 Illumination for case study II: proportion of operating load in pre- and post-action year.

benefit of appliance efficiency can be realized through better strategic management of operational behavior of the efficient appliances.

Conclusion

Energy technology matters in economic development. Thinking holistically about socio-economic systems is key to capturing the complex dynamic between energy technology and energy use. Integration of global as well as national policy is key to influencing technology development, deployment, energy pricing, and energy intensity. It is necessary that technological progress achieves increasingly higher levels of efficiency in resource use and/or waste reduction. Technological progress is expected to keep pushing forward limits to growth that might arise from a finite stock of natural resources. Left to itself, the gap between inventions and their adoption in social practice can occupy a very long time, as has been the case in the Indian cement industry so far. However, target-oriented strategically managed energy efficient technological deployment policy can make technology diffusion faster. However, this cannot happen through unilateral or bilateral national efforts alone. Given that energy efficient technology is a "genre," there have to be multi-sector, multinational, and multidisciplinary efforts simultaneously to achieve global sustainability. Sustainability cannot be achieved by an isolated national effort. Also technology deployment policy needs to be strategically integrated with education, research and training, and capacity building to make transformation faster through socio-cultural embedding and operational behavior change.

Notes

1. Corresponding Author: Department of Economics, Jadavpur University, 188 Raja S.C. Mullick Road, Kolkata 700032. Contact: +91 9836007382, joyashreeju@gmail.com.
2. http://en.wikipedia.org/wiki/Communications_in_India.

References

Abbott, R. J. 1992. Plant Invasions, Interspecific Hybridization and the Evolution of New Plant Taxa. *Trends in Ecology and Evolution* 7, 12: 401–405.

Albrecht, J., D. François, and K. Schoors. 2000. A Shapley Decomposition of Carbon Emissions Without Residuals. *Energy Policy* 30, 9: 727–736.

Allen, R. 1983. Collective Invention. *Journal of Economic Behaviour and Organization* 4, 1: 1–24.

Ang, B. W., and F. Zhang. 1999. Inter-Regional Comparisons of Energy-Related CO_2 Emissions Using the Decomposition Technique. *Energy* 24, 4: 297–305.

Anson, S., and K. Turner. 2009. Rebound and Disinvestment Effects in Refined Oil Consumption and Supply Resulting from an Increase in Energy Efficiency in the Scottish Commercial Transport Sector. *Energy Policy* 37, 9: 3608–3620.

Arthur, B. W. 1988. Self-Reinforcing Mechanisms in Economics. In P. W. Anderson, K. J. Arrow, and D. Pines, eds. *The Economy as an Evolving Complex System*. Reading, MA: Addison-Wesley.

Ausubel, J. H. 1995. Technical Progress and Climate Change. *Energy Policy* 23, 4–5: 411–416.

Babiker, M. H., J. M. Reilly, M. Mayer, *et al.* 2001. *The MIT Emissions Prediction and Policy Analysis (EPPA) Model: Revisions, Sensitivities, and Comparisons of Results*. Report 71, MIT Joint Program on the Science and Policy of Global Change. Cambridge, MA: MIT Press.

Barro, R. J., and X. Sala-i-Martin. 1992. Convergence. *Journal of Political Economy* 100, 2: 223–251.

Baumol, W. J. 1994. *Entrepreneurship, Management and the Structure of Payoffs*. Cambridge, MA: MIT Press.

Berkhout, P. H. G., J. C. Muskens, and J. W. Velthuijsen. 2000. Defining the Rebound Effect. *Energy Policy* 28, 6–7: 425–432.

Berndt, E. R., and G. C. Watkins. 1981. Energy Prices and Productivity Trends in the Canadian Manufacturing Sector 1957–76: Some Exploratory Results. Study for the Economic Council of Canada.

Binswanger, M. 2001. Technological Progress and Sustainable Development: What About the Rebound Effect? *Ecological Economics* 36: 119–132.

Boyd, G. A., D. A. Hanson, and T. Sterner. 1988. Decomposition of Changes in Energy Intensity: A Comparison of the Divisia Index and Other Methods. *Energy Economics* 10, 4: 309–312.

Brookes, L. 1990. The Greenhouse Effect: The Fallacies in the Energy Efficiency Solution. *Energy Policy* 18: 199–201.

Christensen, L. R., D. W. Jorgenson, and L. J. Lau. 1971. Conjugate Duality and the Transcendental Logarithmic Production Function. *Econometrica* 39, 4: 255–256.

Clark, N. 2002. Innovation Systems, Institutional Change and the New Knowledge Market: Implications for Third World Agricultural Development. *Economics of Innovation and New Technology* 11, 4–5: 353–368.

Dasgupta, M., and J. Roy. 2001. Estimation and Analysis of Carbon Dioxide Emissions from Energy Intensive Manufacturing Industries in India. *International Journal of Energy Environment and Economics* 11, 3: 165–179.

Dasgupta, M., and J. Roy. 2002. Energy Consumption in India: An Indicator Analysis. *Development Alternatives* October: 12–13.

Dasgupta, S. 2010. Understanding Productivity Growth and Climate Change Mitigation Potential of Iron and Steel Industries in India. MPhil Dissertation. Department of Economics, Jadavpur University, Kolkata.

Dasgupta, S., F. Salm, and J. Roy. 2011. Designing PAT as a Climate Policy in India: Issues Learnt from EU-ETS. Presented at Sixth Biennial Conference of Indian Society of Ecological Economics, October 20–22, 2011, Hyderabad.

David, P. A. 1975. *Technological Choice, Innovation and Economic Growth*. Cambridge: Cambridge University Press.

Denison, E. 1972. Some Major Issues in Productivity Analysis: An Examination of the Estimates by Jorgenson and Griliches. *Survey of Current Business* 49, 5: 1–27.

Denison, E. 1979. Explanations of Declining Productivity Growth. *Survey of Current Business* 59 (August): 1–24.

Denison, E. 1985. *Trends in American Economic Growth, 1929–1982*. Washington, DC: Brookings Institution.

Duro, J. A., and E. Padilla. 2006. International Inequalities in Per Capita CO_2 Emissions: A Decomposition Methodology by Kaya Factors. *Energy Economics* 28, 2: 170–187.

Ethridge, D. 1973. The Inclusion of Wastes in the Theory of the Firm. *Journal of Political Economy* 81, 6: 1430–1441.

Farla, J., K. Blok, and L. Schipper. 1997. Energy Efficiency Developments in the Pulp and Paper Industry: A Cross-Country Comparison Using Physical Production Data. *Energy Policy* 25, 7–9: 745–758.

Geels, F. W., and J. Schot. 2011. The Dynamics of Transitions: A Socio-Technical Perspective in Transitions to Sustainable Development – New Directions. In J. Grin, J. Rotmans, and J. Schot, eds. *The Study of Long Term Transformative Change*. London: Routledge.

Ghosh, D., and J. Roy. 2011. *Approach to Energy Efficiency Among Micro, Small and Medium Enterprises in India: Results of a Field Survey*. Working Paper 08/11. Vienna: UNIDO.

Girod, B. 2009. Integration of Rebound Effects into LCA. PhD Thesis, Swiss Federal Institute of Technology, Zurich.

Goldar, B., and A. Kumari. 2003. Import Liberalization and Productivity Growth in Indian Manufacturing Industries in the 1990s. *The Developing Economies* 41, 4: 436–460.

Goldar, B. 2010. *Energy Intensity of Indian Manufacturing Firms: Effect of Energy Prices, Technology and Firm Characteristics*. Delhi: University of Delhi, Institute of Economic Growth.

Golove, W. H., and L. J. Schipper. 1996. Long-Term Trends in U.S. Manufacturing Energy Consumption and Carbon Dioxide Emissions. *Energy* 21, 7–8: 683–692.

Golove, W. H., and L. J. Schipper. 1997. Restraining Carbon Emissions: Measuring Energy Use and Efficiency in the USA. *Energy Policy* 25, 7–9: 803–812.

Greene, D. L. 1992. Vehicle Use and Fuel-Economy: How Big Is the Rebound Effect? *Energy Journal* 13, 1: 117–143.

Greening, L. A., D. L. Greene, and C. Difiglio. 2000. Energy Efficiency and Consumption. The Rebound Effect – a Survey. *Energy Policy* 28: 389–401.

Greening, L. A., and D. L. Greene. 1998. *Energy Use, Technical Effciency, and the Rebound Effect: A Review of the Literature*. Oak Ridge, TN: Oak Ridge National Laboratory, Center for Transportation Analysis.

Greening, L.A., and M. Khrusch. 1996. Modeling the Processes of Technological Innovation and Diffusion: An Overview of Issues. White Paper, Climate Policies and Programs Division, Office of Economy and Environment. Washington, DC: US Environmental Protection Agency.

Grin, J. 2011. The Politics of Transition Governance. Conceptual Understanding and Implications for Transition Management. *International Journal of Sustainable Development* 14.

Herring, H., and R. Roy. 2007. Technological Innovation, Energy Efficient Design and the Rebound Effect. *Technovation* 27, 4: 194–203.

Hogan, W. W., and D. W. Jorgenson. 1991. Productivity Trends and the Costs of Reducing Carbon Dioxide Emissions. *Energy Journal* 12: 67–85.

Howarth, R. B., and B. Andersson 1993. Market Barriers to Energy Efficiency. *Energy Economics* 15, 4: 262–272.

INCCA. 2010. *India: Greenhouse Gas Emission 2007*. Indian Network for Climate Change Assessment. Delhi: Ministry of Environment and Forests, Government of India.

IPCC. 2007. *Climate Change 2007: Synthesis Report, Summary for Climate Change*. Geneva: Intergovernmental Panel on Climate Change.

Jorgenson, D. W., and B. M. Fraumeni. 1981. Relative Prices and Technical Change. In E. R. Berndt and B. Field, eds. *Modeling and Measuring Natural Resource Substitution*. Cambridge, MA: MIT Press, pp. 17–47.

Jungbluth, N., R. Steiner, and R. Frischknecht. 2007. Graue Treibhausemissionen der Schweiz 1990–2004. Umwelt-Wissen Nr. UW-0711. Bern: Bundesamt für Umwelt.

Kanbur, R., and L. Squire. 1999. *The Evolution of Thinking About Poverty: Exploring the Interactions*. Washington, DC: World Bank.

Katz, M. L., and C. Shapiro. 1985. Network Externalities, Competition, and Compatibility. *American Economic Review* 75, 3: 424–440.

Kawase, R., Y. Matsuoka, and J. Fujino. 2006. Decomposition Analysis of CO_2 Emission in Long-Term Climate Stabilization Scenarios. *Energy Policy* 34, 15: 2113–2122.

Kaya, Y. 1990. Impact of Carbon Dioxide Emission Control on GNP Growth: Interpretation of Proposed Scenarios. Paper presented to the IPCC Energy and Industry Subgroup, Response Strategies Working Group, Paris (mimeo).

Kaya, Y., and K. Yokobori. 1993. *Environment, Energy, and Economy: Strategies for Sustainability*. Tokyo: United Nations University Press.

Khazzoom, D. J. 1980. Economic Implications of Mandated Efficiency in Standards for Household Appliances. *Energy Journal* 1, 4: 21–39.

Khazzoom, D. J. 1987. Energy Saving Resulting from the Adoption of More Efficient Appliances. *Energy Journal* 8: 85–89.

Khazzoom, D. J. 1989. Energy Saving Resulting from the Adoption of More Efficient Appliances: A Rejoinder. *Energy Journal* 10: 157–166.

Khazzoom, J. D., and S. Miller. 1982. Economic Implications of Mandated Efficiency Standards for Household Appliances: Response to Besen and Johnson's Comments. *Energy Journal* 3, 1: 117–124.

Kim, Y., and Worrell, E. 2002. International Comparison of CO_2 Emissions Trends in the Iron and Steel Industry. *Energy Policy* 30: 827–838.

Liaskas, K., G. Mavrotas, M. Mandaraka, and D. Diakoulaki. 2000. Decomposition of Industrial CO_2 Emissions: The Case of European Union. *Energy Economics* 22, 4: 383–394.

Lovins, A. B. 1988. Energy Saving Resulting from the Adaptation of More Efficient Appliances: Another View. *Energy Journal* 9, 2: 155–162.

Manne, A. S., and R. G. Richels. 1990. CO_2 Emission Limits: An Economic Cost Analysis for the USA. *Energy Journal* 11, 2: 51–74.

Manne, A. S., and R. G. Richels. 1992. Buying Greenhouse Insurance: The Economic Costs of CO_2 Emissions Limits. Cambridge, MA: MIT Press.

Ministry of Power, Government of India. 2012. PAT Notification. *Gazette of India*, Extraordinary, March 30, Part II Section 3, Sub-Section (ii).

NAPCC. 2008. National Action Plan on Climate Change. Government of India, Prime Minister's Council on Climate Change.

Nelson, N., and S. Wright. 1994. Participation and Power. In N.Nelson and S. Wright, eds. *Power and Participatory Development: Theory and Practice*. London: IT Publications.

Pradhan, G., and K. Barik. 1998. Fluctuating Total Factor Productivity in India: Evidence from Selected Polluting Industries. *Economic and Political Weekly* 33, 9: M25–M30.

Reitler,W., M. Rudolph, and M. Schaefer. 1987. Analysis of the Factors Influencing Energy Consumption in Industry: A Revised Method. *Energy Economics* 9, 3: 145–148.

Rotmans, J., R. Kemp, M. B. A. van Asselt, *et al*. 2000. *Transitions and Transition Management: The Case of a Low-Emission Energy Supply*. Maastricht: ICIS.

Roy, J. 1992. *Demand for Energy in Indian Manufacturing Industries*. Delhi: Daya Publishing.

Roy, J. 2000. The Rebound Effect: Some Empirical Evidence from India. *Energy Policy* 28: 433–438.

Roy, J. 2007. Delinking Economic Growth from GHG Emission Through Energy Efficiency Route: How Far Are We in India? *The Bulletin of Energy Efficiency* 7 (Annual Issue).

Roy, J. 2010. Iron and Steel Sectoral Approach to the Mitigation of Climate Change: Perform Achieve and Trade in India. Briefing Paper, Climate Strategies, UK.

Roy, J., and S. Pal. 2009. Lifestyle and Climate Change. *Current Opinion in Environmental Sustainability* 1, 2: 192–200.

Roy, J., J. Sathaye, A. Sanstad, *et al*. 1999. Productivity Trends in Indian Energy Intensive Manufacturing Industries. *Energy Journal* 20, 3: 33–61.

Roy, J., B. Roy, S. Dasgupta, and D. Chakravarty. 2011. *Business Model to Promote Energy Efficiency: Use of Appliances in Domestic Sector in West Bengal – A Baseline Study*. Energy Conservation and Commercialization Project, Phase III (ECO III).

Roy, J., A. H. Sanstad, J. A. Sathaye, and R. Khaddaria. 2006. Substitution and Price Elasticity Estimates Using Inter-Country Pooled Data in a Translog Cost Model. *Energy Economics* (Special Issue).

Sanne, C. 2000. Dealing with Environmental Savings in a Dynamical Economy. How to Stop Chasing Your Tail in the Pursuit of Sustainability. *Energy Policy* 28, 6–7: 487–495.

Sanstad, A. H., J. Roy, and J. A. Sathaye. 2006. Estimating Energy-Augmenting Technological Change in Developing Country Industries. *Energy Economics* 28, 5–6: 720–729.

Sarkar, S., and J. Roy. 1995. Interfuel Substitution During Post Oil Embargo Period – Case Study of Two Energy Intensive Manufacturing Industries in India. *Indian Economic Journal* (Oct.–Dec.): 33–46.

Sathaye, J., J. Roy, R. Khaddaria, and S. Das. 2005. *Reducing Electricity Deficit Through Energy Efficiency in India: An Evaluation of Macroeconomic Benefits*. Berkeley, CA: Lawrence Berkeley National Laboratory.

Saunders, H. D. 1992. The Khazzoom-Brookes Postulate and Neoclassical Growth. *Energy Journal* 13, 4: 131–148.

Saxena, A. 2010. Best Practices and Technologies for Energy Efficiency in Indian Cement Industry. http://beeindia.in/seminar/document/2010/, BEE-A Saxena, accessed November 30, 2012.

Schipper, L., and M. Grubb. 2000. On the Rebound? Feedback Between Energy Intensities and Energy Uses in IEA Countries. *Energy Policy* 28: 433–438.

Schipper, L. J., S. Meyers, R. Howarth, and R. Steiner. 1992. *Energy Efficiency and Human Activity: Past Trends. Future Prospects*. Cambridge: Cambridge University Press.

Schipper, L. J., F. Unander, C. Marie, *et al.* 1998. The Road from Kyoto: The Evolution of Carbon Dioxide Emissions from Energy Use in IEA Countries. In *Proceedings of the 1998 Summer Study on Energy Efficiency in Buildings, August 23–27, 1998*. Washington, DC: American Council for an Energy-Efficient Economy.

Sheerin, J. C. 1992. Energy and Economic interaction in Thailand. *Energy Journal* 13, 1: 145–156.

Sheinbaum, C., and L. Rodríguez. 1997. Recent Trends in Mexican Industrial Energy Use and Their Impact on Carbon Dioxide Emissions. *Energy Policy* 25, 7–9: 825–831.

Solow, R. M. 1956. A Contribution to the Theory of Economic Growth. *Quarterly Journal of Economics* 70, 1: 65–94.

Sorrell, S., and J. Dimitropoulos. 2008. The Rebound Effect: Microeconomic Definitions, Limitations and Extensions. *Ecological Economics* 65, 3: 636–649.

Stern, N. 2007. *The Economics of Climate Change*. Cambridge: Cambridge University Press.

Sun, J. 1999. Decomposition of Aggregate CO_2 Emissions in the OECD: 1960–1995. *Energy Journal* 20, 3: 147–155.

Tomson, R. 1988. *The Path to Mechanized Shoe Production in the United States*. Chapel Hill: University of North Carolina Press.

Torvanger, A. 1991. Manufacturing Sector Carbon Dioxide Emissions in Nine OECD Countries, 1973–1987. *Energy Economics* 13: 168–186.

Turner, K. 2009. Negative Rebound and Disinvestment Effects in Response to an Improvement in Energy Efficiency in the UK Economy. *Energy Economics* 31, 5: 648–666.

UNIDO. 2011. *Industrial Development Report 2011: Industrial Energy Efficiency for Sustainable Wealth Creation*. Vienna: UNIDO.

Wang, C., J. Chen, and J. Zou. 2005. Decomposition of Energy-Related CO_2 Emission in China: 1957–2000. *Energy* 30, 1: 73–83.

Weizsacker, V. E., A. Lovins, and L. Lovins. 1998. Factor Four: Doubling Wealth – Halving Resource Use. London: Earthscan Publications.

Worrell, E., J. A. Laitner, M. Ruth, and H. Finman. 2003. Productivity Benefits of Industrial Energy Efficiency Measures. *Energy* 28, 11: 1081–1098.

Worrell, E., L. Price, N. Martin, *et al.* 1997. Energy Intensity in the Iron and Steel Industry: A Comparison of Physical and Economic Indicators. *Energy Policy* 25, 7–9: 727–744.

Part V Global Energy and Sustainability

Chapter 18

Regulation, Economic Instruments, and Sustainable Energy

Neil Gunningham[1]

Introduction

In 2008 the International Energy Agency (IEA) called for an energy revolution, involving radical action by governments at national and local levels, and through participation in coordinated international mechanisms. Fundamental to achieving such a revolution is the role of governments, and the policy instruments that they use to achieve their objectives. For it is governments that "hold the key to changing the mix of energy investment" and it is the policy and regulatory frameworks that they establish, nationally and internationally, that "will determine whether investment and consumption decisions are steered towards low-carbon options" (IEA 2008: 41).

This chapter explores the roles of a range of policy instruments that might be introduced by governments to facilitate a transition to a low carbon economy, with a focus on the energy sector (as distinct from transportation or deforestation). It is driven by the questions: what works, when, and why?

Since there are many different types of policy measure that can be used to reduce emissions, considerable selectivity is necessary within the space constraints of this chapter. The actual selection is shaped by the insight that, notwithstanding the multitude of policy instruments available, "all policies designed to promote lower carbon emissions must somehow either provide incentives (carrots) to abate, or disincentives (sticks) to emit CO_2, or both" (Banks 2011: 5). Recognizing this, the present focus is on policies whose aim is to penalize emissions or give an incentive for abatement and that also have a material impact on a country's emissions or impose significant total costs (Productivity Commission 2011).

Applying the above criteria, the chapter will address a range of policy tools that have been developed to advance progress toward achieving a low carbon economy. These include economic instruments such as tradable emission rights, taxes on electricity production or consumption, subsidies for low emissions technology, prescriptive regulation and permitting regimes, feed-in tariffs for renewable energy production, and renewable energy mandates.

The Handbook of Global Energy Policy, First Edition. Edited by Andreas Goldthau.
© 2013 John Wiley & Sons, Ltd. Published 2013 by John Wiley & Sons, Ltd.

Such instruments have been developed principally at national or sub-national level (with the European Union providing a rare regional exception, primarily in terms of its Emissions Trading System, labeling schemes, and regional targets). Put differently, energy sector reform has so far been addressed primarily *within* the nation state and it is reform at this level that is the subject of the chapter. This is not to downplay the role of international collaboration and coordination, which in some respects is likely to be crucial to energy sector reform. Questions such as how to encourage technological innovation on a Manhattan Project scale, how to overcome intellectual property constraints and effectively disseminate energy innovations, or how to support developing countries to address energy poverty without exacerbating climate change are all fundamentally questions of governance which can only be effectively addressed at global (and to a lesser extent, regional) level. And of course global and regional targets (such as the European Union's Climate and Energy Package) can serve to drive national policy initiatives. But global energy governance is a large and complex topic in its own right (Gunningham 2012) and beyond the scope of the present chapter.

There are some further constraints on the scope of the discussion below. Given the multiple strategies adopted by different nation states, it is not possible to provide a country by country analysis of where they are applied. Neither is it possible to identify any single template or formula for achieving a sustainable energy future, for there are many credible routes that might be followed and much will depend on a country's stage of development, the region in which it is located, and country-specific characteristics.

Rather, the aim is to provide a broad-brush approach, highlighting trends and developments, indicating obstacles and opportunities, and providing examples of successes and failures. As such, the aim of this chapter is to provide an overview of the role that regulation and economic instruments can provide in sustainable energy development, rather than a fine-grained analysis.[2]

Energy Policy in the Nation State: Toward an Energy Revolution?

Overall, energy sector reform within the nation state is at a relatively early stage. Indeed, a 2009 report that reviewed energy policy in a sample of eight countries found "the absence of a long-term low-carbon policy framework or coherent set of policies" (Global Climate Network 2009: 4). However, the situation is changing. Governments must increasingly compete for dwindling fossil-fuel supplies, mindful that future energy demand is also likely to grow substantially and that a profligate use of energy will generate considerable energy *in*security. Nevertheless, sometimes energy security considerations are in tension with those of climate change mitigation, and militate against measures directed toward achieving low carbon outcomes (Helm 2007; Mitchell 2008). However, in others, these pressures are prompting a concern to harness renewable energy and energy efficiency (and sometimes, nuclear power). And in a minority of countries (principally but not exclusively in Western Europe)[3] climate change mitigation has also risen up the political agenda, inclining decision-makers to take a more proactive approach. In contrast, in the developing world – where an estimated 1.4 billion people (20% of the global population) lack access to electricity and some 2.7 billion people rely on the traditional use of biomass for cooking (IEA 2010c: 16–17) – the most important drivers of change are economic and social development.

As energy policy moves to centre stage, many countries (at the time of writing, almost 100) have set national targets, most commonly in terms of renewable energy as a percentage of total energy generation. Perhaps the best known is the European Union's commitment to the goal of 20% renewable energy in final energy use by 2020 (in conjunction with National Renewable Energy Action Plans elaborating on how member states will reach these targets). Also of note is UN Secretary-General Ban Ki-moon's vision statement for 2030: ensuring universal access to modern energy services; doubling the rate of improvement in energy efficiency; and doubling the share of renewables in the global energy mix (Ban 2011: 4).

Some countries have also set more concrete targets in terms of various technologies.[4] In some cases, such targets are genuine aspirations that may help galvanize action on the ground, while in others they may prove to be largely symbolic. But at their best, they focus attention and resources, form an integral part of planning, provide motivation, and support the achievement of policy objectives.

More important for present purposes is the question: what tools and strategies (hereafter "carbon policies") might most appropriately be invoked to achieve such targets? Here the choice is considerable, but the principal carbon policies can conventionally be divided into two categories: economic instruments and regulation.

Economic Incentives

According to mainstream environmental economics, environmental degradation in general, and climate change in particular, results from a failure to fully value environmental endowments in the market. In this paradigm, it is because no price emerges to reflect their scarcity value that the market fails to ration scarce assets efficiently. As regards greenhouse gas (GHG) emissions, the conventional solution is to put a price on energy-related CO_2 emissions either by determining the amount of carbon pollution that the environment can assimilate, allocating rights to those emissions, and letting the market determine the appropriate price, or by introducing market signals by charging appropriately for the use of such scarce resources (e.g., through a tax or charge on carbon).

Economists argue that economic instruments are usually more cost effective than direct regulation, in large part because they give energy producers and consumers more flexibility as to how they achieve resource productivity and prevent carbon pollution.

The full range of economic instruments that might be deployed in energy policy is extensive, and ranges from the removal of perverse incentives, through the use of taxes, charges, and market creation, to the use of liability rules, and includes both property rights and price-based instruments. The remainder of this section focuses on the most influential types of instruments which have been applied to advance the cause of sustainable energy: emission trading schemes, fiscal instruments, and public finance mechanisms. These strategies have been applied both to encourage energy efficiency and to increase the development and uptake of renewable energy. With the former, the emphasis is on facilitating "win-wins" in the areas where the biggest gains seem plausible, while in the latter it is with advancing technologies and stimulating markets.

Emission Trading Schemes

Much collective energy has been invested in seeking international agreement on climate change mitigation embodied in some form of international agreement. Such efforts have so far proved unsuccessful. The principal mechanism contemplated to deliver emission reduction targets under a climate change agreement is to place a price on carbon and to rely on market mechanisms to achieve agreed reductions. Raising the price of high carbon goods and services is likely to result in a decrease in their use, an increasing focus on energy efficiency, a shift toward existing low carbon fuels and technologies and a stimulus to the development of new ones. The favored mechanisms through which to increase the carbon price are a series of Emissions Trading Schemes (perhaps ultimately merging into regional or a global scheme), a tax on the use of carbon, or some hybrid mechanism involving elements of both (Antes *et al.* 2011; Hoel 2010; Kerr 2000; McKibbin and Wilcoxen 2008). What is common to all such mechanisms is the central role of economic incentives, with the market arguably providing the most flexible and least costly mechanism for delivering agreed carbon reductions. It is the first of these options which has gained most support internationally and which is the subject of this section.

Even in the absence of an international agreement, a growing number of countries (and at regional level, the European Union) have sought to put a price on carbon via an emissions trading scheme. The European Union has led in terms of its European Union Emissions Trading System but is not alone in this regard. New Zealand has also developed such a scheme, and Japan and South Korea are also contemplating going down this path (Anger 2008; Tyler *et al.* 2009). Australia, while initially introducing fixed price permits in 2012, will move to a floating price within three to five years. And China is also contemplating a pilot Emissions Trading Scheme in some provinces. At sub-national level, the developing Regional Greenhouse Gas Initiative incorporates 10 States in the northeast of the US (but lacks a binding cap) while the Eastern Climate Initiative extends to seven US States and four Canadian provinces. The most advanced sub-national emissions trading scheme is that of California, which will operate from 2012.

There is a voluminous literature arguing the virtues of emissions trading schemes as a cost-effective means of curbing GHG emissions[5] (and a much smaller one warning of their shortcomings: e.g., Hanemann 2010). To date, the evidence suggests that such schemes are capable of performing relatively well (though they have confronted numerous problems in the design and early implementation stages: Ellerman *et al.* 2010).

However, as various reports have pointed out, such mechanisms (even assuming that the price is sufficiently high to substantially shape behavior) will not be sufficient in and of themselves to achieve a transformation of the energy sector in the constrained time period which the science suggests remains available for effective mitigation.[6]

There are a number of reasons why this is the case. Some of these relate to limitations in the design of particular schemes while others concern the inherent limitations of such incentive-based mechanisms. As regards the former, the prices generated by carbon markets may be too low (or insufficiently stable) to send effective signals to major carbon emitters to shift to lower emitting technologies. With regard to the latter, the incentives provided by carbon markets may be "fuzzy" – involving such market imperfections as monitoring, enforcement, and asymmetric information problems, all of which may constrain some emitters from responding to price signals, or worse, may generate fraud, speculation, or rent seeking (OECD 2009: 20–21).[7]

But beyond the above imperfections lie other limitations which are more fundamental. In particular, as the OECD (2009: 20–21) points out, "carbon pricing does not address the large market failures undermining R&D in climate mitigation, such as incompatibility with existing infrastructure and weak intellectual property rights protection." Nor will market mechanisms such as emissions trading be sufficient to encourage and facilitate the development and dissemination of low carbon technologies within the relatively short time frame available to us (Sims 2009: chap. 4).[8]

Of particular importance will be infrastructure changes (not least, the development of smart grids) in conjunction with policy approaches that stimulate investment in low carbon technologies (including large-scale research and development projects which only government funding is likely to make viable),[9] together with fast tracking of embryonic renewable technologies and intellectual property rules that encourage the development *and* facilitate the transfer of such technologies.

Fiscal Incentives

Fiscal instruments can be defined as "all economic instruments of a fiscal nature – that is via the use of the fiscal system – as well as direct subsidies that provide *incentives* to shift from environmental harmful activities towards cleaner and more sustainable alternatives" (Kosonen and Nicodème 2009: 2). Fiscal instruments are used to encourage environmentally responsible behavior through full (or partial) cost pricing of consumption or production. That is, rather than establishing property rights over common or unpriced resources,[10] this approach involves establishing prices on them, as a different way to internalize externalities. But like property-based rights, such instruments have an advantage over conventional regulation in that they allow firms to make individual choices about energy performance free from outside interference and in doing so give them an incentive to innovate and reduce carbon pollution at the lowest cost. As such, in principle at least, they provide a cost-effective means of promoting energy policy goals. There are two main categories of fiscal instruments: tax instruments and subsidies.

As regards the former, tax credits and similar incentives can be important in facilitating start-up of projects involving new technologies, where upfront costs are high, and long-term benefits uncertain. For example, under the US American Recovery and Reinvestment Act of 2009, tax credits have been used to promote other aspects of sustainable energy policy. There are provisions for a residential energy property credit (for homeowners who make energy efficient improvements) and residential energy efficient property credits (to encourage residential alternative energy equipment). Other countries, such as India, promote renewable energy use by providing accelerated depreciation for small hydropower and biomass and preferential tax rates for other renewable energy projects.[11]

Many tax credits or exemptions focus on achieving energy efficiency in terms of lighting and heating equipment and in electric household appliances, since these are perceived to be areas where gains are likely to be most substantial. For example consumers in Italy are provided with a tax credit for purchases of energy efficient refrigerators and freezers while France provides a credit for purchase of condensing boilers (Kosonen and Nicodème 2009: 22).

Fiscal incentives have also been the predominant policy mechanism for encouraging the development of renewable energy with regard to heating and cooling, though far more with respect to the former than the latter. Chief amongst these have been capital rebates, grants, variable consumption rates, production incentives, and tax credits (IPCC 2011: section 11.5.5.1).

Less common are taxes on fossil fuels, including fossil fuels used to generate electricity, although such may be found in India and Japan, and differential electricity taxes (imposed at rates based on generation technology), although the latter may be found in the UK in the form of the Climate Change Levy (imposed on non-residential users of electricity but exempting generation from renewables).

In contrast, there has been relatively little policy interest in demand-side abatement (due to lower electricity consumption) and estimates suggest, with the limited exceptions of Germany and the UK, that it has played a relatively minor role in the energy policies of a range of developed countries (Productivity Commission 2011: 98–99).

In evaluating economic instruments such as the above, it is important to distinguish between theory and practice, and to be conscious of how far market imperfections may impede the effective operation of the market. For example, as the IEA has pointed out in terms of renewable energy, these include: **imperfect information** (e.g., when a consumer buys an appliance but there is insufficient or inaccurate information provided on the energy performance of the product); **principal–agent problems** (e.g., landlords are commonly responsible for buying electrical appliances such as refrigerators for their properties but it is only the tenant who will benefit from reduced electricity bills from buying energy efficient products); and **behavioral failures** (e.g., personal decisions made by consumers that appear not to be economically rational) (IEA 2010a; IEA 2011b; Productivity Commission 2011).

The severity of such impediments will vary with the context, and in some circumstances they can be overcome or substantially mitigated (e.g., by mandating product labeling – a regulatory mechanism – to overcome informational asymmetries). However, in others these impediments will be so severe as to suggest that other types of instruments (usually regulation) may be more appropriate. This is the case for example with regards to energy efficiency in buildings.

Subsidies are the second principal type of fiscal instrument. The use of subsidies is deplored by some environmental economists because such mechanisms prevent the "internalization of externalities" and are for this reason inefficient, and because they are a drain on public revenue. Treasury departments often oppose such incentives because of the difficulties in classifying products and practices which might attract the subsidy, and because they have the potential to lock in current technologies at the expense of as yet unknown alternative solutions.

Nevertheless, subsidies may have merit because of imperfect information, deficiencies in capital markets, or the public good characteristics of technology. In consequence, "there may be insufficient investment in the production and dissemination of environmentally friendly technology, technique and products. If such market failures are pervasive, then an argument can be made for government subsidies for the development and use of such technologies and products" (Pearson 2000: 162). In these circumstances, the international experience suggests that subsidies, suitably designed and targeted *can* in some circumstances act as an effective spur to energy efficiency and in advancing low carbon technologies (IPCC 2011: chap. 11).

Direct capital renewable energy subsidies are widely used by governments for a diversity of purposes from facilitating the provision of large-scale renewable energy generation capacity (e.g., wind farms) to micro-level projects under which householders and small businesses are the main recipients (e.g., facilitating the installation of solar PV cells). Such subsidies are often provided in addition to other financial incentives for renewable generation (e.g., feed-in tariffs) (Productivity Commission 2011: 26). However, while some direct subsidy and tax credits programs have proven cost effective in the sense that their benefits exceed the costs to the society (Kosenen and Nicodème 2009: 25), this is far from being the case across the board and much depends upon the particular context and technology. For example, subsidizing solar PV is currently a relatively costly abatement option (Productivity Commission 2011: 80).

Subsidies have also been widely used to promote energy efficiency. For example, under the aegis of the Car Allowance Rebate System (popularly dubbed "cash-for-clunkers") the US government provided $3 billion of subsidies in July and August of 2009 to consumers opting to trade in an old vehicle for a newer, more fuel-efficient vehicle, with doubtful efficacy (Sivak and Shoettle 2009; Yacobucci and Canis 2010).

Subsidies have been employed to greater effect in China, which has shown an impressive capacity for improvement in energy efficiency. The nation achieved a 19.06% reduction in energy consumption per unit of GDP in the 11th Five Year Plan ending in 2010 and has ambitious targets under the 12th Five Year Plan. Amongst the arsenal of Chinese energy efficiency policies canvassed by Zhou *et al.* (2010) in their study of the area was a suite of major subsidy programs, including the flagship Ten Key Projects program, which offered financial incentives for enterprise and local government authorities willing to embark on investment in major energy efficiency improvements. Under government supervision, participants received 60% of the necessary capital upfront, with the remainder payable on achievement of the budgeted energy savings (World Resources Institute 2009: 2). The four most significant of these projects were the renovation of coal-fired boilers, district level combined heat and power, oil conservation and substitution, and energy efficiency in buildings. These programs alone were estimated by the Chinese National Development and Reform Commission as likely to deliver some 40% of the national efficiency target under the 11th Five Year Plan (NDRC 2004). It might be questioned, however, whether an extensive subsidy program like the Ten Key Projects would satisfy a cost-benefit analysis *outside* of a developing country context.

Overall, a mixed picture of the value of subsidies in delivering energy efficiency and renewable energy goals emerges. While some energy subsidies can be justified, there are numerous others that cannot. And *perverse energy incentives* – subsidies (often hidden) of fossil fuels – are estimated to have caused wasteful consumption that totaled $312 billion in 2009 (IEA 2010b). As the G20 and others continue to emphasize, the removal (or more realistically, the reduction) of such subsidies should be very high on the list of sustainable energy policy reforms. Doing so would enhance energy security, reduce GHG emissions and air pollution, *and* bring economic benefits (IEA 2010b: 13).

How can we assess the costs and benefits of direct fiscal incentives in their various forms? In terms of energy efficiency, perhaps the most comprehensive study is that conducted within the European Commission in 2008 (BIO Intelligence Service 2008). On the basis of examining four different products and four different EU Member States, Kosonen and Nicodème (2009: 25) note that this study concluded: (1) in case of increases of energy taxes the benefits exceed the costs by a relatively wide margin and energy taxation thus appears to be a cost-effective way of improving energy efficiency in the economy; (2) subsidies have also a considerable potential in generating energy savings exceeding that of energy taxation substantially *in some cases*; (3) in contrast, tax credits to manufacturers are costly policies (emphasis added).

In terms of renewable energy, evaluation is more difficult because "investment support instruments like investment grants, rebates and tax policies are difficult to measure as they are generally used as supplementary policy tools" (IPCC 2011: 40). In general, as the Intergovernmental Panel on Climate Change has pointed out (IPCC 2011: 40): "countries that have relied heavily on tax-based incentives have often struggled with unstable or insufficient markets for wind power or biogas [and that] generally, tax credits work best in countries where there are numerous profitable, tax-paying private sector firms that are in a position to take advantage of them."

More broadly, fiscal instruments have the particular attraction that they are generally more efficient than regulation because they rely on incentives rather than direction and prescription. Moreover, they can raise revenues for the state which in turn can be deployed either for R&D or to reduce distorting taxes in other areas. But as we will see below, they tend to be most effective when combined with other policy instruments to compensate for informational or market failures, a matter to which we return in the concluding section.

Public Finance Mechanisms

Public finance mechanisms involve "public support for which a financial return is expected (loans, equity) or financial liability is incurred (guarantee)" (IPCC 2011: 883). For example, preferential loan schemes involving low or zero interest loans can reduce the cost of borrowing to invest in low emissions generation technologies, as can loan guarantees whereby the government takes the default risk (Productivity Commission 2011: 26). Also significant are public procurement policies (providing preference for the purchase of renewable technologies) and investment policies (providing financing in return for equity ownership in a renewable energy initiative). UNEP has developed a comprehensive taxonomy of the various types of policy measures in the public finance space (UNEP 2008: esp. Part II).

The most obvious financing mechanism is of course government provision of debt finance on terms more favorable than the private sector could furnish, termed "soft" loans (IEA 2007: 64–65; IPCC 2011: 893). Various preferential loan schemes can be identified both in developed and developing countries. Such schemes can provide support at the micro-level (as with the off-grid and decentralized solar program providing low interest loans for small solar power generation in India), or at the middle range (as with Brazil's use of preferential loans as an incentive for the use of wind, small hydropower, and biomass) (Productivity Commission 2011: 26). On a large scale, of course, the efficacy of simple debt financing by public authorities will depend on the ready availability of large amounts of capital.

Public finance is particularly important in facilitating the development of renewable energy technologies which may have difficulty accessing commercial financing given the level of risk involved. Here public finance mechanisms play important roles in mobilizing or leveraging commercial investment and in "indirectly creat[ing] scaled up and commercially sustainable markets for these technologies" (IPCC 2011: 892). More broadly, public finance is an indispensable source of the investment that will be needed to meet GHG emissions reduction targets and keep global warming below the 2 degree "guardrail." As the IPCC (2011) points out, public finance mechanisms play a particularly important role in developing countries, where an immature commercial financial sector is usually less than willing to provide the sort of financing the renewable energy initiatives require.

The type and structure of public financial support appropriate in a given case will depend upon the ability of private institutions to finance emissions abatement projects in the absence of public support (though it should be noted that the relevant "public" authority might in fact be a regional or international development institution such as the World Bank rather than a domestic government: IPCC 2011: 893; UNEP 2008). Public finance is most cost effective when it leverages extant private sector capacity. In UNEP's (2008: 6) assessment of public finance mechanisms, it noted the considerable multiplier effect that well-directed public financial assistance can deliver.

In a country with very underdeveloped financial markets, then, an appropriate course of action might be for the state to provide a line of credit to commercial financial institutions to be lent on to provide the bulk of funding for major renewable energy projects. UNEP-Centre (2012: 27) gives the example of Thailand's Energy Efficiency Revolving Fund, under which the Thai government provides capital to Thai banks to be lent at their discretion to worthy projects. Only the principal sum is repaid into state coffers and the capital is then used to finance further projects (see APEC 2005).

A range of other measures are possible depending on the particular weaknesses of private actors in a given jurisdiction. The state might involve itself in the provision of venture capital, for instance, or specific monetary prizes to incentivize R&D. Where private capacity exists but is constrained by a lack of expertise, the state might provide technical assistance to key market actors.

Evaluating public finance mechanisms is extremely challenging and for this reason the United Nations Environment Program commissioned a report *Evaluating Clean Energy Public Finance Mechanisms* (Irbaris and Climate Bonds Initiative 2011) which proposed a methodology to evaluate the performance of different public finance mechanisms and their suitability for replication in other conditions and policy contexts. The report suggests that the clean energy policy, regulatory framework, and investment climate in the target country will be key variables.

Regulation

While regulatory measures are in theory less flexible and less efficient than market mechanisms, whether this is the case in practice depends on numerous factors, including the particular type of regulation and the degree to which the theoretical advantages of market mechanisms are not realized in practice because of various market failures or weaknesses such as those discussed above.

More broadly, the relative effectiveness and efficiency of regulation as compared with economic instruments will vary substantially with the context, including the characteristics of the entities being regulated and the capacity of the state to implement and enforce the particular instruments being introduced. China, for example, given its distinctive command economy, has relied particularly heavily on regulation (coupled with targets) to promote low carbon technology development and deployment, as with the 11th Five Year Plan's energy intensity target and specific regulations requiring closure of inefficient industrial capacity.[12]

Numerous forms of regulation might be invoked to bring about a transition to a low carbon economy. The following section discusses the most important but is by no means exhaustive. It begins by making the conventional distinction between technology and performance standards before examining in more detail some of the most important energy regulation policy instruments that have been introduced successfully in a variety of jurisdictions.

Technology and Performance Standards

Technology-based standards typically require the use of specified equipment, processes or procedures while performance-based standards "specify allowable levels of pollutant emissions or allowable emission rates, but leaving the specific methods of achieving those levels to the regulated entities" (Aldy and Stavins forthcoming). Performance standards are widely seen as preferable to technology-based standards because they prescribe outcomes to be achieved but not *how* they might be achieved and accordingly are more flexible.

Not least, performance standards have served to limit energy demand by removing inefficient products from the market (for example products which fail to meet a minimum energy efficiency standard) and by promoting more efficient alternatives. California in particular has been an exemplar of "what can be achieved in energy efficiency through relentless pressure on mundane product standards" (Aldy and Stavins forthcoming) such as those mandating more energy-saving appliances and buildings. Similarly, some US States have specified limits to the emissions intensity of new electricity generators.

In some contexts the particular virtue of performance-based regulation has been in overcoming the sorts of principal–agent problems, informational asymmetries, and bounded rationality effects that can seriously impede the effectiveness of economic instruments as described above. For example, building codes mandating energy efficiency requirements for new buildings have been applied in numerous countries and have demonstrably achieved impressive levels of energy efficiency (Stern 2006: 382) in a way that economic incentives cannot because of the principal–agent problems described earlier.

Notwithstanding their perceived inflexibility, technology-based standards have also figured prominently in the energy policies of many nation states. For example, while renewable portfolio standards that mandate generators, suppliers or consumers to satisfy specified renewable energy targets (which themselves are often ratcheted up over time) may often be performance-based, others specify *how* targets are to be achieved.

Mandatory renewable energy targets which require a proportion of electricity to be generated using *specified* renewable technologies are not uncommon. Such targets have been set in a number of countries including Australia, Germany, the UK, Japan, and South Korea, while at sub-national level 41 US States have renewable targets, many of them mandatory (Productivity Commission 2011: xviii). More modestly, the requirement in the UK that new coal-fired power stations over a certain capacity must be "carbon capture ready" is also a technology-specific standard. Finally, some building codes require the installation of particular renewable energy heat or power technologies, often in conjunction with efficiency investments.

In some circumstances at least, the conventional assumption of mainstream economists that technology-based standards are inefficient is overstated or incorrect (Latin 1984: 1271). Overall, the evidence confirms that the imposition of technology standards can accelerate the deployment of best technologies (Gunningham *et al.* 2003: chap. 3), but for the most part, such standards have been slow to develop, particularly in the developing world, where economic pressures to opt for "cheap but dirty" technologies often trump climate change mitigation.[13]

Feed-in Tariffs and Tradable Certificate Schemes

Some types of energy regulation have proven more successful than others, but often it is only by trial and error that it becomes apparent into which category a particular instrument falls. Such has been the case regarding whether renewable energy support should best take the form of a tradable certificate scheme or whether feed-in tariffs should be preferred (Lauber 2005). The former are the most common instruments for delivering on renewable energy targets in developed countries. Conventionally, "tradable certificates are issued to renewable electricity generators for the units of electricity they produce. An obligation is placed on generators or electricity retailers to surrender these certificates to a regulator to meet the renewable energy target. Renewable generators receive the market price for the electricity they produce, and earn an additional subsidy by selling the certificate to a retailer or generator with obligations under the scheme" (Productivity Commission 2011: 23).

Feed-in tariffs in contrast seek to accelerate investment in renewable energy technologies by paying a guaranteed tariff to large renewable energy producers (e.g., wind farms, biogas producers) and/or small ones (e.g., domestic solar PV). The general aim is to set a price for each individual renewable technology, reflecting its particular level of costs to produce and providing sufficient incentive for such production. The overall aim is to provide cost-based compensation to renewable energy producers, and sufficient certainty through long-term contracts to make their investments commercially viable. Such tariffs have been adopted in a number of countries including Germany, Japan, South Korea, and the UK. China and India also operate national and sub-national schemes (Productivity Commission 2011: 25).

Contrary to conventional wisdom, there is evidence suggesting that tradable certificates have led to higher prices per kWh, to less deployment, less innovation, and hence lower dynamic efficiency (Lauber 2006) and that feed-in tariffs may be a better option. Germany, in particular, has successfully used feed-in tariffs to support investments in wind, solar, and biomass and through them has achieved a demonstrably accelerated growth in the use of these various forms of renewable energy.[14] However, it is not alone in this regard.[15] Indeed a 2011 IPCC report suggests that feed-in tariffs at national level have achieved dramatic increases in renewable capacity and that they are a key enabling policy for promoting renewable energy (IPCC 2011: chap. 11).

However, the success of feed-in tariffs is by no means guaranteed and its effectiveness has been demonstrated more clearly than its efficiency.[16] To the extent it does succeed, the principal reasons would appear to be "the combination of long-term fixed price or premium payments, network connections, and guaranteed purchase of all RE electricity generated" (IPCC 2011: 23). On the other hand, rapidly falling prices for photovoltaics are presenting a problem for feed-in tariffs, albeit one that may prove only to be temporary, and may yet be capable of remedy without the need to stop such programs entirely.

It is also the case that "inevitably [feed-in tariff] rates are set at higher levels than would be necessary to induce the least-cost mix of renewables and the overall resource cost of using a particular level of renewables will be higher than under a [renewable energy certificate] scheme" (Productivity Commission 2011: 81). For reasons such as these, some jurisdictions continue to favor tradable certificates.

Other Regulatory Initiatives

While the above are the most widely used carbon policies, there are a number of others that have had significant success albeit on a limited scale or only in individual jurisdictions. Of particular note is China's "Large Substitute for Small" policy which encourages the decommissioning of small, energy-inefficient thermal power plants and their replacement with larger, more efficient ones. This policy, self-evidently, is likely to be cost effective, reducing generation costs and carbon emissions and promoting local environmental objectives more generally.

There is also a role played by "informational regulation," which involves the state encouraging or requiring the provision of information about energy impacts but *without* directly requiring a change in those practices. Rather, this approach relies upon incentives and public opinion as the mechanisms to bring about improved energy performance. In terms of carbon policy, its most important manifestation is product labeling and certification with regard to energy performance, an approach that has been adopted in numerous countries, in part because of the relative ease of measuring energy and in quantifying benefits and costs. Such initiatives may involve endorsement or warning labels, or may provide consumers with comparative information with regard to energy consumption, cost, and attributes of a product – popularly termed "eco-labeling."

Over and beyond those instruments that have been directly invoked to drive carbon policy, regulation also plays a key role in underpinning or facilitating other sustainable energy initiatives. For example, regulation is important for the indirect but nevertheless crucial role it can play in making particular technologies viable (as in the case of carbon capture and storage and nuclear power.

Another dimension of a low carbon economy will be to shift from centralized to distributed forms of electricity generation. This cannot be achieved simply by matching a particular technology to a particular need, but will involve numerous institutional factors, including, in particular, an appropriate regulatory regime (see Zerriffi 2010). Not least, a suitably designed grid (and local grid and off-grid renewable energy sources) would involve the deployment of micro-generational capacity on a considerable scale, implying the need not only for mechanisms that render such sources cost effective and a "smart grid"[17] but also for appropriate regulatory frameworks that provide incentives to consumers and energy suppliers to promote diffusion *and* that provide priority grid access to renewable energy sources.

Finally, it should be noted that regulation in major markets (particularly where such markets are environmental leaders, as with the European Union and California) can also have a ripple effect, since other countries, to export to these markets, must meet their standards. This, as David Vogel (1995) has demonstrated, is part of the wider process of "trading up" whereby regulatory standards are ratcheted upward. For example, many global firms adopt EU standards globally because if they meet these most stringent standards "they can be marketed anywhere in the world" (Vogel forthcoming). Such pressures have demonstrably forced "changes in how industries around the world make plastics, electronics, toys, cosmetics and furniture" (Cone 2005) and may soon, in conjunction with the efforts of the EU to actively globalize its standards, have a similar impact with regard to energy efficiency standards for various appliances.

Conclusion

No single policy instrument or set of instruments (whether it be economic incentives or regulation) can claim superiority in terms of contributing to a sustainable energy future. While economic instruments in principle have the edge over regulation in terms of their flexibility and consequently their efficiency, this is not always the case in practice. Sometimes market imperfections are so great that regulation is a demonstrably preferable alternative. But even where this is the case, selecting the most appropriate type of regulation is context specific and it cannot be said that particular types of tools (e.g., performance standards) are invariably superior to others (technology standards).

Nevertheless, there is much that we do know about what works and when, in terms of sustainable energy policy. We know that emissions trading schemes can be relatively cost effective, though much depends upon how they are designed and implemented (see Ellerman *et al.* 2010) but that they will be insufficient in and of themselves to bring about the needed "energy revolution." And we know that a range of other measures, such as mandatory renewable energy targets, feed-in tariffs, energy efficiency measures, and capital subsidies for constructing or installing renewable energy technologies can also make important contributions (IEA 2011b).

On the other hand, some policies supporting small-scale renewable generation can be relatively costly. This is particularly the case with subsidies for solar photovoltaic systems, and often little abatement has resulted (Productivity Commission 2011: chap. 4). Supply-side policies on product prices also seem to have had only a modest impact, with the notable exception of electricity prices in Germany and the UK (Productivity Commission 2011: chap. 6). Other policies such as the use of government funds to leverage more private finance for energy projects, as under current European Union initiatives (Buchan 2011), are still at too early a stage to be confident of their relative merits.

Crucially, complementary combinations of policy instruments are likely to work better than "stand alone" tools (Buchan 2011). For example, consumers may have insufficient information with regard to the energy-saving capacity of a particular appliance, in which case informational regulation via energy labeling may usefully complement tax instruments which reduce the cost of such appliances. Numerous country-specific examples illustrate the value of developing context-specific policy mixes. For example, to promote renewable energy, South Korea promotes feed-in tariff tax exemptions for dividends in combination with long-term loans for manufacturing facilities, China combines various clean energy policies such as feed-in-tariffs for wind with integrated solar PV, while Japan's solar roof program combines gradually declining rebates with net metering, low interest loans, and public education.[18]

However, it is equally the case that some instrument combinations can be counter-productive. For example Australia managed to develop state and territory feed-in tariffs that overlapped completely with a renewable energy target and not only did not lead to any additional abatement but added to the total financial cost of meeting the target and indeed "could have actually led to higher emissions than if there had been no FIT schemes" (Productivity Commission 2011: 83).

Beyond this, which policies might sensibly form part of the policy mix will depend substantially upon each country's individual energy profile (energy importer/importer, opportunities for harnessing particular energy sources, technological capabilities, economic circumstances, etc.) as well as upon the political, economic, and social constraints within which its government must make decisions.[19]

But governments have too infrequently taken advantage of the knowledge that we do possess and have often failed to implement the lowest cost options first, choosing instead various relatively high-cost options (Productivity Commission 2011: 151). For example, it is widely accepted that the building sector offers the highest potential level of reductions, but only in the Scandinavian countries, Germany, and France have considerable resources been devoted to achieving substantial energy efficiencies in this sector. And where governments have acted, equity considerations have often been downplayed, notwithstanding that in the developing world in particular, energy poverty is a compelling issue.

To what extent this situation will change and greater policy learning and transfer will take place as energy policy networks develop, remains to be seen. Certainly there is considerable potential for modeling and for strengthening best practice on a transnational level through such emerging networks as the International Confederation of Energy Regulators, but whether these opportunities will be grasped is an open question.

Frequently too, when countries have developed a policy mix, this has not been done strategically. Rather, there is much overlap and inconsistency with different levels of government sometimes supporting the same project, or with overlaps between different policies even within the same level of government. For example, Germany continues to support renewable electricity by various measures notwithstanding being part of the European Union's emissions trading scheme. This has the effect of reducing the emissions reduction burden of other EU countries, lowering ETS permit prices, and increasing emissions in other EU countries at Germany's expense (Traber and Kemfert 2011).

Finally, individual countries have often paid insufficient attention to overcoming the well-known obstacles to achieving more efficient and effective sustainable energy outcomes, not least informational and market failures, a shortage of technical skills, inadequate capacity within government, and insufficient financing (IPCC 2011: chap. 11).

Notes

1. The author gratefully acknowledges the excellent research assistance of David Rowe.
2. While this chapter draws from a diversity of sources, particular mention must be made of the *Special Report on Renewable Energy Sources and Climate Change Mitigation* (IPCC 2011). This report addresses comprehensively, and with considerable sophistication, much of the policy terrain of sustainable energy.
3. See in particular, South Korea's Framework Act for Low Carbon Green Growth 2010.
4. See generally UN General Assembly (2011).
5. For an overview see Garnaut (2008).
6. See for example US Government Accountability Office (2009). This is not to deny the importance of energy taxes or to suggest that they do not play an important role but rather that they are necessary but not sufficient (see for example Schmidt *et al.* 2011).
7. See also Sachs (2009) arguing that barriers to effective market signals include principal–agent divergence of interests, high implicit discount rates used in purchase of energy-using products, inadequate information on energy pricing and usage by individuals, and lack of incentives for utilities to undertake investments in efficiency measures.
8. See also Sims (2009).
9. See further Aldy and Pizer (2008: 21), pointing out that R&D generates benefits that the innovator cannot fully appropriate.
10. Property rights, for example, underpin emissions trading schemes, providing incentives for businesses to reduce their emissions by clarifying their rights to and responsibilities with regard to units of carbon, which can then be traded.
11. See UN General Assembly (2011).
12. However there are now examples of economic incentives including reforms to coal-fired electricity tariffs to promote efficiency and these are of growing importance. See Watson *et al.* (2011: 52).
13. A good example is Indonesia. See PT Media (2008).
14. However, it would appear that policies requiring priority access to the grid for renewables, priority purchase of generation from renewable resources, and differential tariffs based on the cost of generation plus a reasonable profit, are essential. See United Nations (2009: 60).
15. See IPCC (2011: section 11. 5) citing studies suggesting that some feed-in tariffs have been effective and efficient in promoting renewable electricity.
16. For example Germany's feed-in tariff regime is relatively costly and may also lead to "carbon leakage." See Traber and Kemfert (2011: 33–41).
17. The development of a "smart grid" (capable of predicting and responding intelligently to the behavior and actions of all electric power users connected to it) will be of particular importance in facilitating energy reform. See Farhangi (2010).
18. See generally UN General Assembly (2011).
19. See also IEA (2011a), which provides guidance on how to assess the need for supplementary policies for energy efficiency and renewable energy with existing carbon pricing.

References

Aldy, Joseph, and William Pizer. 2008. *Issues in Developing US Climate Change Policy.* Washington, DC: Resources for the Future.

Aldy, Joseph, and Robert Stavins. Forthcoming. The Promise and Problems of Pricing Carbon: Theory and Practice. *Journal of Environment and Development*.

Anger, Niels. 2008. Emissions Trading Beyond Europe: Linking Schemes in a Post-Kyoto World. *Energy Economics* 30, 4: 2028–2049.

Antes, Ralf, Bernd Hansjürgens, Peter Letmathe, and Stefan Pickl, eds. 2011. *Emissions Trading: Institutional Design, Decision Making and Corporate Strategies*. 2nd edn. Dordrecht: Springer.

APEC. 2005. *Thailand's Energy Efficiency Revolving Fund: A Case Study*. Asia-Pacific Economic Cooperation Energy Working Group.

Ban, Ki-moon. 2011. *Sustainable Energy for All: A Vision Statement by Ban Ki-moon, Secretary-General of the United Nations*. New York: United Nations.

Banks, Gary. 2011. Comparing Carbon Policies Internationally: The "Challenges." Presentation to the BCA/AIGN Carbon Pricing Forum, Parliament House, Canberra, March 23.

BIO Intelligence Service. 2008. *A Study on the Costs and Benefits Associated with the Use of Tax Incentives to Promote the Manufacturing of More and Better Energy-efficient Appliances and Equipment and the Consumer Purchasing of these Products*. Copenhagen Economics for European Commission, DG TAXUD.

Buchan, David. 2011. *Expanding the European Dimension in Energy Policy: The Commission's Latest Initiatives*. Oxford: Oxford Institute for Energy Studies.

Cone, Marla. 2005. Europe's Rules Forcing US Firms to Clean Up. *Los Angeles Times*, May 16.

Ellerman, A. Denny, Frank J. Convery, and Christian de Perthuis. 2010. *Pricing Carbon: The European Union Emissions Trading Scheme*. Cambridge: Cambridge University Press.

Farhangi, Hassan. 2010. The Path of the Smart Grid. *IEEE Power & Energy Magazine* 8, 1: 18–28.

Garnaut, Ross. 2008. *Emissions Trading Scheme Discussion Paper*. Canberra: Australian Government.

Global Climate Network. 2009. *Breaking Through on Technology: Overcoming the Barriers to the Development and Wide Deployment of Low-carbon Technology*. Washington, DC and London: Center for American Progress and Global Climate Network.

Gunningham, Neil. 2012. Confronting the Challenge of Energy Governance. *Transnational Environmental Law* 1, 1: 119–135.

Gunningham, Neil, Robert Kagan, and Dorothy Thornton. 2003. *Shades of Green: Business, Regulation and Environment*. Palo Alto, CA: Stanford University Press.

Hanemann, Michael. 2010. Cap-and-trade: A Sufficient or Necessary Condition for Emission Reduction? *Oxford Review of Economic Policy* 26, 2: 225–252.

Helm, Dieter, ed. 2007. *The New Energy Paradigm*. Oxford: Oxford University Press.

Hoel, Michael. 2010. *Climate Change and Carbon Tax Expectations*. CESifo Working Paper Series 2966. Munich: CESifo.

IEA. 2007. *Renewables for Heating and Cooling: Untapped Potential*. Paris: International Energy Agency.

IEA. 2008. *World Energy Outlook: 2008*. Paris: International Energy Agency.

IEA. 2010a. *Combining Policy Instruments for Least-Cost Climate Mitigation Strategies*. Paris: International Energy Agency.

IEA. 2010b. *World Energy Outlook: 2010*. Paris: International Energy Agency.

IEA. 2010c. *Energy Poverty: How to Make Modern Energy Access Universal?* Paris: International Energy Agency.

IEA. 2011a. Energy Efficiency Policy and Carbon Pricing. Paris: International Energy Agency.

IEA. 2011b. *Summing Up the Parts: Combining Policy Instruments for Least-Cost Climate Mitigation Strategies*. Paris: International Energy Agency.

IPCC. 2011. *IPCC Special Report on Renewable Energy Sources and Climate Change Mitigation*. UK: Cambridge University Press.

Irbaris and Climate Bonds Initiative. 2011. *Evaluating Clean Energy Public Finance Mechanisms*. Montpelier, VT: UNEP SEF Alliance.

Kerr, Suzi, ed. 2000. *Global Emissions Trading: Key Issues for Industrialized Countries*. Cheltenham, UK: Edward Elgar Publishing.

Kosonen, Katri, and Gaëten Nicodème. 2009. *The Role of Fiscal Instruments in Environmental Policy*. European Commission: Directorate-General for Taxation and Customs Union.

Latin, Howard. 1984. Ideal v. Real Regulatory Efficiency: Implementation of Uniform Standards and Fine Tuning Reforms. *Stanford Law Review* 37: 1267–1332.

Lauber, Volkmar. 2005. Renewable Energy at the Level of the European Union. In Danyel Reiche, ed. *Handbook of Renewable Energies in the European Union*. Frankfurt: Peter Lang.

Lauber, Volkmar. 2006. Tradeable Certificate Schemes and Feed-in Tariffs: Expectation versus Performance. In Volkmar Lauber, ed. *Switching to Renewable Power*, chap. 12. London: Earthscan.

Macintosh, Andrew. 2011. Searching for Public Benefits in Solar Subsidies: A Case Study on the Australian Government's Residential Photovoltaic Rebate Program. *Energy Policy* 39: 3199–3209.

McKibbin, Warwick, and Peter Wilcoxen. 2008. *Building on Kyoto: Towards a realistic Global Climate Agreement*. Working Papers in International Economics 3.08. Sydney: Lowy Institute.

Mitchell, Catherine. 2008. *The Political Economy of Sustainable Energy*. London: Palgrave Macmillan.

NDRC. 2004. *Medium and Long-Term Energy Conservation Plan*. Beijing: National Development and Reform Commission.

OECD. 2009. *The Economics of Climate Change Mitigation*. Paris: OECD.

Pearson, Charles. 2000. *Economics and the Global Environment*. Cambridge: Cambridge University Press.

Productivity Commission. 2011. *Carbon Emission Policies in Key Economies: Research Report*. Canberra: Australian Government.

PT Media. 2008. *Comprehensive Study on Crash Program Progress and National Electricity Business Opportunity, 2008–2015*. Jakarta: PT Media Data Riset.

Sachs, Noah. 2009. Greening Demand: Energy Consumption and US Climate Policy. *Duke Environmental Law and Policy Forum* 19: 295–320.

Schmidt, Sigurd Naess, Eske Stig Hansen, Janatan Tops, *et al*. 2011. *Innovation of Energy Technologies: The Role of Taxes*. Copenhagen Economics for European Commission.

Sims, Ralph. 2009. Can Energy Technologies Provide Energy Security and Climate Change Mitigation? In S. Stec and B. Baraj, eds. *Energy and Environmental Challenges to Security*. Dordrecht: Springer, pp. 283–305.

Sivak, Michael, and Brandon Schoettle. 2009. *The Effect of the "Cash for Clunkers" Program on the Overall Fuel Economy of Purchased New Vehicles*. Ann Arbor: University of Michigan Transportation Research Institute.

Stern, Nicholas. 2006. *Stern Review on the Economics of Climate Change*. London: HM Treasury.

Traber, Thure, and Claudia Kemfert. 2011. Refunding ETS Proceeds to Spur the Diffusion of Renewable Energies: An Analysis Based on the Dynamic Oligopolistic Electricity Market Model EMELIE. *Utilities Policy* 19, 1: 33–41.

Tyler, Emily, Michelle du Toit, and Zelda Dunn. 2009. *Emissions Trading as a Policy Option for Greenhouse Gas Mitigation in South Africa*. Cape Town: Energy Research Centre.

United Nations. 2009. *World Economic and Social Survey 2009: Promoting Development, Saving the Planet*. New York: United Nations.

UNEP. 2008. *Public Finance Mechanisms to Mobilise Investment in Climate Change Mitigation*. Nairobi: United Nations Environment Programme.

UNEP-Centre. 2012. *Case Study: The Thai Energy Efficiency Revolving Fund*. Frankfurt: Frankfurt School UNEP Collaborating Centre for Climate and Sustainable Energy Finance, available at http://fs-unep-centre.org/sites/default/files/publications/fs-unepthaieerffinal2012_0.pdf, accessed December 11, 2012.

UN General Assembly. 2011. *Promotion of New and Renewable Sources of Energy: Report to the Secretary General*, New York: United Nations.

US Government Accountability Office. 2009.Testimony Before the Subcommittee on Energy and Environment, Committee on Energy and Commerce, House of Representatives: Observations on the Potential Role of Carbon Offsets in Climate Change Legislation. Statement of John

Stephenson, Director Natural Resources and Environment, March 5. Washington, DC: United States Government Accountability Office.

Vogel, David. 1995 *Trading Up: Consumer and Environmental Regulation in a Global Economy.* Cambridge, MA: Harvard University Press.

Vogel, David. Forthcoming. The Transatlantic Shift in Health, Safety and Environmental Risk Regulation, 1960 to 2010. In *The Politics of Precaution: Regulating Health, Safety and Environmental Risks in Europe and the United States.* Princeton, NJ: Princeton University Press.

Watson, Jim, Rob Byrne, Michele Stua, *et al.* 2011. *UK–China Collaborative Study on Low Carbon Technology Transfer.* University of Sussex (UK), Sussex Energy Group.

World Resources Institute. 2009. China's Ten Key Energy Efficiency Projects. Washington, DC: WRI.

Yacobucci, Brent, and Bill Canis. 2010. *Accelerated Vehicle Retirement for Fuel Economcy: "Cash for Clunkers."* Washington, DC: Congressional Research Service.

Zerriffi, Hisham. 2010. *Rural Electrification: Strategies for Distributed Distribution.* Dordrecht: Springer.

Zhou, Nan, Mark Levine, and Lynn Price. 2010. Overview of Current Energy Efficiency Policies in China. *Energy Policy* 38, 11: 1–37.

Chapter 19

The Role of Regulation in Integrating Renewable Energy: The EU Electricity Sector

Jaap Jansen and Adriaan van der Welle

Introduction

The Climate Change challenge is looming large with potentially catastrophic consequences for the earth's human population, flora, and fauna. The Earth Summit in 1992 at Rio de Janeiro, that brought together an unprecedented number of heads of state from all over the world, concluded that the most affluent countries have the obligation to shoulder the first and highest burdens to adequately address this issue of paramount global significance. The European Union (EU) has responded to the climate change by adopting a set of quite ambitious policy objectives to bring down greenhouse gas (GHG) emissions in 2020 by 20% with respect to 1990, raise the share of renewables in final energy consumption to 20% in 2020, and realize energy savings of 20% in 2020 compared to an official baseline energy consumption level. The European Commission and several EU member states have gone further and have set an indicative GHG emissions reduction target for the EU of 80–95% for year 2050 compared to emissions in 1990. Furthermore, the EU heads of state stipulated in 2011 that by 2014 the national and regional electricity systems will have to merge into one truly EU-wide electricity system. This ambition is challenging indeed, given the currently still insufficiently interconnected power systems and a multitude of non-harmonized national regulations and grid codes.

According to the Energy Roadmap 2050 outlined by the European Commission in 2011, the power sector will have to play a key role in achieving the 2050 GHG emissions reduction target. A GHG emissions reduction in the power sector over the corresponding period of 90–95% is required, whilst at the same time the share of electricity in final energy consumption is to expand substantially. A phase-out of fossil fuels in passenger transportation and domestic heating systems would have to occur to a large extent, possibly by way of electrification.

To make this happen, significant improvement of energy efficiency in energy conversion, distribution, and end use is required, along with strong penetration of low carbon generation technologies in the electricity supply mix. It is envisaged that variable renewables such as wind and solar power will have to contribute quite significantly, with

The Handbook of Global Energy Policy, First Edition. Edited by Andreas Goldthau.
© 2013 John Wiley & Sons, Ltd. Published 2013 by John Wiley & Sons, Ltd.

nuclear and/or coal and gas with carbon capture and storage having to account for the remainder. Hence it can be expected that future European electricity supply will come to a large extent from less controllable renewable resources. Furthermore, the power sector will be more capital intensive, warranting more front-loaded and hence more risky investment finance.

Besides limiting the environmental impact of energy supply by notably limiting carbon emissions reduction, European energy policy is oriented around two other main objectives: a competitive, single internal energy market which is to yield affordable energy prices, and security of supply.

The three principal policy dimensions are sometimes conflicting. For instance, European policy-makers strongly stimulate the penetration of electricity production from renewable energy sources (RES-E). As such this may help to decarbonize the European economy. Yet this also implies higher market price volatility and more frequent high-amplitude fluctuations in network flows with, *ceteris paribus*, negative consequences for both competitiveness and short/medium-term security of supply.

This chapter focuses on some major regulatory reforms and challenges facing European policy-makers in fostering and enabling a fast rising share of renewables in EU electricity supply. It will address some of the most important conflicts between the drive for affordable electricity prices and a secure electricity supply on the one hand, and a renewable (sustainable) electricity supply on the other. Pursuance of the low carbon energy road map will require smart institutional framing of, and regulations governing, the European power sector. Quite a few controversial issues have to be resolved. In this chapter some of the following issues will be discussed. How can the European electricity system contain the huge transition costs in per unit of energy terms and in so doing mitigate surging affordability issues? How can the European electricity system ensure supply adequacy: is it possible in an "energy only" market framework or is there a need for a capacity mechanism? Either way, can market-based solutions be found? How can the European electricity system contain the huge transition costs in per unit of energy terms and in so doing mitigate surging affordability issues? Do national support schemes for renewable energy need to give way to the internal energy market concept? Is there a need for adjustment of congestion management methods within the EU? Does the prevailing national network planning approach have to convert into a European approach? Do the European experiences hold out lessons for electricity systems elsewhere in the world? The next sections will address these questions.

The Internal Energy Market and Renewable Electricity in The EU

This section begins by shedding light on the ongoing European integration process of a multitude of national power markets into one single EU-wide power market. Next the various policy instruments, applied within the EU to stimulate the market uptake of renewable generation, are set out. Finally some key aspects of the regulatory framework for use of the grid by European renewable generators are explained.

The Internal Energy Market

The EU electricity supply sector is undergoing a long evolutionary process that should lead to one internal energy market (IEM). The starting point was a fragmented situation in the early 1990s, characterized by closed national electricity markets with typically one public state electricity company or a multitude of regional public electricity companies, taking care of both generation, transmission, and distribution, as well as delivery to the final customers. By 2007, in each EU member state all electricity customers were to have the freedom to choose their supplier within the framework of a liberalized electricity market. Great progress has been made to that effect. Still in most member states supply-side market concentration is rather high, with just a few vertically integrated companies accounting for 90% or more of the market. In several member states the majority of electricity customers is still serviced by one company, the (former) state electricity company.

The so-called Third Package of EU legislation regarding the gas and electricity markets was adopted in 2009. A key issue in this package is the unbundling of vertically integrated electricity companies.[1] Commercial activities to be undertaken in a liberalized market framework, production and supply, should be disconnected from transmission and distribution as network services are conceived as forming a natural monopoly by the operators of networks. The maximum tariff rates for network transport services are set by national regulating agencies. EU legislation allows three unbundling options:

- *total unbundling*, the most radical alternative, in which the vertically integrated companies divest their transmission networks to an independent actor operating and owning the transmission networks;
- *mandating an independent system operator* (ISO) to operate the transmission networks of integrated power companies; in the so-called deep ISO approach the transmission networks are sold to a transmission owner (TO) who is fully independent from the ISO;
- *legal unbundling* of high-voltage transmission networks: vertically integrated power companies are to rearrange their structure and operate their transmission networks through a separate legal entity. The transmission network operator (TSO) company is to be separated by "Chinese walls" from the subsidiary companies taking care of commercial activities such as power generation and/or retail supply.

An example of a member state having transposed EU legislation into national legislation in accordance with the first alternative is the Netherlands, where even distribution networks have to be unbundled. Yet in several other member states the unbundling requirement has been implemented rather poorly to date, with a quite dominant position of the former state electricity company. Moreover, at odds with EU legislation, electricity retail prices are still regulated in quite a few member states.

The Third Package on EU gas and electricity markets envisages a further coordination of the institutional and regulatory frameworks within the EU electricity sector. This and intervening market framework developments are to enable the emergence of a genuine IEM for the European electricity sector. Currently, by way of the so-called Regional Initiatives, national electricity markets are evolving into six EU regional electricity markets. Early in 2011 EU heads of government decided that the IEM is to be realized by 2014. A key role in this process is played by ENTSO-E/G.[2] One of the key tasks of ENTSO-E is to draft harmonized network codes, with special reference to cross-border aspects (European Union 2009b: Art. 8.1/2). Moreover, the EU Agency for the Cooperation of Energy Regulators (ACER), is to guide on behalf of the European Commission the process leading to compatible national network regulation and network codes, notably regarding network issues of cross-border significance. To that effect, ACER and other relevant EU bodies have formulated a so-called Target Model that should be implemented by 2014.

This encompasses *among others* a harmonized approach for:

- coordinated calculations of transfer capacity (TC) for interconnectors as a basis for implicit (bundled) allocation of cross-border transfer capacity and approval of cross-border bilateral energy trades on the basis of a common grid model (CGM);
- single price coupling (SPC) for day-ahead markets all over Europe, where one single matching algorithm is able to establish prices and volumes across all borders between intra-European market areas;
- implicit continuous inter-regional and ultimately pan-European capacity allocation for cross-border intra-day markets;
- cross-border balancing markets with a harmonized approach (e.g., with respect to gate closure time).

Policies to Stimulate the Market Uptake of Renewables

In 2009 the renewables (RES) directive, on the promotion of the use of energy from renewable sources, was adopted.[3] This EU directive sets each member state a mandatory target for its share of renewables in final energy consumption in 2020. It provides a framework for renewable energy policies and measures in the EU member states. The corresponding RES target for the whole EU is 20%. At the time of drafting of the RES directive no consensus could be reached on an EU-wide approach to market stimulation of renewables. Therefore, subject to complying with targets and framework conditions stipulated in the directive and other EU legislation, member states have the prerogative to design and implement policies and measures of their own liking.

Indeed the EU member states pursue rather divergent aims with national renewable energy policy. These range from ambitious national industrial policies, decarbonization policies, social cohesion in peripheral regions, to plain cost-effectiveness in complying with the national target of the member state concerned set by the RES directive. As a result, divergent national support mechanisms have been put in place, effectively fragmenting the renewable segment of the EU electricity market by national borders. Feed-in tariffs (FIT) and feed-in premiums (FIP) schemes are most popular as the main stimulation mechanisms. These comprise pre-set, technology-specific preferential production prices (FIT) and production subsidies (FIP) respectively. Typically, if with some exceptions, renewable generators benefiting from FIT are entitled to inject their production into the public grid for free and without regard to the cost impact this has for system operators and other network users. Other defined grid stakeholders, e.g., transmission system operators, have to market the renewable power concerned. Eventually, the small-scale end users have to pay the lion's share of the bill, whereas industrial customers are typically largely relieved so as to foster their international competitiveness. This is for instance the case in Germany. Conversely, renewable generators benefiting from FIP tend to be responsible themselves for marketing of the power they produce. Also, FIP-receiving renewable generators do have to shoulder the cost of imbalances to the power system, when they deviate from their production schedules which they had to notify to their TSOs beforehand. Hence, typically renewable generators receiving FIP are exposed to market risk, while their counterparts receiving FIT are shielded from market risk.

Consequently, market integration of FIP beneficiaries as compared to FIT beneficiaries tends to be more optimal. Also grid integration of FIP beneficiaries is typically better. For these reasons, two large member states, Spain and Germany, have changed their initial FIT system into an optional FIT-FIP system. In these countries eligible renewable generators can choose between receiving pre-set FIT or FIP rates respectively. As the governments concerned wish to stimulate a transition from FIT toward FIP, they tend to set more attractive technology-specific FIP rates than FIT rates. Based on prior experience gained with windfall profit issues, since recently FIP rates are being subjected to adjustment mechanisms to the evolution of wholesale prices.

Originally, FIT and FIP systems were open-ended. Yet recent financial constraints have triggered ad hoc volume caps or ad hoc downward price interventions, especially for the high cost technologies. For example, since 2011 the German government has introduced several ad hoc subsidy cuts for PV following the quite remarkable effectiveness of German feed-in tariffs for this technology: in years 2010 and 2011 7.4 GW and 7.5 GW of new PV capacity has been installed in Germany. This has put strong upward strain on the final electricity prices for German households and prompted questions about the efficiency of the German support instrument. In Spain FIP subsidy to eligible PV installations has even been discontinued altogether recently.

A minority of EU member states have opted for renewable quota systems (RQS) as their main support instrument. Typically, RQS-benefiting renewable generators receive a RQS certificate for each MWh fed into the grid. They sell the certificates, originated on their behalf, on the RQS certificate market to traders or directly to electricity suppliers. The latter have to comply with the pre-set renewable target by submitting a number of certificates that accords with this target. RQS beneficiaries are responsible themselves for marketing the power they produce and for making arrangements to settle the cost of system imbalances for which they are accountable. Such arrangements may include a contractual shift of balancing responsibility to another system user, for instance by way of a power purchase agreement. Well-designed RQS systems may foster cost-efficient deployment of RES potentials. Yet, unless properly addressed, promising high cost emerging technologies might be locked out. So far, most RQS systems in the EU have suffered from major design flaws (e.g., UK, Italy) or from small, illiquid certificates markets (Flanders, Wallonia). In contrast, the Swedish RQS system has proved quite successful in meeting its effectiveness and efficiency objectives.

Although so far harmonization of renewable support systems turned out to be a bridge too far, the RES directive encompasses so-called cooperation mechanisms, intended to foster cross-border efficiency of investments in renewable generation capacity. So far, the utilization of these cooperation mechanisms has been rather subdued. The most significant development to date is the joint Norwegian-Swedish RQS support scheme, implemented from the beginning of 2012.

The Regulatory Framework on Grid Use by Renewable Generators

The RES directive mandates member states to ensure among others:

1. Clear definition and coordination of authorization, certification, and licensing procedures with transparent timetables for determining planning and building applications.
2. The use of minimum levels of energy from renewable sources in new buildings and in existing buildings that are subject to major renovation, with public buildings fulfilling an exemplary role in these respects.
3. To develop transmission and distribution grid infrastructure, intelligent network, and storage facilities in the electricity system, in order to allow secure operation of the electricity system as it accommodates the further development of RES-E,

including interconnection between member states and between member states and third countries.

4. To accelerate authorization procedures for grid infrastructure and coordinate grid infrastructure along with streamlined administrative and planning procedures.
5. To ensure, subject to maintenance of (transparent and non-discriminatory) reliability and safety standards:
   a. Guaranteed transmission and distribution of RES-E.
   b. Priority or guaranteed access of RES-E to the grid-system.
   c. Priority dispatch of RES-E installations by TSOs based on transparent and non-discriminatory criteria.
6. To minimize the curtailment of RES-E and, if significant RES-E curtailment measures occur, TSOs concerned are to report these and to indicate measures to prevent inappropriate curtailments.
7. TSOs and DSOs are to set up and make public their standard rules relating to the bearing and sharing of costs of grid extension and reinforcement and grid code rules, necessary for feeding RES-E into the interconnected grid. These rules should take particular account of all the costs and benefits, including location-specific ones, of connecting RES-E producers to the grid. Where appropriate, TSOs and DSOs might be required to bear, in part or in full, the aforementioned grid extension and reinforcement costs.
8. Streamlining of the handling of requests by RES-E producers for grid connection with clear timetables and comprehensive and detailed estimate of connection costs.
9. Charging of transmission and distribution tariffs is not to discriminate against RES-E producers and is to account for realizable cost benefits resulting from grid connection of RES-E producers.

The transposition of the RES directive into the national legislation of many member states boils down to the following:

- Application of "shallow" or even "supra-shallow" connection charges; in the latter case, e.g., the costs of interconnecting offshore wind farms at sea to the onshore grid are "socialised," paid for by mainly small-scale electricity users through grid charges on top of their energy bills.
- Very low or even zero use-of-system charges for generators, including RES-E generators.
- Priority dispatch for RES-E generators in congested grid areas.

On the other hand, in some member states quite a few complaints are filed by (potential) RES-E investors, who claim to be facing discriminatory treatment of their grid connection requests. This especially appears to apply in member states with a quasi-monopolist state grid company, notably but not only in the new member states who have acceded to the EU since May 2004.

We observe that achievement of the 2020 RES deployment objectives has been the leading consideration in shaping the RES directive. This consideration has taken precedence over considerations of EU competitiveness regarding the cost of delivered energy and completion of a genuine IEM. The RES directive allows and at times even mandates a preferential market risk reduction treatment for operators of (selected) power plants from RES. This is not only confined to allowing state-specific support benefits to facilitate the market uptake of energy, but also applies to allowance of preferential grid connection and use in grid-constrained situations, along with cross-subsidization by other grid users. Such an approach may have certain merits to kick-start RES-E market deployment. Yet the disproportionally rising cost of facilitating steeply increasing variable RES-E segments and the rising prominence of affordability issues are poised to warrant a radical recast of the 2009 RES directive in due course. In 2012, the number of disconnected households because of payment arrears has been reported to have risen substantially in several EU member states; affordability is poised to gain prominence on the agenda of European policy-makers.[4]

Security of Supply

Security of supply has different time scales. To ensure reliability of supply, power demand has to be matched by power supply on a second-by-second basis. Supply adequacy regards the capacity of generation plants as well as transmission and distribution infrastructure to meet system peak demand. Supply adequacy needs to be considered quite some time ahead as the construction of generation plants including pre-construction preparations takes several years, depending on the generation technology, whilst T&D system expansions can take 10 years or more. As will be explained below, market failures are at the base of adequacy and reliability of supply.

Addressing the Challenge of Integrating Variable Renewables

Variable renewable energy technologies (VRE: wind, solar PV, wave, and tidal energy) represent additional effort in terms of their integration into existing power systems.[5] Balancing power systems with large shares of VRE poses a severe challenge, due to limitations to the predictability and high fluctuations of VRE power injections.

Existing flexible resources to manage fluctuations in power demand or supply are:

- dispatching flexible conventional power plants;
- electricity storage;
- demand-side management;
- interconnections to neighboring power markets.

Variability and uncertainty of VRE are greater than on the demand side. It is generally easier to predict fluctuations in demand than in VRE supply. Can the use of existing flexible resources be enhanced efficiently to balance increasing variability resulting from VRE deployment?

System operators have vast experience with responding to variability in demand and contingencies by ramping flexible sources up or down. When fast response is required, the operators will call upon the most flexible resources, which are:

- power plants designed for peaking (e.g., open-cycle gas turbines, hydropower plants);
- storage facilities (e.g., pumped hydro) and in some cases:
- interconnections and
- contracted demand-side management (load shedding).

The IEA (2011) has designed a crude four-step method, FAST (flexibility assessment) to assess what share of VRE is possible with effective use of existing flexible resources on four time scales within a 36 hours balancing time frame (36 hours, 6 hours, 1 hour, 15 minutes):

1. Assess the Technical Flexible Resource, i.e., the maximum flexible ability of the four flexible resources to ramp up or down over the balancing time frame. Network constraints are disregarded.
2. Assess (qualify) the extent to which certain attributes of the power area in question will constrain the availability of the technical resources. This yields the Available Flexible Resource.
3. Calculate the maximum Flexibility Requirement of the system, i.e., a combination of fluctuations in demand and VRE output (the net load), and contingencies. Allow for smoothing through geographical and VRE technology spread when a strong grid is in place.
4. The Present VRE Penetration Potential (PVP) is a function of the requirement for flexibility and the available flexible resource (in %).

The pioneering IEA study (2011) has only looked at transmission level VRE power plants. Network constraints are disregarded in the PVP quantification. Also, complementarity of demand and VRE output fluctuations is disregarded (net load fluctuations might therefore have been overestimated to some extent). Furthermore, no allowance has been made for smoothing effects nor for the opportunity to curtail VRE output.

With due regard to these caveats, the IEA study tentatively established in a number of case studies that present-day power systems tend to be able to integrate much larger shares of renewables than previously thought, ranging from 19% in Japan to 63% in Denmark. The study has some noteworthy recommendations to facilitate the integration of large volumes of power from weather-determined variable renewable sources:

1. Assess and mitigate without delay weaknesses in the existing transmission network.
2. Markets should be (re)configured so that the full flexible resource is able to respond in time to assist in balancing. Power markets should incorporate mechanisms that enable sufficient response from supply-side and demand-side flexibility assets. Markets overly relying on supply of electricity locked up in long-term bilateral contracts find it harder to balance variability and uncertainty. Such contracts hamper the use of flexibility assets closer to real time.
3. Operation of (notably) existing mid-merit plants must remain economic, when socio-economically optimal. Increasing VRE penetration tends to depress electricity prices, pushing conventional plants out of the market. Lesser operating hours and increasing wear and tear costs due to increased cycling may lead to early retirement. New market mechanisms might be needed to prevent a possible shortfall in flexible resources. We revert to this issue in the next section.
4. Owners of flexible resources need incentives additional to that of a fluctuating electricity price to prompt them to offer the full extent of their flexibility to the market, also allowing for increasing wear and tear cost of start-ups, shut-downs, and ramping. New mechanisms might be needed to prompt slower assets to respond to flexibility needs forecasted 36 hours ahead.
5. TSOs making use of the best available forecasting tools of VRE plant output combined with more dynamic power-trading (e.g., short gate closure time) and planning of system operation, can make more efficient use of the flexible resource.

6. Seek to expand power markets: larger power markets with VRE resources widely distributed over a strong grid will see a lesser requirement for flexibility.[6]
7. Merging balancing areas enables more smoothing (lessening flexibility needs) and sharing of flexibility resources.
8. *Last but not least*: expensive new capacity measures should be considered a last resort, taken only after optimizing the availability of existing flexible resources.

Based on a review of existing studies and contingent on local conditions, the IEA estimates the balancing cost for wind power of $1–7 USD/MWh at a penetration rate of 20% of average electricity demand. Other additional system costs, not included in these figures, relate to transmission and support to supply adequacy.

Supply Adequacy

The optimal investment level in flexible generation capacity will be impacted by the increasing deployment of renewable generation. Especially in predominantly thermal power systems with an increasing share of electricity from renewables with marginal costs close to zero, like wind and solar, less electricity will be produced by conventional coal- and gas-fired power plants. Hence, in energy-only markets[7] running hours of existing power plants will decrease substantially, which means that they have to recover their fixed costs during few hours with high (peak) demand and low availability of wind and solar resources. When annual running hours go down, fewer hours occur when prices in excess of short-term marginal costs occur. Then full recovery of fixed costs becomes more problematic and the business case of investment in new flexible capacity diminishes. Consequently, larger price spikes and lower, even more negative price valleys are set to occur in wholesale markets; overall price volatility will increase.

This market price variability leads to an important issue. On the one hand, since flexible power plants are required for system security as well as supply adequacy purposes (i.e., both for balancing supply and demand and for provision of system flexibility), price spikes are needed to offer generators the opportunity to recover their investments ("scarcity rents"). On the other hand, larger price spikes in wholesale markets are probably unacceptable for politicians. Hence, they may either react by (more restrictive) price caps on wholesale market prices or by other types of government interventions. However, such interventions limit scarcity rents and give rise to pressure on conventional power producers to withdraw their generating assets from the system. Furthermore, such interventions are likely to induce a lack of future investment in new conventional but flexible generation facilities. The upshot is a capacity scarcity problem.

This has triggered the debate about the introduction (extension) of alternative remuneration schemes for conventional power generation in Europe. Those mechanisms usually remunerate conventional generators for the reservation of (part of) their generation capacity for utilization during system peaks. A distinction can be made between capacity payments and capacity requirements. For the former, the regulator sets a price for capacity and the market determines the amount of available capacity, whereas for the latter the regulator determines beforehand the total amount of capacity that has to be made available, with the market fixing the price for making available the pre-set aggregate capacity level.

In many electricity markets in the US (especially on the East Coast, including PJM, ISO-NE, NYISO) as well as in some EU member states (Spain, Ireland) capacity remuneration mechanisms have already been in use for a long time even before the uptake of renewables started. The deployment of capacity markets in the US was (partly) motivated by legal regulatory requirements to mitigate unreasonable exercise of market power in their absence. Hence, price hikes with associated high scarcity rents at times of capacity scarcity triggered rent-distorting regulatory intervention (e.g., price caps). In Spain, capacity markets have been set up for partly the same reason. Besides mitigation of market power of large incumbent generators, allowance is being made in Spain for cost recovery of (stranded) assets dating from the pre-liberalization era. Moreover, due to its rather isolated location in the western rim of Europe, Ireland has small interconnection capacity with neighboring countries, which diminishes possibilities for import during plant failures (Roques 2008).

Yet any debate about additional government interventions should not be limited to the introduction of capacity remuneration mechanisms only. *First*, supply adequacy is not only influenced by measures aimed at supply adequacy itself but also by other security of supply dimensions, such as participation in congestion management and balancing market design (Roques 2008). Several academics as well as stakeholders (including Eurelectric 2011) deem that, when policy-makers consider changes to electricity wholesale market design, the focus should not only be put on energy-only markets versus markets with both remuneration of capacity and energy produced. The analysis scope should be broader. The income of conventional producers is also affected by institutional conditions and the market rules for provision of reserves, participation in congestion management, network planning standards, and balancing market design. This means that it is also important to adapt conditions and market rules that unnecessarily impede the flexibility of the power system and hence increase its price variability. Such conditions include generation support schemes and rules for preferential market access (during congestion) which limit the responsiveness of market participants (e.g., renewable generators) to market prices.

Second, the extraordinary market price variability in the power market is related to market failures; *non-storability of electricity* requires that supply and demand have to be matched on a second-by-second basis. Generators do not have an incentive to invest in enough production capacity that accounts for many uncertainties about supply and demand since they do not face the full costs of a disruption. They only face the costs of electricity not sold, not the resulting costs for society (i.e., the value of lost load). Furthermore, a lack of information exists due to *demand inelasticity* related to the absence of real-time metering and billing for small consumers, as a result of which a large group of consumers does not pay the time- and location-dependent spot price, but rather a price averaged over a certain period. Consequently, electricity consumers such as households do not usually instantaneously face high prices during periods with high price levels nor price variability. Household electricity demand thus typically does not react at all (or at best with a large time lag) to changing market conditions.

For both market failures, reserve capacity (and security of supply in general) has public good characteristics. This relates to the fact that for technical and economic reasons it is not possible to curtail all customers individually from using capacity, even when some of them are not paying for the cost of keeping that reserve capacity available for them when they need it. This follows from the non-excludability nature of reserve capacity. In many cases there is thus free-riding of electricity consumers on reserve capacity. In this respect, policies aimed at increasing demand response seem quite valuable as demand response mitigates the aforementioned market failures and hence the large price variability.

If despite the measures outlined above, the price variability is still unacceptably high, capacity remuneration schemes are an option to be considered. Yet two points are then important to heed. First, such schemes should *only* be aimed at potential flexibility providers, that is providers ramping up when system demand peaks, and/or ramping down when demand valleys occur. Besides flexible generating capacity this notably also includes demand response. *Non-flexible generation capacity should not be remunerated for availability.* Second, after the massive uptake of variable renewables in a certain power system, there is a wider array of power system needs than before. This implies that capacity remuneration mechanisms deployed in several countries to date may not be future-proof anymore and hence unfit for replication elsewhere around the globe. The availability of sufficient generation capacity for meeting peak demand is not enough. For the integration of renewables in electricity systems, additional flexibility for accommodating low-demand/high-supply situations is key.

Network Planning in the European Union

This section emphasizes the need for improved network planning procedures in Europe, as escalating delays put realization of Europe's ambitious energy and environment goals in the balance. Europe's electricity systems are in a transition process, mainly in response to policy aimed at de-carbonization. It also relates to the development of an IEM within Europe and to policies aimed at maintaining and enhancing security of supply.

These developments drive the need for additional investments in network extension and improved operational network management.[8] Renewable generation, especially wind, is often furthest located from areas with highest demand. This implies that electricity has to be transported over larger distances. Furthermore, the higher variability and lower predictability of renewable generation increases variability and lowers predictability of power flows. This, in turn, requires grid adaptations. Moreover, renewable energy production of variable resources (in particular wind and solar PV) is not load following like conventional power plants, increasing the demand for power exchanges across borders in large areas.

Difficulties in the realization of network extensions are impeding the realization of appropriate levels of grid infrastructure in Europe and are commonly considered the largest hurdle for the transition to power systems with large shares of renewable electricity in 2050. First, lack of public acceptance for grid extensions in densely populated Europe increases the length of permitting procedures for grid extensions significantly to typically 10 years or more. Mainly for this reason, at least 33% of important grid infrastructure projects within Europe is reported to be materially delayed (ENTSO-E 2012). Secondly, the low network investment level can be explained by a set of inappropriate institutional conditions including suboptimal investment incentives for TSOs.[9] This is primarily related to the fragmented, bilateral network-planning methodologies across Europe. These do not adequately allow for existing interdependencies between national power systems and cross-border externalities in meshed grids.

When electricity is transported between countries, power in an AC network flows according to physical laws (Kirchhoff laws). Hence, part of the network flow will not be similar to commercial transactions between countries ("contract paths") but makes a detour through neighboring power systems. These types of flows are called parallel or loop flows. Loop flows can originate both from (new) interconnections which span borders as well as from internal transmission lines within countries.

The network externality arising from loop flows is not taken into account in current bilateral network-planning methodologies within the EU. Hence, EU member states considering transmission network extensions (including interconnectors) do not tend to allow for positive or negative effects on other member states. As a result, to date, typically projects with negative benefits for one potentially implementing country but with overall net benefits for Europe as a whole are not realized. The European Commission deems that reformed procedures aiming at more appropriate network planning require proper coordination of bilateral policies.

Despite the Third Energy Package legislation, hitherto projects are considered to be of European significance dependent on opinions and considerations of the country (or countries in case of interconnectors) concerned (ENTSO-E 2010). Hence a uniform framework for the assessment of proposed infrastructure investments with cross-border impacts within Europe is missing. Therefore, the European Commission proposes a regulation to cover this gap which is currently under debate (European Commission 2011). The Commission foresees the application of a common cost-benefit methodology to identify *projects of common interest* (PCIs). Projects that qualify for the PCI label are eligible to favorable rules for faster permitting procedures and advantageous regulatory treatment.

PCIs for electricity and gas are identified in three steps. First, regional groups consisting of representatives of the member states, national regulatory authorities, TSOs, project promoters, ACER, ENTSO-E, and the Commission have to draw up a list of proposed PCIs using the common cost-benefit methodology. Second, ACER has to submit the list with an opinion to the European Commission. Third, the latter has to decide whether or not to adopt the list.

The sketched procedure has some similarities with the US where uniform cost-benefit analyses are typically set up by market area (CAISO, PJM) and project selection also takes place at regional level (FERC 2011). It seems a promising way forward when an objective and common cost-benefit methodology for project selection can be put in place.

The proposed regulation (European Commission 2011) also aims at the realization of faster permitting procedures for PCIs with the introduction of both a one-stop-shop approach for obtaining permits by project developers and sanctions in case of project delays which are insufficiently justified. The latter measures address the other important hurdle for the realization of an appropriate level of grid infrastructure in Europe.

Congestion Management

This section discusses the need for adjustment in market design of EU power markets with regard to congestion management. It begins with alternative approaches to congestion management, moves on to incompatibilities between current EU legislation regarding the internal electricity market and legislation regarding the promotion of renewables, and ends with nodal pricing. Nodal pricing is already being applied quite successfully in the US. It certainly holds out an attractive ultimate solution direction for Europe.

Congestion Management Methods

Because of difficulties in realizing network expansions and an increase in lines with low utilization rates resulting from rising feed-in of electricity from intermittent renewable energy sources,[10] there is an urgent need to improve utilization levels of existing transmission lines. Therefore, the deployment of congestion management is likely to increase. Furthermore, the application of smart grids and demand response will lower the costs of congestion management and drive wider application of the latter.

Table 19.1   Classification of congestion management methods.

Preventive Method (before gate closure)	Curative Methods (near real time, after gate closure)
Implicit auctions	
Explicit auctions	
Counter trade	Counter trade
	Unilateral/Joint cross-border dispatch

Two basic congestion management methods can be distinguished. That is, *preventive* methods which aim at resolving congestion during the operational planning phase before day-ahead gate closure, and *curative* methods which aim at resolving unexpected congestion after gate closure during the day of operation (ETSO 2005). Preventive methods can be subdivided into mechanisms that assign the available capacity based on economic principles (market-based mechanisms) and mechanisms that assign the capacity based on other criteria (distributive mechanisms, i.e., priority access and pro rata) (de Jong 2009).

Curative methods are relatively expensive compared to preventive methods. This relates to the fact that most available resources – both the production and consumption side – are likely to be already deployed earlier in time, either through the day-ahead market or by energy contracting. Therefore, curative methods are only allowed in case lower-cost measures cannot be applied (Congestion Management Guidelines, Regulation (EC) 714/2009, Annex I, Article 1.3 (European Union 2009b)). Hence curative methods are typically only a remedy of last resort. Table 19.1 provides a more detailed classification of preventive and curative congestion management methods.

Implicit and explicit auctions are preventive methods as they allocate only the available transmission capacity. Capacity and energy can be traded either separately (explicit auctions) or combined (implicit auctions). In contrast, in the case of cross-border redispatching the TSOs concerned may resolve congestion through direct intervention in the generation dispatch after gate closure of the day-ahead market. They do so by reducing the output of generators downstream of the cross-border constraint and increasing the output of generators upstream of the constraint, without the requirement of any economic evaluation. As redispatching is not market-based, it is mainly used as a curative method in cross-border congestion management. Also countertrading is usually deployed as a curative method[11] and is generally considered a subset of redispatching. If congestion occurs after market clearing, the TSOs involved create a market to remove the congestion; constrained-on generators are requested to reduce production and constrained-off generators to increase production, based on bids submitted to the TSOs.

Countertrading and redispatching are generally considered less efficient than congestion management methods which take into account network constraints directly in the electricity market design (Dijk and Willems 2011). Countertrading gives the wrong signals for entry and exit of power plants and allows for gaming of redispatch, which may result in high congestion costs. Generators in export constrained zones can play the "inc-dec" (incremental-decremental) game by selling such an amount of electricity at the day-ahead market that they will receive a payment not to produce afterwards during congestion management. Likewise, generators in import constrained zones can play the game by not producing in the day-ahead market and waiting until the TSO requests them afterwards to produce during congestion management. Hence, the European target model for the day-ahead market is market coupling by using auctions.[12]

Conflict Between Internal Market Legislation and Renewable Energy Legislation

On the one hand, market-based mechanisms are mandated by legislation aimed at the achievement of an internal energy market for electricity within Europe (see Regulation (EC) 714/2009, Article 16 and the Congestion Management Guidelines (Annex I, Article 2.1) (European Union 2009b)). On the other hand, legislation on the promotion of renewable energy (European Union 2009a) mandates or at least recommends non-market-based, distributive mechanisms to be transposed in national legislations of the EU member states. These include priority access to the power network system for renewable generators in general, whilst curtailment of power injections by renewable generators in the case of locally congested network situations is strongly dissuaded in the absence of network emergency situations. Given that renewable power is being mainstreamed and commanding a fast increasing share in the EU power mix, it would appear that the renewable energy legislation will have to be adjusted on these scores to bring it in line with the principles of the IEM. At the same time, subsidies and existing biases in favor of incumbent conventional generators have to be phased out as well.

Market Clearing In Europe: Why Nodal Pricing Is a Solution

Implicit auctions are generally preferred as they are considered to be most efficient for congestion management in short time frames. In case of a network constraint, an implicit price is attached to network capacity. Prices for network capacity differ with the network granularity that is considered. In case of *zonal pricing* network capacity is priced for a set of nodes, while in case of *nodal pricing* network capacity is priced for each separate node. This seems rather a technical issue. Yet its implications are of much wider economic significance. The reason is that one average price for a set of nodes does not yield the optimal price for separate nodes. Hence uniform pricing for different nodes incentivizes inefficient producer and consumer decisions. As a result, the network is not used to its full capacity and benefits from the existing network infrastructure are lower than possible.

Currently, whereas the US has implemented nodal pricing because of weaker network infrastructure and hence more network congestion, Europe still has country-based zonal pricing (van der Welle *et al.* 2011). However, Europe has an increasing need for a more efficient congestion management system, as pointed out before. More efficient congestion management procedures can be very valuable: both for the integration of larger amounts of electricity from fluctuating renewable sources of energy in electricity networks and for the realization of one internal energy market for electricity in Europe.

To date, in Europe interconnector capacity is allocated to the market based on the contract path paradigm; as long as there is capacity available on the contract path of the commercial transaction proposed, the transaction is accepted. Actual physical flows resulting from commercial transactions are not taken into account in the capacity allocation phases but accounted for in the available trading capacity (ATC). A transition from an ATC-based to a flow-based capacity allocation which takes into account the physical consequences of proposed commercial transactions is under preparation, especially for interconnections in meshed grids (ACER 2011). Although this is an important step forward, a number of significant challenges will remain which are mainly related to the zonal pricing market clearing structure in Europe. These challenges relate to differences between intra-zonal and inter-zonal congestion management schemes that are likely to result in increasing intra-zonal congestion costs which exceed inter-zonal congestion costs, as well as lengthy and time intensive renegotiations of required periodic zone adjustments (van der Welle *et al.* 2011).

For the reasons just stated, in the US several market areas (PJM, ERCOT, CAISO) shifted to nodal pricing (Baldick *et al.*, 2011; Leuthold *et al.* 2008: Neuhoff and Hobbs, 2011). Experiences gained in the US with nodal pricing has been rather satisfactory so far from a perspective of market functioning. The foreseen increase of generation from variable renewable sources makes such a shift all the more attractive for serious consideration. Intermittent generation will be an increasingly important driver of more frequent changes of congestion patterns in the time frame to 2050. These frequent changes would require frequent adjustment of price zones which can have dramatic consequences for power companies, rendering frequent zonal changes (politically) impossible. Such adjustments are not necessary with nodal pricing, since nodal prices reflect the opportunity costs of transmission capacity. Furthermore, nodal pricing has some important benefits, including notably better price and investment signals to consumers and generators respectively, as well as higher transparency about price formation and required network investments.

Nevertheless, the introduction of nodal pricing meets several, often political, objections in Europe. Some governments, like the German government, consider the introduction of nodal pricing as detrimental to market competition and market liquidity (Frontier Economics and Consentec 2011). On the other hand, if physical constraints are not properly taken into account in the market framework, traders and investors will experience market interventions with concomitant complexity and unpredictability as well as risks of strategic behavior by incumbent market players. Experiences from the US suggest that market liquidity was not affected fundamentally by the introduction of nodal pricing.

Additionally, the introduction of nodal pricing can have important distributional consequences, especially on generators in export-constrained areas and load located in import-constrained areas. As wind turbines are often located further from load than conventional generation facilities, they are more often located in export-constrained areas and hence will earn less revenues. One way to compensate losers is to implement Financial Transmission Rights (FTRs). FTRs are an important instrument for market parties to hedge locational price differences in time. However, since FTRs are often defined for blocks of sequential hours they are a suboptimal hedge for locational price differences for generation of intermittent renewable energy sources, as the latter are exposed to risks when they do not produce. A real solution for the latter problem has not yet been found; it remains an issue for further research. Overall, the social welfare benefits which can be realized due to efficiency and transparency gains of nodal pricing seem likely to more than offset potential losses from the introduction of nodal pricing in Europe.

Relevancy for Emerging and Developing Economies

The EU is global frontrunner in the promotion of renewable generators and the decarbonization of the power sector with high ambitions for the medium term (2020) and the long term (2050). This implies huge challenges related to the market and network integration of power from fluctuating renewable sources. Moreover, tremendous financing as well as regulatory hurdles and other non-financial implementation hurdles are still to be overcome to achieve the ambitious policy goals for the European power system. Emerging economies and developing countries can learn important lessons from European successes and setbacks in these regards. To date, power utilities in many developing countries are vertically integrated with a mandate for both generation, owning and operating networks, and delivery of power. The evolution of the EU regulatory framework sets a useful benchmark for those developing countries seeking to liberalize their power sector.

We have pointed out that whilst the EU can be regarded as frontrunner in rendering its power system more environmentally sustainable and has a better developed network infrastructure, on a number of counts the EU can learn from the US, such as the introduction of nodal pricing and making the demand side more responsive. Furthermore, the EU is an interesting laboratory for instrumentation of market stimulation of renewable electricity generation. What can be learned is that already at an early stage deployment of market-oriented stimulation instruments, such as feed-in premiums with built-in ex-post adjustment to power prices or (at a more advanced stage and for not too small power markets) a certificates-based renewable quota system, can appreciably bring down market and system integration costs. Moreover, feed-in stimulation instruments (FIT, FIP) should not be open-ended for notably the higher cost technology categories. Furthermore, market design reforms need to stimulate flexibility resources to be offered in the market-place to the maximum cost-effective extent possible. The option to introduce a capacity remuneration mechanism should be resorted to only when all flexibility enhancing design features do not prove to be adequate enough. And if introduced, such a mechanism should be designed in a market-based fashion and only rewarding availability of flexible resources.

A final observation can be made regarding technology transfer for the integration of renewable power. Organizations such as IRENA (International Renewable Energy Agency) with its head office in Abu Dhabi, the Paris-based IEA (International Energy Agency), the World Bank, the EBRD, and other regional development banks need to rise to the occasion. Indeed, they have to adjust their priorities toward playing a (more) important role in knowledge transfer to quicken the cost-effective transition toward a more sustainable power sector in emerging and developing economies.

Notes

1. Regarding the electricity sector, key Third Package legislation adopted in 2009 concerns Directive 2009/72/EC stipulating framework conditions for the internal electricity market, Regulation 713/2009 establishing ACER (Agency for the Cooperation of Energy Regulators), and Regulation 714/2009 on access to the network for cross-border exchanges in electricity.
2. ENTSO-E stands for European Network of Transmission System Operators for Electricity, whilst the letter G in ENTSO-G stands for Gas.
3. Directive 2009/28/EC, ultimately replacing two previous directives on renewable electricity (Directive 2001/77/EC) and biofuels (Directive 2003/30/EC).
4. Not only member states in the southern and eastern EU rim, but in central and western Europe as well, such as Germany. See two articles in *Die Welt*: Teure Energie: Hunderttausende Haushalten wird der Strom gesperrt, February 21, 2012, and Wegen Netzausbau. Deutschen droht Strompreis-Erhöhung, March 29, 2012.
5. This section is based on IEA 2011.
6. But adequate arrangements need to be made for efficient treatment of network constraints, see below.
7. In such wholesale power markets, conventional generators are only remunerated for delivery of energy, not for making available (flexible) capacity.

8. This is related to the physical characteristics of electricity. Non-storability of electricity means that supply and demand have to be balanced instantaneously, limiting possibilities for storage, while demand inelasticity limits possibilities for shifting consumption over time.

9. In Europe, investments are mainly made by regulated TSOs as room for merchant investments is limited by EU regulation.

10. Low utilization rates generally mean that the benefits of additional network capacity do not outweigh the congestion costs they prevent. Current network utilization rates are already low due to network security rules (N-1 criteria) and low controllability of AC power flows. Higher variability and lower predictability of electricity from RES-E tend to decrease utilization rates even further.

11. Although it can also be deployed as a preventive method, see ETSO (2005).

12. In the process of further elaborating the target model, implicit auctions have been selected as most appropriate for the day-ahead market time frame.

References

ACER. 2011. *Framework Guidelines on Capacity Allocation and Congestion*. Management for Electricity. FG-20110E-002. Ljubljana, July 29.

Baldick, Ross, James Bushnell, Benjamin Hobbs, and Frank Wolak. 2011. *Optimal Charging Arrangements for Energy Transmission: Final Report*. Report prepared for and commissioned by Project TransmiT. Great Britain Office of Gas & Electricity Markets, May 1.

de Jong, H. 2009. Towards a Single European Electricity Market – A Structured Approach to Regulatory Mode Decision-Making. PhD, Delft University of Technology.

Dijk, Justin, and Bert Willems. 2011. The Effect of Counter-Trading on Competition in Electricity Markets. *Energy Policy* 39: 1764–1773.

ENTSO-E. 2010. *Ten Year Network Development Plan 2010–2020*. Brussels, June.

ENTSO-E. 2012. *Ten Year Network Development Plan 2012*. Draft for public consultation. Brussels, March.

ETSO. 2005. *An Evaluation of Preventive Countertrade as a Means to Guarantee Firm Transmission Capacity*. Background paper. Brussels, April.

Eurelectric. 2011. *RES Integration and Market Design: Are Capacity Remuneration Mechanisms Needed to Ensure Generation Adequacy?* Brussels: Eurelectric.

European Commission. 2011. *Proposal for a Regulation of the European Parliament and of the Council on Guidelines for Trans-European Energy Infrastructure and Repealing Decision No 1364/2006/EC*. COM (2011) 658 final. Brussels.

European Union. 2009a. *Directive 2009/28/EC of the European Parliament and the Council of 23 April 2009 on the Promotion of the Use of Energy from Renewable Sources and Amending and Subsequently Repealing Directives 2001/77/EC and 2003/30/EC*. OJ L 140/16. Brussels.

European Union. 2009b. *Regulation (EC) No 714/2009 of the European Parliament and of the Council of 13 July 2009 on Conditions for Access to the Network for Cross-Border Exchanges in Electricity and Repealing Regulation (EC) No 1228/2003*. OJ L 211. Brussels.

FERC. 2011. Transmission Planning and Cost Allocation by Transmission Owning and Operating Public Utilities. Docket No. RM10-23-000; Order No. 1000, July 21. Washington, DC: Federal Energy Regulatory Commission.

Frontier Economics and Consentec. 2011. *Relevance of Established National Bidding Areas for European Power Market Integration – An Approach to Welfare Oriented Evaluation*. Report prepared for Bundesnetzagentur. London.

IEA. 2011. *Harnessing Variable Renewables. A Guide to the Balancing Challenge*. Paris: International Energy Agency.

Leuthold, Florian, Hannes Weigt, and Christian von Hirschhausen. 2008. Efficient Pricing for European Electricity Networks – The Theory of Nodal Pricing Applied to Feeding-in Wind in Germany. *Utilities Policy* 16: 284–291.

Neuhoff, Karsten, and Benjamin Hobbs. 2011. *Congestion Management in European Power Networks: Criteria to Assess the Available Options*. Discussion Paper 1161. Berlin: DIW (German Institute for Economic Research).

Roques, Fabian. 2008. Market Design for Generation Adequacy: Healing Causes Rather Than Symptoms. *Utilities Policy* 16: 171–183.

Van der Welle, Adriaan, Jeroen de Joode, Karina Veum, *et al.* 2011. *Socio-Economic Approaches for Integration of Renewable Energy Sources into Grid Infrastructures*. D5.1 of FP7 Susplan project. Petten, Netherlands: ECN.

Chapter 20

Global Climate Governance and Energy Choices

Fariborz Zelli, Philipp Pattberg, Hannes Stephan, and Harro van Asselt

Introduction

This chapter analyzes the increasingly fragmented climate change governance architecture (Biermann *et al.* 2009; Keohane and Victor 2011) and how it relates to global energy choices. When referring to a fragmented governance architecture, we understand different institutional approaches to be situated along a continuum ranging from international and public, to public–private or private interventions. Some are related to international agreements and norms and thus fall under a shadow of hierarchy, while others are situated in the realm of non-hierarchical steering without any overarching authority (Pattberg and Stripple 2008). A complete picture of the climate–energy nexus will therefore require mapping institutions and policies beyond the United Nations Framework Convention on Climate Change (UNFCCC) and its Kyoto Protocol, henceforth referred to as the international climate regime. When speaking of energy choices, we largely refer to technological choices. We examine to what extent the global climate regime and other climate-related institutions intend to influence the proliferation and use of clean energy technologies, as a means for countries to transform their energy sectors.

Anthropogenic climate change caused by emissions of greenhouse gases into the Earth's atmosphere is basically a function of population, economic growth (affluence, measured in GDP per person), and technology (carbon dioxide, CO_2, emitted per unit of GDP). Historic decrease in population growth has been slow but substantial, resulting in an average of 1% increase per year over the past few decades with a further decrease to 0.7% growth over the past few years. While world GDP per person has on average increased by 1–2% per year, technological advances have resulted in an average 1% decline in CO_2 emissions per dollar GDP. The net result has been an average increase of CO_2 emissions of 1% annually over the twentieth century (Dessler and Parson 2010: 126–127). Given the political contestation about far-reaching measures to curb population growth and a broad political consensus about the necessity for further economic growth, reducing emission intensity per unit GDP and CO_2 intensity per unit of energy seem to be the most feasible strategies for mitigating climate change (ignoring for now options to reduce CO_2

The Handbook of Global Energy Policy, First Edition. Edited by Andreas Goldthau.
© 2013 John Wiley & Sons, Ltd. Published 2013 by John Wiley & Sons, Ltd.

emissions from non-industrial activities and other greenhouse gases such as methane as well as geo-engineering scenarios).

Assuming a political will to limit global mean temperature increase to 2 °C above pre-industrial levels (corresponding with a total future emission budget of 750 Gt CO_2-eq by 2050) by means of efficiency improvements, and assuming a world population of 9 billion by 2050 with an average income growth of 1.4% per year (an extrapolation of the current growth rate), the carbon intensity of goods and services would have to decrease from currently around 770 g/$ economic output to 36 g/$, a 21-fold efficiency improvement over the current level (Hoffmann 2011). While this target is not impossible, compared to historic efficiency gains the scope of the challenge is indeed huge. After a decrease of around 1.5% in 2009 due to the world economic recession, global emissions from fuel combustion have increased by 5.3% in 2010 (IEA 2011) while improvements in carbon intensity of goods and services have been lagging behind the background trend of 1.7% increase by almost 1% in 2009 (Friedlingstein *et al.* 2010). Against these trends, the challenge of decoupling economic growth from energy use while rapidly expanding access to energy services in the non-OECD world is of immense proportions. Close to 90% of increase in global energy demand is expected to occur in these countries and approximately $25 trillion of investment in energy will be required (IEA 2011) while current financial support for low carbon energy infrastructures covers only a small fraction of the expected additional costs (World Bank 2010).

Against this background, questions of energy policy in the context of climate change mitigation (both reducing the carbon intensity of production and increasing the efficiency of energy production) are paramount. While global energy policy is concerned with a broad range of questions, including supply security, decreasing energy poverty, questions of domestic energy governance, and sustainability (Dubash and Florini 2011), we focus only on the narrow climate–energy nexus in this chapter. We start by discussing the role of the international climate regime in global energy policy, then turn to global climate change governance beyond the international regime, before analyzing the inter-linkages between different institutional arrangements, within and beyond the climate change arena, explicitly addressing the intersections with the other three dimensions of this edited volume – markets, security, and development. We conclude with a brief discussion of options for improved management of institutional interlinkages and a number of future research challenges.

The International Climate Change Regime and Energy Policy

This section provides an overview of the role of the international climate regime in global energy policy. After a brief discussion of those provisions of the UNFCCC and Kyoto Protocol that are relevant for countries' energy choices, the section discusses how the international climate regime has sought to promote research, development, and deployment of clean energy technologies. We contend that while the effectiveness of the UNFCCC in promoting clean technology development and transfer can be debated, an elaborate system addressing various aspects of clean technology cooperation has been put in place.

Although the international climate regime was not created specifically with a view to governing energy supply and demand, it is nevertheless of key importance given the fact that about two-thirds of global greenhouse gas emissions stem from energy use. As noted by Dubash and Florini (2011: 14), a "comprehensive global climate agreement organized around explicit national carbon caps would be transformative and become a de facto global energy governance regime." However, the Kyoto Protocol's emission reduction targets only cover developed countries, and thereby do not directly seek to alter the energy policy choices of developing countries. Furthermore, the international climate regime is marked by a continuing transformation toward a "bottom-up" approach to policy-making, leaving the determination of climate change mitigation actions largely up to individual countries. These considerations decrease the relevance of the international climate regime in the development of global energy policy. Nevertheless, the regime comprises several mechanisms relevant from the perspective of energy policy.

Promoting Clean Energy Technology Cooperation

The UNFCCC requires its Parties to "[p]romote and cooperate in the development, application and diffusion, including transfer, of technologies, practices and processes that control, reduce or prevent anthropogenic emissions of greenhouse gases" (Article 4.1(c)). Furthermore, the UNFCCC contains commitments related to technology transfer, determining that developed countries need to provide new and additional financial resources to developing countries. The Kyoto Protocol expands on these provisions, introducing the term "environmentally sound technologies" to encompass not only the "hardware" (equipment, etc.) but also the "software" (know-how) (Yamin and Depledge 2004: 306–307).

Implementation of the technology provisions received a boost following a UNFCCC decision adopted in 2001. To analyze and identify ways to facilitate and advance technology transfer activities, the decision established the Expert Group on Technology Transfer (EGTT). The EGTT was mandated to focus on five themes: (i) technology needs and needs assessments; (ii) technology information; (iii); enabling environments; (iv) capacity building; and (v) mechanisms for technology transfer (UNFCCC 2002a). It established a work program related to these five themes and regularly analyzed country information provided by parties to the climate treaties. Furthermore, in recent years, the EGTT has also discussed innovative options for financing technology (e.g., UNFCCC 2007).

Assessing the effectiveness of the UNFCCC in achieving the transfer of clean energy technologies on the whole is fraught with difficulties. A recent report by the UNFCCC Secretariat provides some insights, noting that the overall technology transfer framework can be deemed effective, but remarking that private sector engagement needs to be improved, and that the level of financial support for the development and transfer of technologies is inadequate (UNFCCC 2010: 47–48).

Technology played a prominent part in the negotiations on an international climate agreement beyond 2012. In 2007, "enhanced action on technology development and transfer" became one of the four building blocks under the Bali Action Plan (UNFCCC 2007). An initial agreement on a new Technology Mechanism was part of the 2009 Copenhagen Accord and was fleshed out a year later in the Cancún Agreements. The Cancún decision on technology development and transfer specified that the mechanism is to consist of a Technology Executive Committee and a Climate Technology Centre and Network (UNFCCC 2011). The Committee's broad mandate includes the identification of technology needs, addressing barriers to technology development and transfer, cooperation with international technology initiatives, and providing guidance and recommendations to enhance the effectiveness of technology development and transfer. The Climate Technology Centre's purpose is to facilitate and coordinate a wider network of technology initiatives at different levels of governance. The centre and the network are intended to provide a flexible framework to support technology development and deployment, especially in developing countries. The mandate of the centre and the network includes the provision of assistance for identifying technology needs, implementing technologies, and capacity building. In addition, the centre and the network are expected to foster cooperation between public and private institutions on technology development and transfer. With the introduction of the Technology Mechanism, the EGTT ceased to exist.

The Clean Development Mechanism and Clean Energy Technologies

The transfer of clean energy technologies also plays an indirect, yet very important role in the Clean Development Mechanism (CDM). Under the CDM, developed countries can form voluntary partnerships with developing countries to undertake greenhouse gas emission reduction projects, with the dual purpose of achieving sustainable development in the host country while contributing to cost-effective mitigation targets for the developed countries. The Kyoto Protocol does not explicitly mention clean energy technology transfer as a goal of the CDM, but a follow-up decision does so (UNFCCC 2002b: preamble).

Although a majority of CDM projects operate in the area of renewable energy, most of the credits are generated through projects that reduce emissions of industrial greenhouse gases with high global warming potential, such as hydrofluorocarbons (HFCs) and nitrous oxide (N_2O). For instance, while in March 2012 67% of the CDM projects were in the area of renewable energy, compared to 1.7% for large-scale industrial gas projects, the latter accounted for 68% of the credits issued, compared to only 18% for renewables. Energy efficiency – both in terms of absolute numbers and in terms of credits issued – scores even less (Fenhann 2012). We observe a number of barriers to widespread use of the CDM in the promotion of renewable energy and energy efficiency. First, the CDM's "additionality" requirement, which means that a project has to go beyond business-as-usual, acts as a barrier. Since energy efficiency projects often pay for themselves through reduced energy costs over time, and both (small-scale) renewable energy and energy efficiency projects typically generate few credits, this makes it difficult to prove that these projects would not have happened without the CDM (Driesen 2006). In contrast, it is quite easy to prove additionality for end-of-pipe projects involving industrial gases such as HFCs and N_2O, especially when there are no national regulations on these gases and when CDM credits deliver the only return on investment. A second barrier to renewable energy and energy efficiency projects is that, in general, these projects require more investment per generated carbon credit than alternative projects (Matschoss 2007: 119). As a result, renewable energy and energy efficiency projects are largely "crowded out" by the low-cost, high-credit projects.

Recent years have seen a burgeoning literature on the (potential) contribution of the CDM to technology transfer, fueled in part by the claim that it "is currently the strongest mechanism for technology transfer under the UNFCCC" (Schneider *et al.* 2008: 2936). The findings of these studies depend on the data sources used (e.g., project design documents drafted by CDM project developers, or independently collected data) and the definition of technology transfer (e.g., equipment only, and/or associated know-how). While studies using a wide definition of technology transfer (including also the transfer of hardware only) tend to provide cautiously optimistic assessments of the CDM's potential (e.g., de Coninck *et al.* 2007: 455; Haites *et al.* 2006: 346), a more comprehensive definition of technology transfer (including software) led to the conclusion that the CDM's contribution "can at best be regarded as minimal" (Das 2011: 28). What the various studies have in common, however, is that they show that the CDM works better for some clean technologies than for others, and that there are differences in the rates of technology transfer among developing countries (Dechezleprêtre *et al.* 2009: 710; Seres *et al.* 2009: 4924).

Dealing with Fossil Fuel Producers

Another way in which the international climate regime is of relevance for global energy policy is in how it accounts for the position of fossil fuel producing countries. The UNFCCC itself acknowledges the special circumstances of countries "whose economies are highly dependent on income generated from the production, processing and export, and/or on consumption of fossil fuels" (Article 4.8(h)). Similarly, the Kyoto Protocol seeks to ensure that the "adverse effects" of climate policies and measures implemented in developed countries are minimized (Articles 2.3 and 3.14). These provisions have formed the basis for claims by fossil fuel producing countries to receive compensation for expected losses in export revenues due to climate change mitigation measures. As we will describe in more detail below, several countries belonging to the Organization of Petroleum Exporting Countries (OPEC), in particular Saudi Arabia, have been notorious for their obstructive role in the climate negotiations, either seeking to delay the implementation of new policies, or to link such policies to the compensation discussion (Depledge 2008; Dessai 2004). Not surprisingly, such compensation has faced heavy opposition from developed countries, and, as a result, no compensation mechanism to support fossil fuel producers in their economic diversification has been put in place.

Climate Governance Beyond the International Climate Regime

In this section, we first analyze the role of international organizations and UN institutions outside the international climate regime, which – through their funding practices, programmatic activities, and ideational leadership – have a significant impact on energy choices around the world. This also includes the activities, related to global energy and sustainability, of major UN institutions such as the United Nations Development Programme (UNDP), the United Nations Environment Programme (UNEP), and multilateral financial institutions such as the World Bank. Second, we discuss in more detail the role and relevance of so-called club arrangements for global climate and energy governance. Specific attention is paid to informal clubs such as the G8 and G20. Third, this section analyzes the contribution of public–private partnerships to global energy policy, e.g., the more than 350 public–private partnerships for sustainable development that have emerged as a result of the 2002 World Summit on Sustainable Development in Johannesburg. Finally, we briefly discuss the relevance of transnational governance arrangements for global energy policy.

Formal Organizations

International organizations play a crucial role in supporting the global trend toward low carbon economic development, particularly by promoting the construction of sustainable energy infrastructures and by helping countries to exploit untapped potentials for energy efficiency. The expectation of growing flows of energy and climate-related funding has attracted interest from many international organizations (Newell 2011: 97), ranging from long-established UN bodies to energy-related organizations and multilateral development banks.

UNEP lists the implementation of renewable energy systems among its six sub-programs. Given funding constraints, UNEP's operational activities in the energy arena have centered on creating public–private networks (see below for more details), such as the "Renewable Energy Policy Network for the 21st Century" (REN 21). UNEP's vision of spearheading the global transition toward the "Green Economy" highlights renewable energy, energy efficiency, and cleaner transport as most critical areas for long-term investment (UNEP 2011). This message chimes with its annual publication of *Global Trends in Renewable Energy Investment* which documents the rapidly growing scale of investments and seeks to galvanize the interest of the financial sector.

UNEP's larger sister agency, UNDP, sees energy primarily through the prism of sustainable development. In 2010, UNDP supported 14 energy efficiency and 30 renewable energy projects and spent $508 million (approx. 11% of its total expenditure) on "managing energy and the environment for sustainable development" (UNDP 2011: 6).

While sustainable energy has only recently moved to the centre of the policy debate on climate change, the International Energy Agency (IEA) has sought to mainstream its relevant expertise since the mid-1990s. Founded in 1974 as an OECD response to the oil crisis, the IEA's agenda has significantly broadened over time. Since the early 1990s, the organization has muted its erstwhile endorsement of fossil fuels and nuclear power and, by 2005, the issues of energy security and sustainability had been declared top priorities (Van de Graaf and Lesage 2009). This means that the IEA's traditional functions of policy advice, information sharing, and technology transfer are now also supporting the worldwide uptake of renewable energy and energy efficiency. Furthermore, the regular participation of the IEA's Executive Director in G8 summits has translated into an enhanced status, greater policy influence, and additional funds. The decision to engage in more capacity-building and to expand its publications on technology-specific road maps will cement the IEA's position as a major font of expertise for the low carbon energy transition (Florini 2011).

However, the establishment of the International Renewable Energy Agency (IRENA) in 2009, with now 88 member countries (as of April 2012), signifies that being a global coordinator of sustainable energy will not be an exclusive competence. With staff numbers and budgets evenly matched (at around $25 million per year) and overlapping functions, much depends on whether a recently signed partnership agreement between IRENA and the IEA ensures a fruitful division of labor (Van de Graaf 2013). Early indications are that IRENA is actively pursuing its goal of becoming the global authority on renewable energy and that it is building alliances with the major public–private networks in this area (e.g., REN 21 and the Renewable Energy and Energy Efficiency Partnership).

Whatever the important facilitative and catalytic functions of the above institutions, they arguably compare modestly with the influence on energy choices exerted by the World Bank and regional multilateral development banks. First, countries and private investors seeking to access their funds are subject to particular ideas and guidelines regarding appropriate technologies, regulations, policies, and methods of service delivery (Nakhooda 2011). As Ferrey (2010: 113) puts it, "[w]hat the World Bank does and supports is a critical starting point." The World Bank not only shapes the lending practices of other regional multilateral development banks, but also those of national export banks and private banks in the industrialized world. Second, in terms of funding, the Bank's Climate Investments Funds (CIFs) established in 2008 represent "more public finance than has ever before been dedicated to climate change" (Newell 2011: 97). The

Clean Technology Fund ($4.5 billion) offers concessional loans for both transfer and deployment of low carbon technologies. It is expected to leverage another $36 billion in co-financing. The Strategic Climate Fund ($2 billion) also has a dedicated program for the scaling-up of renewable energy (World Bank 2011).

At the same time, the Bank is not so much an independent organization as a "financial aid conduit" for projects deemed appropriate by its member states (Ferrey 2010: 112). Under pressure from civil society organizations and some governments, the Bank is clearly trying to integrate climate mitigation objectives more systematically in its projects, but – in the context of the increasing weight of emerging economies in the Bank's voting arrangements – there has been limited success in assimilating these concerns into mainstream energy finance lending (Nakhooda 2011: 127), which accounted for $5.8 billion in 2011.

Club Arrangements

Since the mid-2000s, various club arrangements have become engaged in governing the climate–energy nexus. Club arrangements generally refer to governance initiatives involving "small groups of pivotal nations" (Victor 2009: 342). This raises questions about *how many* countries should be involved and *which* countries should be included (Eckersley 2012). Countries have generally been considered pivotal because of their greenhouse gas emissions profile or their economic power, meaning that most clubs have at least included the US, China, and India, although membership of the clubs is variegated. The rationale behind club arrangements is presumably that it is generally easier to get to an effective agreement by including only a limited number of players compared to the multilateral process under the international climate regime that is considered to be cumbersome (Naím 2009; Victor 2009). Various examples of such a mini-lateral approach already exist.

The Group of Eight (G8) was not established in 1975 with a view to addressing climate change, but following the 2005 G8 summit in Gleneagles, Scotland, climate change has been a recurring issue on its agenda. The Gleneagles summit produced the Gleneagles Plan of Action on Climate Change, Clean Energy and Sustainable Development, which invigorated the G8's activities on energy, and kick-started its involvement in climate change issues. The renewed focus on energy and climate change resulted in various pledges and actions (Van de Graaf and Westphal 2011). Notable pledges were the goal to reduce global emissions by at least half by 2050 (at the G8 summit in Hokkaido, Japan in 2008); the specification to achieve 80% or more of these reductions in developed countries; and acknowledgment of the goal to keep the global temperature increase below 2 °C above pre-industrial levels (both at the G8 summit in L'Aquila, Italy in 2009). In addition to these broad (and rather symbolic) pledges, the G8 has also established more practical initiatives, notably the International Partnership for Energy Efficiency Cooperation at the Hokkaido summit (Lesage *et al.* 2010).

At the Heiligendamm summit in 2007, the G8 also established a (mainly informal) dialogue with five other countries – China, India, Brazil, Mexico, and South Africa – known as the G8+5. While energy, and in particular energy efficiency, were on the agenda of this dialogue, the momentum behind the initiative has waned after the UNFCCC Copenhagen conference in 2009 (Bausch and Mehling 2011: 28–30), and with the emergence of the Group of 20 (G20). The G20 is a coalition of large economies (including both developed and developing countries), primarily focused on international finance and economic development, which is increasingly looking to take over the role of the G8 in coordinating international economic policy. From a climate change perspective, the most notable development in this forum has been that political leaders pledged to phase out fossil fuel subsidies in the medium term at the 2009 Pittsburgh summit (Van de Graaf and Westphal 2011).

Another mini-lateral initiative is the Major Economies Process on Energy Security and Climate Change launched by US President George W. Bush in 2007, which has been continued as the Major Economies Forum (MEF) by President Obama. The MEF brings together 17 major economies from the developed and developing world, operating rather like the G20 but with a focus solely on climate change and energy issues. Although initially seen as a diversion from the UNFCCC process, it later was explicitly stated to be supportive of the international climate regime. The MEF has focused primarily on technology cooperation, with one of its main actions being the establishment of a Global Partnership to advance the development and deployment of climate-friendly technologies. As part of the MEF, activities to advance the use of clean energy technologies were also pursued through several Clean Energy Ministerials. Whether and how the MEF (and the Clean Energy Ministerials) will be continued remains to be seen: the initiatives are mainly driven by the US, and given the lack of domestic progress there, the interest in the initiatives seems to be decreasing (Bausch and Mehling 2011: 23–25).

Public–Private Partnerships on Climate Change and Energy

Scholars observe that intergovernmental attempts to address the multiple challenges of global climate change have been gradually complemented by transnational – border-crossing and non-state-based – forms of governance. In this context, authors have argued that transnational governance goes beyond more established forms of transnational relations, as it is geared toward authoritative steering toward public goals (Abbott 2012; Andonova *et al.* 2009; Pattberg 2010; Pattberg and Stripple 2008).

As one concrete manifestation of this broader trend, public–private partnerships – that is, networks of different societal actors, including governments, international organizations, corporations, research institutions, and civil society organizations – have become a cornerstone of the current global environmental order, both in discursive and material terms. At the United Nations level, partnerships have been endorsed by the former Secretary-General Kofi Annan through the establishment of the Global Compact, a voluntary partnership between corporations and the UN, as well as through the so-called type-2 outcomes concluded by governments at the UN World Summit on Sustainable Development (WSSD) in Johannesburg in 2002 that institutionalizes public–private partnerships in issues areas ranging from biodiversity to energy (Pattberg *et al.* 2012).

Out of the 340 partnerships registered with the UNCSD in early 2012, 46 have a primary focus on energy issues. An in-depth study of these energy partnerships (Szulecki *et al.* 2011) concludes that the majority is not fulfilling the high expectations that were placed on them in terms of effectiveness. While many partnerships are non-operational or have very little traceable output, those that are assessed to be effective resemble international organizations in terms of organizational structure. While this line of research has painted a rather gloomy picture of the overall sample of WSSD partnerships in terms of contributing to the much needed sustainability transition, it also acknowledges that individual partnerships can make a distinct contribution to sustainable development. In answering the question why some partnerships perform better than others, Pattberg *et al.* (2012) highlight the importance of organizational structure, resources, and powerful actors. In more detail, the authors conclude that in terms of problem-solving effectiveness, there is something like an ideal model or best practice of partnerships for sustainable development (2012: 106–109).

The Renewable Energy and Energy Efficiency Partnership (REEEP) is a prime example of the larger universe of public–private partnerships devised and established around the 2002 summit and the ideal model referred to above. As an open-ended initiative to facilitate multi-stakeholder cooperation in the renewable energy, climate change, and sustainable development sector, REEEP is a cooperative platform for more than 3500 members and 250 registered partners, among them 45 governmental actors (both national and subnational), including all G8 members except Russia, 180 private entities, and 6 international organizations. With an annual budget of just over $7,800,000, and $16,450,000 of available funds, it is one of the largest public–private partnerships for sustainable development. The partnership has an international secretariat (administrative office), eight regional secretariats, and two additional local focal points (North Africa and West Africa). REEEP is implementing activities and programs in 57 countries.

REEEP represents a market-oriented group of actors working for sustainable develop-ment. Its intention is to facilitate the exchange of technologies, identify and remove policy and regulatory barriers in the renewable energy market (and also create such markets in the first place if they do not exist), and provide information for various stakeholders, including the general public. The partnership is foremost a platform for communication between the partners, and a means to streamline the idea of renewable energy into efforts of informing and educating the wider public. It is therefore both a deregulatory and reg-ulatory enterprise, aimed at the removal of state and regional barriers for the renewable energy market, yet at the same time devoted to regulation within this relatively new and rapidly growing sector. The membership remains open, and the number of partners is constantly growing. In 2012 more than 33% of the governmental partners were Euro-pean, 31% were from Asia (with 6 separate regional governments in India), 18% were American states, and 11% were from Africa, along with Australia and New Zealand.

In addition, a number of multilateral public–private arrangements on sustainable energy technologies were planned or launched in the early 2000s, seeking to provide complementary or alternative institutional routes to the UNFCCC, especially given the uncertainty, at the time, on whether the Kyoto Protocol would ever enter into force (van Asselt and Karlsson-Vinkhuyzen 2009). Some of these forums focus on a limited range of technologies, e.g., on carbon sequestration or bioenergy. These include, for instance, the Carbon Sequestration Leadership Forum, established in June 2003, and the Global Methane Initiative (November 2004). Other initiatives cover multiple technologies, e.g., the Asia-Pacific Partnership on Clean Development and Climate (APP), which started in January 2006 (McGee and Taplin 2006), but concluded five years later.

All of these initiatives are mini-lateral and come close to the club-style composition of the G20 or the MEF. With few exceptions they include members from the Umbrella Group (a US-led loose coalition of non-EU developed countries, often acting as laggards in the UN climate regime), the EU, and emerging economies like China or India, while small-island developing states and least developed countries are excluded. Additionally, the initiatives envisage an important role for business actors, primarily through the implementation of specific projects.

Transnational Climate Governance by Disclosure

In addition to a wide range of firm- and industry-level emissions reduction schemes and market-building approaches, a number of transnational networks have emerged that only indirectly aim at the reduction of greenhouse gas emissions, but rather focus on creating the necessary information and transparency for societal actors to assess corporate responses to climate change and thereby induce lasting behavioral change. Often these schemes are supported by institutional investors that have begun to include sustainability in their investment decisions. These benchmarking processes create a global competition among business actors to address climate change as a serious limitation to their profit-making activities. The emerging information-based governance schemes effectively institutionalize new norms at the transnational level, for example the norm to disclose corporate carbon emissions (in addition to the country-based reporting of the climate convention) (Florini and Saleem 2011: 144–145).

For instance, the Investors Network on Climate Risk (INCR) is a case of investor-driven governance that may impact on energy and climate-related business development through internalizing climate change externalities (INCR 2008). Other examples include the Joint Oil Data Initiative, the Global Reporting Initiative, the Electricity Governance Initiative, and the Extractive Industries Transparency Initiative (see Quiroz and Bauer, this volume).

Interlinkages Between the Global Climate Regime and Other Global Governance Institutions

Building on the previous sections and their mapping of international energy-related institutions within and beyond the international climate change regime, this section focuses on the relations between these institutions. An emerging strand of literature has come to address the institutional patchwork of global energy governance (cf. Dubash and Florini 2011), describing it as "chaotic, incoherent, fragmented, incomplete, illogical or inefficient" (Cherp *et al.* 2011: 76). A more systematic, concept-driven approach to inter-institutional relations on energy is largely missing, with a few exceptions (Bradshaw 2010; Cherp *et al.* 2011; Colgan *et al.* 2011; Goldthau, Introduction, this volume). The literature on global environmental governance is more advanced in this regard, establishing concepts like institutional interaction (Oberthür and Gehring 2006), fragmentation (Biermann *et al.* 2009; Zelli 2011a), and complexes (Keohane and Victor 2011; Oberthür and Stokke 2011).

One concept that may lend itself to the analysis of global energy policy is a three-fold typology by Stokke (2001: 10–23) who distinguishes: utilitarian interlinkages where one institution alters the costs or benefits of another institution; ideational interlinkages where one institution influences another through learning processes; and normative interlinkages where one institution affects compliance with the norms of another institution (cf. McGee and Taplin 2006: 180). We employ Stokke's typology to briefly characterize some of the interlinkages between the climate regime and other institutions in the four dimensions of global energy policy that structure this handbook (Cherp *et al.* 2011; Dubash and Florini 2011; Goldthau, Introduction, this volume): sustainability, markets, security, and development.

The relations between the UNFCCC and formal organizations in the realm of sustainability and renewable energy could generally be portrayed as an ideational interlinkage. Even the relation with the IEA has changed from strong tensions, due to the agency's original bias toward fossil and nuclear industries, to one of mutual learning. After its telling absence in the early stages of climate negotiations, the IEA has eventually come to feed its expertise on energy technologies into climate summits. Likewise, the agency has broadened its climate-related work since 2005, albeit primarily incentivized by the G8 summit in Gleneagles (Van de Graaf and Lesage 2009: 304–305). Nonetheless, there is still a conflictive utilitarian side to this interlinkage. After all, the climate regime architecture was designed to profoundly restructure energy choices around the world through its restrictions on carbon emissions and concomitant price increase for traditional energy carriers. Some of the European countries that advocate this role for the climate regime consequently pushed for the creation of IRENA as a renewables counterpart to the IEA (Van de Graaf 2013).

Likewise, interlinkages between the international climate regime and club or public–private arrangements over energy issues are characterized by both synergistic and conflictive features. There are supportive utilitarian and ideational overlaps wherever such arrangements have provided their members with additional incentives and awareness to advance their low carbon development paths. The G8+5 with the Gleneagles Process and G20 are cases in point here. As described in further detail above, recent summit declarations in Heiligendamm 2007, L'Aquila 2009, and Pittsburgh 2009 endorsed the UNFCCC process and included soft commitments for phasing out inefficient fossil energy subsidies, and for avoiding an average global temperature rise of more than 2 °C above pre-industrial levels. Similar stimuli toward global energy transition – and respective financing efforts – came from the Clean Energy Ministerial at the 2009 Copenhagen climate summit, the MEF, and from several new public–private partnerships on specific technologies like methane, hydrogen, and carbon sequestration (Florini and Dubash 2011).

On the other hand, observers cautioned against disruptive effects of these various clubs: their non-binding approaches may undermine the climate negotiations' drive toward hard law development (Vihma 2009), and their lack of inclusiveness leaves behind the energy concerns of the majority of developing countries (Biermann *et al.* 2009). This also goes for some of the public–private technology partnerships that evolved in the early 2000s. The now-defunct APP has drawn particular attention from scholars (cf. Karlsson-Vinkhuyzen and van Asselt 2009). In terms of a negative utilitarian interlinkage, the APP's approach with non-binding nationally determined intensity targets may reduce the enthusiasm for the more ambitious multilateral objectives of the UN process. This might also entail disruptive normative effects on the compulsion of core UNFCCC features like the principle of common but differentiated responsibilities (McGee and Taplin 2006; van Asselt 2007).

The relationship between the international climate regime and transnational initiatives can largely be described as synergistic. In normative terms, such initiatives "often fill perceived gaps in the provision of energy governance, which meet specific regulatory needs" (Newell 2011: 102). Moreover, information-based governance schemes like INCR and the Carbon Disclosure Project (CDP) not only institutionalize new norms (Florini and Saleem 2011: 149–150), but also induce energy-related behavioral changes of private actors, thus creating a positive utilitarian interlinkage with the UNFCCC process. Still, many critical voices remain as to potential disruptive effects of "climate capitalism," especially a preference for low-hanging fruits paired with an aversion for potentially risky investments for renewables in poorer developing countries. This criticism however also includes the climate regime itself, in particular, as discussed above, the CDM (Paterson and Newell 2010: 129–140).

As for the dimension of markets, we briefly address three potentially disruptive normative interlinkages between the UNFCCC process and international trade agreements. A highly controversial issue regards carbon border adjustments. A number of actors, including the US Congress and the European Commission, have considered offsetting requirements against carbon-intensive imported goods. Some legal experts hold that such steps could violate WTO law (cf. van Asselt and Brewer 2010), while international relations scholars argue that this very fear might have kept parties from elaborating more ambitious trade- and energy-relevant measures under the international climate regime (Eckersley 2004; Zelli 2011b). A second potential normative conflict regards intellectual property rights (IPRs). In recent climate summits, developing countries led by India and OPEC members called for relaxing provisions of the WTO Agreement on Trade-Related Aspects of Intellectual Property Rights, as these might render the acquisition of renewable energy technologies more costly. Industrialized countries, which host the vast majority of patent-holding companies, rejected this idea, arguing that IPR systems protect innovators and may therefore induce technological research development (Littleton 2008). Third, the climate regime's CDM may collide with some of the hundreds of bilateral investment treaties, and also with regional trade agreements that contain investment provisions. With its restrictions to certain parties and procedures, the CDM could violate the liberal regulations for investors under some of these treaties (Brewer 2004: 7–8). Several agreements meanwhile took steps toward more normative coherence with the climate regime, by including certain environmental safeguards for energy-related investments and activities (Ghosh 2011).

We already touched upon some institutions and alliances that play a role in the realm of energy security, including the IEA and the G8. Unlike these two, OPEC has kept a disruptive relationship with the UNFCCC process. While the ideational clash over values and knowledge has slightly eased (OPEC delegates at least no longer question climate change per se), the issue of adverse impacts of climate policies or response measures is at the core of an ongoing utilitarian conflict over interests. "In essence, OPEC's strategy towards climate policies centers on two main goals: compensation and assistance [...] to diversify away from a natural resource economy" (Goldthau and Witte 2011: 36). From the onset, OPEC countries, regularly advised by the US fossil fuel lobby, made their support for the UNFCCC dependent upon the inclusion of respective clauses. Skilful in positioning themselves as central actors in the G77, they held the issue of response measures "hostage for many years" (Dessai 2004: 25), sidelining non-oil related interests of other group members. While OPEC's dominance in the group has shrunk over the last few years, it still presents a considerable stumbling block for the energy-related development of the climate regime, most recently with its stern opposition to the regulation of emissions from aviation.

Beyond IEA, G8, and OPEC, it becomes difficult to assess the regime's interlinkages with other institutions over energy security, since this governance arena "continues to be deeply entrenched in national political economies" (Florini and Dubash 2011: 2). Over the next years, the growing role of regional agreements like ASEAN might entail closer links to the international climate regime over energy security matters, similar to the observed rapprochement on energy market and investment issues.

To structure the complex energy-related interlinkages of the international climate regime in the realm of development, we distinguish between the closely related issues of low carbon development, climate finance, and adaptation to climate change. Although sustainable development is one of the UNFCCC's core principles (Article 3.4), ideational tensions between development (or rather energy consumption) and sustainability objectives frequently emerged in climate negotiations, most prominently in the ongoing deadlock over burden sharing for limiting greenhouse gas emissions (Dubash and Florini 2011: 9). These tensions somehow resurfaced as turf wars between the UNFCCC and its UN sister agencies over the imprint of climate change on the development agenda. Climate issues were largely subsumed under the "energy" heading at the 2002 World Summit on Sustainable Development, and the UNFCCC secretariat at best played a modest role in the preparations for the Rio+20 summit in 2012. But aside from these rivalries, UNEP, UNDP, and UNFCCC created considerable ideational synergy as norm entrepreneurs for renewable energies and energy efficiency since the late 1990s. Jointly with the CSD, they facilitated interaction and learning processes, providing a "fertile ground for states that wanted to strengthen energy on the UN agenda" (Karlsson-Vinkhuyzen 2010: 191). Further convergence on these matters is reflected in the vibrant cross-institutional rhetoric of a "green economy," notwithstanding the lack of concrete strategies to tackle underlying drivers of energy poverty (Bruggink 2012: 6).

The World Bank's Climate Investment Funds significantly enhanced incentives and opportunities for developing countries to limit their greenhouse gas emissions, thus creating a synergistic utilitarian relationship with the international climate regime. But there are also conflictive aspects to this utilitarian interlinkage, as the Bank largely goes for the commercially most attractive projects that do not show a particular pro-poor focus (Michaelowa and Michaelowa 2011). The donor-oriented voting structure (as opposed to the climate regime's one country-one vote system) further adds to this bias in the Bank's low carbon project portfolio. The new Green Climate Fund is expected to avoid such prioritization and be more in line with energy choices and adaptation goals promoted by the UNFCCC; but this will ultimately depend on its governance structure and allocation criteria.

Finally, adaptation to climate change needs to be factored more appropriately into the nexus between mitigation and energy development. The current ambition of a triple-win solution through mainstreaming has proved unrealistic. In fact, the compromises around "low carbon development" and "green economy" have partly sidelined adaptation concerns. To address this negative utilitarian interlinkage among multilateral climate funds and official development assistance for energy security and access, a sensible division of labor is needed, implying, in the words of Bruggink (2012: 6), "a clear distinction between pro-growth development strategies incorporating mitigation objectives and pro-poor development strategies incorporating adaptation objectives."

In sum, we find considerable variation for the interlinkages involving the climate regime across the four dimensions of global energy policy. This is no surprise given the different actor constellations, objectives, and logics that mark these dimensions, and the relatively feeble ties among them. In the realm of sustainability, the climate regime is largely involved in ideational overlaps on the common goal to tackle climate change through novel energy choices; yet, especially with newcomer institutions, the regime faces competitive utilitarian interlinkages over how exactly this goal shall be achieved. In the fields of markets and security, we find a nearly reversed situation: being originally marked by ideational or normative tensions, some of these interlinkages now bear potential for limited utilitarian convergence and complementarity, e.g., with the IEA and regional trade agreements. The dimension of development offers the most complex picture with both synergistic and conflictive overlaps along several fault lines. This complexity very well reflects the international climate regime's inner balancing act between its sustainability and development features.

Conclusion

This contribution has carefully mapped and assessed current climate change governance and its relation to global energy policy. After a brief discussion of the climate–energy nexus and current emission trends, we have analyzed the global climate governance architecture, including international and transnational institutions, organizations, and arenas. Subsequently, we have discussed the energy-related interlinkages between climate change and other issue areas, namely sustainability, markets, security, and development. This concluding section will provide short suggestions on managing interlinkages before outlining some possible future avenues of research.

At present, an overarching institutional framework like a global energy regime neither seems feasible, given the underlying constellations of interests and power (Newell 2011: 103), nor desirable, as it could never appropriately reflect and harmonize the different objectives and ideas (Cherp *et al.* 2011). A more realistic, albeit ambitious multilateral option would be "joint interplay management" (Oberthür 2009: 375–376), through enhanced inter-institutional cooperation. For instance, in the realm of climate change and energy aid, one could increase earmarking by dividing more clearly between publicly financed poverty prevention aid and carbon market-based green growth aid (Bruggink 2012: 32–35). Such better divisions of labor might lead to a polycentric governance system, "providing stronger interlinkages while preserving the unique and important characteristics of each [governance arena]" (Cherp *et al.* 2011: 86).

Meanwhile, efforts of "unilateral interplay management" taken by individual institutions are more likely to succeed (Oberthür 2009: 375–376). The climate regime could play the role of an "orchestrator" (Abbott *et al.* 2010) within the institutional patchwork: it could help connect the dots by serving as a clearing-house or by coordinating implementation assistance to support country-level transitions toward low carbon societies (Dubash and Florini 2011: 15; van Asselt and Zelli 2012). Likewise, the IEA and the World Bank could do a much better job of integrating climate change and other energy concerns into their work to influence national energy policies (Karlsson-Vinkhuyzen and Kok 2011). As a result, viable options to manage existing conflictive fragmentation should include streamlining the energy policies of various international institutions on the ground.

Based on our mapping of international and transnational climate governance and its implications for global energy policy, we provide some suggestions for future avenues of research on the climate–energy nexus. First, while scholars have started to conceptualize and map the overlap between the climate change and energy governance architectures (e.g., Cherp *et al.* 2011; Colgan *et al.* 2011; Goldthau 2012), more theory-guided and systematic efforts are needed to address the underlying causes of this institutional complexity as well as its consequences, e.g., for institutional effectiveness or equity. Second, while much discussion has focused on the climate regime, and recently also on transnational climate governance, comparatively little is known about the relevance of club governance within the climate–energy nexus. Here, we call for more systematic research on the intricate connections between G8/G20 policy-making, the international climate regime (UNFCCC), and the broad landscape of transnational governance initiatives.

Third, following our discussion of interlinkages, we add an important qualification: our analysis has concentrated on horizontal interlinkages among multilateral institutions (cf. Young 2002). However, in light of the dominance of domestic politics in energy governance, interlinkages of the UN climate regime with national and subnational levels merit much more attention and would render the above overview even more complex. Analyzing these vertical interlinkages is "crucial to understanding what accounts for existing outcomes and where and how reforms may be possible" (Florini and Dubash 2011: 3). In sum, we urgently need more research on the causes and consequences of interlinkages across scales, between climate and energy, and within the energy governance arena.

To conclude, the current climate governance architecture has not been designed to implement a transition toward more sustainable energy. However, given the right strategies and incentives, we believe that climate governance, both international and transnational, can provide an important impetus for the global energy transition.

References

Abbott, Kenneth. 2012. Engaging the Public and the Private in Global Environmental Governance. *International Affairs* 88, 3: 543–564.

Abbott, Kenneth, Philipp Genschel, Duncan Snidal, and Bernhard Zangl. 2010. International Organizations as Orchestrators. Unpublished paper presented at the 7th Pan-European International Relations Conference of the European Consortium for Political Research / Standing Group on International Relations, Stockholm, September 9–11.

Andonova, Liliana B., Michele M. Betsill, and Harriet Bulkeley. 2009. *Transnational Climate Governance. Global Environmental Politics* 9, 2: 52–73.

Bausch, Camilla, and Michael Mehling. 2011. *Addressing the Challenge of Global Climate Mitigation – An Assessment of Existing Venues and Institutions*. Berlin: Friedrich-Ebert Stiftung.

Biermann, Frank, Philipp Pattberg, Harro van Asselt, and Fariborz Zelli. 2009. The Fragmentation of Global Governance Architectures: A Framework for Analysis. *Global Environmental Politics* 9, 4: 14–40.

Bradshaw, Michael J. 2010. Global Energy Dilemmas: A Geographical Perspective. *The Geographical Journal* 176: 275–290.

Brewer, Thomas L. 2004. The WTO and the Kyoto Protocol: Interaction Issues. *Climate Policy* 4: 3–12.

Bruggink, Jos. 2012. *Energy Aid in Times of Climate Change: Designing Climate Compatible Development Strategies*. Publication No. 12-006. Petten: ECN.

Cherp, Aleh, Jessica Jewell, and Andreas Goldthau. 2011. Governing Global Energy: Systems, Transitions, Complexity. *Global Policy* 2: 75–88.

Colgan, Jeff D., Robert O. Keohane, and Thijs Van de Graaf. 2011. Punctuated Equilibrium in the Energy Regime Complex. *The Review of International Organizations* 7, 2: 117–143.

Das, Katsuri. 2011. *Technology Transfer under the Clean Development Mechanism: An Empirical Study of 1000 CDM Projects*. The Governance of Clean Development Working Paper 014. Norwich, UK: University of East Anglia.

de Coninck, Heleen, Frauke Haake, and Nico van der Linden. 2007. Technology Transfer in the Clean Development Mechanism" *Climate Policy* 7: 444–456.

Dechezleprêtre, Antoine, Matthie Glachant, and Yann Ménière. 2009. Technology Transfer by CDM Projects: A Comparison of Brazil, China, India and Mexico. *Energy Policy* 37: 703–711.

Depledge, Joanna. 2008. Striving for No: Saudi Arabia in the Climate Change Regime. *Global Environmental Politics* 8: 9–35.

Dessai, Suraje. 2004. An Analysis of the Role of OPEC as a G77 Member at the UNFCCC. Report for WWF. http://assets.panda.org/downloads/opecfullreportpublic.pdf, accessed March 21, 2012.

Dessler, Andrew, and Edward A. Parson. 2010. *The Science and Politics of Global Climate Change. A Guide to the Debate*. Cambridge: Cambridge University Press.

Driesen, David M. 2006. Links between European Emissions Trading and CDM Credits for Renewable Energy and Energy Efficiency Projects. http://ssrn.com/abstract=881830, accessed April 3, 2012.

Dubash, Navroz K., and Ann Florini. 2011. Mapping Global Energy Governance. *Global Policy* 2(SI): 6–18.

Eckersley, Robyn. 2004. The Big Chill: The WTO and Multilateral Environmental Agreements. *Global Environmental Politics* 4, 2: 24–40.

Eckersley, Robyn. 2012. Moving Forward in the Climate Negotiations: Multilateralism or Minilateralism? *Global Environmental Politics* 12, 2: 24–42.

Fenhann, Jørgen. 2012. CDM Pipeline Overview. http://www.cdmpipeline.org/, accessed April 3, 2012.

Ferrey, Steven. 2010. The Failure of International Global Warming Regulation to Promote Needed Renewable Energy. *Boston College Environmental Affairs Law Review* 37: 67–126.

Florini, Ann. 2011. The International Energy Agency in Global Energy Governance. *Global Policy* 2: 40–50.

Florini, Ann, and Navroz K. Dubash. 2011. Introduction to the Special Issue: Governing Energy in a Fragmented World. *Global Policy* 2(SI): 1–5.

Florini, Ann, and Saleena Saleem. 2011. Information Disclosure in Global Energy Governance. *Global Policy* 2(SI): 144–154.

Friedlingstein, P., R. A. Houghton, G. Marland, *et al.* 2010. Update on CO_2 Emissions. *Nature Geoscience* 3: 811–812.

Ghosh, Arunabha. 2011. Seeking Coherence in Complexity? The Governance of Energy by Trade and Investment Institutions. *Global Policy* 2(SI): 106–119.

Goldthau, Andreas. 2012. From the State to the Market and Back. Policy Implications of Changing Energy Paradigms. *Global Policy* 3, 2: 198–210.

Goldthau, Andreas, and Jan Martin Witte. 2011. Assessing OPEC's Performance in Global Energy. *Global Policy* 2(SI): 31–39.

Haites, Erik, Duan Maosheng, and Stephen Seres. 2006. Technology Transfer by CDM Projects. *Climate Policy* 6: 327–344.

Hoffmann, Ulrich. 2011. *Some Reflections on Climate Change, Green Growth Illusions and Development Space.* UNCTAD Discussion Paper 205. Geneva: UNCTAD.

IEA. 2011. *World Energy Outlook 2011.* Paris: International Energy Agency.

INCR. 2008. *Investor Progress on Climate Risks and Opportunities: Results Achieved since the 2005 Investor Summit on Climate Risk at the United Nations.* Boston: CERES.

Karlsson-Vinkhuyzen, Sylvia I. 2010. The United Nations and Global Energy Governance: Past Challenges, Future Choices" *Global Change, Peace & Security* 2: 175–195.

Karlsson-Vinkhuyzen, S. I., and M. Kok. 2011. Interplay Management in the Climate, Energy and Development Nexus. In Sebastian Oberthür and Olav Schram Stokke, eds. *Managing Institutional Complexity: Regime Interplay and Global Environmental Change.* Cambridge, MA: MIT Press, pp. 285–312.

Karlsson-Vinkhuyzen, Sylvia I., and Harro van Asselt. 2009. Introduction: Exploring and Explaining the Asia-Pacific Partnership on Clean Development and Climate. *International Environmental Agreements* 9, 3: 195–211.

Keohane, Robert O., and David G. Victor. 2011. The Regime Complex for Climate Change. *Perspectives on Politics* 9: 7–23.

Lesage, Dries, Thijs Van de Graaf, and Kirsten Westphal. 2010. G8+5 Collaboration on Energy Efficiency and IPEEC: Shortcut to a Sustainable Future? *Energy Policy* 38: 6419–6427.

Littleton, Matthew. 2008. *The TRIPS Agreement and Transfer of Climate-Change-Related Technologies to Developing Countries.* UN-Doc. No. ST/ESA/2008/DWP/71. New York: UN DESA. http://www.un.org/esa/desa/papers/2008/wp71_2008.pdf, accessed March 20, 2012.

Matschoss, Patrick. 2007. The Programmatic Approach to CDM: Benefits for Energy Efficiency Projects. *Carbon and Climate Law Review* 1: 117–126.

McGee, Jeffrey, and Ros Taplin. 2006. The Asia-Pacific Partnership on Clean Development and Climate. A Complement or Competitor to the Kyoto Protocol? *Global Change, Peace & Security* 18: 173–192.

Michaelowa, Axel, and Katharina Michaelowa. 2011. Climate Business for Poverty Reduction? The Role of the World Bank. *The Review of International Organizations* 6: 259–286.

Naím, Moisés. 2009. Minilateralism. The Magic Number to Get Real International Action. *Foreign Policy* July/August: 5–8.

Nakhooda, Smita. 2011. Asia, the Multilateral Development Banks and Energy Governance. *Global Policy* 2: 120–132.

Newell, Peter. 2011. The Governance of Energy Finance: The Public, the Private, and the Hybrid. *Global Policy* 2: 94–105.

Oberthür, S. 2009. Interplay Management: Enhancing Environmental Policy Integration Among International Institutions. *International Environmental Agreements: Politics, Law and Economics* 9, 4: 371–391.

Oberthür, Sebastian, and Thomas Gehring, eds. 2006. *Institutional Interaction in Global Environmental Governance: Synergy and Conflict among International and EU Policies*. Cambridge, MA: MIT Press.

Oberthür, Sebastian, and Olav S. Stokke, eds. 2011. *Managing Institutional Complexity. Regime Interplay and Global Environmental Change*. Cambridge, MA: MIT Press.

Paterson, Matthew, and Peter Newell. 2010. *Climate Capitalism: Global Warming and the Transformation of the Global Economy*. Cambridge: Cambridge University Press.

Pattberg, Philipp. 2010. Public–Private Partnerships in Global Climate Governance. *Wiley Interdisciplinary Review: Climate Change* 1, 2: 279–287.

Pattberg, Philipp, and Johannes Stripple. 2008. Beyond the Public and Private Divide: Remapping Transnational Climate Governance in the 21st Century. *International Environmental Agreements* 8, 4: 367–388.

Pattberg, Philipp, Frank Biermann, Ayşem Mert, and Sander Chan, eds. 2012. *Public–Private Partnerships for Sustainable Development. Emergence, Influence, and Legitimacy*. Cheltenham, UK: Edward Elgar.

Schneider, Malte, Andreas Holzer, and Volker Hoffmann. 2008. Understanding the CDM's Contribution to Technology Transfer. *Energy Policy* 36: 2930–2938.

Seres, Stephen, Erik Haites, and Kevin Murphy. 2009. Analysis of Technology Transfer in CDM Projects: An Update. *Energy Policy* 37: 4919–4926.

Stokke, Olav S. 2001. *The Interplay of International Regimes: Putting Effectiveness Theory to Work*. Report No. 14/2001. Lysaker, Norway: The Fridtjof Nansen Institute.

Szulecki, Kacper, Philipp Pattberg, and Frank Biermann. 2011. Explaining Variation in the Performance of Energy Partnerships. *Governance: An International Journal of Policy, Administration, and Institutions* 24, 4: 713–736.

UNDP. 2011. *UNDP in Action – Annual Report 2010/2011*. New York: United Nations Development Programme.

UNEP. 2011. *Towards a Green Economy*. Nairobi: United Nations Environment Programme.

UNFCCC. 2002a. Decision 4/CP.7, Development and Transfer of Technologies (Decisions 4/CP.4 and 9/CP.5), UN Doc. FCCC/CP/2001/13/Add.1 (21 January 2002).

UNFCCC. 2002b. Decision 17/CP.7, Modalities and Procedures for a Clean Development Mechanism, as Defined in Article 12 of the Kyoto Protocol, UN Doc. FCCC/CP/2001/13/Add.2 (21 January 2002).

UNFCCC. 2007. *Innovative Options for Financing the Development and Transfer of Technologies*. Bonn: UNFCCC.

UNFCCC. 2010. Report on the Review and Assessment of the Effectiveness of the Implementation of Article 4, Paragraphs 1(c) and 5, of the Convention. Note by the Secretariat. UN Doc. FCCC/SBI/2010/INF.4 (26 May 2010).

UNFCCC. 2011. Decision 1/CP.16, Outcome of the Work of the Ad Hoc Working Group on Long-term Cooperative Action under the Convention, UN Doc. FCCC/CP/2010/7/Add.1 (15 March 2011).

van Asselt, Harro. 2007. From UN-ity to Diversity? The UNFCCC, the Asia-Pacific Partnership, and the Future of International Law on Climate Change. *Carbon and Climate Law Review* 1, 1: 17–28.

van Asselt, Harro, and Thomas L. Brewer. 2010. Addressing Competitiveness and Leakage Concerns in Climate Policy: An Analysis of Border Adjustment Measures in the US and the EU. *Energy Policy* 38: 42–51.

van Asselt, Harro, and Sylvia Karlsson-Vinkhuyzen, eds. 2009. *Exploring and Explaining the Asia-Pacific Partnership on Clean Development and Climate*. Special Issue of *International Environmental Agreements* 9, 3. Dordrecht: Springer.

van Asselt, Harro, and Fariborz Zelli. 2012. *Connect the Dots: Managing the Fragmentation of Global Climate Governance*. Earth System Governance Working Paper 25. Lund and Amsterdam: Earth System Governance Project.

Van de Graaf, Thijs. 2013. Fragmentation in Global Energy Governance: Explaining the Creation of IRENA. *Global Environmental Politics* 13, 3 (forthcoming).

Van de Graaf, Thijs, and Dries Lesage. 2009. The International Energy Agency After 35 years: Reform Needs and Institutional Adaptability. *Review of International Organizations* 4: 293–317.

Van de Graaf, Thijs, and Kirsten Westphal. 2011. The G8 and G20 as Global Steering Committees for Energy: Opportunities and Constraints. *Global Policy* 2(SI): 19–30.

Victor, David. 2009. Plan B for Copenhagen. *Nature* 461: 342–344.

Vihma, Antto. 2009. Friendly Neighbor or Trojan Horse? Assessing the Interaction of Soft Law Initiatives and the UN Climate Regime. *International Environmental Agreements* 9: 239–262.

World Bank. 2010. *World Development Report: Development and Climate Change*. Washington, DC: World Bank.

World Bank. 2011. *Annual Report 2011*. Washington, DC: World Bank.

Yamin, Farhana, and Joanna Depledge. 2004. *The International Climate Change Regime: A Guide to Rules, Institutions and Procedures*. Cambridge: Cambridge University Press.

Young, Oran R. 2002. *The Institutional Dimensions of Environmental Change. Fit, Interplay, and Scale*. Cambridge, MA: MIT Press.

Zelli, Fariborz. 2011a. The Fragmentation of the Climate Governance Architecture. *Wiley Interdisciplinary Reviews: Climate Change* 2: 255–270.

Zelli, Fariborz. 2011b. Regime Conflict and Interplay Management in Global Environmental Governance. In Sebastian Oberthür and Olav S. Stokke, eds. *Managing Institutional Complexity. Regime Interplay and Global Environmental Change*. Cambridge, MA: MIT Press, pp. 199–226.

Chapter 21

The Growing Importance of Carbon Pricing in Energy Markets

Christian Egenhofer

Introduction

The traditional energy challenges such as rapidly increasing demand, increasing production costs, fears about under-investment or increasing government interference and the new assertiveness of "petrol states" is gradually being complemented by the need to fight climate change. Given that energy-related emissions account for about 80% of all greenhouse gas (GHG) emissions in developed economies, the vision of containing the warming of earth temperature to 2 °C by the end of the century as agreed at the climate change negotiations in Copenhagen in 2009 and legally adopted the following year in Cancún will require fundamental changes in energy systems of developed and emerging economies alike. Generally, this will require global GHG emissions reduction by 50% by 2050.

Calculations conclude that in order to reach these aspirations, developed countries would need to cut their own emissions by around 80–95% by 2050. There is no scenario that allows reaching this goal that does not assume an almost total de-carbonization of the power sector including a significant share of around one third to one half of de-carbonized power in the transport sector. Even if one assumes that not all of these ambitious targets will be met in their entirety, de-carbonization will progressively influence energy policies at national, regional, and global level, especially via large-scale electrification, the development of renewable energy sources, reduced oil consumption in the transport sector, a trend toward natural gas in power generation, a more serious and sustainable approach to energy efficiency, and the search for new and low carbon energy technologies, be they clean coal, nuclear or other, yet unknown technologies.

The size of the challenge has revived interest in market-based instruments that offer the prospect of achieving the target at least cost – at least in theory. Ever since the negotiations of the Kyoto Protocol in 1997 the use of market-based instruments has occupied a central role in the international discourse on climate change, despite the fact that some countries are openly hostile to this concept. This has meant that the use of

The Handbook of Global Energy Policy, First Edition. Edited by Andreas Goldthau.
© 2013 John Wiley & Sons, Ltd. Published 2013 by John Wiley & Sons, Ltd.

emissions trading as a tool that up to the negotiations of the Kyoto Protocol was largely restricted to the US has become a credible addition to any government's toolbox.

There is also a second driver for using emissions trading. Given the notorious inertia of the energy sector, some of the domestic reductions required and promised may not be achieved by countries, not least for security of supply reasons. This will require offsetting of domestic emissions by reductions achieved in other countries. Not only can this smooth the investment cycle, it also offers to reach the global target at lower costs and can function as an effective transfer mechanism; effective because once carbon markets are designed they function almost automatically, far more detached from daily politics than the "transfer politics" in international negotiations or development aid.

This, however, is not to say that carbon markets will drive energy markets and energy policy. Far from it; witness the recent difficulties of the US to make progress on a federal carbon market, the Canadian inaction, or the difficult birth of the Australian system. However, carbon markets will increasingly become an important part of governments' energy and climate change toolbox and thereby exert their influence on national, regional, and global energy policy. The speed with which this will happen is not clear at the moment but it is hard to see how global national, regional, and global climate policy and thereby by extension, energy policy discard entirely the use of market-based instruments and therefore emissions trading.

Carbon Market: Snapshot and Mechanics

The emergence of GHG emissions markets was the direct result of the United Nations Framework Convention on Climate Change (UNFCCC) and the Kyoto Protocol, which included three articles that provided for the creation of offsets and the trading of these units. The Kyoto Protocol established that countries with emissions targets, so-called Annex I countries, can buy and sell parts of their emission rights ("assigned amounts" in Kyoto-speak) or emission reduction units from projects with other Annex I countries, and buy certified emissions reductions that are generated from projects in developing countries (so-called non-Annex I countries). Trade can take place in various ways under three mechanisms.

- Clean Development Mechanism (CDM): Article 12 establishes that Annex I countries (and firms in these) can transfer emissions reductions from projects in developing countries through so-called CERs (Certified Emissions Reductions).
- Joint Implementation (JI): Article 6 allows Annex I countries (and firms in these) to transfer reductions in emissions compared to a baseline for individual projects through so-called ERUs (Emissions Reduction Units).
- International emissions trading: under Article 17 Annex I countries (and firms in these) can trade parts of the "assigned amounts" allocated under the Protocol (Assigned Amounts Units or AAUs).

Each mechanism has a different function and rationale, particularly the CDM, which is meant to finance sustainable development and has more elaborate governance provisions. The tradable units from the mechanisms are interchangeable, however, which meant that all three mechanisms worked together in one international system.

The trading mechanisms were meant to become an integral part of and work side by side with other policy measures to provide flexibility to abate emissions where it can be done most cheaply and therefore as a means of reducing compliance costs. Behind this was also the vision of a global carbon market and a single (global) price for carbon. While this has not been accomplished, attempts to do so continue up to this day and many elements of the architecture have been put into place.

A carbon market does exist, however it is fragmented and largely confined to the EU, Japan, and the developing countries. The Kyoto Protocol units have been used for trading and accounting in the EU and Japan. The rest of the world uses emerging, and until now largely voluntary standards, which have little influence on the price for carbon. The voluntary carbon market was principally a result of the decision of the US not to ratify the Kyoto Protocol, accompanied by a long delay by Australia and an official policy of inaction by Canada.

The agreements recently struck during the global climate change negotiations in Durban, South Africa in November/December 2011 have launched a new process on the future of global carbon markets. This process will give an answer as to whether the initial vision of a global carbon market and a single price for carbon, which emerged from the Kyoto Protocol, will prevail or whether the future will see a long period of building through a bottom-up approach, which may or may not lead to a unitary carbon market in the future (e.g., Marcu 2011). Therefore, for the foreseeable future, the two processes will develop in parallel. Irrespective of the outcome, carbon pricing will make itself heard in energy policy as one of the tools to deal with global challenges.

The EU has made a start in 2005 by launching a comprehensive mandatory system. Other systems have been launched by New Zealand and more recently Australia. In the US, despite the collapse of a federal system, northeastern States have created a cap-and-trade system in operation since 2009 while California started in 2012. Discussions are ongoing in Japan, albeit for a long time. Most importantly, there is experimenting in China, notably with the establishment of a measurement, reporting, and verification system under the 12th Five-Year Plan. A number of initiatives in developing and emerging economies under the UN are pointing toward a further proliferation of programs, be they fully fledged or pilots. Finally, the voluntary market is adding further dynamics to the global carbon market.

Cap-and-trade programs are proliferating in many different regions of the world for reducing GHG emissions, or at least as forming a substantial element of climate policy. The CDM and the prospect for new mechanisms have created a constituency that is likely to continue promoting the use of emissions trading. Nowhere has this been more visible than in the EU, where the first major emissions trading system has been running since 2005. In 2010 the global carbon market amounted to a value of around $140 billion (Linacre et al. 2011) although it declined in 2011, mainly due to the fall of EU carbon prices.

The EU Emissions Trading Scheme

In order to meet its obligations under the Kyoto Protocol, the EU has set up the EU Emissions Trading Scheme (ETS), a domestic cap and trade system. Through coverage of currently some 2 billion tonnes of GHG emissions in the countries of the European Economic Area,[1] the EU ETS made up 84% of the global carbon market in 2010 (Linacre et al. 2011). Strictly speaking a regional carbon market, its size means that prices for EU allowances (EUAs) under the ETS are global price setters.[2]

It was designed as a domestic policy, largely protected from the global carbon markets that at the time were seen as likely to emanate from the Kyoto Protocol, such as CDM and JI or International Emission Trading. A principal trigger for this protection from the global carbon market was concern over compliance under the Kyoto Protocol and the Marrakech Accords (Egenhofer 2007). The Kyoto Protocol was seen as far from guaranteeing consistent measuring methodologies or enforcement mechanisms, conditions indispensable to a functioning emissions trading system. This was why the EU ETS could only be guaranteed within a national or regional jurisdiction and not within a more loosely global (UN) framework.

1. Initial Design

The EU ETS is a cap-and-trade system. It limits the total amount of certain GHGs that can be emitted by the sectors covered. To date it covers CO_2 emissions only from large industrial and energy installations from a limited number of sectors,[3] covering some 15,000 installations, which amount to some 40% of total EU GHG emissions. Within this cap, companies receive emissions rights in the form of EUAs, which they can sell to or buy from one another as needed. At the end of each year, companies must surrender enough allowances to cover all their emissions. Otherwise fines are imposed. Unused emissions can be banked for future use. The flexibility associated with the trading system should guarantee that emissions are cut where it costs least to do so. Credits from the Kyoto Protocol's project mechanisms, CDM and JI, can be used for compliance within limits (Lefevere 2006).

The EU ETS had a bumpy start, particularly in its pilot phase (2005–2007) but somewhat carried over into phase 2 (2008–2012). It suffered from a number of teething problems and design flaws, which have been covered extensively in the literature (e.g., Egenhofer 2007; Ellerman et al. 2010; Skjærseth and Wettestad 2010). Most have been addressed by now, notably by a review, completed in 2008, coming into force in the beginning of 2013 (the start of phase 3).

Initial problems were partly the result of the rapidity with which the ETS was adopted, motivated by the EU's desire to show a strong determination to tackle climate change (Skjærseth and Wettestad 2008), while similar carbon pricing systems were largely absent in other economies with which EU industry had to compete. The allocation of allowances by member states on the basis of National Allocation Plans led to a "race to the bottom," i.e., member states were under pressure by industries not to hand out fewer allowances than their EU competitors received (Ellerman et al. 2007, 2010; Kettner et al. 2007). This led to over-allocation, and ultimately to a price collapse.

At the same time, in the absence of a global agreement, concerns over competitiveness and carbon leakage have been high on the agenda. The essential answer by the EU ETS was free allocation. Free allocation constitutes a compensation or a subsidy, potentially creating an incentive to continue producing in Europe. At the same time, grandfathering in the first two phases has led to significant windfall profits. During the period when the EU allowance price was high, free allocation also generated windfall profits, mainly but not only in the power sector. Ellermann et al. 2010, the most authoritative ex-post study conducted so far, conclude that in total the rents were substantial, even at a relatively modest carbon price of €12, and amount to more than €29 billion in windfall profits, although with the caveat of surrounding uncertainties in the calculations.

Some of these issues were addressed in phase 2 as a result of member state cooperation and the European Commission was able to reduce member states' allocation proposals. Throughout both early phases, by and large, the ETS has managed to deliver a carbon price, meaning that the carbon price has managed to enter into boardroom discussions

(Ellerman and Joskow 2008). This also means that the EU ETS has introduced carbon management systems within companies.

2. Review

Experiences from phases 1 and 2 triggered a review in 2008 that led to the adoption of radical changes to the EU ETS which were not even thinkable before its initial adoption in 2005 (Egenhofer *et al.* 2011). The principal element of the new ETS is a single EU-wide cap, which will decrease annually in a linear way by 1.74% starting in 2013. This annual linear reduction factor has no sunset clause.

The revised ETS Directive also foresees EU-wide harmonized allocation rules. Starting from 2013, power companies will have to buy all their emissions allowances at an auction with some temporary exceptions for coal-based poorer member states. For the industrial sectors under the ETS, the auctioning rate will be set at 20% in 2013, increasing to 70% in 2020, with a view to reaching 100% in 2027. The remaining free allowances will be distributed on the basis of EU-wide harmonized benchmarks, set on the basis of the average performance of the 10% most GHG-efficient installations. Industries exposed to significant non-EU competition and thereby potentially subject to carbon leakage, however, will receive 100% of allowances free of charge up to 2020, based on EU-wide product benchmarks set on the basis of the average performance of the 10% most GHG-efficient installations.

Other changes include a partial redistribution of auction rights between member states, restrictions of the total volume of CDM/JI credits, the use of 300 million EU allowances to finance the demonstration of carbon capture and storage (CCS) and innovative renewable technologies, and a general but non-binding commitment from EU member states to spend at least half of the revenues from auctioning to tackle climate change both in the EU and in developing countries, including measures to avoid deforestation and increase afforestation and reforestation in developing countries. Furthermore:

- 12% of the overall auctioning rights will be redistributed to member states with a lower GDP per capita (10%) and those that have undertaken early action (2%).
- The system will be extended to the chemicals and aluminium sectors and to other GHGs, e.g., nitrous oxide from fertilizers and perfluorocarbons from aluminium.
- Member states can financially compensate electro-intensive industries for higher power prices. The European Commission is drawing up EU guidelines as to this end.
- Since 2012 aviation, covering also all incoming and outgoing international flights, is included in the ETS. Similar provisions for maritime transport are likely to be proposed by the European Commission in the course of 2012.

As already in the previous periods, access to project credits under the Kyoto Protocol from outside the EU will be limited. The revised ETS restricts access to no more than 50% of the reductions required in the EU ETS to ensure that emissions reductions will happen in the EU. Leftover CDM/JI credits from 2008–2012 can be used until 2020.

The revised ETS Directive explicitly foresees the possibility for a revision in the case of an international climate change agreement. Depending on the nature of the agreement, this could mean the lowering of the cap, for example if the EU decided to move to a unilateral EU reduction commitment of 30%. This move would trigger a whole number of implementation rules including notably an increase of the linear annual reduction factor of currently 1.74% for the cap allocation, the role of flexible mechanisms, the inclusion of forestry credits, and land use changes. Most importantly, the ETS features an enabling clause for linking the ETS with other regional, national, or sub-national emissions trading programs through mutual recognition of allowances.

At the time of the ETS review, there was a general conviction that the new ETS will be able to cope with the lack of a global climate change agreement, address competitiveness, yet also drive de-carbonization of the EU economy. The 2008/9 economic crisis however has destroyed that confidence by a lowering of EUA prices due to rapid and dramatic decline in economic output. Ever since, EUA prices have been lingering around €5–8 per ton of CO_2 and many analysts believe that prices will remain lower than had been expected.

As a result, the list of those voices calling for some sort of market oversight and price stabilization mechanisms has increased. Many agree that both price stability and a strong carbon price signal are beneficial if not essential. More controversial is the question on the nature of such a mechanism, its organization, and how permanent this should be. This will be an issue high on the agenda throughout 2013 and beyond.

Other OECD Countries

Interest in carbon markets has not been limited to Europe. In other parts of the developed world, numerous initiatives have been developed, with some emissions trading systems being adopted while others still pending.

3. New Zealand

The New Zealand Emissions Trading Scheme started operation in 2008 and is to be fully phased in by 2013.[4] It covers GHG-emitting activities in all major sectors of the economy: forestry, stationary energy, industrial processes, transport fuels, agriculture, synthetic gases, and waste, responsible for some 50% of total emissions. The system, which is mandatory, operates with absolute caps but it foresees a possibility to switch over to an intensity-based system if international climate change agreements in the future include intensity-based approaches. It allows for international offsets. Agriculture is expected to enter in 2015.

4. Australia

Within the context of a comprehensive climate change plan, the Australian government has adopted a cap-and-trade system that covers some 60% of Australia's GHG emissions starting in July 2012. The system starts with a fixed carbon price for the first three years and then transitions to a cap-and-trade program. The program offers the possibility to use credits that comply with the Kyoto Protocol to meet compliance obligations (C2ES 2011).

5. United States

Although both the concept and putting into practice of cap-and-trade and emissions trading systems in general originated in the US (Klaassen 1996), federal carbon pricing appears to be stalled. This became apparent when the US federal cap-and-trade system under the Waxman-Markey legislation that passed the House in June 2009 was abandoned. That legislation had proposed a domestic target of 17% reductions by 2020 based on 2005 with an additional 3% anticipated from forestry offsets.

As a consequence of this federal impasse action has again accelerated at the regional, State, and in some cases local level. The most prominent initiatives are the Regional Greenhouse Gas Initiative (RGGI) in the northeast and the California Global Warming Solutions Act (Assembly Bill 32, or AB 32). RGGI, starting in 2009, is a regional initiative that has capped power sector CO_2 emissions and committed to reduce them by 10% by 2018.[5] By the end of 2011, California adopted a cap-and-trade program applying to all major industrial sources and electricity by 2013, to be expanded in 2015 to the distribution of transport and other fuels. There is a possibility to link the system to the Western Climate Initiative (WCI), which attempts to bring together States in the US and Provinces in Canada to tackle climate change at a regional level explicitly including the prospect of emissions trading. According to C2ES,[6] a well-known Washington think-tank, other States and Provinces as well as Mexican states and tribes are interested in collaborating within WCI.

6. Japan

Japan has operated a voluntary emissions trading scheme since in 2005 that covers CO_2 emissions from fuel consumption, electricity and heat, waste management, and industrial process from over 300 companies. Introduction of a mandatory emissions trading system originally considered for 2013 has been postponed, though discussions are ongoing. In 2010 the Tokyo Metropolitan Government introduced a mandatory cap-and-trade program, essentially covering office space and commercial buildings, areas that are within its authority.

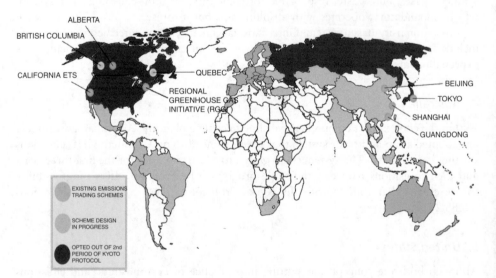

Figure 21.1   Selection of emissions trading programs and developments wordwide.
Source: Prepared by Perspectives in January 2012 and modified by IETA in April 2012.

Emerging Economies

Of the emerging economies, China's plans for emissions trading are the most advanced. The 12th Five-Year Plan on National Economic and Social Development from 2011to 2015 features the objective of gradually setting up a carbon market. More importantly, the Plan also announced a system for measurement, reporting, and verification, which is a precondition for any emissions trading program (Lin *et al.* 2011). While it should be expected that the actual establishment would take many years, the concept seems to be now part of the government's options to tackle climate change. Important elements for the realization of the plan are regional pilots, to be extended at a later stage (Lin *et al.* 2011: 40) and a voluntary carbon market (Babu 2011: 102).

Voluntary Carbon Markets

The decision of the US not to ratify the Kyoto Protocol, accompanied by a long delay by Australia and an official policy of inaction by Canada, led to the emergence of a Voluntary Carbon Market. Main elements are the Voluntary Carbon Standard that emerged as an effort of the business community, mainly in North America, and different trading systems, such as the GHG cap-and-trade and regulatory system emerging in California and the RGGI. By now these can be seen as possible precursors to the creation of mandatory markets elsewhere, for example in China, India, Peru, and Brazil. Although very small compared to the compliance market, with only a 2% share, voluntary markets are expected to shape compliance markets via standards, methodologies, and technical infrastructure (Babu 2011).

The Proliferation of New International Offset Mechanisms

Following the global climate change negotiations in Durban, South Africa in December 2011, there is a new dynamic within the UN both as regards the possibility to agree on a new global framework and the role of offset or market mechanisms as they are also called. While Japan for some time has been exploring the idea of bilateral offset mechanisms, the Durban negotiations opened up the possibilityfor developing this and the concept of new market mechanisms further. Also the EU continues to promote the idea of new or revised flexible mechanisms to advance climate objectives, scaling up existing mechanisms able to achieve real emissions reductions, and possibly other specific objectives such as sustainable development, technology transfer, and financing. In addition, the national, regional, and local emissions trading systems that have been presented all offer the possibility of using offsets for compliance. By introducing a carbon price signal, offsetting or market mechanisms are likely to become stepping stones in a path from a localized carbon price in an individual project, toward cap-and-trade systems within a sector or the entire economy. The possibility of offsetting allows for arbitrage between different offset mechanisms. The emergence of a "'more global" carbon price signal and the convergence of carbon prices eventually becomes a function of eligibility rules for cross-border compliance of the different mechanisms and volumes of offsets.

Clean Development Mechanism

The Clean Development Mechanism (CDM) is the oldest and most successful offset mechanism, which in its heyday has been responsible for up to 10% of the global carbon market. Criticism about complexity, transaction costs, and integrity as well as doubts as to how far scaling up is possible mean that its importance in relative terms is likely to diminish in the future. Nevertheless, there is still scope for improving the CDM. One possible reform option relates to so-called "Programs of Activities," which would transform the CDM from the current project-based mechanisms into a programmatic version that registers a set of activities of the same type under a single umbrella. Scaling up could also be achieved by "sectoral benchmarking" whereby CDM credits could easily be calculated on the basis of a predetermined benchmark for a sector or sub-sector. Such an expansion of the scope to sectoral and programmatic activities would allow the CDM to address more mitigation opportunities. This would however require improvements in efficiency of administration and an increase in the transparency of governance (Fujiwara 2012).

Joint Implementation

Joint Implementation (JI), the other mechanism established by the Kyoto Protocol, has played a much smaller role than the CDM, and both in name and in substance will decline in importance. In particular, it has faced administrative and organizational shortcoming pertaining to the Joint Implementation Supervisory Committee as well as more technical issues such as baseline setting and methodology choices. Existing problems with double-counting have become controversial (see Elsworth and Worthington 2010).

Sectoral Crediting Mechanism

For the EU, most potential is seen in so-called "sectoral crediting." A sectoral crediting mechanism credits emissions reductions from a covered sector against a threshold. The main difference from the CDM is that credits are only granted for reductions that start below a certain threshold, which would be significantly below the business-as-usual trajectory. That would ensure that reductions seen from a global standpoint indeed are additional and not just offsetting those elsewhere. There are ideas to base a sectoral crediting mechanism on so-called no-lose targets. This means that the host country will be rewarded for overperformance in the sector above the threshold but will not be penalised for underperformance. There are a variety of design options for baseline setting. It can be negotiated as part of an international agreement between parties or domestically set on the basis of a sectoral benchmark. The baseline could be expressed in absolute emission levels, carbon intensity, or technology penetration rates (Fujiwara 2009).

Sectoral Trading

Sectoral trading is a cap-and-trade scheme (or alternatively, a baseline and credit program) applied to a whole sector or sub-sector within a country (Fujiwara 2009). This can be done by gradually tightening the negotiated baselines, and converting them into absolute caps. Sectoral trading aims at addressing countries that are not yet ready to take on binding economy-wide targets but are prepared to accept them in key sectors, such as power and industry. Emissions allowances would be allocated to the host country's government, reflecting binding sectoral targets. Governments would be responsible for reducing emissions in particular sectors to a predetermined level, based on national rules such as on allocation or on compliance. Theoretically, sectoral trading, if based on absolute caps, would be simpler with lower transaction costs than sectoral crediting.

Some countries, such as China for example, might prefer this model over sectoral crediting or a scaled up CDM. As sectoral trading is generally seen as a stepping stone to a cap-and-trade system such as the EU ETS, one should expect some sort of "preferential" treatment of credits emanating before it. This would be possible for example by a bilateral agreement between the EU and China, something that has been rumored for some time.

REDD-plus Market

There is a consensus on the importance of providing a value to environmental services such as avoided deforestation. The importance of avoided deforestation has been discussed in detail during the review of the EU ETS and recognized in Article 10(3) of the ETS Directive. Sovereign participation of EU member states and other countries in an international REDD-plus market (Reducing Emissions from Deforestation and forest Degradation) generally appears to be more likely and preferable to linking to the ETS and international carbon markets, irrespective of whether a CDM style (international issuance of credits) or JI style (national issuance of credits) is chosen. Full linking to international carbon markets would first require more clarity of the design of REDD-plus markets, notably addressing questions of permanence, monitoring, reporting and verification, and more generally compliance as well as a solution to the tricky question of how to absorb the expected volumes of credits (O'Sullivan *et al.* 2010).

To date, the link to the EU ETS is the auctioning of EUAs, which will supply EU governments with the necessary funds for sovereign participation. However, current and expected EUA price levels are insufficient for EU finance commitments.

NAMA Crediting

There is a possibility for crediting so-called Nationally Appropriate Mitigation Actions (NAMAs), which are domestic policies of developing or emerging economies to reduce GHG emissions and "merit" international financial and other support. While discussions are continuing, NAMA crediting to date remains very complex, requiring further analysis. Much depends on whether a significant breakthrough on NAMA crediting in the UN negotiation can be achieved. Irrespective, countries most likely will provide support for NAMAs through sovereign climate finance, thereby supporting a carbon price signal.

New Market Mechanisms

The international climate change negotiations in Durban in December 2011 finally opened the road toward the creating of new market mechanisms, possibly two different types: one top-down mechanism under the guidance and authority of the international community within the UN and one bottom-up mechanism that allows countries to develop and implement such approaches in accordance with national circumstances (Marcu 2012). This has reinvigorated the discussions on many of the mechanisms that have been described above as well as new ones; for example, bilateral offset credit, sectoral crediting, or NAMAs.

Implications of a Patchwork of Regional Emissions Markets

In the absence of a global framework and a global carbon market that generates one or at least a reasonably consistent carbon price signal across major economies, the plethora of regional, national, or local emissions trading systems will continue to generate different carbon constraints in different countries. As a result, carbon shadow prices are not equivalent. However, as long as the primary instrument of climate change policy remains emissions trading, respective carbon markets can be linked (see Stavins 2011). This can happen formally or informally. Formal linking occurs when emissions trading systems recognize other systems' emissions allowances or credits as compliance tools. Informal linking occurs when market participants of different emissions trading schemes search for arbitrage possibilities between different carbon markets or commodities. As long as domestic or regional emissions trading schemes allow for the use of credits from such projects, and there is sufficient volume, carbon prices will converge and therefore become more visible (Jaffe *et al.* 2009; Mehling and Haites 2009).

"Formal" Linking

From the beginning, the EU ETS Directive allowed for linking the ETS with other emissions trading schemes by international agreement. This provision has been strengthened following the review. In Article 25 the "new" ETS foresees different types of linking arrangements, via international agreement or through reciprocal commitments applied through domestic policies. The latter provision is innovative, both internally and internationally, as it would allow schemes to be linked through administrative decisions. In essence, this could mean that over time non-EU emissions trading schemes could be linked to the EU ETS, using the EU ETS as a "docking station" for the global carbon market. Other regional systems foresee similar provisions.

Emissions trading systems in different jurisdictions exhibit widely divergent design features. This is an expected result of the domestic political dynamics. Nonetheless, linking schemes does not necessarily run into fundamental problems as long as technical fixes such as gateways or restrictions are put into place. Such fixes, however, generally reduce the efficiency through additional transaction costs, market fragmentation, or perverse effects (Delink *et al.* 2010; Ellis and Tirpak 2006).

We should however expect political obstacles to linking as a result of potential distributional impacts (see Bode 2003 for Germany). When two schemes are linked, the market price will be higher than the pre-link price in one of the trading schemes and lower than the pre-link price in the other zone, thereby creating winners and losers. The winners will be net sellers in the low-price scheme, as the price will go up for them, and net buyers in the high-price scheme, as price will go down for them. The reverse is true for net buyers in the low-price scheme and net sellers in the high-price scheme.

Another option for an emerging global carbon market would be to move toward sectoral agreement on an international scale (Sterk 2011) as opposed to linking domestic schemes that include a variety of sectors. This would have the benefits of combining similar sectors or "carbon commodities" with similar characteristics.

"Informal" Linking

All systems currently in operation allow for international offsets such as CDM or JI, even if the domestic political economy typically "demands" additional quality and quantity restrictions. This has been the case for the EU ETS but also for the Australian, New Zealand, and the US State systems.

Given that emissions trading programs other that the ETS are not yet or barely operational, at this stage it is difficult to estimate exact volumes of demand for offset credits, except for the EU. The EU – both for the ETS and non-ETS sectors – is expected to require somewhat more than 300 Mt CO_2e through 2012. For the period until 2020 estimates range between 1750 to 2100 Mt CO_2e for the EU's unilateral 20% reduction target and between 2550 and 3800 for a possible 30% reduction target. In the "Roadmap 2050," the European Commission (2011) estimates that a 25% reduction by 2020 can be achieved by full and effective implementation of the Energy Efficiency Plan and the legally-binding renewables targets, while only a 30% reduction target would generate additional demand for post-2012 credits or offsets from non-Annex 1 countries (see Fujiwara 2012 for details).

Even if for the time being there is no demand for large volumes of credits in the EU due to demand reduction as a result of the recession, interest in the further development of existing and new mechanisms remains strong. This can to an extent be explained by the overarching interest in progress toward a global carbon market and price.

An Unexpected Ally for Carbon Pricing: Carbon Import Taxes

Attempts to scale up and reform existing flexible mechanisms such as the CDM and create new ones to gradually create a global carbon price may be supported by trade policy. This for example could happen if the EU and/or the US would impose an import tax on the content (including the embedded carbon) of CO_2 of all goods imported from countries that do not have their own cap-and-trade system or equivalent measures. While there is strong reluctance in the EU to even discuss carbon import taxes, to date it is has been an indispensable pillar of any US proposal for a cap-and-trade system, also in the most recent one from Waxman-Markey. From a purely economic perspective, this would be a straightforward way to move toward a global "shadow" carbon price even in the rest of the world. It transfers carbon pricing, at least partially, via trade flows, even to those parts of the world where governments have so far refrained from imposing domestic measures of any magnitude. Thereby it creates a mechanism that enforces the pass-through of carbon costs across the globe. A key effect of such a tariff is that it would always lower global emissions. There are solutions to issues such as WTO compatibility and equity, the latter for example through rebating (Gros *et al.* 2010; Gros and Egenhofer 2011). The EU's decision to include international aviation into the EU ETS is an application of the concept of border measures, although in a small section of the economy only.

Conclusion

Although the 2009 Copenhagen climate change negotiations failed to make decisive progress toward a global framework of climate change, it documented the consensus that a carbon constraint is accepted globally. This is the meaning of enshrining the 2 °C target into the Copenhagen Accord and the Cancún Agreements, which gave the Accord legal force under international law. At the same time, Copenhagen made clear that there will be no "top-down" global legal framework with absolute emissions ceilings any time soon. Instead, the future of the international climate change regime will be built upon domestic policies that emerge in a bottom-up fashion. Emissions trading in the form of cap-and-trade systems will be an essential part of these approaches. Indeed, since Copenhagen, new cap-and-trade systems have come into operation, notably in Australia, California, and Tokyo, thereby complementing the existing system in the EU, the Northeastern US, and New Zealand. Since then both China and the Republic of Korea have announced their intention to develop national cap-and-trade systems.

It is fair to assume that in the next decade additional emissions trading systems will be implemented while existing ones are enlarged. For example it is possible that the California system becomes a Western initiative including both US States and Canadian provinces and possibly Mexican states. It is also perceivable that the RGGI system in the Northeast US might include Canadian provinces. The EU ETS will almost certainly expand into Southeast Europe and may even include former Soviet Republics. The issue of a federal US program will reemerge after the 2012 US elections. Discussions in Japan continue.

All existing and planned systems foresee international offsets. This offers arbitrage possibilities for market participants, which will contribute to generate a more universal carbon price signal. At the same time, there are possibilities for linking the different emissions markets – be they regional, national, or local – in a formal way. Linking provisions are part of the design. Nevertheless, we should not expect formal linking to proceed quickly. We have learned that emissions trading systems are controversial at best, requiring careful crafting of compromises to accommodate different interests. Formal linking also means that design options from other systems are imported wholesale, which risks upsetting the carefully achieved balance. Nevertheless, in the medium and long term the economic argument of the promise of higher efficiency through wider and deeper markets will be too hard to ignore.

In the meantime however the carbon price will continue to emerge, first locally then by convergence into an increasingly robust signal. As this happens, the energy sectors worldwide will no longer be able to ignore the carbon price. Implementation of a cap-and-trade system in the US – the original home of emissions trading – and/or in China will alter the dynamics fundamentally.

The EU ETS offers a good showcase of how a domestic ET market can be put together. Although not all lessons are transferable, the study of the EU ETS and how it came together can enhance the understanding of how the carbon market might develop.

Notes

1. The EEA covers the EU plus Norway, Iceland, and Liechtenstein, which are closely associated to the EU's internal market and obliged to adopt most of the EU's economic regulation.
2. With demand from those countries that have ratified the Kyoto Protocol fast decreasing, the EU ETS will become – at least temporarily – an even more important component of the global carbon market.
3. These sectors include electricity and heat generation, cement production, pulp and paper production, refining, coke ovens, iron and steel, glass, ceramics, and paper and board, amounting to some 15,000 installations.
4. See Egenhofer and Georgiev (2010: 43); Government of New Zealand, http://www.c limatechange.govt.nz/emissions-trading-scheme/about/questions-and-answers.html#units, accessed December 3, 2012.
5. RGGI includes the States of Connecticut, Delaware, Maine, Maryland, Massachusetts, New Hampshire, New York, Rhode Island, and Vermont. Several other US States and Canadian provinces act as observers.
6. Formerly the Pew Center on Global Climate Change.

References

Babu, Nityanandam Yuvaraj Dinesh. 2011. Voluntary Market: Future Perspective. In *Progressing Towards Post-2012 Carbon Markets*. Roskilde: UNEP Risoe Centre, pp. 101–111.

Bode, S. 2003. *Implications of Linking National Emissions Trading Schemes Prior to the Start of the First Commitment Period of the Kyoto Protocol*. Discussion Paper 214. Hamburg: Hamburg Institute of International Economics (HWWI).

C2ES. 2011. *Australia's Carbon Pricing Mechanism*. Arlington, VA: C2ES.

Delink, Rob, Stephanie Jamet, Jean Chateau, and Roman Duval. 2010. *Towards Global Carbon Pricing: Direct and Indirect Linking of Carbon Markets*. Paris: OECD.

Egenhofer, Christian. 2007. The Making of the EU Emissions Trading Scheme: Status, Prospects and Implications for Business. *European Management Journal* 25, 6: 453–463.

Egenhofer, Christian, and Anton Georgiev. 2010. *Benchmarking in the EU: Lessons from the EU Emissions Trading System for the Global Climate Change Agenda*. Brussels: Centre for European Policy Studies.

Egenhofer, Christian, Monica Alessi, Anton Georgiev, and Noriko Fujiwara. 2011. *The EU Emissions Trading Scheme and Climate Policy towards 2050*. Brussels: Centre for European Policy Studies.

Ellerman, A. Denny, and Paul Joskow. 2008. *The European Union's Emissions Trading System in Perspective*. Arlington, VA: Pew Center on Global Climate Change.

Ellerman, A. Denny, Barbara Buchner, and Carlo Carraro. 2007. *Allocation in the European Emissions Trading Scheme: Rights, Rents and Fairness*. Cambridge: Cambridge University Press.

Ellerman, A. Denny, Frank Convery, and Christian de Perthuis. 2010, *Pricing Carbon: The European Union Emissions Trading Scheme*. Cambridge: Cambridge University Press.

Ellis, Jane, and Dennis Tirpak. 2006. *Linking GHG Emissions Trading Schemes and Markets*. Paris: OECD/IEA.

Elsworth, Rob, and Bryony Worthington. 2010. *E R Who? Joint Implementation and the EU Emissions Trading System*. London: Sandbag.

European Commission. 2011. *A Roadmap for Moving to a Competitive Low Carbon Economy in 2050*. Luxemburg: Office for Official Publications of the EU.

Fujiwara, Noriko. 2009. *Flexible Mechanisms in Support of a New Climate Change Regime: The CDM and Beyond*. Brussels: Centre for European Policy Studies.

Fujiwara, Noriko. 2012. *Post-2012 Carbon Markets*. Brussels: Centre for European Policy Studies.

Gros, Daniel, and Christian Egenhofer. 2011. The Case for Taxing Carbon at the Border. *Climate Policy* 11, 5: 1212–1225.

Gros, Daniel, Christian Egenhofer, Noriko Fujiwara, *et al.* 2010. *Climate Change and Trade: Taxing Carbon at the Border?* Brussels: Centre for European Policy Studies.

Jaffe, Judson, Mathew Ranson, and Robert N. Stavins. 2009. Linking Tradable Permit Systems: A Key Element of Emerging International Climate Policy Architecture. *Ecological Law Quarterly* 39: 789–808.

Kettner, Claudia, Angela Köppl, Stefan Schleicher, and Georg Thenius. 2007. *Stringency and Distribution in the EU Emissions Trading Scheme – The 2005 Evidence*. Nota di Lavora 22.2007. Milan: Fondazione Eni Enrico Mattei.

Klaassen, Geert. 1996. *Acid Rain and Environmental Degradation: The Economics of Emissions Trading*. Cheltenham, UK: Edward Elgar.

Lefevere, Jürgen. 2006. The EU ETS Linking Directive Explained. In Jos Delbeke, ed. *The EU Greenhosue Gas Emissisons Trading Scheme*. Deventer: Claeys & Casteels, pp. 117–151.

Lin, Wei, Hongbo Chen, and Jia Liang. 2011. China Carbon Market. In *Progressing Towards Post-2012 Carbon Markets*. Roskilde: UNEP Risoe Centre, pp. 37–47.

Linacre, Nicolas, Alexandre Kossoy, and Philippe Ambrosi. 2011. *State and Trends of the Carbon Market 2011*. Washington, DC: World Bank.

Marcu, Andrei. 2011. The Durban Outcome. A Post-2012 Framework Approach for Green House Gas Markets. In *Progressing Towards Post-2012 Carbon Markets*. Roskilde: UNEP Risoe Centre, pp. 127–138.

Marcu, Andrei. 2012. *Expanding Carbon Markets Through New Market-Based Mechanisms*. Brussels: Centre for European Policy Studies.

Mehling, Michael, and Erik Haites. 2009. Mechanisms for Linking Emissions Trading Schemes. *Climate Policy* 9, 2: 169–184.

O'Sullivan, Robert, Charlotte Streck, Timothy Pearson, *et al*. 2010. *Engaging the Private Sector in the Potential Generation of Carbon Credits from REDD+: An Analysis of Issues*. Report to the UK Department for International Development (DFID), Climate Focus.

Skjærseth, J. B., and J. Wettestad. 2008. *EU Emissions Trading*. Aldershot: Ashgate.

Skjærseth, J. B., and J. Wettestad. 2010. Fixing the EU Emissions Trading System? Understanding the Post-2012 Changes. *Global Environmental Politics* 10, 4: 101–123.

Stavins, Robert. 2011. The National Context of U.S. State Policies for a Global Common Problem. In *Progressing Towards Post-2012 Carbon Markets*. Roskilde: UNEP Risoe Centre, pp. 49–57.

Sterk, Wolfgang. 2011. Sectoral Approaches as a Way Forward for the Carbon Market? In *Progressing Towards Post-2012 Carbon Markets*. Roskilde: UNEP Risoe Centre, pp. 113–125.

Chapter 22

The Influence of Energy Policy on Strategic Choices for Renewable Energy Investment

Rolf Wüstenhagen and Emanuela Menichetti[1]

Introduction

Renewable energy technologies and energy efficiency can help secure energy supplies at lower risk compared to conventional energy sources. This requires significant initial investment, though. The International Energy Agency estimates that in order to reach the 2 °C reduction scenario, worldwide annual investment in renewables alone should be $235 bn until 2020 (IEA 2012). In times of constrained government budgets, relying on public funding alone may not be sufficient to bring about the necessary initial investment to reap the benefits of renewables and energy efficiency. The question then becomes how can policy frameworks be designed to improve the flow of private capital into these sectors? In order to address this question, it is essential to gain a deeper understanding of how energy investors make strategic choices.

Over the past decade we have seen significant growth in clean energy investment, often supported by favorable policy frameworks. On the other hand, policy has not only created opportunities, but also posed risks for renewable energy investors. How do investors take decisions in the light of those risks and opportunities? How much of their decision process can best be explained by traditional economic models of full rationality, and what is the role of bounded rationality, path dependence, and other "behavioral" factors that influence return expectations and risk perception? And what can policy-makers learn from insights about investor decision-making to design even better policies?

While addressing this question could focus on a wide range of aspects of decision-making, we propose to pay particular attention to *strategic choices*. Strategic choices are characterized by one-off, new, ambiguous, and complex decision contexts; they require resource commitment (or the decision not to commit), and they are not easily reversible (Bansal 2005; Eisenhardt and Zbaracki 1992; Mintzberg *et al.* 1976). When it comes to energy investment, not all decisions made by financial, corporate, or retail investors are equally important. Whether a venture capitalist decides to set up yet another fund in biotechnology or endeavors to start a new fund in clean energy has more impact

The Handbook of Global Energy Policy, First Edition. Edited by Andreas Goldthau.
© 2013 John Wiley & Sons, Ltd. Published 2013 by John Wiley & Sons, Ltd.

than where he makes an incremental investment in an existing fund. An oil company's decision to enter or exit the solar industry has more far-reaching consequences than another company's decision to extend its manufacturing facilities by one more assembly line. A utility's decision whether to invest in a new coal-fired power plant or an offshore wind park determines output for decades to come. Where consumers choose to invest in a new house relative to their place of work determines their transport-related energy consumption (and the possibility to satisfy this demand with renewable energy) for several years. The strategic portfolio allocation of public research funding agencies among conventional and renewable energy technologies carries more weight than support for one particular research project or another, and – due to path dependences – such allocation decisions are not easily reversed ever after. Properly understanding the determinants of such far-reaching, strategic choices will be particularly helpful in creating effective frameworks for renewable energy investment.

Current Insights on the Policy–Investment Nexus

Status of Global Investment in Renewable Energy

Over the last few years, investment in renewable energy technologies has steadily increased in both developed and developing countries. According to the Intergovernmental Panel on Climate Change (IPCC 2011a), renewable energy accounted for 12.9% of global primary energy supply in 2008, with the largest share coming from biomass. While new renewable energy sources such as wind and solar energy only account for a small fraction of global energy supply, they have recently experienced significant growth, especially in countries with active renewable energy policies. Denmark has seen the share of wind power grow to nearly 25% of the country's electricity supply since the 1990s and wants to increase it further to 49.5% by 2020. Germany has increased the share of renewable energy from 3.1 to 20.0% of electricity supply and from 2.1 to 10.4% of heat supply between 1990 and 2011 (BMU 2012). China has seen strong domestic growth in the wind energy sector, with installed capacity nearly doubling from 25.8 to 44.7 GW in 2010 alone, overtaking the US as a global market leader for wind energy (GWEC 2011). Apart from energy policies, technological improvement and cost reductions have been a strong driver for growth. For instance, the project cost for onshore wind energy has decreased by around a factor of three between 1982 and 2002 (Wiser and Bolinger 2008), and the cost of solar photovoltaic (PV) modules has decreased by a factor of nine, with a 75% cost reduction in 2008–2011 alone (IEA 2012). Costs are expected to decrease further as a result of technology development, deployment, and economies of scale.

Investment in renewable energy was fairly limited until the early 2000s. According to Bloomberg (2011), total investment in clean energy amounted to $52 billion USD in 2004. Since then, investment in clean energy has recorded substantial growth, reaching $180 billion in 2008. In the light of the financial crisis, growth in renewable energy investment almost came to a standstill in 2009, but rebounded in 2010 with an annual growth rate of about 30%. While the global financial crisis seems to have had limited overall impact on the renewable energy investment community, it led to some significant structural shifts (UNEP 2010). While for example new investment in US windpower showed a sharp decline in 2009 and 2010, the Chinese market continued to exhibit high growth rates. Also, investment in PVs boomed in 2010, with a particular focus on new manufacturing capacity in China and new installations in Germany, which saw 7.4 GW of new capacity added in 2010 (BMU 2012), a 75% increase over 2009 levels.

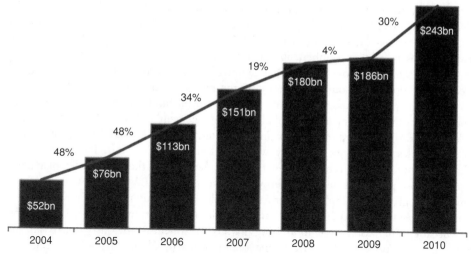

Figure 22.1 Global new investment in clean energy, 2004–2010.
Source: Bloomberg (2011).

While government used to be the most important source of funding a decade ago, private investments have now become the largest source of capital for renewable energy projects. This growth is the result of two factors: on the one hand, technology improvement has led to increased reliability and declining costs of many renewable energy options, and on the other hand, renewable energy policies have successfully created new market opportunities, which in turn spurred private sector investment.

Wind energy has been the engine of growth in renewable energy investments for more than a decade now. While the global financial crisis has slowed down that growth, wind power generation capacity continued to grow by over 30% in 2009 and by 24% in 2010 (GWEC 2011). In the European Union (EU), new wind generation capacity in 2009 reached over 10 GW, with an annual average market growth of 23% over the last 15 years (EWEA 2010). Wind installations represented 39% of total generating capacity added in Europe in 2009, more than any other power generation technology. Investment in EU wind farms in 2009 totaled €13 billion, of which €11.5 billion was in onshore and €1.5 billion in offshore wind energy projects. In 2010, another 9.9 GW of new wind power capacity were installed in Europe, indicating the first year in which new installations were slightly below the previous year (EWEA 2011).

Although smaller than wind in absolute values, investments in PVs has been experiencing dynamic growth. PV is the fastest growing renewable energy technology, with a 14-times capacity increase in the last 10 years, and an annual average growth of 30% over the same period. By 2050, PVs as well as concentrated solar power (CSP) are each expected to be in the same order of magnitude as wind energy in terms of contribution to electricity generation and greenhouse gas reductions (Frankl and Philibert 2009).

The Future of Renewable Energy Investment in a Carbon-Constrained World

In a future carbon-constrained world, investment in renewable energy is supposed to show further growth. The IPCC's recent Special Report on Renewable Energy Sources estimates that global investments in renewable power generation technologies will range from $1360 to $5100 billion in this decade, and from $1490 to $7180 billion in the decade 2021 to 2030, with the higher values being consistent with a stabilization of CO_2 concentration at 450 ppm (IPCC 2011a). While five to seven trillion dollars in two decades is a large number, annual investment is equivalent to less than 1% of global GDP, even in those climate change mitigation scenarios that assume renewables become the dominant source of energy by the middle of this century, and equivalent to less than 5% of the assets currently held by pension funds, mutual funds, insurance funds, and other players in the global fund management industry (McKinsey 2011). Also, the IEA (2012) estimates that approximately 80% of the additional investment in renewables and energy efficiency will be offset by fuel savings. It may be noted that these numbers for renewable energy investment are in the same order of magnitude as those published on total energy investment by the IEA (2003), who estimated investment in energy-supply infrastructure worldwide (including conventional energy sources) over the period 2001–2030 at $16 trillion (which translates to $5300 billion per decade). Therefore, while mobilizing private investment is obviously not trivial, the true challenge policy-makers are facing is not primarily about "paying a green premium" but one of influencing strategic choices of those investors who will deploy capital anyway, and are selecting between opportunities in conventional and renewable energy projects.

Scaling up sustainable energy investments requires important changes in the social and institutional context (Krewitt et al. 2007). As pointed out by Grubb, "technological advances, and in some cases breakthroughs, are certainly needed: but the revolution required is one of attitudes" (Grubb 1990: 716). This change in attitudes concerns a multitude of actors, from policy-makers to citizens, industry and market operators, and investors. Managing social acceptance of renewable energy innovation, and especially acceptance by the financial community, may therefore be an important success factor for future energy and climate policies (Wüstenhagen et al. 2007).

Linking Renewable Energy Investment and Energy Policy

In measuring the effectiveness of renewable energy policies, most of the literature has used installed capacity as the dependent variable rather than explicitly addressing investment. For example, a large number of country-level case studies have been carried out across different geographies, renewable energy technologies, and policy instruments (Breukers and Wolsink 2007; Jacobsson and Lauber 2006; Lipp 2007; Toke et al. 2008; Wüstenhagen and Bilharz 2006), and asked whether some policy instruments (for example, price-driven vs. quantity-driven) work better than others in bringing about new capacity. Much of this literature suggests that the answer may be: "it depends" (IPCC 2011b). For example, economic modelers would argue that trading schemes such as Renewable Portfolio Standards (RPS) may be an effective way of minimizing cost (Palmer and Burtraw 2005), while a more interdisciplinary stream of energy policy research has pointed out that their implementation in real life suffers from limitations like market power and transaction cost (Bergek and Jacobsson 2010; Jacobsson et al. 2009; Jensen and Skytte 2002; Menanteau et al. 2003; Verbruggen 2004). On the other hand, feed-in tariffs have brought about significant new capacity in Germany, but they have also been criticized for their cost implications, especially in the case of PV (Frondel et al. 2008), and their implementation in other countries shows a variety of outcomes (Campoccia et al. 2009; Lüthi 2010; Rowlands 2005). Apparently, the devil is in the details of policy implementation (Dinica 2006; Menanteau et al. 2003; Ringel 2006).

What is different then about using investment rather than capacity as the dependent variable of policy analysis? Can any additional insights be gained by moving from tracking megawatts to dollars (Usher 2008)? A possible answer to this question relates to a recent stream in the energy policy literature that highlights the importance of risk in policy design. Variations in policy outcomes, these authors suggest, are strongly influenced by variations in the level of risk that different policies imply for investors. Therefore, while two policy instruments, such as feed-in tariffs and green certificate systems, might be expected to lead to similar outcomes in economic models without proper consideration of investment risk, one may actually outperform the other if risk is accounted for. This has been offered as an explanation for why feed-in tariffs have resulted in higher levels of new renewable energy capacity than green certificate systems (Mitchell *et al.* 2006). Lower risk translates into lower financing cost for renewable energy projects by affecting investors' cost of capital (De Jager and Rathmann 2008; Langniss 1999; Wiser and Pickle 1998). Policies that effectively reduce (perceived) risk for investors are therefore more likely to result in large-scale deployment of renewable energy. In this chapter, we add to previous calls for including an investor perspective in assessing the effectiveness of energy policies (Dinica 2006; Gross *et al.* 2010; Hamilton 2009; IEA 2003). The empirical evidence about how policies and their risk are actually perceived by investors and project developers has been limited so far (e.g., Bürer and Wüstenhagen 2008, 2009; Lüthi and Wüstenhagen 2012).

An interesting feature of looking at investment rather than installed capacity is that it works almost like a time machine, or a crystal ball (Usher 2008). By measuring today's investment in project finance, researchers can gain insights into tomorrow's installed capacities. Today's investment in manufacturing capacity for renewable energy equipment (like wind turbines or solar cells) gives an early indication for which installations are to be expected in 2–5 years. And today's investment in venture capital is a precursor of technologies that will be deployed 5–10 years down the road. Finally, moving from revealed to stated preferences can stretch this effect a little further on the time axis. Especially in the renewable energy sector, where markets are facing dynamic growth in recent years only, an analysis of investment (decisions) rather than – or in addition to – the traditional ex-post analysis of installed capacities, can therefore help to alleviate the inherent problem of data availability and allow policy-makers to take informed decisions with the benefit of foresight.

Conceptualizing Strategic Choices for Renewable Energy Investment

Renewable Energy Investment as a Function of Risk, Return, and Policy

When trying to understand what determines current levels of renewable energy investment, a basic model is to represent investments as a function of risk, return, and policy (cf. Figure 22.2). Risk and return have long been established as fundamental determinants of investment in finance theory. Investors, so the argument goes, rationally weigh the levels of risk and return of possible investment opportunities, and will pick those opportunities that provide the best return for a given level of risk. Another way of putting this is that investors compare investment opportunities by looking at their risk-adjusted returns. In energy, investment opportunities in renewables tend to be at a disadvantage compared to conventional energy because of environmental externalities. Therefore, there is a case for energy policy to correct those externalities. The effect of such policies on investment, in the basic model, is to make the risk–return equation more favorable for renewable

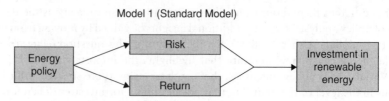

Figure 22.2   A simple model of renewable energy policy and investment.

energy investors, for example by increasing the returns for renewable energy investment (e.g., through feed-in tariffs) or by reducing the risk (e.g., through loan guarantees). The relative influence of policy versus "pure" risk–return considerations on renewable energy investment is subject to some debate in the literature. While some observers see policy as the essential driver of RE investment (IPCC 2011b), others, especially in the investment community, emphasize the role of private capital that is seeking opportunities with or without policies (e.g., some of the venture capital investors interviewed by Wüstenhagen and Teppo 2006). Despite the nuances, there seems to be agreement that policy is – at least for the time being – one of the important drivers, but to reach the order of magnitudes of investments outlined above, markets cannot be driven by policy alone to infinity. Therefore, further investigating the subtleties of the policy–investment nexus is an endeavor worth undertaking.

The following section will drill deeper into some of the drivers of renewable energy investment, including but not limited to policy. This will ultimately result in a more sophisticated conceptual model for investigating strategic choices for renewable energy investment (cf. Figure 22.4), which serves as a starting point to identify promising avenues for further research.

Portfolio Aspects

A first extension of the basic risk–return model of investment comes from portfolio theory. Markowitz (1952) was the first to highlight the concept of portfolio diversification, noting that risk can be reduced by combining different assets. Therefore, there is a systematic difference between the risk–return ratio of one single investment and that of a portfolio of investments. This is important for renewable energy investment on two levels.

First, adding renewable power generation assets to a portfolio of conventional power generation assets may provide a diversification effect. Traditional engineering/economic models of valuing power generation assets, as they are employed by most electric utilities, do not account for this effect, hence potentially undervaluing renewables (Awerbuch 2000a, 2000b, 2004; Bhattacharya and Kojima 2012). For financial investors, on the other hand, the idea of portfolio diversification is deeply embedded in the way they assess opportunities. This might explain why some financial investors, such as insurance companies and pension funds, have become more proactive than utilities when it comes to investing in renewable energy assets.

Second, there is diversification among different renewables: the risk–return ratio of a wind park combined with a set of solar power generation facilities is likely to be more favorable than one of the two alone. Laurikka (2008) calls this the diversification of plant-specific risk. Failure to properly value diversification effects may result in underinvestment in renewables.

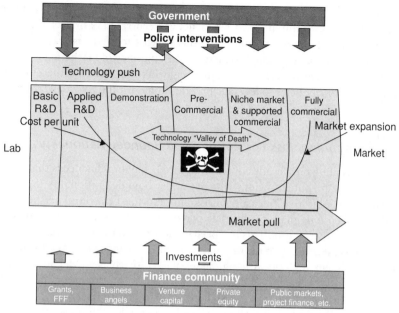

Figure 22.3 Segmentation of financial investors along the innovation chain.
Source: Bürer and Wüstenhagen (2009); Grubb (2004).

What are the implications of portfolio diversification effects for policy that intends to stimulate renewable energy investment? First, failure to value portfolio effects may constitute one of the cognitive barriers to renewable energy investment. Second, since financial investors tend to be more experienced in valuing portfolio effects than incumbent utilities, policy-makers aiming at increasing levels of RE investment may be well advised to target their policies at a wide range of investors, and not just incumbent players in electricity markets. Third, discussions about the cost-effectiveness of renewable energy policies that focus on "additional cost" based on a comparison of additional cost per kilowatt hour may be missing an important part of the equation, and should reconsider their key metrics to capture the value of portfolio diversification (and avoided fuel price risk).

Investor Heterogeneity and the Segmentation of Policies

Another aspect where Figure 22.2 depicted an oversimplified model of reality was that investment decisions about renewable energy are not taken by one type of financial actor alone, but instead there is heterogeneity among investor types. This is true for investors along the different stages of the innovation chain (see Figure 22.3): for example, venture capitalists investing in early-stage technology firms require different policies (Bürer and Wüstenhagen 2009) than project financiers who deploy mature technology (Lüthi and Wüstenhagen 2012). When it comes to large-scale deployment of renewables, the number and diversity of investors that need to be targeted increases further. For example, corporate (e.g., electric utilities), financial (e.g., insurance companies, pension funds), and retail investors (e.g., homeowners) all invest in PVs, but they are likely to differ in their policy preferences. While empirical evidence is rare so far, it might be worth systematically investigating such differences, which might refer to required rates of return, preferences for initial down-payments vs. recurring tax breaks, etc.

Borrowing from marketing terminology, there seems to be a case for segmentation. Just as consumers can be segmented to increase the efficiency of marketing efforts, thinking about ways to identify relevant investor segments may increase the efficiency and effectiveness of public policies to leverage private capital for the growth of the renewable energy market.

The Role of Cognition, Risk Perceptions, and Bounded Rationality

An important insight from decades of research into behavioral finance is that investment decisions are made by human beings who act under bounded rationality (Simon 1955). Rather than some "objective" measure of risk and return, a behavioral perspective would suggest that *perceptions* matter, and that perceptions of risk and expected return are influenced by cognitive factors. This has been demonstrated for a wide range of decisions under uncertainty, starting with the seminal work of Kahneman and Tversky in the 1970s (Tversky and Kahneman 1974). Among the set of cognitive biases (McFadden 2001) that have been identified by behavioral finance scholars are anchoring-and-adjustment, availability, representativeness, and status quo biases (Barnes 1984; Katz 1992; Pitz and Sachs 1984; Samuelson and Zeckhauser 1988). These biases can lead to conservatism in adjusting to new information (Kahneman 2003; Tversky and Kahneman 1974) and to decisions in which losses are weighted differently than gains (Kahneman and Tversky 1979). Pursuing this line of reasoning further, the behavioral economics and finance literature has investigated how real financial markets deviate from classical economic models, including investor behavior in stock markets (Chan and Lakonishok 2004; Jordan and Kaas 2002; Lakonishok *et al.* 1994), currency speculation (Bikhchandani *et al.* 1992; Bikchandani and Sharma 2001; Froot *et al.* 1992), managerial decision-making (McNamara and Bromiley 1999), and the energy sector (Masini and Menichetti 2012).

What are the implications for renewable energy investment and policy? If one accepts that investors act under bounded rationality, what ultimately matters is the impact of policies on perceived levels of risk and expected returns. While actual risk and return certainly still matter, not all investors will have all the information available to comprehensively judge the actual levels of risk and return before taking an investment decision in renewable energy. Therefore, in addition to trying to lower actual risk or increasing actual return, it is important for policy-makers to manage expectations. Risk–return perceptions can, for example, be negatively influenced by frequent policy changes or unclear targets. The value of long-term policy stability, which may not be entirely obvious from classical economic models assuming full rationality, becomes evident from a bounded rationality perspective. Similarly, green public procurement strategies may have positive effects on private sector investment in renewables by reducing perceived risk and adding a stamp of credibility to renewable energy technologies. Policy-makers can also try to address cognitive barriers directly, for example by investing in education and training for financial decision-makers, or by collaborating with opinion leaders in the financial community.

Path Dependence in Energy Investments

One important implication of a bounded rationality perspective is the existence of path dependence (Goldstone 1998; North 1990), i.e., that past events have an impact on present choices. For example, Wüstenhagen and Teppo (2006) have identified evidence for path dependence in the venture capital market, which slowed down the flow of venture capital investments into the newly emerging renewable energy sector. Lovio *et al.* (2011) discuss effects of path dependence on Finland's attempts to diversify its economy away from fossil fuels. Path dependence can also lead to lock-in, and it has been suggested that the world is currently locked into a high-carbon economy (Unruh 2000), which is difficult to transcend (Unruh 2002). Path dependence may not only occur on the level of industries or innovation systems, but also on the firm level. For example, past investments in fossil fuels may influence the risk–return perception of decision-makers in oil companies, leading them to see more opportunities on their previous path than in the less familiar territory of renewable energies (see Pinkse and van den Buuse 2012).

What are the implications of path dependence on policy-makers' efforts to stimulate investment in renewable energy? A first recommendation would be for policy-makers to be sensitive for history, in that new technologies like renewables do not enter the market in a vacuum, but instead there are vested interests of incumbents and risk–return perceptions based on past experience that determine today's choices. Therefore, simply providing a "level playing field" for renewables and expecting the invisible hand of the market to result in an optimal allocation of capital may not be enough, as boundedly rational actors, in case of doubt, will still stick to their past patterns of investment. Systems tend not to switch from one path to another by themselves, but need some initial impulse to get a new path started. As a consequence, while some authors view generous feed-in tariffs as wasteful overfunding for renewables, a path-dependence perspective may suggest that this is the kind of initial impulse that is needed to overcome inertia in the financial system. Just like building a new road needs some initial investment, path creation in the energy system is not for free.

Conclusion: A More Nuanced Picture of Renewable Energy Policy and Investment

Taking the considerations of the previous sections together, there seems to be a case for a more nuanced picture of the antecedents of renewable energy investment. As a starting point, risk and return are important drivers of investment decisions. Therefore, policy-makers aiming at an increased share of renewable energy should do what they can to reduce risk and provide adequate returns. Creating a level playing field, and helping the market to value positive externalities of renewables, is also important. However, the story does not end here. We live in a world of bounded rationality, and therefore perceptions matter, and policy needs to take such perceptions into account. Surveying investor attitudes and preferences can help to identify which risks are perceived as particularly relevant, and therefore help policy-makers to prioritize their efforts. On the other hand, especially when it comes to long-term decisions, investor preferences should obviously be regarded as complementing, rather than being a substitute for, strategic choices made by policy-makers themselves. Portfolio effects and the idea of diversification add another layer to the understanding of investment choices, because investment in more than just one asset has different implications. Finally, not all investors are the same, and similar investment opportunities are valued differently by different investors. These differences are driven by rational aspects such as the effects of portfolio diversification, but also have a boundedly rational component, for example in the form of path dependence and prior investment choices. To conclude, we suggest that an effective policy mix is based on a

Model 2 (Extended Model)

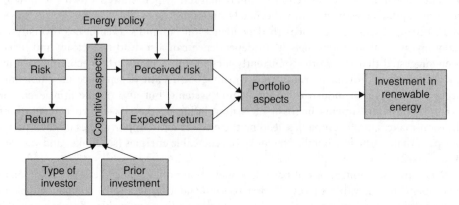

Figure 22.4   A more nuanced model of renewable energy policy and investment.

thorough understanding of investor realities, including cognitive factors, and includes segmentation.

Further Research

The topic of strategic choices for renewable energy investment is an emerging area of research with high policy relevance. We outline a couple of possible avenues for further research below.

First, there is a recurring theme of understanding appropriate measures for risk, return, and the resulting cost of capital for renewable energy investments across different geographies and asset classes. As shown by, for example, Sadorsky (2012) and Donovan and Nuñez (2012), renewable energy investments can be either more or less risky than other asset classes in different countries. Moreover, those studies demonstrate that investment risk in the renewable energy sector can indeed be quantified, but measures are highly influenced by assumptions, both input assumptions to the asset pricing model and, systematically, the assumptions that support the economic logic of the model itself (e.g., the capital asset pricing model). One conclusion then is that policy-makers should become more familiar with the tools that investors use to set the cost of capital for renewable energy investments, but the publications referenced above also identify the limitations of using a purely rational framework to understand strategic investment decision-making in this sector. On a methodological level, the findings of these early studies are limited by a relatively short history of renewable energy investments. Future research can benefit from improved data availability due to the maturing nature of the renewable energy industry, and explore the suitability of different asset pricing models in explaining firm and industry betas.

Second, apart from understanding risk and return on the firm or project level, several authors highlight the importance of portfolio effects (Awerbuch 2004; Bhattacharya and Kojima 2012; Fuss et al. 2012). Future research should further investigate the usefulness of portfolio theory to assess the value of renewables in energy systems, for example as a means to reduce fuel price risk. A topical issue could also be to investigate how the decision of some governments to phase out nuclear energy after the Fukushima accident in 2011 influences the composition of the optimal power generation portfolio, and what the implications are for investment in different renewable energy technologies.

A third area for future research is to further investigate the idea of investor segmentation as a starting point for developing "tailor-made" policies. Given the heterogeneity of investors along the innovation chain (Figure 22.3) as well as with regard to other dimensions like firm size (incumbents vs. entrepreneurial new entrants) and geography (industrialized vs. developing countries), more empirical research is needed to explore which policies work well across the entire spectrum, and which ones may be effective for one segment of investors but not for another, both with regard to investor attitudes as well as actual investment behavior.

Moving from the world of "rational" finance into behavioral approaches, the theme of real-world decision processes of investors under bounded rationality seems worth further exploration. Future research in this area could take one of two philosophical approaches: it could either try to detect cognitive biases and follow the tradition of behavioral economists to "prove irrationality," as for example in the case of status quo bias. Alternatively, it could follow the approach of scholars in the heuristics and intuitive decision-making camp, which would argue that rationality is a relative term, and if actual investor behavior deviates from the predictions of textbook economics, this may not necessarily be an indication of a lack of rationality, but rather a mis-specification of such rationality on behalf of the researcher. An example would be Goldstein and Gigerenzer's (2009) work demonstrating that simple rules of thumb ("fast and frugal heuristics") often work just as well as more complex decision procedures, especially in the context of limited information and limited processing time – arguably two conditions that are relevant in the context of a newly emerging industry such as renewable energy.

Whatever the philosophical approach pursued, a bounded rationality perspective could shed light on the role of cognitive aspects shaping risk perceptions and return expectations. What are the cognitive processes that lead some investors to perceive investment opportunities in renewable energy as too risky, whereas others evaluate the same opportunity as attractive, as in the case of oil industry incumbents vs. other new entrants to the solar industry? What are the determinants of path dependence in energy investments? Can for example prior investment, affective influences, group dynamics, organizational culture or network effects explain observed patterns of investment better than pure risk–return considerations? Are risk-assessment tools and return metrics that have been used to judge conventional energy projects appropriately capturing the costs and benefits of renewable energies? And once this has been explored, how can policy build on this knowledge to devise more effective regulatory frameworks in an evolutionary and boundedly rational world? What is, for example, the relative importance of "symbolic" policies compared to the actual monetary value of incentives? Interdisciplinary collaboration between scholars in political science, finance, marketing, and economic psychology may be useful to come up with new insights here, although there are obviously limits as to how far investor preferences can form an adequate basis for long-term policy-making.

Given this wide array of possible areas for further research on strategic choices for renewable energy investment, we hope that readers will find this chapter an inspiring starting point to think about the multiple interdependencies of energy policy, finance, and behavioral sciences. We are confident that future research in this area will provide valuable further insights in how to reach the substantial investment levels needed to successfully manage the transition into a renewable energy future.

Note

1. This chapter is a revised version of a paper by the same authors which appeared as the introduction to an *Energy Policy* special issue on "Strategic Choices for Renewable Energy Investment" (Wüstenhagen and Menichetti 2012).

References

Awerbuch, S. 2000a. Getting it Right: The Real Cost Impacts of a Renewables Portfolio Standard. *Public Utilities Fortnightly*, February 15.

Awerbuch, S. 2000b. Investing in Photovoltaics: Risk, Accounting, and the Value of New Technology. *Energy Policy* 28: 1023–1035.

Awerbuch, S. 2004. *Portfolio-Based Electricity Generation Planning: Implications for Renewables and Energy Security*. London and Paris: REEEP/UNEP.

Bansal, P. 2005. Responsible Strategic Decision Making. *Proceedings of the International Association for Business and Society* 16: 57–62.

Barnes, J. H. 1984. Cognitive Biases and Their Impact on Strategic Planning. *Strategic Management Journal* 5: 129–137.

Bergek, A., and S. Jacobsson. 2010. Are Tradable Green Certificates a Cost-Efficient Policy Driving Technical Change or a Rent-Generating Machine? Lessons from Sweden 2003–2008. *Energy Policy* 38: 1255–1271.

Bhattacharya, A., and S. Kojima. 2012. Power Sector Investment Risk and Renewable Energy: A Japanese Case Study Using Portfolio Risk Optimization Method. *Energy Policy* 40: 69–80.

Bikhchandani, S., and S. Sharma. 2001. Herd Behavior in Financial Markets. *IMF Staff Papers* 47, 3: 279–310.

Bikhchandani, S., D. Hirshleifer, and I. Welch. 1992. A Theory of Fads, Fashions, Customs and Cultural Change as Informational Cascades. *Journal of Political Economy* 100: 992–1026.

Bloomberg. 2011. Bloomberg New Energy Finance Summit: Results Book 2011. London. www.bnefsummit.com, accessed December 3, 2012.

BMU. 2012. Erneuerbare Energien 2011. Berlin: Bundes Ministerium für Umwelt.

Breukers, S., and M. Wolsink. 2007. Wind Power Implementation in Changing Institutional Landscapes: An International Comparison. *Energy Policy* 35: 2737–2750.

Bürer, M. J., and R. Wüstenhagen. 2008. Cleantech Venture Investors and Energy Policy Risk: An Exploratory Analysis of Regulatory Risk Management Strategies. In R. Wüstenhagen, J. Hamschmidt, S. Sharma, and M. Starik, *Sustainable Innovation and Entrepreneurship*. Cheltenham, UK: Edward Elgar Publishing, pp. 290–309.

Bürer, M. J., and R. Wüstenhagen. 2009. Which Renewable Energy Policy Is a Venture Capitalist's Best Friend? Empirical Evidence from a Survey of International Cleantech Investors. *Energy Policy* 37: 4997–5006.

Campoccia, A., L. Dusonchet, E. Telaretti, and G. Zizzo. 2009. Comparative Analysis of Different Supporting Measures for the Production of Electrical Energy by Solar PV and Wind Systems: Four Representative European Cases. *Solar Energy* 83: 287–297.

Chan, L. K. C., and J. Lakonishok. 2004. Value and Growth Investing: Review and Update. *Financial Analysts Journal* 60, 1: 71–86.

De Jager, D., and M. Rathmann. 2008. Policy Instrument Design to Reduce Financing Costs in Renewable Energy Technology Projects. Utrecht: Ecofys.

Dinica, V. 2006. Support Systems for the Diffusion of Renewable Energy Technologies – An Investor Perspective. *Energy Policy* 34: 461–480.

Donovan, C., and L. Nuñez. 2012. Figuring What's Fair: The Cost of Equity Capital for Renewable Energy in Emerging Markets. *Energy Policy* 40: 49–58.

Eisenhardt, K. M., and M. J. Zbaracki. 1992. Strategic Decision Making. *Strategic Management Journal* 13: 17–37.

EWEA. 2010. *Wind in Power. 2009 European Statistics*. Brussels: European Wind Energy Association.

EWEA. 2011. *Wind in Power. 2010 European Statistics*. Brussels: European Wind Energy Association.

Frankl, P., and C. Philibert. 2009. Critical Role of Renewable Energy to Climate Change Mitigation. Presented at COP 15 IEA Day Side Event, Copenhagen, December 16.

Frondel, M., N. Ritter, and C. M. Schmidt. 2008. Germany's Solar Cell Promotion: Dark Clouds on the Horizon. *Energy Policy* 36: 4198–4204.

Froot, K., D. Scharfstein, and J. Stein. 1992. Herd on the Street: Informational Efficiencies in a Market with Short-Term Speculation. *Journal of Finance* 47: 1461–1484.

Fuss, S., J. Szolgayová, N. Khabarov, and N. Obersteiner. 2012. Renewables and Climate Change Mitigation: Irreversible Energy Investment Under Uncertainty and Portfolio Effects. *Energy Policy* 40: 59–68.

Goldstein, D. G., and G. Gigerenzer. 2009. Fast and Frugal Forecasting. *International Journal of Forecasting* 25: 760–772.

Goldstone, J. A. 1998. Initial Conditions, General Laws, Path Dependence, and Explanation in Historical Sociology. *American Journal of Sociology* 104, 3: 829–845.

Gross, R., W. Blyth, and P. Heptonstall. 2010. Risks, Revenues and Investment in Electricity Generation: Why Policy Needs to Look Beyond Costs. *Energy Economics* 32: 796–804.

Grubb, M. J. 1990. The Cinderella Options. A Study of Modernized Renewable Energy Technologies. Part 2 – Political and Policy Analysis. *Energy Policy* 18, 8: 711–725.

Grubb, M. J. 2004. Technology Innovation and Climate Policy: An Overview of Issues and Options. *Keio Economic Studies* 41, 2: 103–132.

GWEC. 2011. *Global Wind Report. Annual Market Update 2010*. Brussels: Global Wind Energy Council.

Hamilton, K. 2009. *Unlocking Finance for Clean Energy: The Need for "Investment Grade" Policy*. Energy, Environment and Development Programme Paper no. 09/04. London: Chatham House, Renewable Energy Finance Project.

IEA. 2003. *World Energy Investment Outlook*. Paris: International Energy Agency.

IEA. 2012. *Tracking Clean Energy Progress*. Energy Technology Perspectives 2012 excerpt as IEA input to the Clean Energy Ministerial. Paris: International Energy Agency.

IPCC. 2011a. Summary for Policymakers. In *IPCC Special Report on Renewable Energy Sources and Climate Change Mitigation*. Cambridge: Cambridge University Press.

IPCC. 2011b. Policy, Financing and Implementation. In *IPCC Special Report on Renewable Energy Sources and Climate Change Mitigation*. Cambridge: Cambridge University Press.

Jacobsson, S., and V. Lauber. 2006. The Politics and Policy of Energy System Transformation – Explaining the German Diffusion of Renewable Energy Technology. *Energy Policy* 34: 256–276.

Jacobsson, S., A. Bergek, D. Finon, *et al.* 2009. EU Renewable Energy Support Policy: Faith or Facts? *Energy Policy* 37: 2143–2146.

Jensen, S. G., and K. Skytte. 2002. Interactions Between the Power and Green Certificate Markets. *Energy Policy* 30: 425–435.

Jordan, J., and K. P. Kaas. 2002. Advertising in the Mutual Fund Business: The Role of Judgmental Heuristics in Private Investors' Evaluation of Risk and Return. *Journal of Financial Services Marketing* 7, 2: 129–140.

Kahneman, D. 2003. Maps of Bounded Rationality: Psychology for Behavioral Economics. *American Economic Review* 93, 5: 1449–1475.

Kahneman, D., and A. Tversky. 1979. Prospect Theory: An Analysis of Decisions Under Risk. *Econometrica* 47, 2: 263–291.

Katz, J. A. 1992. A Psychosocial Cognitive Model of Employment Status Choice. *Entrepreneurship Theory and Practice* 17, 1: 29–37.

Krewitt, W., S. Simon, W. Graus, *et al.* 2007. The 2 °C Scenario – A Sustainable World Energy Perspective. *Energy Policy* 35: 4969–4980.

Lakonishok, J., A. Shleifer, and R. W. Vishny. 1994. Contrarian Investment, Extrapolation, and Risk. *Journal of Finance* 49, 5: 1541–1578.

Langniss, O., ed. 1999. *Financing Renewable Energy Systems*. Stuttgart: Deutsche Forschungsanstalt für Luft- und Raumfahrt.

Laurikka, H. 2008. A Case Study on Risk and Return Implications of Emissions Trading in Power Generation Investments. In R. Antes *et al.*, eds. *Emissions Trading*. Heidelberg: Springer, pp. 133–147.

Lipp, J. 2007. Lessons for Effective Renewable Electricity Policy from Denmark, Germany and the United Kingdom. *Energy Policy* 35: 5481–5495.

Lovio, R., P. Mickwitz, and E. Heiskanen. 2011. Path Dependence, Path Creation and Creative Destruction in the Evolution of Energy Systems. In R. Wüstenhagen and R. Wuebker, eds. *Handbook of Research on Energy Entrepreneurship*. Cheltenham, UK: Edward Elgar Publishing, pp. 274–304.

Lüthi, S. 2010. Effective Deployment of Photovoltaics in the Mediterranean Countries: Balancing Policy Risk and Return. *Solar Energy* 84: 1059–1071.

Lüthi, S., and R. Wüstenhagen. 2012. The Price of Policy Risk – Empirical Insights from Choice Experiments with European Photovoltaic Project Developers. *Energy Economics* 34: 1001–1011.

Markowitz, H. M. 1952. Portfolio Selection. *Journal of Finance* 7, 1: 77–91.

Masini, A., and E. Menichetti. 2012. The Impact of Behavioural Factors in the Renewable Energy Investment Decision Making Process: Conceptual Framework and Empirical Findings. *Energy Policy* 40: 28–38.

McFadden, D. 2001. Economic Choices. *American Economic Review* 91, 3: 351–378.

McKinsey. 2011. *Mapping Global Capital Markets 2011*. McKinsey Global Institute.

McNamara, G., and P. Bromiley. 1999. Risk and Return in Organization Decision Making. *Academy of Management Journal* 42, 3: 330–339.

Menanteau, P., D. Finon, and M.-L. Lamy. 2003. Prices Versus Quantities: Choosing Policies For Promoting the Development of Renewable Energy. *Energy Policy* 31: 799–812.

Mintzberg, H., D. Raisinghani, and A. Theoret. 1976. The Structure of "Unstructured" Decision Processes. *Administrative Science Quarterly* 21: 246–275.

Mitchell, C., D. Bauknecht, and P. M. Connor. 2006. Effectiveness Through Risk Reduction: A Comparison of the Renewable Obligation in England and Wales and the Feed-in System in Germany. *Energy Policy* 34: 297–305.

North, D. C. 1990. *Institutions, Institutional Change and Economic Performance*. Cambridge: Cambridge University Press.

Palmer, K., and D. Burtraw. 2005. Cost-Effectiveness of Renewable Electricity Policies. *Energy Economics* 27: 873–894.

Pinkse, J., and D. van den Buuse. 2012. The Development and Commercialization of Solar PV Technology in the Oil Industry. *Energy Policy* 40: 11–20.

Pitz, G. F., and N. J. Sachs. 1984. Judgment and Decision: Theory and Application. *Annual Review of Psychology* 35: 139–163.

Ringel, M. 2006. Fostering the Use of Renewable Energies in the European Union: The Race Between Feed-in Tariffs and Green Certificates. *Renewable Energy* 31: 1–17.

Rowlands, I. H. 2005. Envisaging Feed-in Tariffs for Solar Photovoltaic Electricity: European Lessons for Canada. *Renewable and Sustainable Energy Reviews* 9: 51–68.

Sadorsky, P. 2012. Modeling Renewable Energy Company Risk. *Energy Policy* 40: 39–48.

Samuelson, W., and R. Zeckhauser. 1988. Status Quo Bias in Decision Making. *Journal of Risk and Uncertainty* 1, 1: 7–59.

Simon, H. A. 1955. A Behavioral Model of Rational Choice. *Quarterly Journal of Economics* 69, 1: 99–118.

Toke, D., S. Breukers, and M. Wolsink. 2008. Wind Power Deployment Outcomes: How Can We Account for the Differences? *Renewable and Sustainable Energy Reviews* 12: 1129–1147.

Tversky, A., and D. Kahneman. 1974. Judgment Under Uncertainty: Heuristics and Biases. *Science* 185 (4157): 1124–1131.

UNEP. 2010. *Global Trends in Sustainable Energy Investment 2010.* Paris: United Nations Environment Programme.

Unruh, G. 2000. Understanding Carbon Lock-in. *Energy Policy* 28, 12: 817–830.

Unruh, G. 2002. Escaping Carbon Lock-in. *Energy Policy* 30, 4: 317–325.

Usher, E. 2008. Global Investment in the Renewable Energy Sector. In O. Hohmeyer and T. Trittin, eds. *IPCC Scoping Meeting on Renewable Energy Sources.* Geneva: IPCC, pp. 147–154.

Verbruggen, A. 2004. Tradable Green Certificates in Flanders (Belgium). *Energy Policy* 32: 165–176.

Wiser, R., and M. Bolinger. 2008. *Annual Report on U.S. Wind Power Installation, Cost, and Performance Trends: 2007.* Washington, DC: US Department of Energy.

Wiser, R., and S. Pickle. 1998. Financing Investments in Renewable Energy: The Impacts of Policy Design. *Renewable and Sustainable Energy Reviews* 2: 361–386.

Wüstenhagen, R., and M. Bilharz. 2006. Green Energy Market Development in Germany: Effective Public Policy and Emerging Customer Demand. *Energy Policy* 34: 1681–1696.

Wüstenhagen, R., and E. Menichetti. 2012. Strategic Choices for Renewable Energy Investment: Conceptual Framework and Opportunities for Further Research. *Energy Policy* 40: 1–10.

Wüstenhagen, R. and T. Teppo. 2006. Do Venture Capitalists Really Invest in Good Industries? Risk-Return Perceptions and Path Dependence in the Emerging European Energy VC Market. *International Journal of Technology Management* 34, 1/2: 63–87.

Wüstenhagen, R., M. Wolsink, and M. J. Bürer. 2007. Social Acceptance of Renewable Energy Innovation: An Introduction to the Concept. *Energy Policy* 35: 2683–2691.

Part VI Regional Perspectives on Global Energy

The Handbook of Global Energy Policy, First Edition. Edited by Andreas Goldthau.
© 2013 John Wiley & Sons, Ltd. Published 2013 by John Wiley & Sons, Ltd.

Chapter 23

Global Energy Policy: A View from China

Alvin Lin, Fuqiang Yang, and Jason Portner[1]

Introduction

Over the past decade, China's expansive energy development and utilization has had an increasingly large impact on the global energy market. In 2011, China's energy production reached 3.18 billion tons of coal equivalent (tce) and energy consumption reached 3.48 billion tce (National Bureau of Statistics 2012a), representing 21.3% of global energy consumption (BP 2012). By 2009, China had become the world's largest producer and consumer of energy (IEA 2010). In addition to fueling its rapid economic growth, China's growing energy consumption has made it increasingly dependent on foreign countries for its energy. By 2011, 55.2% of China's oil, 21.6% of its natural gas, and 5.3% of its coal supply came from foreign sources (China Energy Research Society 2011). Although China's growing energy imports have led to booming international energy markets, they have also made securing a reliable energy supply one of China's key concerns.

China's reliance on coal to fuel its industry and provide electricity and heating for its population has led to severe environmental pollution, public health problems, and logistical challenges in transporting coal to demand centers. In 2011, China's coal consumption increased more than any time in history, reaching 3.6 billion tons and representing roughly 70% of its total energy consumption (National Bureau of Statistics 2012b). Realizing the challenges posed by its excessive coal consumption, China's government has made sustainable development a key objective of its energy policy. During the 11th Five Year Plan (2006–2010), China established a set of strengthened energy efficiency and renewable energy policies to increase its use of clean energy, including for the first time establishing a national target to reduce its energy intensity (the amount of energy consumed per unit GDP) by 20%. These policies are continuing to be strengthened in the current 12th Five Year Plan (2011–2015).

China's growing energy consumption, fueled by its industrialization and urbanization, also has severe implications for global climate change. As early as 2006, China became the largest carbon emitting country in the world (PBL 2007). At the current rate of emissions

The Handbook of Global Energy Policy, First Edition. Edited by Andreas Goldthau.
© 2013 John Wiley & Sons, Ltd. Published 2013 by John Wiley & Sons, Ltd.

growth, by 2020 it could emit as much carbon as the US and the European Union (EU) combined. China's government has realized the impact that climate change could have on its agricultural production, water resources, and coastal cities, and is actively seeking to transition to low carbon development through investment in clean energy and exploration of policy mechanisms such as a carbon tax or carbon cap-and-trade system. China's 12th Five Year Plan is seen as its greenest yet, including an unprecedented number of environmentally focused targets and a strengthened commitment to energy savings and low carbon development (Yu and Elsworth 2012). China's future policy choices, including reform of its energy policy-making bodies to more effectively coordinate energy policy, have important implications for global energy and climate change policy.

The Costs of China's Growing Energy Use

In 2010, China surpassed Japan as the second largest economy in the world after the US. This achievement was the result of 30 years of industrialization, globalization, and marketization following the reform and opening of China's economy in the 1980s. This rapid economic growth, along with the improvement of living standards associated with it, was made possible through the provision of an abundant energy supply. At the same time, China's economic model has been one of high input, low output, and low efficiency. China accounted for 9.3% of global GDP and 20.3% of global energy consumption in 2010 (China Energy Research Society 2010). Its current economic model depends upon excessive energy consumption to maintain rapid GDP growth.

The combined effects of expansion in infrastructure and real estate, urbanization and industrialization, and skyrocketing imports and exports following China's entry into the World Trade Organization in 2001 caused China's economic development to shift to a more energy-intensive pattern. From 1980 to 2002, China's energy demand grew at less than half the rate of its GDP, a result of "strict oversight of industrial energy use, financial incentives for energy-efficiency investments, provision of information and other energy efficiency services through over 200 energy conservation service centers spread throughout China, energy-efficiency education and training, and research, development, and demonstration programs" (Price et al. 2011). From 2002 to 2005, however, this pattern of a reduction in energy intensity reversed, with China's average 5% annual energy intensity reduction from 1980 to 2002 changing to an average 5% annual energy intensity increase, visible in Figure 23.1.

This rapid increase in energy intensity was largely a result of concerted growth of China's energy-intensive heavy industries, such as iron and steel, chemicals, and cement (Rosen and Houser 2007). China's energy-intensive industries play a dominant role in its energy structure, with heavy industry increasing from 61% of total industrial output in 2001 to 71% by 2010 (National Bureau of Statistics 2002, 2011). The growth in energy intensity led the government to focus on building coal-fired power plants to improve the supply of electricity. From 2001 to 2011, China embarked on a period of rapid construction of electricity supply sources, with an average of 100 GW of installed capacity constructed each year (National Bureau of Statistics 2002, 2011).

The Chinese government aimed to reverse the trend of energy overconsumption in the 11th Five Year Plan (2006–2010), setting for the first time a target to reduce energy intensity by 20%. This national energy intensity reduction target was allocated down to and made mandatory for provincial and local governments and large state-owned enterprises, with provincial and enterprise leaders evaluated on their performance in meeting

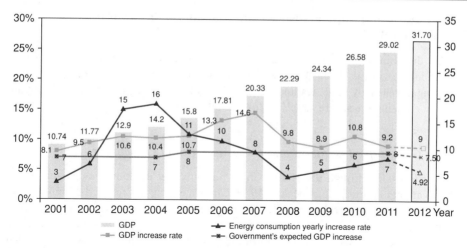

Figure 23.1 GDP and energy consumption growth rates (in trillion RMB adjusted for year 2000). Source: National Bureau of Statistics (2011).

their respective assigned targets. The government established a Top 1000 Enterprises program to focus specifically on the energy saving actions of the largest 1000 industrial enterprises, and established central and provincial energy efficiency subsidies for a Ten Key Energy Saving Projects program focused on industrial and building efficiency (Price *et al.* 2011). By the end of 2010, China had achieved an energy intensity reduction of 19.1%, saving 630 million tons of coal and avoiding about 1.55 billion tons of CO_2 emissions (Qi Ye *et al.* 2012).

During the 11th Five Year Plan, China shut down about 71 GW of smaller, less efficient thermal power plants, replacing them with larger, more efficient power plants; as of 2011, the average efficiency of Chinese coal-fired power plants reached 330 grams of coal per kWh. China also shut down outdated, inefficient manufacturing capacity in cement, steel, chemicals, and other heavy industries.

Even though China's energy conservation efforts have achieved significant results, however, it has proven difficult to change China's economic development model, which continues to be dominated by energy-intensive heavy industry and energy overconsumption. China in 2011 still relied on coal for 69.7% of its total energy consumption. Oil and natural gas accounted for roughly 22.8% of total energy consumption, while hydropower, nuclear power, wind, and other renewable energies combined accounted for just 7.5% of total energy consumption (National Bureau of Statistics 2011, 2012b). Such a high carbon energy structure has led to low efficiency, severe pollution, high CO_2 emissions, and a high mortality rate in the coal-mining industry.

On the supply side, China has made significant efforts to decarbonize its energy supply, including passing a renewable energy law in 2005 and establishing feed-in tariffs and other subsidies to encourage the expansion of wind power and other renewables. This has resulted in China installing 62.4 GW of wind power by the end of 2011 (grid and non-grid connected), the highest installed capacity in the world (Xinhua 2012c). However, challenges in connecting wind farms to the grid, integrating wind electricity into the grid, and improving and coordinating renewables policy and incentives remain.

China's Increasing Dependence on and Influence in International Energy Markets

China's distribution of energy resources can be characterized as rich in coal, short in oil, and lacking in gas. The gap between domestic supply and demand for oil and gas is becoming increasingly large, leading China to a greater reliance on imports and a greater focus on ensuring the secure supply of foreign oil and gas.

China's Growing Oil Demand and the International Oil Market

During the 1960s through the 1980s, China not only met its domestic demand for oil, but was an oil exporter. At its peak in 1986, China's oil exports reached 33 million tons (National Bureau of Statistics 1990). This changed in 1993 when China became a net oil importer due to a rapid increase in domestic oil demand. China's oil imports have increased since then, with China importing 250 million tons of oil in 2011, equivalent to 55.2% of its total oil consumption (China Petrochemical News Net 2012). China by 2011 surpassed the US in its dependence on foreign oil to come directly behind Japan, India, and the EU (BP 2011; China Energy Research Society 2010; First Caijing 2012).

Although China's consumption of oil as a proportion of total energy consumption has decreased from 21.8% in 2001 to 19% in 2010, absolute consumption of oil has continued to increase during the same period, reaching 10.7% of global oil consumption in 2010 (BP 2011). This increase has largely been driven by rapid growth in car sales, with China's car production and sales surpassing the US in 2009 (China Daily 2009, 2010). In 2011, the number of private vehicles in China increased 20.4% to roughly 79 million (National Bureau of Statistics 2012b). Further, the automotive market still has immense growth potential, with China averaging only 13 cars per 100 urban households in 2010 (National Bureau of Statistics 2011). By 2020, one scenario estimates that China will consume 600–650 million tons of oil per year, with its reliance on foreign oil rising to 65–70% (China State Council Development Research Center et al. 2009).

To meet China's increasing oil demand, China's three major state-owned oil enterprises (PetroChina, Sinopec, and CNOOC) have greatly expanded their search for overseas oil resources and their participation in overseas markets. They have engaged in overseas mergers and acquisitions of oil companies and oilfields, and expanded their operations to include exploration and development of oil and gas, production and sales, pipeline transportation, and refining. China's major oil enterprises currently have cooperation projects in nearly every part of the world with oil resources, as well as the production rights for about 85 million tons of oil (China News 2012).

As the gap between supply and demand for domestic oil and gas continues to increase, Chinese oil companies are forced to weigh the pros and cons of expanding overseas. Although Chinese oil companies have gained from direct investment in and access to overseas oil and gas resources, they have been unable to eliminate all of the political and economic risks associated with oil imports. China's oil enterprises are also expanding their domestic offshore drilling, which carries with it more technically challenging conditions and risks to marine environments, as was evident in the 2011 oil spill in the Bohai Bay.

Given China's status as a latecomer to international oil markets, many of the world's oil- and gas-rich areas have already been occupied and operated for years by the multinational oil firms of developed countries. Chinese oil firms have therefore concentrated on oilfields in Africa, South America, and Central Asia that have higher political risks and lower commercial appeal.

Table 23.1 China's top 10 sources of crude oil, 2011.

Country	Import Volume (tons)	Proportion of Total Imports (%)	Increase from 2010 (%)
Saudi Arabia	50,277,700	19.8	+12.61
Angola	31,149,700	12.3	−20.9
Iran	27,756,600	10.9	+30.19
Russia	19,724,500	7.8	+29.42
Oman	18,153,200	7.2	+14.4
Iraq	13,773,600	5.4	+22.57
Sudan	12,989,300	5.1	+3.1
Venezuela	11,517,700	4.5	+52.66
Kazakhstan	11,211,000	4.4	+11.51
Kuwait	9,541,500	3.8	−2.94

Source: China Oil News (2012).

According to China's General Administration of Customs, China imported a total of 254 million tons of crude oil in 2011, representing a 6% increase from 2010. The 10 countries listed in Table 23.1 account for 81.2% of China's total imports, with Middle Eastern countries accounting for 47% of China's imported oil (China Oil News 2012). China has sought to diversify its oil supply to strengthen its energy security, with Venezuela, Iran, and Russia constituting the countries with the greatest percentage increase in oil exports to China in 2011. Saudi Arabia, however, still accounts for 20% of China's total oil imports, and China faces challenges in further diversifying its oil imports.

Saudi Arabia, Russia, Canada, and Venezuela are all rich in oil resources, holding respectively 19.1%, 5.6%, 2.5%, and 12.9% of the world's remaining recoverable oil reserves. Together, the four countries play a decisive role in the global oil supply, accounting for 40.1% of the world's remaining recoverable reserves (China Energy Research Society 2010). They are attractive as oil exporters to China given that they are relatively stable politically, not concentrated in a single region (meaning it is unlikely that trade would be significantly interrupted from multiple sources), and share friendly relations with China.

China's rise as a major oil importer coincides with significant changes in the structure and dynamic of the global oil market. Since the 2008 financial crisis, the world's economic growth centers and demand for oil and gas have shifted toward Asia and the five BRICS countries. China, India, Japan, Singapore, South Korea, and Taiwan currently account for 75% of Middle Eastern oil exports and 45% of global oil exports, and this proportion is steadily increasing (China Energy Research Society 2010). In contrast, European oil and gas imports have decreased, and in recent years they have plummeted. Similarly, the US, seeking to strengthen the security of its energy supplies and its energy independence, has been rapidly developing shale gas and domestic oil resources, with its reliance on foreign oil decreasing to 53.5% (First Caijing 2012). China, on the other hand, has focused on securing a consistent and diverse oil supply to limit its vulnerability to market fluctuations and disruptions in international oil supplies, and to support its growing consumption of oil.

To further shield itself from disruptions in international oil supplies, China is also developing its strategic petroleum reserves. Current reserves, however, are still insufficient

100 million cubic meters

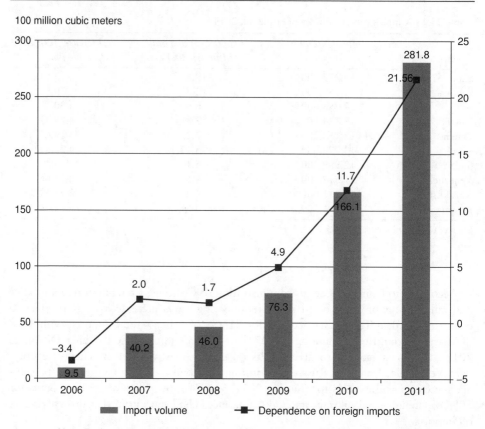

Figure 23.2 China's natural gas imports and dependence on foreign sources, 2006–2011.
Source: China Energy Statistics 2010 Annual Report; China Customs Statistics.

to meet this purpose. As of 2012, China's strategic petroleum reserves amount to roughly 40 days' worth of oil imports, significantly less than the International Energy Agency requirement of 90 days (Cui 2012; IEA n.d.). China is continuing to build its strategic petroleum reserves, with the goal of reaching 90 days of oil imports by 2020 (Xinhua 2012a).

China's Expansion of Domestic and Foreign Natural Gas Consumption

China has approximately 2.45 trillion cubic meters of proved conventional natural gas reserves, and as of the end of 2011, 202 billion cubic meters of remaining technologically recoverable coal bed methane reserves (China Energy Research Society 2010; Ministry of Land and Resources 2012). In addition, China's initial estimate of its geologically recoverable shale gas reserves is 25 trillion cubic meters (National Energy Administration 2012b). Natural gas, however, is still a relatively small part of China's energy portfolio, with natural gas production and consumption respectively representing only 4.1% (95 billion cubic meters) and 4.2% (106 billion cubic meters) of total energy in 2010 (National Bureau of Statistics 2011).

China has only one-seventh the natural gas reserves and consumes one-sixth the natural gas of the US, where natural gas accounts for 22% of total energy consumption (China Energy Research Society 2010). Increasing production and consumption of natural gas is a key aspect of changing China's heavily carbonized portfolio.

To satisfy its rapidly increasing natural gas consumption, China will need to utilize both its domestic market and the international markets for conventional and unconventional natural gas. The National Development and Reform Commission (NDRC) projects that by 2015, China's natural gas consumption will have reached 260 billion cubic meters, necessitating the import of 90 billion cubic meters. By 2030, China is expected to consume 500 billion cubic meters of natural gas, about what the EU consumes today, half of which will be from foreign sources (Wang Xiuqiang 2012).

China's shale gas and coal bed methane have great potential for development. Although China possesses abundant coal bed methane resources, in 2010 it utilized only about 10 billion cubic meters of surface extracted and underground coal bed methane after having developed the resource for 15 years (National Energy Administration 2012a). If China is to reach its 12th Five Year Plan goal of producing 30 billion cubic meters of coal bed methane per year by 2015, it will need to further develop and scale-up coal bed methane development technology (National Energy Administration 2012a).

China has set a goal of verifying 600 billion cubic meters of geological and 200 billion cubic meters of recoverable shale gas reserves by 2015, and it has set shale gas production goals of 6 billion cubic meters of shale gas per year by 2015, ramping up to 60–100 billion cubic meters per year by 2020 (National Energy Administration 2012b). China, however, faces numerous challenges as it attempts to meet these goals. China's shale gas is mostly in mountainous and remote districts and requires large amounts of water to extract, in addition to large-scale equipment for fracturing. China has not yet mastered newer fracturing technology and depends on joint ventures to obtain new technology (National Energy Administration 2012b). Given the large amounts of water required for hydraulic fracturing and the pollution that results from the use of fracturing chemicals, China will need to carefully consider the technologies and regulations required to develop shale gas in a manner that will protect its water resources.

China also plans to expand its natural gas imports, which in 2011 already exceeded 30 billion cubic meters (Ministry of Commerce 2012). Its natural gas imports principally include pipeline natural gas from Turkmenistan, Uzbekistan, and Kazakhstan, and coastal liquefied natural gas imports from Australia, Indonesia, Malaysia, Qatar, Yemen, and Russia (BP 2011). China is also currently negotiating the importation of pipeline natural gas from Russia. Further diversification would better shield China's natural gas supply from the myriad of possible economic fluctuations.

China's oil and gas companies, however, have little incentive to develop natural gas given the current pricing system, in which prices are set by the government. Retail prices for natural gas are fixed at low rates for consumers, particularly residential users, in order to avoid inflation; these prices are much less than the cost for companies to provide the gas. For example, it cost companies 3.5 RMB per cubic meter to deliver gas from Central Asia to Beijing along the Second West–East Gas Pipeline, considering the wellhead price and pipeline transmission costs. With retail gas prices fixed at 2.05 RMB per cubic meter in Beijing, companies were actually losing 1.45 RMB per cubic meter of gas sold (Wang Xiuqiang 2012). Reform in pricing is therefore needed to let the market function more efficiently and to give domestic enterprises more incentive to develop and sell natural gas. This is especially true for more expensive natural gas imports and for unconventional natural gas due to the higher cost of development and extraction (IEA 2012).

Table 23.2   China's top 7 sources of imported natural gas, 2010. (All sources are LNG, except for Turkmenistan which is PNG (pipeline natural gas).)

Country	Import Volume (billion cubic meters)
Australia	5.21
Turkmenistan	3.55 (PNG)
Indonesia	2.45
Malaysia	1.68
Qatar	1.61
Yemen	0.70
Russia	0.51

Source: BP (2011).

Coal and Electricity Imports, Nuclear, Hydropower, and Renewables Continue to Grow

In addition to China's growing imports of oil and natural gas, China's growing energy consumption has meant that it is continuing to seek foreign sources of coal, and to expand nuclear power, hydropower, wind, and solar, with ever greater influence on and interaction with the global market.

China in 2009 became a net importer of coal, and after just a few years reached 167 million tons of coal imports, primarily from Vietnam, Indonesia, South Africa, and Australia (NDRC 2012). Due to the high energy needs of China's southeast coastal areas and the challenges of transporting coal domestically, the price of domestic coal has increased to more than 800 RMB per ton in these regions, making the price of imported coal more competitive (State Coal Mine Safety Supervision Bureau 2011). Parts of Canada and the US, which have been reducing their consumption of coal and lessening its environmental impacts, are now preparing for coal exports to China. Additionally, China is actively entering into coal mining development in Mongolia, and coal imports from Mongolia are therefore expected to increase (Thermal Energy Net 2012).

China is also planning to greatly expand its use of nuclear power in the next decade. By 2020, China is expected to develop nuclear power from the current 12 GW to at least 58 GW. After the Fukushima nuclear accident in 2011, China carried out safety reviews of its civilian nuclear installations, mandated requirements for improvements, and suspended approvals of new nuclear power projects until the State Council approved plans for nuclear safety and nuclear development in October 2012.

China will soon become a net importer of electricity. In the mid term it is expected to purchase thermal power from Russia to supply its northeast region, hydroelectric from Mongolia, and coal-fired power from Southeast Asia. China recently signed a contract with Russia to buy 100 billion kilowatt hours of thermal power (China Energy Net 2012). The State Grid Corporation of China is also looking beyond China to become a stockholder of Spanish and Australian power companies.

China's hydroelectric power equipment and construction ability is among the largest and most developed in the world, with Chinese hydroelectric companies operating in many countries. Chinese companies are also constructing coal power facilities abroad, keeping the cost of construction lower than 4500 RMB per kilowatt of power (Zhang 2011). In the renewable energy market, China's solar PV cells represent 50% of the international market, with roughly 90% of domestically manufactured solar PV cells exported abroad (US Department of Commerce 2011). Chinese wind turbine manufacturers have also been seeking to export abroad, and have reduced the cost of wind power to

3500 RMB per kilowatt, enabling wind power to compete with coal in wind resource-rich areas (Zhang 2012).

China's Motivations, Policies, and Challenges in Addressing Climate Change and Pursuing Clean Energy

In the last three decades, China has been able to provide a secure energy supply to sustain the fast pace of its economic development. However, the government has recognized that the high investment costs, resource consumption, pollution emissions, and low efficiency and production of its current economic development pattern cannot sustainably continue. China has also recognized the threat that severe climate change will pose to its economy and its people, and is taking actions domestically and on the international stage to address climate change.

China's Motivations for Addressing Climate Change

China's status as the largest CO_2 emitting country in the world means that it has faced increasing pressure internationally to take action to address its emissions. By 2011, China's CO_2 emissions had risen to roughly 7.7 billion tons (22% of global emissions), with per capita emissions reaching 5.7 tons. Unless the pace of emissions growth is reduced, by 2020 China will emit 28–30% of global CO_2 emissions (Qi Ye *et al.* 2012).

Another important question is when China's absolute greenhouse gas (GHG) emissions will peak and then begin to reduce. In an analysis of China's carbon emissions scenarios through 2050, a number of Chinese research institutions determined that at the current rate, carbon emissions will peak by about 2030 (China State Council Development Research Center *et al.* 2009). Under this scenario, even as China reduces its carbon intensity at a faster rate, total emissions will continue to rise relatively quickly. It is thus necessary to intensify emissions reduction efforts to ensure that China's emissions peak as early as possible, by 2020 or 2025. An earlier peak in China's CO_2 emissions would contribute greatly to minimizing the potentially severe impacts of global climate change.

The major motivating force behind China's efforts to address climate change, however, is domestic, since China will be among the countries most affected by climate change. According to the Second National Climate Change Assessment Report, China's average temperature increased by 1.38 degrees C from 1951 to 2009. This temperature rise has led to a loss of more than 10% of China's glaciers, and will greatly exacerbate China's already severely strained water resources and threaten one-third of the population's water supplies. Further, the report projects that sea levels around China's coasts will continue to rise for 30 years, and that a temperature rise of 2.5 degrees C would reduce China's grain production by at least 20% (Science Publishing 2011).

Strengthening Policies for Energy Efficiency, Renewable Energy, and Other Low-Emissions Clean Energy Technology

As part of the current 12th Five Year Plan (2011–2015), China aims to develop a "resource conserving, environmentally friendly society" (NDRC 2011a). The Twelfth Five Year Plan builds upon the 20% energy intensity reduction target set during the 11th Five Year Plan (2006–2010) by putting in place targets to reduce energy intensity (the amount of energy consumed per unit GDP) by 16% and carbon intensity (the amount of CO_2 emitted per unit GDP) by 17%. Non-fossil energy sources (i.e., nuclear and renewables) are to make up at least 11.4% of total energy by 2015.

These targets are in line with the climate targets announced by the Chinese government in 2009 at the UN Climate Change conference in Copenhagen to reduce carbon intensity by 40–45% from 2005 levels and increase non-fossil energy to 15% of total energy consumption by 2020 (Qi Ye *et al.* 2012). The government has also set targets for reducing absolute emissions of the following key environmental pollutants during the 12th Five Year Plan: Chemical Oxygen Demand (COD) and SO_2 (pollutants that were targeted in the 11th Five Year Plan) are to be reduced by 8% by 2015, and ammonia nitrogen and NOx emissions are to be reduced by 10% by 2015.

Significantly, the government set the target for annual average GDP growth during the 12th Five Year Plan at a relatively low 7%, in an effort to send a signal to local governments that the pursuit of unrestrained GDP growth at the cost of the environment is no longer favored. Local officials will be evaluated in their performance in meeting energy and carbon intensity reduction targets as well as the other pollutant emissions reduction targets.

The Chinese government will build upon the energy efficiency programs begun in the 11th Five Year Plan by, for example, expanding the Top 1000 Enterprises program to the top 10,000 energy consuming enterprises. China is also investing heavily in energy efficiency and renewables: during the 11th Five Year Plan period, it invested 1730 billion RMB and 860 billion RMB, respectively, in new and renewable energy and energy efficiency (Qi Ye *et al.* 2012). China is also expanding upon provincial demand-side management (DSM) energy efficiency programs, passing a national DSM regulation in 2010 that will require grid companies to participate in demand-side energy efficiency programs (Finamore 2010). Even after the energy savings achieved during the 11th Five Year Plan, China still has great potential for energy savings, given its advantage as a late developer.

China's passage of a Renewable Energy Law in 2005 played an important role in supporting the rapid growth of its renewables industries, establishing national targets for renewable energy, setting a renewable energy surcharge on electricity to fund payments to renewable generation sources, and requiring that grid companies connect and purchase electricity from renewable sources (Schuman 2010).

Wind power in particular expanded rapidly during the 11th Five Year Plan period. At the beginning of the period in 2006, China had just 2.6 GW of installed wind capacity. Over the next four years, China more than doubled its wind installed capacity each year. By the end of 2011, China had installed a cumulative 65 GW of wind power capacity, with 45 GW connected to the grid, making it the country with the largest wind installed capacity in the world, ahead of the US. China has transitioned from a concession program, in which individual wind projects were placed for bidding by project developers, to a national feed-in tariff program begun in 2009, in which all wind projects are paid predetermined on-grid electricity rates based on their geographic region. Although there remain significant challenges to integrating greater amounts of wind energy into the grid, including establishing stronger on-grid connection standards for wind farms, improving turbine technology, and improving management of wind farms, wind power has become a notable component of China's energy portfolio.

In a similar manner, China has also begun to support greater amounts of solar PV installation domestically. By the end of 2011, China had installed a cumulative 3 GW of solar PV installations. It also established in 2011 a national feed-in tariff for electricity generated from solar PV, and plans to greatly expand domestic installation of solar PV modules, rather than continuing to manufacture primarily for export. China has set ambitious plans for future wind and solar PV growth, with wind and solar PV installed capacity targeted to reach 100 GW and 15 GW, respectively, by 2015, and 200 GW and 50 GW by 2020.

China's hydropower installed capacity is expected to reach over 350 GW by 2020, pushing upon the limits of China's hydropower resources. China is also developing other forms of renewable energy, with biomass and geothermal energy respectively reaching 5.12 million and 28,000 kilowatts of installed capacity in 2010, and production of solar hot water heaters reaching 49 million square meters and accounting for roughly 80% of the world's total production (China Energy Research Society 2011).

Additionally, China is pursuing development of technologies to reduce emissions, including Carbon Capture and Storage (CCS) technology to capture and store the CO_2 from coal-fired power plants, electric vehicles, ultra-high voltage electricity transmission lines for transmitting power across long distances, and battery storage technology. China has identified clean technologies as strategic industries for future development. With regard to CCS, China is building a large, coal-fired Integrated Gasification Combined Cycle (IGCC) power plant demonstration project known as GreenGen in Tianjin that will eventually use CCS technology. Given that coal will still represent a significant portion of China's total energy consumption in the future (estimated at around 35% by 2050), it is sensible to explore the technological and economic feasibility of storing CO_2 emissions from coal-fired power plants underground.

Finally, the government in 2010 also established low carbon development pilots in five provinces (Guangdong, Liaoning, Hubei, Shaanxi, and Yunnan) and eight cities (Tianjin, Chongqing, Shenzhen, Xiamen, Nanchang, Guiyang, Baoding, and Hangzhou) (NDRC 2010). These pilots are meant to serve as examples for low carbon development policies that can then be transferred to other cities and provinces.

China's Exploration of a Carbon Tax and Carbon Market Mechanisms

Although the government achieved important successes in saving energy and promoting renewable energy during the 11th Five Year Plan, it has determined that mandatory regulatory tools are insufficient, and has made preparations to begin implementation of a carbon tax and carbon market mechanisms during the 12th Five Year Plan. Such regulatory tools will provide an important complement to existing clean energy policies, helping to send a signal to the market that could reverse persistent trends toward heavy industry and extensive fossil fuel consumption, while providing a revenue source for the government that can be directed toward clean energy.

In 2007, the government included a carbon tax in its agenda and various government financial and environmental research institutes have conducted research for policy-makers regarding the design and effect of a carbon tax. Current plans would be for the carbon tax to start at a relatively low rate, such as 10 RMB per ton of CO_2, and ramp up, with covered sectors including coal, oil, and natural gas (Caijing 2012; Ministry of Finance 2009). Revenues from the tax would be invested in clean energy projects, and the tax could be implemented as soon as the middle or latter part of the 12th Five Year Plan period (Caijing 2012).

The Chinese government is also piloting the use of carbon cap-and-trade markets, establishing at the end of 2011 a pilot program for developing carbon markets in two provinces (Guangdong and Hubei) and five cities (Beijing, Chongqing, Shanghai, Shenzhen, and Tianjin) (NDRC 2011b; Yu and Elsworth 2012). The development of these sub-national carbon markets during the 12th Five Year Plan are meant to provide experience and learning for the development of a national carbon cap-and-trade market in the 13th Five Year Plan. China is currently developing carbon markets and a carbon tax as complementary policies that need not be mutually exclusive.

However, a number of challenges remain to implementing carbon markets in China, including issues surrounding the design of the markets, such as "the allocation of allowances, market behaviour, trading regulations and responsibilities of parties as well as legal infrastructure" (Yu and Elsworth 2012). The carbon price must be set neither too low nor too high, so that it can send a consistent market signal (Lin and Yang 2012). Another key task, fundamental to the functioning of any carbon market, is the development of a robust monitoring, reporting, and verification system to ensure that carbon credits that are allocated and traded in a carbon market scheme indeed represent actual emissions. Despite these challenges, China's efforts to establish carbon markets are likely to spur on the global trend toward setting emissions limits on GHG emissions through market mechanisms.

Addressing China's Growing Coal Consumption Through a National Coal Consumption Control Plan

As the world's largest consumer of coal, China faces significant challenges in reducing its coal consumption and utilization. Although China has only 14% of the world's recoverable coal reserves, it produces nearly half of the world's coal output (China Energy Research Society 2010). China consumed 1.4 billion tons of coal in 2000; coal consumption has grown to 3.48 billion tons of coal in 2011, accounting for half of the world's total coal consumption, and growth continues to gain momentum. If China's coal consumption continues at its current rate, it will reach over 5 billion metric tons by 2020, a level that is unsustainable for China's local environment and for addressing global climate change.

As an accessible and available energy source, coal has been important in boosting local economies. On the other hand, the process of mining, transportation, transfer, and combustion of coal causes significant harm to the environment and public health, and is a key contributor to climate change. Indeed, one analysis of coal's true cost in China found that if the external costs are internalized, the price of coal would increase by 70–80%; it further found that the environmental and public health costs from coal consumption in China were equivalent to roughly 7.1% of GDP (Mao *et al.* 2008).

In light of China's increasing coal consumption and coal's severe impact on the environment and public health, it is important that the government establish policies that can limit and reduce future consumption and scale up alternatives to coal. China's existing energy efficiency, renewables, and low carbon development policies can help form the basis of a coal cap. China's energy and carbon intensity targets, and pollutant emissions reduction targets, also align with the establishment of a coal consumption cap.

China's policy-makers are already considering the establishment of total energy consumption targets and regional coal consumption targets. As of 2012, the central government is evaluating the establishment of a national energy consumption target of no more than 4.2 billion tons of coal equivalent by 2015, which would be allocated down to provinces and local governments, but would be more flexible than a mandatory target (China Daily 2012). Setting total energy consumption caps would send a stronger signal to local governments that controlling energy use is a priority.

The Ministry of Environment Protection will establish regional coal consumption caps in three regions and 10 city agglomerations to address severe air pollution conditions, particularly PM 2.5, which became an issue of public concern in 2011 (Xinhua 2012b). In these pilot areas, the coal consumption of projects will be evaluated as part of their environmental impact assessments (Xinhua 2012b). Beijing, Tianjin, and Zhejiang province have plans to set coal consumption targets during the 12th Five Year Plan, with Beijing planning to limit coal consumption to 15–20 million tons by 2015, from the 27 million tons consumed in 2010 (Xinhua 2012b). Further, the NDRC in its 12th Five Year Plan for Coal Industry Development stated that it will implement controls on the supply and demand of coal (NDRC 2012). These projects serve as foundations for implementing regional targets and systems that will be necessary under a country-wide coal cap.

In its 12th Five Year Plan for Coal Industry Development, the Chinese government has set a goal of limiting coal production to 3.9 billion tons by 2015 from 3.6 billion tons in 2010 (NDRC 2012). Coal consumption and total energy consumption targets, however, have consistently been exceeded in previous Five Year Plans, necessitating the use of a more focused coal consumption cap policy. Establishing a national coal consumption control policy would help China to reach its GHG emissions peak earlier, and have multiple benefits for the environment and public health. Setting a coal cap will require both top-down government planning and design and bottom-up public and enterprise involvement.

Reforming China's Energy Regulatory System and Market Structure

Improving the regulatory institutions for and market structure of China's energy sector is an important task for developing a more coordinated and rational energy sector. In March 2008, China restructured its national energy policy-making institutions, establishing a National Energy Commission with leaders from 17 ministries and agencies and led by the Prime Minister, and a National Energy Agency (NEA) under the NDRC, which was meant to develop a more coordinated approach to energy planning and policy-making across ministries (Downs 2008). The establishment of an NEA fell short of calls for a Ministry of Energy, which would have had the bureaucratic authority to effectively regulate equivalent-level ministries and state-owned enterprises. Notably, the NEA lacks the power to set energy prices, which remains the purview of the NDRC Pricing Department, given the importance of energy prices to other macro-economic policies such as combating inflation (Downs 2008). The lack of pricing authority has similarly limited the ability of institutions such as the State Electricity Regulatory Commission (SERC) to effectively regulate their sectors.

There thus exists a need for further reform of the energy planning and regulatory system, through the establishment of a Ministry of Energy that has overall authority for planning and policy regarding energy supply, energy conservation, and carbon emission reductions, and can guide and coordinate inter-ministerial energy policy. China's energy system reform should emphasize the ability to adapt to climate change, the sustainable use of energy resources, and environmental protection (Li 2012).

With regard to climate change policy, China has a National Leading Working Group on Addressing Climate Change, headed by the Prime Minister, to coordinate inter-ministerial policy, and a Climate Change Department within the NDRC, which leads climate policy-making. However, China has no local government equivalent bodies for climate change, which limits the effectiveness of implementation of climate change response policies at the local level.

Other energy regulatory reforms that are necessary include the strengthening of the nuclear safety regulatory body, by establishing an independent Nuclear Safety Regulatory Commission with sufficient independence, responsibility, and authority to regulate nuclear safety; providing the existing SERC with the authority to regulate natural gas pipelines and markets; and rationalizing pricing of energy, including raising electricity prices to better account for the real cost of production (coal-fired generators currently suffer a loss since they purchase coal at market prices but must sell at government-determined market prices), and adopting real-time and tiered electricity pricing to incentivize energy-saving.

Conclusion

China's energy policies and energy use and production have an increasingly significant impact on international energy and climate policy and energy markets. China is both the largest energy and coal consumer and carbon emitter in the world, and at the same time, it is actively promoting policies to transition to a low carbon and clean energy development model supported by increased energy efficiency. China's actions and achievements over the next decade will have important implications for its own domestic energy and environmental situation and the state of the global environment.

Note

1. The authors give their sincere thanks to Angela Shen for her significant contribution to this paper.

References

BP. 2011. *BP Statistical Review of World Energy*, June 2011. London: BP.

BP. 2012. *BP Statistical Review of World Energy*, June 2012. London: BP.

Caijing. 2012. Carbon Tax Planned as an Independent Tax (Chinese). http://economy.caijing.com .cn/2012-01-05/111590186.html.

China Daily. 2009. China Surpasses US Auto Market in H1 Sales. http://www.chinadaily.com .cn/china/2009-07/09/content_8404128.htm.

China Daily. 2010. China is Now World Champion in Car Production. http://www.chinadaily.com .cn/business/2010-02/03/content_9420521.htm.

China Daily. 2012. Energy Use May be Capped for 2015. http://www.china.org.cn/business/2012- 05/03/content_25288835.htm.

China Energy Net. 2012. Development of China-Russia Energy Cooperation Diversifying, Electricity Cooperation Continuing to Heat Up (Chinese). http://finance.qq.com/a/20120427/ 000599.htm.

China Energy Research Society. 2010. *Energy Policy Research* (Chinese). Beijing: China Energy Research Society.

China Energy Research Society. 2011. *China Energy Development Report* (Chinese). Beijing: China Energy Research Society.

China News. 2012. Last Year Chinese Enterprise Oversees Oil and Gas Rights Reached 85 Million Tons of Oil Equivalent (Chinese). http://finance.chinanews.com/ny/2012-02-09/3657672.shtml.

China Oil News. 2012. China Crude Oil Import Country and Regional Statistics (Chinese). http://oilinfo.cnpc.com.cn/ypxx/ypsc/tjsj/yy/.

China Petrochemical News Net. 2012. China Must Strengthen Energy Independence (Chinese). http://www.china5e.com/show.php?contentid=216128#weblog.

China State Council Development Research Center, National Development and Reform Commission Energy Research Institute, and Tsinghua University Nuclear Energy and New Energy Technology Research Institute. 2009. 2050 China Low Carbon Development Scenario Research (Chinese). *2050 China Energy and CO$_2$ Emissions Report*. Beijing: Science Press.

Cui, Carolyn. 2012. China Seen Bolstering Oil Reserves. *The Wall Street Journal*, April 11.

Downs, Erica S. 2008. China's "New" Energy Administration: China's National Energy Administration Will Struggle to Manage the Energy Sector Effectively. *China Business Review*, November–December: 42–45.

Finamore, Barbara. 2010. Taking Action to Meet its Climate Pledge – China Enacts National Energy Efficiency DSM Regulations to Dramatically Scale Up Investments in Energy Efficiency. http://switchboard.nrdc.org/blogs/bfinamore/taking_action_to_meet_its_clim.html.

First Caijing. 2012. America Has Achieved "Energy Independence," What Should China Do?" (Chinese). http://finance.qq.com/a/20120229/000298.htm.

IEA. n.d. Oil Markets and Energy Preparedness. http://www.iea.org/about/ome.htm.

IEA. 2010. *World Energy Outlook 2010*. Paris: International Energy Agency.

IEA. 2012. *Gas Pricing and Regulation: China's Challenges and IEA Experience*. Paris: International Energy Agency.

Li, Ting. 2012. Discussion of Systemic Reform for China's Energy Management System (Chinese). *Energy Review* 62.

Lin, Alvin and Fuqiang, Yang. 2012. Design Tips for a Carbon Market. China Dialogue. http://www.chinadialogue.net/article/show/single/en/4797-Design-tips-for-a-carbon-market.

Mao, Yushi, Hong Sheng, and Fuqiang Yang. 2008. *The True Cost of Coal* (Chinese). Beijing: Coal Industry Publishing.

Ministry of Commerce of the People's Republic of China. 2012. Natural Gas Industry Promotion Has Led to Significant Growth in the Oil Industry (Chinese). http://www.mofcom.gov.cn/aarticle/hyxx/fuwu/201203/20120307997202.html.

Ministry of Finance, Research Institute for Fiscal Science. 2009. Analysis of the Necessity and Feasibility of a National Carbon Tax (Chinese). http://finance.ifeng.com/roll/20090923/1273279.shtml.

Ministry of Land and Resources. 2012. Xu Dacun: 2011 National Coal Bed Methane Exploration Newly Discovered Geologically Reserves Total 142.174 Billion Cubic Meters (Chinese). http://www.mlr.gov.cn/wszb/2012/sytrq/zhibozhaiyao/201202/t20120223_1066530.htm.

National Bureau of Statistics. 1990. *1990 China Statistical Yearbook*. Beijing: China Statistics Press.

National Bureau of Statistics. 2002. *2002 China Statistical Yearbook*. Beijing: China Statistics Press.

National Bureau of Statistics. 2011. *2011 China Statistical Yearbook*. Beijing: China Statistics Press.

National Bureau of Statistics. 2012a. *2012 China Statistical Yearbook*. Beijing: China Statistics Press.

National Bureau of Statistics. 2012b. *2011 National Economic and Societal Development Statistical Report* (Chinese). National Bureau of Statistics of China.

NDRC. 2010. Notice from the National Development and Reform Commission Regarding Establishment of Low Carbon Provinces and Low Carbon Cities Pilots (Chinese). National Development and Reform Commission, Climate Change Department, No. 1587.

NDRC. 2011a. *Outline of the 12th Five Year Plan for Economic and Social Development* (Chinese). National Development and Reform Commission.

NDRC. 2011b. Notice from the NDRC Office on Establishing Carbon Emissions Rights Trading Pilot Work (Chinese). National Development and Reform Commission, Climate Change Department, No. 2601.

NDRC. 2012. *Twelfth Five Year Plan for Coal Industry Development* (Chinese). National Development and Reform Commission.

National Energy Administration. 2012a. *12th Five Year Plan for Coal Bed Methane Development* (Chinese). National Energy Administration.

National Energy Administration. 2012b. *Shale Gas Development Plan (2011–2015)* (Chinese). National Energy Administration.

PBL. 2007. China Now No. 1 in CO_2 Emissions; USA in Second Position. PBL Netherlands Environmental Assessment Agency. http://www.pbl.nl/en/dossiers/Climatechange/moreinfo/ChinanownolinCO2emissionsUSAinsecondposition

Price, Lynn, Mark D. Levine, Nan Zhou, *et al.* 2011. Assessment of China's Energy-Saving and Emission-Reduction Accomplishments and Opportunities During the 11th Five Year Plan. *Energy Policy* 39, 4: 2165–2178.

Qi, Ye, *et al.* 2012. *Blue Book of Low-Carbon Development: Annual Review of Low-Carbon Development in China (2011–2012)* (Chinese). Tsinghua University Center for Climate Policy Analysis. Beijing: Social Sciences Academic Press.

Rosen, Daniel, and Trevor Houser. 2007. *China Energy: A Guide for the Perplexed*. Washington, DC: Peterson Institute for International Economics.

Schuman, Sara. 2010. *Improving China's Existing Renewable Energy Legal Framework: Lessons from the International and Domestic Experience*. Beijing: Natural Resources Defense Council.

Science Publishing. 2011. *Second National Climate Change Assessment Report* (Chinese). Beijing: Science Publishing.

State Coal Mine Safety Supervision Bureau. 2011. *China Coal Industry Yearbook 2010* (Chinese). Beijing: Coal Information Research Institute.

Thermal Energy Net. 2012. China in 2012 Will Remain a Net Importer of Coal Chinese. http://news.bjx.com.cn/html/20120120/337810.shtml.

US Department of Commerce International Trade Administration. 2011. Renewable Energy – Solar and Wind. http://export.gov/china/doingbizinchina/eg_cn_025864.asp.

Wang, Xiuqiang. 2012. Tong Xiaoguang: The Main Problem with Restricting Imported Natural Gas is Gas Prices (Chinese). Sina Finance. http://finance.sina.com.cn/roll/20120307/041311530070.shtml.

Xinhua. 2012a. Iran Oil Ban Triggers Chinese Oil Reserve Concerns. Xinhua News Agency, February 21. http://www.china.org.cn/world/2012-02/21/content_24695371.htm.

Xinhua. 2012b. 8 Chinese Cities Air Does Not Meet Standards, Photochemical Smog Is Severe (Chinese). Xinhua News Agency, May 14. http://news.xinhuanet.com/society/2012-05/14/c_111941222.htm.

Xinhua. 2012c. China World's Wind Power Leader: New Figures. Xinhua News Agency, March 23. http://news.xinhuanet.com/english/china/2012-03/23/c_131485088.htm.

Yu, George, and Rob Elsworth. 2012. *Turning the Tanker: China's Changing Economic Imperatives and Its Tentative Look to Emissions Trading*. London: Sandbag Climate Campaign.

Zhang, Guobao. 2012. The Price of Wind Energy Has Already Dropped to 3500 RMB Per Kilowatt Hour (Chinese). http://finance.ifeng.com/stock/roll/20120310/5730778.shtml.

Zhang, Hui. 2011. The First One Gigawatt Coal Fired Power Generation Program Has Been Launched at Caofeidian (Chinese). http://ts.yzdsb.com.cn/system/2011/08/25/011389441.shtml.

Chapter 24

Dismounting the Subsidy Tiger: A Case Study of India's Fuel Pricing Policies

Sudha Mahalingam

Introduction

As India races towards its manifest destiny of becoming a developed country, its predominant concerns remain focused on its ability to access adequate energy resources to sustain the growth momentum. Commercial energy consumption has grown in tandem with the economy although it trails GDP growth marginally.[1] Efforts to de-link the Indian economy from energy consumption have had limited impact, partly because of the nature of India's chosen growth paradigm but also because of contradictory policy signals.[2] While the service sector, relatively less energy-intensive, has long overtaken agriculture and manufacturing, other energy-intensive industries like automobiles, steel, cement, and construction, have been registering impressive growth rates.[3] Besides, service sector-led growth has overturned traditional wisdom primarily because of the energy-profligate lifestyles of its increasingly prosperous and numerous workforce.

Apart from availability and accessibility of adequate energy supplies, concerns that are common to most fuel-importing nations, India, starting from a low energy base, has the additional responsibility of providing a modicum of equity in access to commercial energy to its billion plus population. At least 300 million Indians subsist on less than a dollar a day, and as such, cannot afford commercial energy.[4] More than half the rural households do not even have access to electricity[5] and even those that do, have intermittent supplies. According to the 2011 Census, 377 million Indians now live in urban areas and this poses its own set of problems to the government, which has to provide a modicum of basic human needs such as cooking fuel and at least a single light bulb at affordable prices to the economically disadvantaged. Equity apart, environmental considerations also dictate that clean cooking fuels are made available at affordable prices.

Integrated Energy Policy, the first comprehensive policy document prepared by India's Planning Commission and adopted by the government, defines energy security to include access to "lifeline" energy needs to its citizens, regardless of their ability to pay, a definition that is unique and inclusive.[6] In fact, this single statement in the document reflects the state's acknowledgement of its responsibility to provide energy to meet basic human

The Handbook of Global Energy Policy, First Edition. Edited by Andreas Goldthau.
© 2013 John Wiley & Sons, Ltd. Published 2013 by John Wiley & Sons, Ltd.

needs, quite unrelated to the ability of its citizens to pay for it. In that sense, the Indian State has perhaps made a distinction between energy that meets basic human needs versus energy that satisfies human wants.

This distinction, though, becomes less focused when translated into policy and even more blurred in the implementation. Policy distortions and muddled implementation have led to runaway fiscal deficits, part of which goes to fund energy subsidies.[7] The democratic feature of the polity has restricted the maneuverability of the government of the day. The growing trend in favor of coalition politics and greater say for regional parties has exacerbated the situation, limiting the options of the state to follow policies that are sensible and sustainable.[8]

This paper attempts to examine India's energy conundrum through the prism of fuel subsidies. The country's developmental aspirations and achievements have gradually but steadily pushed its economy towards greater integration with the global energy market-place. Yet, such integration has at best been half-hearted, considering that the policy-makers have simultaneously attempted to shield its citizens from the vagaries and volatil-ity of the global energy markets through the instrument of fuel subsidies. This conflicting and contradictory approach to energy policy has locked the country into an increas-ingly unsustainable subsidy paradigm. It is noteworthy that this half-hearted approach to markets cuts across fuels and sectors, be they coal, electricity, or hydrocarbons.

But nowhere is it more egregious and devastating than in the case of hydrocarbons. In fact it is the hydrocarbon sector that is the fastest growing segment in India's energy basket. Currently accounting for 41%, it is growing at 5.5% per annum, driven largely by the rapid growth of transportation industry, but also by India's growing population which needs to be supplied with cooking fuel. With three out of every four barrels of crude consumed in the country coming from imports, the hydrocarbon sector has crossed the Rubicon, as it were, in terms of integration with global markets. For that very reason, the current paradigm in the oil sector is even less sustainable than in coal or electricity, both more inward-looking. Therefore, this paper limits itself to examining the pricing policies in the petroleum sector which have locked the country into an unsustainable trajectory.

Evolution of India's Petroleum Industry

In 1947, at the time of Independence from British rule, there was no domestic production except marginal quantities of oil produced in the northeastern state of Assam by Assam Oil and Oil India Limited, the latter a joint venture between Burmah Oil and the Gov-ernment of India (GoI). This oil was refined at Digboi refinery in Assam and consumed locally. The rest of the country relied on imported products which were retailed by three multinationals, Burmah Shell, Caltex, and Standard Oil, and a plethora of private traders.

Initially, petroleum product prices were fixed by government committees on import-parity basis. An oil pool account was set up to cushion the impact of volatile product prices. However, as global prices hardened, the oil pool account was insufficient to cushion the domestic market. Between 1960 and 1969, three government committees progressively reduced import parity prices to keep fuel affordable to the masses.

Meanwhile, in 1956, the Oil and Natural Gas Commission (ONGC) was set up with Soviet technology and assistance to engage in domestic exploration and production activities.[9] By the 1970s, ONGC also found substantial reserves of crude in Cambay basin in Western India and in Bombay High, an off-shore basin, also in Western India. Yet, India had to import substantial quantities of crude and oil products to supplement domestic production.

Between 1974 and 1976, in the wake of the oil crisis of 1973, the Indian government nationalized the three foreign oil companies, all subsidiaries of major MNCs. Shell became Bharat Petroleum Corporation Ltd (BPCL) and Esso-Caltex, Hindustan Petroleum Corporation Ltd (HPCL), both state-owned monopolies. A third, Indian Oil Corporation (IOC), already in existence since 1959, was given the express mandate to set up refineries and retail outlets. All three Oil Marketing Companies (OMCs) went about setting up refineries, with the help of technology from the socialist countries. The OMCs also set up marketing outlets all over the country, with a tacit understanding not to tread on each other's toes. The three OMCs were envisaged as integrated refining and marketing operations while ONGC and OIL (Oil India Limited, a National Oil Company engaged in exploration and production in northeast India) confined their operations to exploration and production. Thus the Indian petroleum sector came to be the monopoly of National Oil Companies in the entire hydrocarbon value chain.

Moving from State Monopoly to Market

Concomitant with nationalization, GoI moved to regulate petroleum product prices through a policy instrument called Administered Pricing Mechanism (APM). The upstream oil companies received prices fixed on cost-plus basis wherein the oil companies earned a predetermined return on capital in addition to all their costs being covered, for the crude they supplied. A similar dispensation prevailed for downstream integrated refining marketing companies.

Meanwhile, domestic crude production began to plateau at around 32 million tonnes, consumption increased rapidly, necessitating import of crude oil, but also oil products from time to time. While these were imported at global prices, domestic market prices were controlled through APM and kept at rates that were considered affordable to the consuming public.

The Indian economy launched a massive reform process in 1991. At this point, GoI decided to move certain petroleum products out of the APM regime. In 1974, lubricants were moved to an import-parity price regime. In 1998, all the petroleum products other than five were moved out of APM to an import-parity pricing regime. The five exempted fuels were Motor Spirit (MS), High Speed Diesel (HSD), Kerosene, Liquefied Petroleum Gas (LPG), and Aviation Turbine Fuel (ATF). It is noteworthy that these five products constituted nearly 80% of the Indian basket. Thus, government control over pricing remained despite marginal measures to free up the market. In 2001, ATF was moved to import parity leaving only the four mass-consumed products under the APM regime.

In November 1997, GoI decided to open up the entire hydrocarbon sector to private investors, based on the recommendation of a high-powered government committee.[10] FDI was allowed in exploration, refining, and marketing, although foreign equity participation in the refining segment was restricted to 49%. Two large and versatile new refineries, both export oriented, came up with domestic private capital, on the back of huge incentives and tax concessions. From 1999 onwards, upstream oil acreages – both onshore and offshore – were offered to investors for exploration through structured policies offering production-sharing contracts through a transparent competitive bidding process.

In 2002, GoI officially dismantled the APM regime for petroleum products. Henceforth, domestic product prices would be at par with import prices. Import parity pricing was to be applied to all petroleum products except kerosene and LPG, the two cooking fuels. While the prices of these two fuels were also notionally fixed at import parity, they were to be sold at a discount in the domestic market and GoI would reimburse the discount to OMCs through a transparent budgetary subsidy. Private investors with certain threshold investments in the sector were also allowed to retail petroleum products in the Indian market. Reliance, Essar, and Shell set up retail outlets and began to compete with OMCs for market share for all products other than kerosene and LPG which continued to be retailed by OMCs. In fact, although Reliance and Essar were envisaged as fully export-oriented refineries and as such, had availed tax concessions, they decided to enter the domestic market because of the differential between export prices and import parity pricing offered by the new policy. Within a year, private retailers had captured as much as 18% share of the domestic market for MS and diesel.

In back to back reforms, the price paid to upstream oil companies for their crude was also raised to import parity prices. This increased the cost of raw materials to OMCs, but since they were to realize import parity prices on two mass-consumed fuels, MS and diesel, and would be fully compensated through a budgetary subvention for underpricing cooking fuels, the situation did not seem too alarming. Global crude prices in March were hovering around $24 per barrel and import-parity prices of petroleum products were not high enough to ring alarm bells.

The move to deregulate prices signaled India's integration with global oil markets wherein domestic fuel prices would reflect global prices. Considering that imports constituted a significant proportion of total consumption even at this time, it was logical that domestic consumer prices should reflect this reality.

Simultaneously, in a conscious decision, the government also levied heavy taxes on these fuels – more than 100% on MS – on the rationale that the subsidies on LPG and kerosene would be met from these taxes. These taxes, duties, cess, and other charges were ad valorem both on crude and products.[11] As crude prices soared to over $100 per barrel, the government had to bow to public pressure and allow marginal cuts in the tax rates. As on March 31, 2012 the tax component in the pump price of petrol accounted for 41% of which central taxes accounted for 24% and State taxes 17%. In the case of diesel, these are 7% and 11% respectively.[12] In fact, these taxes constitute a substantial chunk of the tax revenues of both the central government and the States. Some local bodies also began to levy octroi or entry tax and claimed their own share of petroleum revenues which became quite significant over a period of time.

The Subsidy Juggernaut

In an unforeseen and unfortunate development from an importing country's perspective, the deregulation of fuel prices in the Indian market coincided with an upward thrust in global crude prices owing to a variety of factors. From around $23.65 per barrel of Indian basket crude in March 2002, prices went up to $26.65 in 2003, $39.21 in 2005, $55.72 in 2006, $79.25 in 2008, $83.57 in 2009, $85.09 in 2011, and to more than $115 in 2012.[13] The initial increases were triggered by the Iraq war and the consequent tightening of the oil market. Subsequently, peak oil fears, terror premium, and speculative premium kept prices on a firm upward spiral except for 2010 when the price declined slightly to $69.76 a barrel.[14]

An insidious, and from the government's own viewpoint not unwelcome outcome of rising crude prices, was the concomitant increase in government revenues. Taxes on petroleum were on ad valorem basis and thus served up a bonanza for the exchequer. While this was welcomed by the government, beyond a certain inflexion point, high diesel prices resulting from high crude prices and the ad valorem taxes on them started impacting the general inflation rate so much so that the government had to step in to moderate prices.

The original policy decision envisaged freedom to OMCs for periodically revising the retail prices of MS and HSD in tune with changes in global crude prices. But as inflation began to pinch the economy, the invisible hand of the government began to restrain OMCs from increasing the prices of MS and diesel to align them with import parity. In the first year, MS and diesel prices were revised 15 times, but subsequently, both the magnitude of revision and the periodicity declined. From 2004 onwards, the invisible hand especially targeted diesel, which had a steep inflationary impact.[15] Eventually, diesel prices stagnated far below import parity while MS prices increased modestly, but also lagged behind import parity. The public sector OMCs managed primarily because their refining costs were lower and what they lost in terms of not charging import parity for retail products was made up for by import parity for refinery gate prices. Thus, by an admixture of subsidies and cross-subsidies, the government had created a complex web which would soon ensnare the OMCs.

That the government of the day was a coalition of many political parties whose base was regional, exacerbated the situation.[16] The regional parties, strident in their opposition to any price rise at the best of times, became recalcitrant during election times. With 25 States making up the Indian Union, there was always some election round the corner, restraining the GoI from raising fuel prices so as not to upset the vote banks in that State.[17]

As crude prices went up and the coalition partners became even more strident in their opposition to price increases, the government began to control MS and diesel prices directly. In fact, price increases began to be announced by the Ministry of Petroleum and Natural Gas and not by the OMCs. The gap between import parity price and domestic market price steadily widened. Very often, even when OMCs were allowed to raise fuel prices, coalition partners forced the government to roll-back the price rise, partially or wholly.[18] The government agonized over weeks before raising petrol or diesel prices and it was debated intensely in the media, giving enough time to allow coalition partners to mobilize public opinion against the proposed price rise.

Meanwhile, from FY 2005 onwards, GoI had also decided that it would foot only a third of the difference through direct government subsidy, while another third was to be borne by upstream oil companies like ONGC which enjoyed windfall profits from import parity price for the crude it produced. The OMCs were asked to bear the last third of the difference between domestic market price and trade parity price. OMCs began to show up an entry called "under-recoveries" in their balance sheets. The rationale for the terminology was that the OMCs were not making losses, since their refining margins made up for the difference between domestic market price and import parity price. As long as OMCs covered their refining and retailing costs from the sale proceeds plus two thirds subsidy payouts by both government and upstream companies, they managed to hobble along.

However, with crude prices spiraling above $100 a barrel, OMCs' gross refinery margins were no longer sufficient to cover even the one-third subsidy which they were expected to absorb, primarily because the subsidy, originally meant only for cooking fuels, had begun to spill over to diesel and MS as well. In 2006, import parity pricing was discarded in favor of trade parity pricing on the recommendation of an official committee.[19] The rationale was that if 20% of petroleum products are exported, trade parity should comprise 80% import parity plus 20% export parity. This further impacted the revenues of the OMCs since trade parity price is less than import parity price.

The cumulative unmet under-recoveries of all the three OMCs on account of MS and diesel are estimated at Rs. 3693 billion up to FY 2011. The cumulative combined unmet under-recoveries on account of all four fuels up to FY 2011 are estimated at Rs. 7819 billion ($17.15 billion).[20] For individual products, the current (2012) under-recovery is Rs. 14.29/litre of diesel, Rs. 31.03/litre of kerosene, and Rs. 570.68/LPG cylinder.[21]

OMCs are being increasingly driven to a situation where their under-recoveries are threatening to translate into actual losses, seriously impacting their ability to carry on their activities. Until FY 2011, all the OMCs reported a cumulative Profit After Tax of 1.69% of their collective turnover.[22] However, the spike in crude prices since then threatens to drive these companies into the red. IOC, the biggest of the OMCs, has reported a net loss of Rs. 74.85 billion for the two quarters of FY 2011 ending September 2011, seriously jeopardizing its ability to source crude from the international market or even the ability to raise debt to carry on its activities.

By all indications, it appears that the Indian government has trapped itself and its OMCs into an unstoppable and unsustainable subsidy juggernaut. That India is a federal democracy where coalition governments are here to stay is, no doubt, a major contributing factor in bringing about this unsustainable predicament.

Fuel Subsidies: Devil in the Details

Exempting cooking fuels from price decontrol and instead funding the difference in prices through the general budget is almost irresistible to populist regimes. First, it satisfies the canon that any subsidies should be funded from revenues generated within the sector. Second, the government could justify its stand that heavy taxation sent out the right price signals to consumers. Third, being indirect taxes, these were easy to collect and leakages were minimal. Fourth, petroleum fuels have low price elasticity and as such, taxing them assured revenue collections for the exchequer. Fifth, taxing select segments of consumers, in this case, the luxury segment that used MS to drive private cars and diesel (the latter used in primarily for freight whose inflationary impact is not readily apparent to the general public) and returning part of it as subsidy on cooking fuel – a welfare measure – invests the government of the day with an aura of munificence. The government is seen as discharging its responsibility to provide clean cooking fuel at affordable prices to an increasingly urbanizing India.

However, the policy-makers erred on two counts. First, they failed to identify and target the fuel subsidies at the vulnerable sections of society. Cooking fuel subsidies were made available to every household, including those that could well have paid its economic cost. Thus, every household with an LPG connection became automatically eligible for subsidized LPG cylinders, although LPG can be safely termed a middle-class fuel. LPG subsidy, largely pre-empted by the urban middle class, was redundant from the start and this fuel should have been kept out of the ambit of subsidies.

Second, the government lost focus, partly because of coalition pressures, but also because of populist tendencies, expanding the scope of the subsidy from cooking fuels to first diesel and then to MS as well, the latter a fuel used by the relatively rich. In a sense, the government got carried away. In fact, the situation would not have deteriorated to such alarming levels had the government stuck to its initial policy of subsidizing only cooking fuels. Over the years, cooking fuel subsidies became a smaller and smaller proportion of total fuel subsidies. In 2011, the share of cooking fuel subsidies in the total under-recoveries stood at just 47%.

Petroleum Sector as Milch Cow

A relatively inelastic demand for petroleum products has made them an attractive source of tax revenue for governments. Coupled with ease of collection, petroleum taxes have become the single largest source of indirect tax revenues for the governments at the Center, State, as well as some local bodies. Local bodies like Brihanmumbai Municipal Corporation (Greater Mumbai) earn substantial revenues from octroi and entry tax on petroleum products entering its jurisdiction.

Table 24.1 shows the revenues to the Central government, generated from sale of petroleum products from 2002 to 2011, as also the subsidies paid out by the Central exchequer. The revenues have been segregated into duties and taxes on the one hand and dividends, corporate taxes, and royalties on the other.

In 2003, nearly half of tax revenues of the GoI came from the petroleum sector. Apart from the Central government, the State governments also found the petroleum sector a lucrative source of revenues. State taxes included Sales Tax, royalties payable to the States, dividends to State governments, etc. Even the municipalities often found the petroleum sector an easy and ready source of revenue. They began levying octroi, entry tax, etc. and this became the major revenue source for many municipalities like Brihanmumbai Municipal Corporation. However, rarely were State governments and municipalities called upon to fund the subsidies, although occasionally they were persuaded to cut the rates of their levies.

A close look at the figures in Table 24.1 shows that the Central government alone collected taxes and duties far in excess of the subsidies it paid out to OMCs with much publicity, all with an eye on the vote bank. In the first two years after APM dismantling, the Central government did foot the entire subsidy bill on account of LPG and kerosene. However, in 2004, a policy decision was made to rope in upstream oil companies to share the subsidy burden. After all, they had been given import parity price for the crude they produced and as such were enjoying windfall profits, since their cost of production remained low relative to prices charged. The government was to pay only a third of the subsidy on petroleum fuels with the remainder being borne equally by upstream oil companies and the OMCs themselves. Initially, even the one-third share of subsidy to be paid by the government was funded through off-budget bonds with an average maturity of 5–7 years. This implied postponing the subsidy burden to be borne by future governments while the present government was free to use the petroleum tax revenues for other purposes. Only from FY 2009 did the GoI begin dispensing subsidy in cash.

The beleaguered OMCs, who had to raise funds from the market, often did so by deeply discounting the oil bonds. However, after much persuasion by OMCs, the government decided to reimburse them in cash through the general budget. Media reports frequently played up the subsidy element in the budget, but paid little attention to the tax and other revenues generated by the sector.

Perverse Outcomes

While the politically-driven pricing policies in the petroleum sector have had egregiously deleterious impact on the finances of the OMCs in the form of mounting "under-recoveries" and incipient actual losses, these are by no means the only outcomes. There are other unintended consequences, perverse and often more damaging, primarily because they have virtually tied the government in knots and launched it into a vicious circle. These have ranged from mass diversion of subsidized fuels to unintended beneficiaries

Table 24.1 Central Government of India revenues versus subsidies in the petroleum sector (Rs. billion). * = revised estimates.

Year	2002	2003	2004	2005	2006	2007	2008	2009	2010
Contribution to Central Exchequer, (customs duty, excise duty, royalties, cess, corporate tax, dividend, etc.)	645.95	691.95	776.92	693.47	782.29	848.33	799.80	820.98	1111.72*
Contribution to Central Exchequer only from duties, taxes, and cess	465.33	501.07	551.77	631.43	718.93	783.77	705.57	717.67	1026.17*
Cooking fuel subsidies paid by Central Government	22.96	40.79	29.30	26.62	25.24	26.41	26.88	27.70	29.05*
Total petroleum subsidies paid by Central Government	52.55	63.51	29.56	26.83	26.99	28.20	28.52	149.51	383.86*
Total under-recoveries of OMCs	99.26	155.66	207.72	272.92	311.08	372.66	485.13	343.91	441.61*

Source: Compiled by the author from various sources.

such as diesel car owners, proliferation of diesel vehicles including SUVs, adulteration of diesel with subsidized kerosene meant for the poor, and its attendant impact on fuel efficiency of diesel vehicles, diversion of subsidized LPG to commercial uses and worse, cheap LPG crowding out piped gas to domestic households, etc.

Foremost among the unintended consequences is the inability of the government – Central, State, and local – to wean itself away from substantial, easy, and assured revenues that accrue from the petroleum sector. Hydrocarbon taxes account for a third of total revenues from customs duty and excise collected by the GoI, 12–30% of the tax revenues of States, and the major chunk of revenues of some local bodies. Substituting these levies with other sources of revenue will prove to be extremely challenging especially as avenues for further taxation get squeezed. Meanwhile, Central government's revenue expenditures (food subsidies, fertilizer subsidies, interest payments, pensions, and salaries) have been steadily increasing, so much so that these account for 38% of total government spending in 2011. In short, having treated the hydrocarbon sector as a milk cow, it becomes increasingly difficult for the government to make a tactical retreat. The government seems to have locked itself into a vicious circle where high tax rates on soaring crude and product prices lead to higher fuel prices for end consumers, which, in turn, necessitates even higher fuel subsidies to keep OMCs afloat.

Second, the government had unwittingly mounted a tiger when in 2002 it decided to offer subsidized cooking fuel indiscriminately to every household, instead of targeting vulnerable sections of society. Now these relatively affluent and influential urban households which have become used to subsidized cooking fuel, constitute a veritable vote bank. This segment is also articulate, and uses the mainstream media to maximum advantage to state its case against withdrawal of subsidies. In the event, dismounting the fuel subsidy tiger is sure to cost the government its incumbency. In a coalition government, it seems well-nigh impossible. The costs of withdrawing the subsidy will be identified with and borne by the leading party in the coalition whereas its coalition partners can oppose such withdrawal with profit. This has been tested time and time again when coalition partners have often forced the government to roll back fuel price increases for populist gains.

Third, keeping diesel prices down by not allowing OMCs to raise them in tandem with international trends has led to unforeseen shifts in the market-place. From the time the government started controlling diesel prices, car manufacturers have been flooding the Indian market with diesel-fueled cars. Initially these addressed the urban taxicab sector, but gradually they began to encroach upon private cars. Even luxury brands have come up with diesel-fueled models to entice fuel-price sensitive consumers. In fact, one study found that 40% of the diesel used in the country is by diesel cars.[23] Cheap diesel, primarily meant for freight, has also led to indiscriminate increase in truck-borne traffic as opposed to rail-borne freight, a more economical way to transport goods. Indian highways are perpetually clogged with truck traffic, endangering the environment as well as human lives, not to speak of the jumps in diesel consumption in recent years.[24] The share of diesel in the fuel basket stood at 43.7% in FY 2011, up from 35.19% in FY 2002.[25] Diesel car output growth has outpaced growth of petrol-driven private cars so much so that the diesel automobile lobby is threatening to become a forceful voice in ensuring that diesel remains a subsidized fuel.[26]

As for MS, the government does not even have a fig-leaf to justify its reluctance to allow OMCs to raise prices in tune with global prices. It is crass populism at play. Subsidizing MS has led to unbridled increases in the number of petrol-driven private vehicles even as public transportation has taken a back seat. Road infrastructure is severely stressed and road accidents kill thousands of people every year.

The fourth unanticipated outcome of kerosene subsidy has been large-scale adulteration of diesel with subsidized kerosene. Subsidized kerosene, the cooking fuel used by poor households, is distributed through the Public Distribution System to holders of ration cards on a predetermined quota linked to household size. At Rs. 27 per litre, the price differential between subsidized kerosene and diesel is indeed very significant, pushing the former into diesel tanks of cars, lorries, and trucks. Kerosene when mixed with diesel defies easy detection. The diversion, that continues to date, has been traced to the level of dealers even before kerosene reaches PDS outlets and at PDS outlets before ration card holders can claim their quota. Chemical markers used to distinguish subsidized PDS kerosene have been neutralized with counter-chemicals by an ingenious mafia that has fattened at the expense of the poor as well as of the OMCs. An official committee reports that the diversion accounts for 40% of all kerosene consumed in the country.[27]

Even subsidized LPG intended as domestic cooking fuel gets diverted to commercial eateries and establishments including kitchens of star hotels and fancy restaurants. There has been a proliferation of duplicate LPG connections to households with attendant spare cylinders available for diversion. Corruption and nepotism on the part of politicians and the oil bureaucracy have played an important role in facilitating duplicate or multiple connections to the same households, especially in urban areas, whereas a large population in rural areas remains outside the pale of commercial energy. Ingenious ways have been devised to outwit the OMCs which have diligently designed different sizes of LPG cylinders for domestic and commercial uses. Illegal and risky transfer of LPG from domestic to commercial cylinders is rampant and a common sight in most cities and towns. Such diversion often happens at the dealer or distributor level, without even the knowledge of the householder who unwittingly facilitates it by not drawing her quota regularly.

LPG subsidies alone cost the exchequer Rs. 19.74 billion in FY 2010.[28] Together, kerosene and LPG subsidies cost the exchequer Rs. 29 billion in FY 2010. Modest as these are, compared to the revenues collected from the sector, they have had a perverse impact on the spread of clean fuels in cities. Around 25 cities and towns in India have local gas distribution networks which bring piped cooking gas to households. Subsidized LPG has all but killed the market for piped cooking gas since it has to be priced below the former. India's domestic gas production has declined drastically in recent months – it is not within the scope of this paper to discuss the reasons behind this development – and as such, domestic piped gas has to be supplied increasingly from imported LNG. With a sizeable price differential between domestically produced natural gas and imported LNG, piped cooking gas is bound to be more expensive than subsidized LPG. There are no subsidies on piped natural gas, although some companies resort to cross-subsidies to keep cooking fuel prices low. Nevertheless, with rising global LNG prices, the price of piped cooking gas is expected to rise, seriously threatening its acceptability in a market used to subsidized LPG. In a sense, the government has unwittingly capped the price of cooking gas at the level of subsidized LPG, a perverse outcome not envisaged hitherto.

Dismounting the Subsidy Tiger

Apart from the political infeasibility of dismounting the subsidy tiger, the government has unwittingly created a chain of unintended beneficiaries because of the huge corruption potential the pricing policies have unleashed. The beneficiaries range from politicians to bureaucrats, both in government and the oil industry – dealers, transporters, retailers, and others all of whom are likely to resist stoutly any attempt to set right the distortions. It has already created a huge vested interest in an urban middle class increasingly used to subsidized fuel that it does not deserve. That this class is also the most articulate and vocal does not help matters much. In the event, it seems extremely difficult, even unlikely that the government will ever be able to dismount the pricing tiger.

In the last few years, the government has finally acknowledged its failure to identify beneficiaries and target the subsidies at them. Technology may come to the rescue in the form of biometric identification of targeted beneficiaries. The government has put in place a project that would issue a unique identification number to each resident in the country. Smart cards embedded in advance with the monthly/annual subsidy amount and married to biometric identification in the form of thumb impressions, are being issued to intended beneficiaries – those below a poverty line – on a pilot basis. There is no product differentiation such as different weight cylinders as in LPG or blue markers as in kerosene. Instead, every petroleum product will be priced at market rates and available from any retail shops, not necessarily PDS outlets. Only the targeted beneficiary can use the smart card using his/her biometric identity, to buy the product from the market. The smart card is akin to a debit card where the beneficiary pays only part of the price and the rest is debited from the smart card already embedded with the subsidy amount. The challenges to implementing the smart card subsidy dispensation system are myriad, but hopefully, technology will find a way out.

While the introduction of smart cards might bring down diversion or misuse of subsidy, it is unlikely to get the government off the subsidy tiger. Limiting subsidies only to smart card holders below a poverty line and excluding the middle classes will be a huge challenge to any government desiring to be voted back to power in the next elections. Yet, if ever there was a chance of surgical correction of the subsidy canker, this might be it. The government can legitimately claim it has taken care of the deserving and the needy.

The sooner this is done, the easier it would be for the government to effect the surgical correction. At current global prices of LNG, piped cooking gas can still be supplied at prices similar to those of subsidized LPG. Or, the government could signal a modest rise in the price of LPG so that piped cooking gas remains competitive. Meanwhile, it can launch a scheme to immediately withdraw LPG connections to those households supplied with piped cooking gas. An independent statutory regulator for the sector is currently engaged in the roll-out of city gas distribution (CGD) networks. A smart government would lend its weight behind this move and allocate domestic natural gas on a priority basis[29] to CGD so that piped gas reaches as many households as possible. If subsidized LPG to these newly connected households is withdrawn simultaneously, perhaps the government can control the subsidy outgo on LPG to some extent. Better still, these newly released LPG connections can find their way to rural areas where non-commercial fuels dominate, in which case, the subsidy component will remain, but may now be better targeted.

Finally, any resource not priced right tends to be used profligately and even wastefully. Subsidized fuels have encouraged Indians to adopt lifestyles that are not energy-efficient. Efforts by the Petroleum Conservation Research Agency and Bureau of Energy Efficiency to persuade consumers to use efficient gadgets have had only a modest impact compared to a much more effective instrument like price signal. The private automobile sector has witnessed explosive growth rates in recent years. A globalized world, brought into living rooms by satellite television, offers tantalizingly accessible lifestyle choices to an aspirational population, choices that are often energy-intensive.

Conclusion: Implications for Global Energy Policy

India, which was hobbling along leisurely at the Hindu rate of growth for decades, has finally emerged from its lethargy after liberalization of its economy. In the last two decades, nearly a third of its population has finally tasted the fruits of development and is hungry for more. The others have been impatiently watching from the sidelines for their turn. The growth momentum has to be sustained if the legitimate aspirations of its young and eager population are to be fulfilled. Thus, by all indications, India is locked into an ineluctable fast growth paradigm that also happens to be energy-intensive.

While that is daunting in itself, the fact that India's middle class has got used to subsidized fuels which it neither needs nor deserves, makes it that much more difficult for a democratic government to wean them away from these subsidies and force them to pay economic costs. Dismounting the subsidy tiger would be difficult enough in any democratic regime, but in a federal democratic coalition with a disproportionate say for regional political parties, it seems well-nigh impossible.

The inability to dismount the subsidy tiger has even more far-reaching implications. It will severely restrict the country's ability to switch to renewables, the latter being more expensive than fossil fuels, even with government subsidies. It might be a while before renewables can compete with conventional fuels in the market-place, but even when they do, it is doubtful if the Indian consumer, grown complacent on subsidized fuels, will be ready to switch to renewables. For a climate-stressed world, a rapidly developing India with an expanding and youthful population riding on an energy-intensive, fossil-fueled growth trajectory could have serious consequences.

It is evident that India's current growth paradigm is unsustainable. Any course correction will necessarily spell pain. A top-down policy-driven surgical correction seems unlikely for the reasons outlined above. Yet the sheer unsustainability of the current paradigm is bound to devise a self-correction mechanism. Gazing into the crystal ball, one can perhaps predict a situation where, unable to support the subsidies, the GoI might just retreat into inaction, leaving OMCs to fend for themselves. R. S. Butola, current CEO of the IOC, is already on record to say that there could be a supply crunch for petroleum products since the company's balance sheets may not be able to fund crude purchases at levels required to keep the markets fully supplied.[30]

Shortages in the supply of fuels, currently non-existent, could become a regular feature in the not-too-distant future if fuel prices are not increased substantially and immediately. Where price signals are weak or non-existent, supply signals will take over, forcing consumers to either cut down consumption or look for alternatives which are bound to be a lot more expensive or both. It could spawn a flourishing in black markets for petroleum fuels with its own set of attendant problems. A similar situation is already unfolding in the electricity sector where the grid is unable to meet the demand fully and thus forces consumers to set up captive generation based on liquid fuels. There has been a proliferation in captive power plants set up by industries as well as residential condominiums that use diesel generators, despite the steep costs involved. In the not-too-distant future, consumers of petroleum fuels will either reduce their consumption or will eventually switch over to solar-powered battery-operated automobiles, CNG-driven vehicles, or even solar cookers, etc. The latter will entail steep costs and/or the inconvenience associated with such technologies.[31] Domestically produced biodiesel or ethanol, alternatives to hydrocarbon fuels, cannot displace the latter primarily because growing crops that make these fuels has been found to be too input- and water-intensive, compromising their economics. Besides, in a densely populated country, they entail diversion of food acreages to grow these crops, a solution that is unlikely to be accepted because of its far-reaching implications for food security.

In the near term, subsidized fuels will encourage profligate consumption and proliferation of private transport. Import-intensity will aggravate and the import bill will pre-empt funds that could be invested in infrastructure or social development. Populist governments will divert funds from education and public health to pay for imported oil and will attempt to keep the subsidies going as long as they can. In the medium term, painful and cathartic transformation will be inevitable. When the Indian economy either slows down or even collapses, it is bound to have a ripple effect on global markets.

Subsidies will also hamper the introduction of fuel-efficient technologies. India's carbon-intensity, now hiding behind average per capita, will increase disproportionately for the urban rich and the middle classes. Because of their sheer numbers – 300 million and growing – it might have a significant impact on global efforts to contain carbon emissions. Any efforts to tax carbon will be met with stout resistance by a population spoilt by cheap fuels. A populous, young, and aspirational India locked into a subsidy trap can slow down collective efforts to tackle global warming.

To conclude, like all public policy, the policy to provide cooking fuels at affordable prices to an increasingly urbanizing population began with laudable intent, but got distorted along the way. It is a case study that holds valuable lessons for all aspiring economies in Asia and elsewhere, who have launched themselves onto a path of rapid development that is at once energy-intensive and import-dependent. Integration into the global market-place is ineluctable especially for a predominantly imported commodity life fuel. Short-sighted and undiscriminating pricing and subsidy policies and misguided implementation could lead to severe distortions in the economy and seriously jeopardize the ability of governments to sustain the growth momentum.

Notes

1. According to the Planning Commission of India, GDP growth rates of 9% would require energy consumption growth of 6.5–7%. http://planningcommission.nic.in/plans/planrel/12appdrft/approach_12plan.pdf, p. 9, accessed April 28, 2012.
2. On the one hand, India is introducing more fuel-efficient cars and advocating petroleum conservation, but on the other, encourages policies that promote the use of private cars against public transport. India's Automotive Mission Plan 2006–2016 envisages an increase in the country's automotive industry turnover from $35 to $145 billion USD accounting for 10% of GDP and providing employment to 25 million people by 2016.
3. Advance estimates of the India Economic Survey 2012 put the share of service sector in the Indian economy at 56.3% for FY 2011 (excluding construction). Automobiles grew at 17.6%, steel at 9.6%, and construction 8.1% during the first three quarters of FY 2011 (compiled from various sources).
4. 60% of Population in Most States Below Poverty Line. *Times of India*, April 29, 2012.
5. http://www.cea.nic.in/reports/national_elec_policy.pdf, accessed April 29, 2012.
6. http://planningcommission.nic.in/reports/genrep/rep_intengy.pdf (now adopted as policy by the GoI), accessed April 29, 2012.
7. Revised Budget Estimates for FY 2011 shows fiscal deficit at 5.9% of GDP and Budget Estimates for FY 2012 at 5.1% of GDP. Hydrocarbon fuel subsidies funded from the general budget account for 8.48% of fiscal deficit in 2012. http://indiabudget.nic.in/ub2012-13/bag/bag1.pdf, accessed April 29, 2012.
8. It is the responsibility of the GoI to provide cooking fuels. Yet, the national party leading the coalition at the Center will have to heed the voices of other coalition partners who

represent regional parties. On several occasions in the past, the Central government run by a coalition has had to roll back increases in the prices of cooking fuels in response to the demand of coalition partners from the States.

9. ONGC became a corporation in 1993.

10. Report of the Strategic Planning Group on Restructuring of Oil Industries, Ministry of Petroleum and Natural Gas, Government of India, September 1996.

11. Since 2010, the rates have been changed to part ad valorem and part fixed.

12. Petroleum Planning and Analysis Cell, Ministry of Petroleum, GoI, www.ppac.org.in, accessed on April 28, 2012.

13. Petroleum Planning and Analysis Cell (PPAC), Ministry of Petroleum and Natural Gas, Government of India. http://www.ppac.org.in, accessed April 28, 2012.

14. Ibid.

15. Diesel has a weight of 4.67 in the wholesale price index and is the heaviest item in a basket of 670 that constitute WPI. Every increase of one rupee in the price of diesel is estimated to push up WPI by 0.14%. Compiled from answers given in Parliament in response to a question raised by a Member of Parliament on April 24, 2012.

16. In 2002, when APM was dismantled, the ruling coalition led by the BJP had 13 parties in its fold. Subsequently, when the Congress came to power under UPA, it had the support of several parties including the Left Front which supported the government from outside, but vehemently opposed any fuel price increases.

17. Petroleum prices were within the jurisdiction of the GoI and were applicable throughout the country with variations for sales tax and other local levies.

18. From 2002, petrol prices were reduced 15 times and diesel 12 times, to date. A substantial number of these reductions were undertaken because coalition partners and opposition parties demanded a roll-back of price increase.

19. Report of the Committee on Pricing and Taxation of Petroleum Products, GoI (Rangarajan Committee), February 2006.

20. Petroleum Planning and Analysis Cell (PPAC), Ministry of Petroleum and Natural Gas, Government of India. http://www.ppac.org.in, accessed April 28, 2012.

21. Ibid.

22. Ibid.

23. http://articles.timesofindia.indiatimes.com/2012-01-25/pollution/30662379_1_diesel-cars-petrol-car-car-segment.

24. According to government data, 60.4% of diesel is consumed by the road transportation sector in FY 2010. http://petroleum.nic.in/pngstat.pdf, p. 71, accessed April 27, 2012.

25. PPAC.

26. http://www.indianexpress.com/news/maruti-sees-diesel-cars-driving-sales-in-com/928336/, accessed May 11, 2012.

27. Report of the Committee on Pricing and Taxation of Petroleum Products, GoI, February 2006.

28. Ibid.

29. Domestically produced natural gas is allocated by a government committee on priority basis to competing gas-based industries. As such, fertilizer and power top the priority list with CGD taking fourth place.

30. Growing Under-Recoveries Worry Oil Marketing Firms. *The Hindu*, April 4, 2012. http://www.thehindu.com/todays-paper/tp-business/article3274917.ece, accessed April 27, 2012.

31. CNG (compressed natural gas) is the mandated fuel for public transport in the national capital and a fuel of choice for private automobiles in all cities with CGD networks. However, CNG prices will have to be competitive with diesel for consumers to prefer it to the latter.

Chapter 25

The EU's Global Climate and Energy Policies: Gathering or Losing Momentum?

Richard Youngs

Introduction

In recent years, the EU's global energy and climate policies appear to have gathered impressive momentum. The EU has agreed a number of new documents that promise to strengthen Europe's presence in international energy policies. These include an EU Energy 2020 strategy; a communication on external energy security; policy documents outlining a reinforced "climate diplomacy"; and an Energy Roadmap 2050 presenting energy scenarios for the next four decades. This raft of new initiatives was supplemented with a strong performance at the December 2011 UN Climate Change Durban summit. The EU has rarely had such a busy period in the external dimensions of its energy policies.

This chapter assesses the extent of this gathering momentum in the EU's global energy policies. It argues that impressive foundations have been laid for a more coherent and proactive international climate and energy strategy, but that vital issues remain unresolved. Crucially, the EU's leadership in global climate policy is increasingly compromised by tensions between its internal and external policies, as well as between traditional energy security and climate change aims. Internal European cooperation on both climate change and energy market integration serves as the launch-pad for EU global influence; but the lack of clarity in these same internal policies also increasingly detracts from the EU's international projection.

These features reflect policy preferences but also the EU's complex institutional structures. The EU is a *sui generis* actor in energy policy. It is a multilevel organization, with a complex division of energy competences between its supranational bodies and the member states. It relies heavily on a regulatory approach to energy questions. Common EU rules coexist with fiercely independent member state policies, especially in the broader international arena. Some aspects of European global energy policies constitute highly geopolitical paths followed by member state governments. Other aspects have a more institutional character, with outcomes explained by the structure of EU cooperation processes. Crucially, this multi-faceted nature of EU energy deliberation is both a strength and weakness. The EU's policy challenge is to combine the rules-based and

The Handbook of Global Energy Policy, First Edition. Edited by Andreas Goldthau.
© 2013 John Wiley & Sons, Ltd. Published 2013 by John Wiley & Sons, Ltd.

geopolitical approaches in a more reinforcing fashion. At present, the return of very traditional approaches to both domestic economic policy and international energy security threatens to subvert EU global climate leadership.

Domestic Doubts?

Experts and European Commission officials ritually claim that the EU's lead role in climate policies enables it to influence the broader international dimensions of global warming. But are EU climate commitments really exemplary enough to lay the foundations for it to play a lead role in global energy politics?

At the Durban summit in December 2011, the EU pushed hard and on the basis of its proposed accord China, India, and the US finally agreed to emissions targets with legal force, albeit only from 2020. And 35 states agreed to a second round of post-Kyoto commitments. Climate commissioner Connie Hedegaard celebrated: "The EU's strategy worked. When many parties after Cancún said that Durban could only implement decisions taken in Copenhagen and Cancún, the EU wanted more ambition. And got more. We would not take a new Kyoto period unless we got in return a roadmap for the future where all countries must commit."

The EU is by far the largest importer of energy, buying in nearly twice the US's energy import and five times that of China. But it has the lowest energy intensity of all regions (measured as energy supply per unit of GDP) and the highest demand for renewable energy (European Commission 2011d: 7–8). The EU is on track to meet its target to have 20% of its energy generated from renewables by 2020. Two thirds of new generating capacity in the EU now comes from renewable sources. Gradually, the more ambitious target of securing 80% reductions by 2050 has come to dominate policy deliberations. The new Danish government that took office in November 2011 made a commitment to pursue the goal of having the county's entire electricity and heat supply come from renewable sources by 2035. In late 2012, binding EU rules on energy efficiency came into force. The EU has set efficiency targets within public procurement rules.

Low carbon technology now represents a €300 billion market and provides employment to over 3 million workers in Europe. The Commission has supported 12 large-scale pilot projects on carbon capture and storage (CCS). The 2009 EU Energy Programme for Recovery committed €4 billion of investment in infrastructure and interconnections, alongside renewable projects. Of this total, €1 billion went to CCS projects. A first license for the commercial implementation of a CCS project was granted in France in 2011. In 2010, the European Investment Bank (EIB) channeled a record €19 billion of credits to low carbon initiatives, a 20% increase from 2009 and two thirds of all EIB loans in Europe (*Platts EU Energy* 252 (2011)). Across all its various budget lines, by 2010 the Commission was putting €1 billion into "frontier" low carbon research and development.

The UK has launched a Green Investment Bank, with £3 billion capital. In April 2012 the British government launched a new £1 billion scheme for the commercialization of CCS, and another £125 million for research on CCS. For the period 2011–2014, the German government increased its research and development funding for green technologies by 75% over the preceding three-year period, partly in response to the decision to phase out nuclear power generation by 2022 (*Platts EU Energy* 264 (2011): 5). Late in 2011 the EU agreed to make €300 million of Emissions Trading Scheme (ETS) revenues available for CCS and other renewables projects through the EIB.

These all represent significant advances. However, the EU's general performance on climate change policy has been far from faultless. Many observers doubt the logic and impact of the EU's talismanic "20/20/20 by 2020" strategy: its arbitrariness is reflected in the convenience of all the numbers being 20. In May 2010 the Commission produced a review which argued against moving unilaterally from the 20% to a 30% emissions reduction target. European states are not on target to meet their energy efficiency targets. DG Energy has lamented that "the quality of National Energy Efficiency Action Plans, developed by member states since 2008, is disappointing" (European Commission 2011c: 5).

The effect of the economic crisis is such that the 20% emissions target is now achievable without great effort; even the 30% target would not require much additional reform. Moving from 20% to 30% reductions would only cost the EU 0.1% of GDP, according to the Commission. Yet despite all this, the EU promises to increase to a 30% reduction target only if others follow suit. Moreover, this is not a firm or ambitious enough commitment to make states like the US calculate that they would be better off in terms of net welfare gain by increasing their own offers (Bréchet *et al.* 2010).

Some member states, like the UK, have moved to the higher 30% target, in binding fashion. They argue that these moves should be Europeanized if the EU as a whole is to retain credible climate leadership. The UK has been one of the European states most committed to combating climate change but still needs to double its rate of emission reductions to meet its long-term targets. Moreover, the UK lags behind other OECD states in low carbon research and development, which is now at too limited a level to make any notable impact (Bowen and Rydge 2011: 13 and 16). In autumn 2011 the British chancellor, George Osborne, caused waves when he declared that henceforth the UK would seek to move no faster on green commitments than its EU partners, in an effort to conserve jobs.

Respected expert Dieter Helm insists that the EU has only made progress on its emissions targets by dint of the collapse of Soviet-era industries in eastern Europe, a switch from coal to gas, and now the economic recession. Moreover, the EU simply has not constructed the infrastructure for feeding significant amounts of renewable energy into the grid. The EU has focused on subsidies for renewables, but the broader structure of the energy market has not changed, meaning that even current levels of green electricity sit idle unable to get into the grid (Helm 2011). Europe's grid cannot absorb sufficient amounts of renewables-generated power to meet the EU's targets.

Member states' different forms of support for renewables are effectively undercutting any prospect of a single market in green energy. No European government will subsidize green energy in another member state (van Agt 2011). Some analysts fear that the variety of liberalization (or "unbundling") options now available to national energy champions will fragment the European market even more. Renewables-generated electricity is still not traded across borders, holding back its take-off. Proposals for a North Sea supergrid interconnector for wind farms at sea has been held up by national protection of home markets. The Commission has targeted France and the Czech Republic for failing to comply with rules on access for renewable power into national grids.

The recession has eaten into funding for renewables. Italy has announced cuts in solar power incentives. Spain has cut subsidies for solar investments, leaving many companies in severe difficulties. Denmark has gradually reined back on its use of wind farms, as these were proving to be inefficient and of intermittent use. EU officials fret that R&D on renewables has slowed dramatically due to the economic crisis (*Platts EU Energy*

246 (2010): 8). The Commission's flagship research budget, FP7, provides only €2.35 billion to low carbon research out of a €50 billion total allocation. The Commission proposal Horizon 2020 inks in €5.7 billion for renewable research out of a total €80 billion budget for post-FP7 research; both the business community and NGOs criticize this as woefully insufficient. The 2010–2011 Renewable Energy Attractiveness Index ranked China and US in the top positions, displacing EU governments. The share of low carbon power has grown but two thirds of this is still nuclear.

A 2012 mid-term report on the EU energy infrastructure fund reveals that most projects actually under way are for gas power generation; of four eligible CCS projects, three have collapsed. In mid-2011, the Commission started infringement proceedings against nearly all member states for having failed to implement the 2009 Directive on the development of CCS (*Platts EU Energy* 263 (2011): 1). At the end of 2011 Swedish utility Vattenfall canceled its CCS project in Germany.

As analyzed in depth elsewhere in this volume, the EU's much-lauded ETS has not had a dramatic impact on emissions levels. Even in its third phase, the scheme remains well short of full auctioning, which is what is really needed for it to have a major impact. Carbon offsets compromise the scheme's ostensible rationale: nearly all EU states have carbon footprints way in excess of their national reporting, because they simply buy the right to pollute outside EU. Sectors excluded from the ETS still account for 50% of the EU's total emissions. The ETS carbon price fell to an all-time low in April 2012.

Also pertinently, the touted nuclear renaissance is now on hold. A majority of EU member states were considering moving back into nuclear power by 2010. After the Fukushima disaster in April 2011 many backtracked, in particular Germany, Belgium, and Italy, and also non-EU Switzerland. There are exceptions. France is the most nuclear-dependent country in the world and the Czech Republic wants to become a "nuclear superpower." With 58 reactors covering 40% of its energy, France insists on a discourse of "low carbon" rather than "renewables." Nuclear's supporters say it will not only help meet emissions targets but also boost security: uranium supplies are plentiful in stable allies like Canada and Australia. But the trend is now firmly away from nuclear power, despite this making emissions reductions much harder to achieve.

In contrast, high-polluting coal production is booming. Germany, Spain, Poland, and others have been slow to reduce state aid to the coal sector. European states are only on target for their emissions targets because they are relying on coal-based production in China and other markets. They have reduced carbon production on their own territories but have increased carbon consumption, simply importing goods from carbon-rich production. Moreover, Germany and others plan to increase coal usage as they shift away from nuclear. Several German Land-level authorities have sought to extend the operating licenses of coal-fired plants in the wake of the federal government's commitment to phase out nuclear power stations.

In short, the mainstream components of EU environmental policies have advanced but are not without their serious shortcomings. It cannot be said that the EU's commitment to mitigating climate change is so strong that there is a significant or natural spill-over of climate-related considerations into its global policies. The EU regularly claims that the example of its own climate leadership serves as the foundation for an internationalization of its influence in this area of policy. But this domestic–external read-over is not without blemish. The advance of EU environmental policies has certainly been pervasive enough to ensure that other areas of European external policies can no longer remain completely immune from climate change considerations. But neither has their progress been so exemplary as to guarantee a natural externalization of climate primacy across the wider panoply of global energy issues.

Global Climate Funding

This qualified internal–external read-over can be seen in the scale and nature of EU climate funding. European ministers and policy-makers routinely allude to the significant scale of their climate funding commitments. They conceive such generous funding as a concrete contribution to a global presence in energy policy: European climate aid is aimed at helping adaptation in ways that are designed to reduce the strategic knock-on effects of climate change. The Copenhagen summit distinguished between two types of climate finance for developing countries. Fast-start finance enshrines a commitment by developed countries to provide new and additional resources, approaching $30 billion for the period 2010–2012, supposedly with a balanced allocation between adaptation and mitigation. Long-term finance then mobilizes $100 billion a year by 2020 to address the underlying needs of developing countries, and in the context of meaningful mitigation actions and transparency on implementation.[1]

For 2010–2012 the EU's total climate funding contribution was to be €7.2 billion. Of this, €2.2 billion was raised in 2010. This was split between 48% for mitigation, 33% for adaptation, and 16% for reducing emissions from deforestation and forest degradation. Just over half was given in the form of loans, and 48% as grants. Nearly 60% went through multilateral organizations.[2] To date, Denmark, Finland, and Germany have been the most generous proportionate contributors, allocating between 12% and 15% of bilateral aid to climate projects.

British climate aid serves as an example of what such funding is spent on. UK aid has included the inception of an Environmental Transformation Fund; funding for the Global Environment Facility; and a £17 million Climate and Development Knowledge Network designed to enhance developing countries' access to high-quality research and information on climate adaptation. Bangladesh has received the UK's largest country-specific climate change funding, with a budget of £75 million by 2013. Other notable UK initiatives include a £15 million Strategic Climate Institutions Programme in Ethiopia, designed to help build organizational and institutional capacity within the Ethiopian Government, civil society, and the private sector to increase resilience to climate variability, adapt to future climate change, and benefit from the opportunities for low carbon growth.

While regularly trumpeted as a leading edge of EU climate policies, the scale of climate funding has elicited much disappointment. Even many senior officials themselves express frustration with the EU's failure to invest top priority in this area of financial support. Funding for both mitigation and adaptation outside the EU is still subject to limits of a degree that raise doubts over how much importance is really attached to the external dimensions of climate change.

Economists point to the gap in existing climate finance: current allocations stand at around $15 billion a year; the 2020 target is $100 billion a year; $200 billion is required to make any kind of tangible impact (Haites 2011: 967). The amount of climate funding negotiated through the UNFCCC is a tiny percentage of the potential estimated cost of climate change: 0.5% compared to 20% of OECD GDP (Mabey 2009: 5–6). Yet still governments have haggled over the distribution of such funding allocations. Development commissioner Andris Piebalgs acknowledges that the scale of climate financing agreed so far within mainstream Commission development budgets can assist in a minor amount of adaptation at the margins, but not help prepare more anticipatory solutions to the strategic impact of global warming. DG Energy officials recognize there is still a need for more systematic use of research and development budgets to include neighbors in renewables development. Member state officials also acknowledge that dialogue with consumer countries on cooperation in climate aid projects has so far remained pitifully limited.

The EU is still not on target to meet its commitments on climate financing for developing countries. It raised hackles on the first day of the Cancún summit by revealing that half its fast-start funding would take the form of loans and private equity instead of grants. Many environmental campaigners express concerns that the EU still has to demonstrate that it will resist the temptation simply to divert existing development aid. Some accuse the EU and member states of using the climate adaptation agenda as a covert means of introducing new forms of conditionality and even trade barriers. Governments have rejected novel proposals, such as that to ring-fence future taxes on banks for climate adaptation. Italy has been singled out for a particularly poor record in delivering on its promised funding.[3]

Member states still have different ideas on additionality and on reporting criteria. Most governments over-report climate funding to least developing countries. France is guiltiest on this issue of additionality; much of its climate funding commitment simply repackages existing aid projects in disingenuous fashion (Scholz 2010: 2). Member states have pushed up to 50% the share of ETS revenues to be allocated to climate projects, so as to reduce the burden on their own budgets. This represents another dent in the spirit that new climate aid promises should bring additional money to the table, not simply divert resources from other revenue sources.

The balance between internal and external funding for renewables now engenders fierce debate. Most member states express unease over increasing external renewables support relative to funding for projects *within* Europe. Consultations for the new EU Energy 2020 strategy revealed growing doubts on the part of most member states about large-scale funding for adaptation projects in non-EU states. The most common member state position is to argue that the EU should reduce subsidies for renewable energy and adaptation outside Europe and instead channel funds toward internal energy efficiency. Poorer member states express explicit opposition to huge transfers for climate funding to the likes of China, countries growing fast and suffering less economically than many EU governments. Officials advocate much more formal and high-level political backing to sell renewable technology developed in Europe to other consumer countries around the world. This reflects an apparently more mercantile than developmental approach to climate funding.

Global Partnerships for Renewables Technology

The September 2011 Commission communication on security of supply and international cooperation (European Commission 2011a) offers a number of proposals to enhance cooperation with international partners on renewables. These include a range of new international partnership agreements that identify cooperation on renewables as their "primary" aim; an initiative to get other forums such as the G20 to prioritize global rules on renewables development; an enhanced framework for a Mediterranean Solar Plan; the extension of carbon pricing to non-EU states under the rubric of external agreements; the extension of EU initiatives at the level of cities beyond the EU's borders; reciprocity in access to renewables research programs; an extension of the Energy Charter treaty's mandate to include rules on the renewables sector; widening of the EU Energy

Initiative on access to energy in Africa for the first time to include assistance on renewables; a strategic group for International Energy Cooperation made up of member states and Commission representatives; and a database of member states' energy projects in third countries.

Indeed, when the Commission held a public consultation prior to elaborating this communication, the majority of suggestions forwarded by companies, civil society organizations, and experts were focused on means of providing incentives to European actors to support renewables beyond the EU's borders. Proposals included were that external renewables cooperation projects should count toward member states' 20/20/20 targets; that the EU use diplomatic pressure to reduce regulatory uncertainty in renewable sectors in third countries; that the EU should push through its geostrategic tools for greater energy efficiency in ENP states; that the EU should extend feed-in tariffs to European Neighborhood Policy states; and that the EU should combine renewables cooperation and broader economic support for growth (European Commission 2011e).

One of the most visible of initiatives has been EU–China cooperation on low carbon zones and CCS. An EU–China Partnership on Climate Change embraces a range of activities related to clean energy technologies. In 2007, the EIB signed a Climate Change Framework Loan of €500 million to fund climate change mitigation projects in China. A more specific and targeted China–EU Action Plan on Energy Efficiency and Renewable Energies promotes industrial cooperation relevant to protecting the global environment. A biennial EU–China energy conference brings together high-level representatives from European and Chinese industries and governments. The EU–China Clean Development Mechanism (CDM) Facilitation Project aimed to strengthen the role of the CDM to help China's path to sustainable development, until it was wound up in January 2010. The UK leads the EU–China Near-Zero Emissions Coal Initiative, which aims to build demonstration plants in China to test the feasibility of CCS technology on an industrial scale. Phase II of this initiative (2010–2012) examined the site-specific requirements for actual demonstration plants. Phase III will focus on the construction and operation of a commercial demonstration plant in China after 2012.

The EU classifies this cooperation as one of its most notable success stories. Diplomats acknowledge that the need for such coordination with China has placed greater onus on deepening a strategic alliance with Beijing and has relegated the importance of other areas of policy in relation to which the EU and China have for many years not seen eye to eye. It is widely recognized that in climate policy, all other challenges pale alongside the need to cooperate with China on low carbon and CCS. The EU–China CCS initiative is seen as such an exemplary model that the EU has been keen to extend a similar initiative to India.

Other partnerships are also afoot. As a possible harbinger of future alliance-building priorities, Spanish companies are spending heavily to increase uranium supplies from Canada, Kazakhstan, Niger, and Namibia. Scotland is turning to Middle East sovereign wealth funds for renewables investment. The British and German foreign ministers launched a joint initiative in 2011 to encourage Russia to adopt firmer plans on energy efficiency. The EU's Energy Roadmap 2050 calls for a partnership with Russia and Ukraine especially on biomass (European Commission 2011b).

Notwithstanding such initiatives, there is widespread agreement that the EU has so far taken only a few tentative steps in relation to the international development of renewable sources. Six Middle East states are now pursuing nuclear programs; they complain that the EU has not done enough to help them with renewables development. Independent observers are qualified in their judgment of the EU–China CCS program, suggesting that the scale of EU efforts in China have been relatively modest; they also point out that China will not adopt low carbon technologies unless the EU does so in its own coal industry (Burke and Mabey 2011: 30). A British House of Lords report concludes that the pace and depth of cooperation between the EU and China on CCS has been extremely restricted in practice. Only limited funding has been found for phase II of the EU–China CCS initiative, and no funds have been committed for phase III (House of Lords 2010: 55).

There are certainly concerns that European governments have been unduly tempted to use scarce resources to favor indigenous firms rather than helping more international projects. Economists criticize the EU for relying too heavily on subsidies going into green industry development. Subsidies are likely to be beneficial only where countries already possess some existing expertise and infrastructure; they can help deepen a competitive advantage but cannot create it from nothing (Huberty and Zachmann 2011). Companies like Shell have warned that the scale of European governments' domestic subsidies may contribute to a more general unraveling of at least the spirit if not the formal letter of the internal market. And the focus on such large-scale subsidies also undermines prospects for the international extension of a carbon market – which many such companies see as more likely to provide a harmoniously-governed system for tackling climate change than a zero-sum subsidies race. Criticisms are voiced that the extent of European subsidies now weakens the market mechanisms of the ETS and undercuts the EU's credibility when it seeks to encourage non-EU states to buy into the ostensible disciplines of the ETS.

Tensions have arisen with several non-European states over the terms of cooperation over renewables. The frequent complaint from developing countries is that the EU is engaged in a quick grab for large-scale renewable projects oriented toward exporting energy to European markets rather than in a genuine partnership to maximize renewables' potential for host societies too. ActionAid worries that European governments are pumping funds into large-scale, export-oriented renewables projects that are likely merely to worsen local resource scarcity. This is especially the case with Desertec. The latter is budgeted at nearly $600 billion and is driven by German companies who say it will meet 15% of Europe's electricity needs by 2050. This project is seen in particularly negative light in North Africa as cables will take the energy out into EU markets and not supply local demand. Desertec is often held up as symptomatic of an incipient eco-colonialism that could place severe new strains on relations between developed and developing countries.

This is an area where private investors and companies see the EU as too slow and cumbersome. Even those that accept that member states' unity-breaking bilateralism may have been inadvisable in oil and gas, argue that such flexibility is appropriate in the field of renewables as a means of generating competition and getting funding quickly into promising projects. It is widely felt that future policy needs to be flexible rather than predicated primarily on standard forms of EU institutionalized cooperation, which most investors berate as slow-moving and opaque.

Companies like Areva complain that EU-backed investment in basic renewables infrastructure linked to non-European states remains pitifully limited. They warn of approaching bottlenecks restricting renewables exports and imports unless more infrastructure and interconnections are funded and developed very soon. Investors insist that regulatory predictability is still lacking in non-EU states and the EU has wielded limited influence in improving this situation.

The opposite argument is made by organizations like Counter Balance, who admonish the EU for having been overly seduced by high-visibility geoengineering projects. For critics, the solutions these promise are illusory. Indeed, relying heavily on a search for all-conquering technological breakthroughs could create more problems than it solves, to the extent that such a focus diverts policy-makers' attention from getting to the core drivers of resource scarcity. From this perspective, European governments stand accused of colluding too tightly with non-EU regimes on techno-fixes, none of which exhibit convincingly proven potential, rather than targeting the more deep-seated governance pathologies that weave the most menacing links between climate change and geostrategic tension.

Energy Security Versus Climate Change?

A further area of tension has arisen between the EU's climate policies and its approach to energy security. Formally, the EU insists there is no conflict between these two areas of policy. Policy documents conceive energy security through the lens of a longer-term horizon that incorporates renewables and climate-related considerations. Then UK energy minister Chris Huhne argued that the climate security versus energy security debate presented a false dichotomy, to the extent that climate change is likely to disrupt the supplies of oil and gas too.[4]

However, in practice, the way in which the EU has come to attach priority importance to a rather traditional understanding of energy security sits uneasily with its declared climate aims. Part of this is to do with intensified oil and gas diplomacy; part is to do with the focus on non-conventional fossil fuel sources.

Much recent effort has been invested in enhancing the EU's external energy security strategy. The Energy 2020 strategy begins by stating that: "The same collaboration and common purpose that has led to the adoption of the EU's headline energy and climate targets is not yet evident in external energy policy" (European Commission 2011c: 20). It commits the EU to injecting substance into this external dimension of EU energy security coordination. The September 2011 Commission communication (2011a) caught most attention by proposing a new mandatory information exchange on bilateral energy accords and a provision for the Commission to negotiate new energy treaties on behalf of member states (as it had done with Azerbaijan and Turkmenistan on development of the Trans-Caspian pipeline).

The EU has signed a plethora of bilateral energy accords. After an EU–Uzbekistan memorandum of understanding was signed in 2011, all Central Asian states now have such agreements. Under a new EU–Azeri deal, Baku commits to the so-called "southern corridor." Commission president José Manuel Barroso made what was interpreted as a particularly significant visit to Turkmenistan in 2011, in an effort to secure sizeable supplies for the beleaguered Nabucco pipeline project. Indeed, on the back of this visit Turkmenistan committed to supplying Nabucco, EU enticements contrasting with problems experienced under the country's 2007 deal with Moscow to supply northward into Russian networks. The twists and turns of "pipeline diplomacy" have taken up an increasing amount of policy attention in the last two years, as three alternative southern corridor routes (the Nabucco, Trans-Anatolian, and ITGI projects) vie for pre-eminence. The Nordstream pipeline directly connecting Russia with German markets started pumping on November 8, 2011.

Within these accords, policy remains focused on very traditional access issues. The energy NGO Counter Balance argues that the EU is if anything more obsessed with large oil and gas infrastructure projects now and less focused on the broader implications of low carbon than it was in the mid-2000s. This is despite the serious setbacks encountered in all such projects. The proposed Trans-Sahara pipeline has overshot its budget by an estimated $15 billion and with no-one interested in investing that kind of money construction has still not commenced. The opening of the Medgaz pipeline was continuously postponed until 2011, by which time it was running at $1 billion over budget and had to be expensively rescued by EIB loans. And most emblematically, the much-awaited Nabucco project remains stuck: it is still not clear where supplies into the line will come from and, despite a €200 million injection from the European Commission, the consortium has insufficient cash to finish construction work. These pipelines represent the very opposite of the localism needed to mitigate climate change: they entail huge environmental damage, significant energy losses in transmission, and deepen reliance on hydrocarbons.[5]

The EU's third energy liberalization package remained aimed primarily at the Russia challenge; while it did not bring in unbundling of ownership it did require a certain fragmentation of Gazprom operations across the EU, and member states would be less completely autonomous in their bilateral deals with Russia (Barysch 2011). In 2012, headlines were, of course, dominated by the Commission's confrontation with Gazprom along these lines. The EU has offered an "energy roadmap to 2050" to assuage Russian demands for security of demand. While many hope that cooperation with Russia on renewable energy sources can overcome the zero-sum dynamics of gas pipeline politics, this has not become a major strand of the EU–Russia relationship. Standard bilateral deals for gas supply tie-ins continue unabated; RWE and Gazprom signed such a deal in July 2011.

In an attempt to counteract the prevalence of member states' opaque bilateral deals with producer states, in February 2011 EU leaders agreed to share data on third-country energy deals. This was celebrated as an important step forward in guaranteeing transparency. It did not specify what type of information should be revealed and did not empower the Commission to act on information relating to deals that might undermine the spirit of a common EU energy policy. The Commission's communication in September 2011 (2011a) aimed to give a legal basis to this commitment to share information and empower the Commission to act on the information provided on bilateral contracts.

Senior officials now talk enthusiastically of a new emancipation of energy policy from climate policy. The economic crisis and squeeze on competitiveness, combined with a new rise in oil prices, has produced a swing away from the priority attached to climate policy. Experts and policy-makers are increasingly minded to argue in favor of gas and against renewable sources. Long-term energy security is increasingly a matter of the "dash for gas." The UK March 2012 budget provided a significant tax break for oil and gas production in Shetland.

Policy-makers' main concern is now quite clearly with the advent of sizeable shale gas supplies. Unconventional sources have changed the energy panorama. Gas markets now look extremely vibrant. Experts opine that combining traditional and unconventional gases, the world has 300 years of supplies left. Countries like Algeria claim that they have more shale gas than natural gas. The policy priority in this sense is to delink gas from oil markets, by completing the single market in gas infrastructures and linkages.

The US's increasing energy independence should be good for the EU as fungible oil supplies are freed up internationally now that North American demand is decreasing. As unconventional gas supplies have taken the US toward energy independence, this has driven down LNG prices, enticing European buyers. Contrary to increasingly voiced fears, the US will still need to stay involved in policing global energy markets. New energy sources should tilt the balance away from monopoly geopolitics to markets as competition intensifies between a wider range of sources. Yet, the general impression among industry experts is that the EU has been slower to react to this revolution than the US. As gas prices have fallen due to shale output in North America, EU companies are left locked into what now seem extremely overpriced long-term contracts with Gazprom.

The policy focus has shifted away from the question of access to non-European renewable sources to debates over how far shale should be incorporated into the European energy mix. Some experts predict that the high potential for shale extraction in Poland is the factor that definitively kills off the Nabucco project. ExxonMobil and Total have joined forces to explore for shale gas in Poland through large-scale investments. Several other member states have held back on shale gas exploration because of its environmental costs. Most notably, France has prohibited the development of shale reserves. In direct contrast, in July 2011 the UK government decided against restrictions on shale gas drilling.

Unconventional oil has disastrous implications for climate change; unconventional gas is relatively clean but still prolongs the reliance on fossil fuels. The new glut of natural gas has slowed down the drive to renewable energy. And with shale present in many stable, advanced, and friendly countries, the security worries appear less too. Some say this is not disastrous for climate change aims. Industry experts even calculate that using natural gas as a bridging solution would reduce the cost of meeting the EU's 20/20/20 targets relative to the huge subsidies ploughed into wind and solar. And more environmentally friendly drilling techniques are being developed for shale. However, this focus does mean that debates have returned to very traditional questions of the balance between hydrocarbon exploitation and environmental concerns.

The Regulatory Approach: Bad for Climate Policy?

It is impossible to understand EU global energy policies without reference to the EU's distinctive model of energy governance. The EU has set itself the aim of pursuing a range of energy interests through the extension of its own rules and regulations beyond its borders. The EU looks for institutional predictability in neighbors rather than a free market per se. EU officials describe the approach as distinct from neo-liberalism and predicated instead on regulatory reliability.

The EU's basic philosophy is encapsulated by officials' insistence in defining the "European energy space" as extending more widely than the EU itself, spreading across to the Caspian and down to the Sahara. The extension of formal EU energy rules and obligations is enshrined in the so-called Energy Community treaty of 2006, which has been adopted by Western Balkan states, Ukraine, and potentially Turkey. All these states have signed up to abide by EU legal requirements in the management of their energy markets. Under this rubric, for example, in February 2011 Macedonia introduced far-reaching laws to align with EU energy markets. The EU's March 2011 policy document responding to the Arab spring intimated at North Africa also being offered a place in the Energy Community. A majority of member states express support for the idea of extending the Energy Community to both North Africa and the Caucasus.

Some effort is apparent also to incorporate climate change into external relations through this regulatory approach. Wide support exists among member states and the Commission to bring the Renewable Energy Directive (RED) into the Energy Community. DG Energy argues that the RED must expand and change as the "market for renewables is moving from a local to cross-border supply" (European Commission 2011c: 12). The Council of European Energy Regulators welcomes moves to prepare for the integration of the RED into the Energy Community. It argues that the EU needs to support more twinning and capacity-building projects to help new members of the Community implement the RED.

The Emissions Trading Directive is being "externalized" to provide investment certainty for European companies in renewables development in non-EU countries. There is talk of strengthening its rules to impose penalties on third countries for intervening negatively in renewables projects. Under the RED, electricity generated by renewable sources outside the EU can count toward a country's national renewables targets. It is proposed that the geographically wider Energy Charter treaty of 1994 begins to apply its rules to low carbon sources too. Some member states advocate extending the ECT comprehensively to cover renewables. Some officials see relevance for the Middle East and North Africa in using the 20/20/20 targets as a form of experimental governance to galvanize cooperation on climate change mitigation.

Significantly, however, many fear that the regulatory, external governance approach is insufficiently flexible or focused to prioritize renewables development in EU external relations. The focus on regulatory export makes the EU a ponderous actor in the foreign policy dimensions of energy policy. And one implication of this is to compound the difficulties of incorporating climate-related factors into foreign policy planning and initiatives. An underlying concern among some policy-makers has been the need to have a different approach to regulatory convergence in renewables compared to hydrocarbons. They worry that the EU has sought to carry over its basic regulatory model from oil and gas to the renewables sector in a way that is blind to the very different dynamics governing these sectors.

A wide-ranging public consultation held prior to the elaboration of the Energy 2020 strategy revealed growing doubts about the wisdom of the approach based primarily on the export of EU regulations. In these consultations, governments and companies argued for a more direct approach in energy relations with non-EU states, including on the link with supplies of power generated from renewable sources. The French government argued that in the Mediterranean more stress was needed on infrastructure connections than regulations; it wanted Mediterranean states brought into the EU energy market, but on a more pragmatic basis. The main priority should be to push Arab governments on investment protection, especially in renewables. Other member states made similar points: an overly complex set of technical and regulatory convergence criteria are holding up external energy cooperation.

Conclusion

In the last two years, the EU has raised its ambitions in global energy policy. Many impressive new commitments have been introduced. The EU now has a more comprehensive range of instruments at its disposal. On climate change, some of the ghosts of the ill-fated 2009 Copenhagen summit have been laid to rest. The efforts to mainstream climate diplomacy across all areas of EU external relations are commendable. On energy security, a slightly more geopolitical angle has hardened the edges of EU foreign policy. External unity has tightened; the long-standing jibe that the EU has no common external energy policy is no longer entirely fair.

Curiously, however, even as the EU's strategies have gathered momentum, so have new uncertainties filtered into its climate and energy policies. Progress on at least some domestic climate targets has faltered; and the external ramifications of this are apparent. A crisis-compounded strategic introspection increasingly undermines the vitality of external aims; the effect is evident in detailed areas of policy such as climate funding and global renewables partnerships. If anything, in the last two years EU policy commitments in the external dimensions of traditional energy security have advanced further than European global climate policies. The new prominence of unconventional sources of fossil fuels looks set to intensify this trend.

These scenarios capture the difficult policy challenges with which the EU is now grappling. More conceptually, it also raises questions about what kind of energy actor the EU is and should seek to be. The EU has staked out two core pillars to its identity in global energy politics: the primacy of climate diplomacy over "hard security" realpolitik; and the use of its own internal commitments and regulations as the best basis for its international projection. These principles still apply but both are more equivocal. Some degree of flexibility in the EU's approach to climate and energy questions is certainly merited. It is not clear, however, that the current evolution of European strategy represents an enlightened readjustment rather than short-term, ad hoc expediency.

Notes

1. http://ec.europa.eu/clima/policies/finance/index_en.htm.
2. http://ec.europa.eu/clima/publications/docs/spf_startfinance_en.pdf.
3. http://www.europeanvoice.com/article/2010/11/italy-blamed-for-eu-failure-on-climate-ch ange-aid/69421.aspx.
4. Chris Huhne, The Geopolitics of Climate Change. Speech at the Royal United Services Institute, London, July 7, 2011.
5. www.counterbalance-eib.org.

References

Barysch, K. 2011. *The EU and Russia: All Smiles and No Action*. London: Centre for European Reform.

Bowen, Alex, and James Rydge. 2011. *Climate Change Policy in the United Kingdom*. London: London School of Economics, Grantham Research Institute on Climate Change.

Bréchet, Thierry, Johan Eyckmans, François Gerard, *et al.* 2010. The Impact of the Unilateral EU Commitment on the Stability of International Climate Agreements. *Climate Policy* 10: 148–166.

Burke, Tom, and Nick Mabey. 2011. *Europe in the World*. London: E3G.

European Commission. 2011a. *On Security of Supply and International Cooperation – The EU Energy Policy: Engaging With Partners Beyond Our Borders*. COM (2011)539. Brussels.

European Commission. 2011b. *Energy Roadmap 2050*. COM (2011)885/2. Brussels.

European Commission. 2011c. *Energy 2020*. Brussels: DG Energy.

European Commission. 2011d. *Key Facts and Figures on the External Dimension of the EU Energy Policy*. Commission Staff Working Paper SEC(2011)1022. Brussels.

European Commission. 2011e. *Results of the Public Consultation on the External Dimension of the EU Energy Policy*. Commission Staff Working Paper SEC (2011)1023. Brussels.

Haites, E. 2011. Climate Change Finance. *Special edition of Climate Policy* 11, 3: 963–969.

Helm, D. 2011. What Next for EU Energy Policy? In K. Barysch, *Green, Safe, Cheap: Where Next for EU Energy Policy?* London: Centre for European Reform.

House of Lords. 2010. *Stars and Dragons: The EU and China*. European Union Committee, 7th Report. London: The Stationery Office.

Huberty, M., and G. Zachmann. 2011. *Green Exports and the Global Product Space: Prospects for EU Industrial Policy*. Brussels: Bruegel.

Mabey, Nick. 2009. Climate Change and Global Governance. Memo, October. London: E3G.

Scholz, Imme. 2010. *European Climate and Development Financing Before Cancún*. EDC 2020 Opinion 7. Bonn: EADI.

van Agt, C. 2011. The Energy Infrastructure Challenge. In K. Barysch, *Green, Safe, Cheap: Where Next for EU Energy Policy?* London: Centre for European Reform.

Chapter 26

Energy Governance in the United States

Benjamin K. Sovacool and Roman Sidortsov

Introduction

The US occupies a distinct place in the world's energy landscape. It is the largest primary energy consumer in all categories except for coal (US EIA 2012). Many perceive the US as the world policeman aggressively patrolling its energy "beat." Although this assertion may not be as far-fetched in relation to the country's oil interests in the Persian Gulf, applying this metaphor as a whole is inaccurate. A nation's view on a global issue is heavily influenced by international and domestic factors. One can have a long and spirited "chicken or the egg" debate regarding the causal relationship between the complexity of US energy policy, the structure of US energy governance, and domestic and international events. Regardless of the outcome of the debate, energy governance plays a critical role in the nation's outlook on global energy policy. Thus, this chapter explores energy governance in the US by looking at the historic and contemporary foundations of America's energy decision-making.

National energy policy-makers in the US have produced a gargantuan amount of statutes and regulations in the past century. The first half of the twentieth century saw massive federal subsidies and land grants to promote economic growth and advance technological progress through the development of hydroelectric projects, railways, coal mines, oil and gas fields, as well as giveaways of valuable minerals to mining companies (Gulliver and Zillman 2006). The second half of the century witnessed the coming of the "Atomic Age" and efforts to diversify energy production after the crises of the 1970s (Sovacool 2011). In addition, every US Administration starting with President Nixon's has set the agenda of getting rid of dependence on foreign oil.

Two logical questions arise. First, how effective have these efforts been at shaping and implementing energy policy in the US? Second, what have been the core assumptions *behind* energy decision-making within US institutions? This chapter shows that the US has gone through significant changes in its national energy policy regime in the past four decades. The Carter Administration in the 1970s challenged the classic model of heavy-handed utility regulation that had been in place since the 1930s. As a result of

The Handbook of Global Energy Policy, First Edition. Edited by Andreas Goldthau.
© 2013 John Wiley & Sons, Ltd. Published 2013 by John Wiley & Sons, Ltd.

the challenge, federal and State governments saw their power in regulating electricity through monopolies erode. Additionally, once the US ceased to be a net oil-exporting country in March 1971, the federal government no longer sought to control the level of domestic production (Yergin 1991: 544). More changes followed in the 1980s and 1990s when restructuring started, competitive retail electricity markets were introduced, and the government started subsidizing non-petroleum forms of transportation fuel. Changes continued in the 2000s as the portfolio approach or "all of the above" strategy – supporting a wide spectrum of different energy technologies and solutions – took hold. As the next two sections explain, significant changes in the guiding principles of energy governance prompted most of these shifts, and energy has become an even more complicated and disputed area of governance.

Seven Historic Guiding Principles

Until approximately the 1970s, a set of seven guiding principles shaped decisions made about energy in the US. These principles included: (1) the understanding that energy is a public necessity justifiable of government support; (2) commitment to generating abundant supplies of affordable energy; (3) dominance of large-scale centralized energy technologies; (4) increase in supply to satisfy anticipated increases in demand; (5) reliance on technological advancement to overcome scarcity of resources; (6) confidence in expertise and trust in authorities to make energy-related decisions; and (7) understanding that the government should balance competing private property rights to protect public interest.

Energy as a Public Necessity

The notion that energy was a public necessity led to generous governmental support of the power industry through subsidies and grants of monopoly status and other rights that secured comfortable business environments for utilities. However, such support came with the expectation that the industry would satisfy the public necessity in full. For example, public electricity companies historically had a "duty to serve" all customers within their service area. The only real excuses for nonperformance of the duty were the limitations of the available technology. Under the duty to serve, a public electric utility was required to render "safe and adequate service" to all customers within their jurisdiction regardless of all and any foreseeable increases in demand. These utilities were obligated to provide such service to all customers within each service class on equal terms, had to charge "just and reasonable" rates for their services, and could not subject their customers to undue or unjust discrimination. In return for their duty to serve, utilities were granted significant rights. They were given a guaranteed and exclusive customer base. By virtue of a grant of franchise and certificate of public good, convenience, or necessity, utilities were also shielded from competition from other power companies. In addition, they had the right to just and reasonable compensation, including a profit on capital investments made to serve their customers. In many States, public utilities were given the power of eminent domain. Thus, they had the right to take private property for public use (in order to provide adequate service), for just compensation to the owner.

Affordability and Abundance

The US operated under the rationale that in order to make energy a de facto necessity, it must make it abundant and affordable. In his study of the entire history of energy policy-making in the US Melosi (1985) remarked that the lone overarching theme was a dedication to ensuring the abundance of cheap, reliable energy. Throughout the course of the country's economic rise, abundant natural resources (including fossil fuels) enabled the transition from a labor-intensive economy to a capital-intensive one, and generated vast commercial wealth. Melosi concluded that Americans have become "endowed with an abundance of domestic sources of energy and having access to foreign sources" and continue to count on energy supplies to be simultaneously "never-ending and cheap." Clark (1990) made a similar observation in his history on energy use and government policy. He noted that copiousness and price affect the way American people use energy, how firms develop and market it, and how government shapes and implements policies about it. The established and undeniable vision of the American energy policy has been to attain the lowest possible price of electricity per unit while relying exclusively on the fullest exploitation of resources and maximum consumption. This theme was even more pronounced during the early years of oil exploration and extraction in the US. The rush to extract the newly valuable commodity without taking into account the geology of subterranean formations lead to tremendous economic waste and left a trail of broken communities spanning from Pennsylvania to California. Yet in a short term, the "boom and bust" approach channeled by the uniquely American legal rule of capture produced great amounts of oil at a price as low as 10 cents per barrel (Bosselman *et al.* 2006). Inexpensive and plenteous energy, following this paradigm, has attracted more political support in American society than protection of the environment, energy efficiency, or social welfare (Freeman 1973).

Centralized Approach

Several social, political, and technical factors also contributed to the emergence of a large-scale centralized approach to supply of energy, especially electricity. The first steps to centralization were taken during World War I when utilities began interconnecting grids to avoid local power shortages. This trend continued during the Great Depression era when construction of high voltage transmission was viewed as a key component for providing cheap electricity. And cheap electricity was in turn viewed as one of the means of pulling the country from the economic rut. Mega-energy projects were believed to create economies of scale, lowering production costs and generating financial savings that could then be passed on to individual consumers. Prototypical examples of such energy projects in the US include the Hoover Dam in Colorado and the Tennessee Valley Authority's network of 29 hydroelectric dams. In fact, the massive build-up of nuclear power plants in the 1960s propelled by what Bupp and Derian describe as a "Great Bandwagon Market" was largely due to the mirage of economies of scale promised by the nuclear industry with significant help from the US government (Bupp and Derian 1978: 45). The centralization approach featuring massive energy systems appeared to work remarkably well up until the 1970s. Serving as a textbook example of applied economies of scale, the generating capacity of large power plants doubled every 6.5 years from 1930 to 1970. At the same time an average price of electricity in nominal terms dropped from approximately $1 to less than 7 ¢/kWh (see Figure 26.1). The success of the centralized approach was attributed to a number of related assumptions. Those in charge of energy planning believed that power systems should consist of relatively few but large supply and distribution units, and that such units should consist of large, monolithic apparatuses rather than many smaller models (Hirsh 1989).

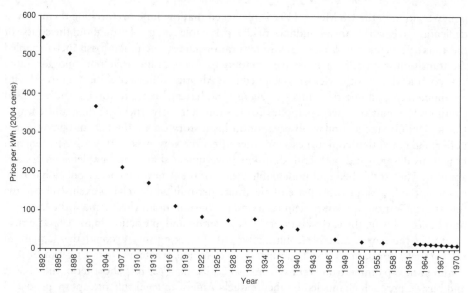

Figure 26.1   Average residential price of electricity in the US, 1892–1970 (adjusted for inflation to year 2004).
Source: Hirsh (1989).

Increasing Supply to Satisfy Demand

Energy decision-makers, especially in the electricity sector, sought to forecast future energy demand, and build sources of supply according to those projections. This approach was premised on the following three assumptions. First, the market-place was the best forum to determine supply and demand and that the efficacy of free markets, not government bureaucrats, should decide how much energy to be generated and consumed. Second, individual customers know their own energy needs best and these individual needs were best satisfied through exchange in the market-place. Third, qualitative variations in energy did not matter, and differences in type, quality, unit scale, and geographic location of a generator and consumer were not as critical as meeting consumer demand for energy regardless of scale. Paradoxically, electricity regulators chose not to learn from the oil and gas industry's disastrous "boom and bust" pattern of development. As a result, overreliance on the "magical" powers of free market led to the build-up of excessive generation capacity. That in turn led to economic inefficiency similar to the "rampant waste of oil and gas" during the pre-market demand prorationing era (Bosselman *et al.* 2006). Even today, some energy decision-makers still firmly believe in the "follow the demand" principle. For example, John Hofmeister, president of Shell Oil Company, recently summarized this view in an eloquently simple manner when he stated that "my job is constantly to look for more fossil fuels, more fossil fuels, and more fossil fuels to try to meet demand."

Technological Optimism

For most of the country's history, Americans have accepted a Promethean view of technology and tended to place their personal needs above that of the environment. Nearly

200 years of industrial growth fueled by cheap energy and abundant natural resources on one side, and an environment that could ostensibly absorb pollution endlessly on another, created a strong belief in the minds of many policy-makers and ordinary citizens that they were, in fact, entitled to conquer Nature. As a result, many Americans formed a worldview that tied energy consumption with economic growth and prosperity. According to this worldview, Americans are entitled to consume as much energy as they wish. And should the nation experience resource constraints, the ever-capable technology will be there to overcome them. Historically, such an outlook can be linked to New England Puritan ideals. The first settlers of the "New World" found their natural surroundings strange and menacing, and their records often referred to nature as something to be dominated and subdued. To aid Americans in this domination, the government placed its confidence in the idea that American ingenuity could solve all problems aided by the great power of technology (Sovacool 2009b; Smith 2009; Winner 1982).

Confidence in the Government as a Decision-Maker

This guiding principle concerns the assumed roles of technical experts, regulators, and consumers. Given the technical complexity of the energy field, Americans by and large agreed that technical experts were the most capable to make decisions about energy and government regulators often deferred to them to choose appropriate technologies. These experts and regulators maintained an active role in energy decision-making. In contrast, consumers played a passive role. They were left to consume energy, not to produce or select it. Lack of public interest, participation, and involvement in the energy sector allowed energy companies and government bureaucrats to sustain their control over an inert system that provided inexpensive energy at a low cost and also modest but stable profits while fueling economic growth.

Protection of Public Interests Over Private Rights

Despite the overreliance on the power of the free market to create a perfect balance between demand and supply of energy, policy-makers in the US recognized the inherent conflict between certain private rights and public interests and, in many cases, gave the latter a clear preference. To recognize the significance of this guiding principle of energy governance, one should take notice of the high level of protection of private property rights in the US. For example, the US unlike the majority of other nations allows for private ownership of subterranean rights. Moreover, the rights of a mineral estate owner (if the estate is divided) usually trump the rights of a surface estate owner. This means that a mineral estate owner can enter private property to develop and produce oil and gas despite objections of the surface rights owner. The mineral estate owner's right to operate on the land is only limited by a duty to reasonably accommodate the needs of the surface owner and pay reasonable compensation for damage to the surface estate (Bosselman *et al.* 2006).

However, the seemingly bullet-proof rights of a mineral estate owner did not hold up against public interest. From the maximum efficiency rate (MER) regulation intended to ensure the most optimal use of oil and gas deposits to zoning and land use regulations enacted to protect urban landscape from looking like streets of Spindletop, Texas in the 1900s, public interests were given significant protection in the US (Laitos and Carr 1998).

Shifts in Seven Historic Principles

The current energy governance system in the US can no longer comfortably rest on the aforementioned seven historic principles. In the last three decades these principles have faced the following challenges: (1) energy is no longer perceived solely as a public necessity, and instead is viewed in at least four other dimensions; (2) efforts to price external environmental and social costs undermine the idea of ever-available low cost energy; (3) the benefits of small-scale and decentralized technologies contest the once unassailable economies of scale foundation of large centralized power systems; (4) "negawatts," the product of energy efficiency, conservation, and demand-side management, reduce the accuracy and value of demand for energy as the benchmark for supply planning; (5) observations, discoveries, and studies about environmental and health impacts of energy production and use have eroded faith in the supremacy of technology; (6) homeowners, community activists, and businesses have started to demand more active roles in decision-making; and (7) departure from large-scale and centralized power systems, economic downturns, and several significant mistakes by government policy-makers have enabled critics to question the public interest foundation of many energy policies.

Energy as Heterogeneous

Energy is no longer solely seen primarily as a public necessity. Several studies challenge this homogeneous viewpoint as they have identified at least five deviating (and often contradicting) perspectives held by analysts, regulators, and consumers. Stern and Aronson (1984), for example, argue that there is no such a thing as an energy user. In fact, they have identified at least five different types of energy users. The first type, the investor, regards energy as a cost that is prudently considered in making financing decisions, and views energy technologies as necessary means to recover costs over their useful life. The consumer thinks of her home, electronic gadgets, and automobile as consumer goods that support her lifestyle. The conformer employs energy technologies to gain access to and affiliation with a particular social group or status. The crusader considers energy use as an expression of self-reliance and environmental stewardship and sees it through ethical lenses. The problem avoider treats energy as an always present fact of life and a potential source of annoyance and inconvenience when it is not readily available, avoiding much thought about it until technologies break down and services cease.

Stern and Aronson's work has been supplemented by various studies of consumer attitudes conducted in the 1980s by the Electric Power Research Institute (summarized in Lutzenhiser 1993). These studies characterized: (1) pleasure and comfort seekers who consume energy (through items like "muscle" cars and heated swimming pools) for the enjoyments that such things bring; (2) appearance conscious conformists who at the time were captivated by large cars and homes (thus needing more energy) to emphasize their social status; (3) lifestyle simplifiers and hassle avoiders who notice energy when it is not available; (4) indifferent consumers who just do not think about the price and consequences of energy use; (5) non-conformists who relish the additional control and try to rely more on their own supplies of energy (solar panels, wind turbines); and (6) conservers who seek to protect the environment through minimizing energy use.

These varying views are not relegated to the 1970s and 1980s, and were confirmed in a recent survey of American attitudes toward energy (Tonn et al. 2009). This study, much like the work before it, identified at least seven different perspectives. "America-firsters" made energy decisions primarily directed at making the country more energy independent; "bottom liners" supported the portfolio of technologies with the lowest cost; "entrepreneurs" supported creativity and research in solving problems through innovation and advancements in technology; "environmentalists" sought to minimize pollution and made decisions based on mitigating or adapting to climate change; "individualists" desired a high quality of life and consumption; "politicians" supported energy systems that accommodated as many interests as possible; and "technophiles" sought big engineering approaches to energy problems.

Constraints on Production, Transportation, Transmission, and Distribution

Once almost universal support for low-cost and plentiful supplies of energy is now being challenged by demands to include the costs of at least some environmental and social externalities associated with producing and moving energy, even if it involves raising its price. Many of such energy-related costs are not reflected in prices and are therefore not borne by consumers; they are instead passed onto society at large or future generations or "externalized." For example, many supporters of coal-fired power plants marvel at the low-cost power that the plants produce. However, they do not include medical bills and lost earnings of miners affected by coal dust that injures and kills thousands of workers per year. Black lung disease alone has resulted in at least $35 billion in health care costs. Coal supporters also rarely talk about the economic cost of coal emissions that cause acid deposition, smog, and climate change. These external costs would easily double the cost of coal if they were reflected in its price. Motor vehicles and gasoline appear inexpensive because their price does not cover the 1.2 million deaths per year and 25–50 million injuries caused by accidents, making them the third largest cause of death and injury in the world. In addition, internal combustion engines emit unhealthy pollutants and particles into the air, contributing to acid rain, ozone depletion, and climate change.

To account for these damages, many economists and even some policy-makers have appealed to the government to increase the price of energy as a matter of objective accounting and ensuring that users receive undistorted price signals related to their energy consumption. This "incorporation of externalities" can be done in many ways and dozens of proposals have been put forth by various groups in the past decade. Rarely a proposal came without an intense discussion about how such externalities ought to be priced and allocated, as well as which mechanisms ought to be employed. Should climate change mitigation or adaptation be the primary focus of such efforts? Should the efforts center on damage costs or avoided costs? Does a carbon tax or a cap-and-trade scheme achieve better results? And should policy-makers opt to rely on feed-in tariffs or renewable portfolio standards? Each proposal has formed into a policy package which in turn has created its own constituency. Such sharp divisions are proving to be ineffective as advocates of pricing externalities often bicker among each other, as well as with those arguing in favor of the status quo.

Yet challenges to the idea of ever-plentiful inexpensive energy come from actors that traditionally have been avid supporters of the same idea. Intense competition for water resources in the Western US has become a fact of life. Power producers have to battle with other industries who are also power users for the precious natural resource. As a result, in many areas in the American West power generation is no longer considered the highest priority use for water resources. Competition for dwindling natural resources and territory has also affected transmission, distribution, and transportation of many forms of energy. From the legal challenges by homeowners regarding siting of high-voltage transmission lines to the opposition to construction of oil and gas pipelines by farmers, the idea of cheap and abundant energy is no longer something that all Americans can get behind. Given the finite natural resources and available territory, as well as the growing US population, these challenges will only become greater, rendering the once powerful guiding principle of US energy governance more obsolete.

Decentralization

The trend toward ever-larger centralized models of energy supply hit significant barriers in the late 1960s and early 1970s. In the electric power sector, efficiencies of thermal power plants plateaued in the 1960s and the scale of power plant equipment climaxed at 1300 MW and then began to decline, as larger units faced technical challenges regarding information processing, materials science, and plant design, producing what one author labeled "technological stasis" (Hirsh 1989).

In his widely acclaimed work, physicist Amory Lovins (1976, 1979a, 1979b) criticized many of the alleged benefits to centralized large-scale production. He classified the centralized approach to energy production as belonging to a "hard path" that features a number of serious shortcomings. Lovins put forth the following arguments in support of his assertion that large energy systems cannot be mass produced. First, centralization demands expensive transmission and distribution networks. Second, they often do not recycle excess thermal energy, making them inefficient. Third, they are much less reliable, and consequently require high-cost reserve capacity. Fourth, they take much longer to construct, and therefore are susceptible to escalated financing costs, misjudged demand forecasts, and wage pressure by unions. To counter these shortcomings, Lovins proposed a "soft" path pushing for energy technologies that were (a) diverse, supplying energy from decentralized sources in smaller quantities; (b) renewable, operating on non-fossil and thus non-exhaustible fuels; (c) simple, or relatively easy to understand; (d) modular, or corresponding in scale to energy needs; and (e) qualitative, or corresponding in energy quality to end-user needs.

Shaping Demand and Influencing Behavior

Vast public and private campaigns to ensure reliability and availability of energy supply during the crises of the 1970s gave birth to a constituency for energy efficiency, conservation, and demand-side management. Originally a grass-roots movement, proponents of this unconventional approach to demand now enjoy support of strong lobbies and groups such as the American Council for an Energy-Efficient Economy (ACEEE) and the Alliance to Save Energy. These groups, armed with the unequivocal support of the hard data, have maintained that energy efficiency measures including substituting fuels, upgrading or retrofitting technology, and changing consumer behavior represent the least expensive, fastest, and most realistic and environmentally friendly approach to addressing energy demand. Public response to the high energy prices of the 1970s gave supporters of energy efficiency plenty of reasons to be optimistic. Individuals insulated and weatherproofed their homes, purchased more fuel-efficient cars, and learned to adjust thermostats to reduce energy consumption. Businesses installed energy management and control systems and retrofitted their buildings with more efficient heating and cooling equipment, achieving a decline of 25% of energy use per square foot of commercial building space. Manufacturing firms replaced their old energy-thirsty equipment with more efficient motors for conveyors, compressors, fans, and pumps and adopted overall more efficient manufacturing processes.

Some electric utilities learned that they could save electricity at a lower cost than producing it. Thus, efficiency could improve cash flow by displacing operating costs, calming investors, and saving consumers money at the same time. The States that adjusted their regulatory regimes to accommodate energy efficiency saw huge economic gains almost immediately. For example, in 1999 Vermont had the second highest electricity rates among seven Northeastern States. In 2005, as a result of State-led energy efficiency initiatives, Vermont electricity prices were the lowest in the region (Sautter *et al.* 2008–2009). Overall, these efforts *saved* more energy nationwide than any single currently available technology, and they did so at a unit cost much less than the value of that energy (Sovacool 2008b).

Technological Pessimism

Forced to confront a burning Cuyahoga River in Ohio, toxic sludge seeping through the ground nearby the Love Canal, New York, and the nuclear accident at Three Mile Island, Pennsylvania, some regulators and citizens have become increasingly skeptical regarding the power of new technologies. Additionally, numerous scientists have repeatedly warned that many ecosystems had reached their assimilative capacity to absorb pollution. Empowering such pessimism are advances from the environmental sciences, including the findings firmly establishing that humans are vitally dependent of on non-human forms of Nature; that pollution of air, water, and land unmistakably puts human health and life in jeopardy; that limits should be set on the exploitation and use of natural resources; and that humans have an obligation to preserve the biosphere for future generations (Sovacool 2009b).

The most serious issue eroding faith in energy optimism is undoubtedly climate change. This reason is crucial enough to turn technological optimists into pessimists. There is mounting evidence that the climate change consequences of "business as usual" in the way we produce and consume energy could be nothing short of catastrophic for the planet. A plethora of studies from the Intergovernmental Panel on Climate Change (IPCC 2007), United Nations (UNDP 1997), and other leading climatologists (Josberger *et al.* 2009; Lackner and Sachs 2004; Rosenzweig *et al.* 2008) have proclaimed that continued emissions of greenhouse gases will likely lead to grave consequences for the planet and species that populate it. Changes in climate will cause acute alterations to the distribution and availability of water resources leading to drinking water shortages for millions of people. Destruction of ecosystems, species, and habitats will continue, and the bleaching of coral reefs and widespread extinction of migratory species will intensify. A significant loss of agricultural and fishery productivity will occur, along with changes in the growing seasons for crops affected by increasing drought in areas with marginal soils that have low buffering potential. Damage from floods and severe storms, especially among coastal areas, will spread geographically and grow in size. Deaths arising from changes in disease vectors, particularly among diseases regulated by temperature and precipitation, will increase. The aggregate impacts of these changes could exceed $13 trillion per year, or 20% of the global GDP, if the more severe scenarios unfold.

Community Control

The levers of energy governance are no longer solely in the grasp of technical experts and bureaucrats. Momentum has partially changed to broaden the roster of decision-makers to include local, State, national, and international communities, as well as individual actors who are not representatives of the prior governance circle. First, correlating with David Orr's "energetics" view, many nongovernmental groups and citizens are arguing for more accessible and democratic modes of energy policy-making to give individuals and communities more power over forms of energy production and use. For example, Sovacool (2009b) examined numerous studies looking at the acceptance of different energy systems. He found that participation and ownership – technologies and systems that had fair forms of permitting and siting and were made available for stakeholders to invest in a given project – played a powerful role in shoring up community support behind energy projects. This transforms the public's role from a passive consumer to an active decision-maker where the newly created "energy democracy" allows consumers to generate all or some of their own energy or at least make more informed decisions about their energy present and future. Several authors have made a strong case in support of such participation, pointing out that it broadens public interest, increases knowledge, promotes equity, and improves overall accountability (Orr 1979).

Resurgence of Private Rights

A combination of the aforementioned challenges, as well as several significant mistakes by government policy-makers, have eroded the primacy of the notion of public interest in the American energy landscape. Energy policy-makers no longer can hide behind the "energy is a public necessity" curtain. Each class or subclass of customers has its own unique demands and levels of flexibility. Some manufacturing facilities can accommodate interruptions in service whereas hospitals and households usually cannot. As a result, the former demand more favorable rate plans whereas the latter are left paying for the "peak premium." Decentralization of electricity supply introduced in the form of "qualified facilities" under the 1978 Public Utility Regulatory Policy Act challenged the "public interest" rationale of mega-projects in favor of often more economically efficient smaller plants. The excessive regulation of natural gas production affirmed by the Supreme Court's opinion in *Philips Petroleum Co. v. Wisconsin* and the following clumsy attempt by the Natural Gas Policy Act of 1978 to encourage production at any cost pushed the pendulum far into another direction.[1]

The departure from the preeminence of public interest as a guiding principle of energy governance has produced mixed results. For example, on a positive side, independent generators were able to enter the market bringing economic efficiency. On a negative side, as the California energy crisis of 2001 demonstrated, overreliance on the free market to strike a perfect balance between private and public interests can lead to disastrous consequences (Bosselman *et al.* 2006).

Current Governance Challenges

These changes to fundamental principles concerning the nature and structure of energy governance in the US have resulted in a plethora of potential problems related to complexity, inconsistency, and vertical and horizontal fragmentation.

Complexity

In the US, decision-makers rely on a vast body of laws and regulations to manage the country's massive energy infrastructure while ensuring appropriate involvement of local, State, and national governments. For example, just one sector, electricity, is governed by more than 44,000 different State and local statutes and regulations. About 240 investor-owned utilities (such as Exelon, Dominion, and American Electric Power) ran three-quarters of the country's total electrical capacity in 2008. In addition to these "power giants," more than 3187 other private utilities provided power services along with 2012 public utilities, 900 cooperatives, 2168 nonutility generating entities, 400 power marketers, and 9 federal utilities (such as the Tennessee Valley Authority and Bonneville Power Administration) (Sovacool 2008b).

Inconsistency

Perhaps this complexity may explain why inconsistencies have occurred in the way energy problems are addressed and resolved. Consider the following example that involves promotion of renewable energy. US national renewable energy policy has lacked internal consistency. Federal research on renewable energy systems concentrated on centralized, large-scale, and utility-owned technologies whereas the legislative effort focused on advancing decentralized, small-scale, and independently-owned technologies. Legislation including the Public Utility Holding Company Act of 1935 (PL 74-333), Wind Energy Systems Act of 1980 (PL 96-345), Renewable Energy Industry Development Act of 1984 (PL 98-370), and provisions of the Energy Policy Act of 1992 (PL 102-486) were allowed to expire or were never fully implemented. The result has contributed to government mandates and lackluster adoption of renewable resources. One study rated renewable energy policy in the US the most inconsistent out of a sample of 17 countries (Haas *et al.* 2008). The study observed that for 16 other countries, the average number of significant revisions to renewable energy policies for the period 1997–2005 was less than 0.6, yet the US managed to change its policies six times.

The same level of inconsistency featuring frequent changes in direction has been common in energy research sponsored by the federal government. Such research has seen volatile changes in year-to-year funding as great as 116% for coal, 84% for petroleum, 64% for natural gas. All major Department of Energy (DOE) research areas and technologies have seen similar ups and downs in funding patterns (Narayanamurti *et al.* 2009: 8.). Gulliver and Zillman (2006) explain the lack of consistency in government research funding through the lens of regulatory capture. Many divisions of the national government have been captured by various energy interests frequently organized in trade associations and represented by powerful lobbying firms. These coalitions zealously protect these interests over those of consumers and the country at large. In some years, their efforts of getting subsidies and research funds are more fruitful than in others, especially when their focus shifts to opposing new regulatory measures.

Vertical Fragmentation

Despite fairly clear delimitation of federal and State jurisdictions over energy matters, an element of vertical fragmentation exists where local, State, and federal actors sometimes duplicate and contradict each other's actions. In some cases such vertical fragmentation has resulted in litigation between States as well as between the States and the federal government.

In other situations the States have tried to "bridge the gap" in areas where they believed the federal government has no or limited authority to act: renewable portfolio standards, greenhouse gas reduction targets, renewable fuel standards, net metering, and carbon trading schemes. Although these attempts have produced sizeable environmental and economic benefits, they have also created problems. Often such unilateral actions have complicated regional, State, and local energy markets, which led to a cacophony of dissenting voices and views negatively affecting public opinion and support for renewable technologies. Consider renewable portfolio standards (RPS), a type of State policy, which requires that States provide a certain percentage of electricity from renewable sources by a specific date (in California all major power utilities must produce 30% of their electricity from renewable resources by 2020). Instead of utilizing a national energy market featuring the same national standards, requirements, and goals, policy-makers (mostly out of necessity) rely on State initiatives that lack such a consistency. As a result, differences among States over what counts as renewable energy, when it has to come online, how large it has to be, where it must be delivered, and how it may be traded clog the renewable energy market like fallen leafs in a gutter. Investors must spend time and resources interpreting and selecting from competing and often arbitrary State statutes and regulations; implementing agencies, both federal and State, must constantly grapple with the perpetually changing and expanding universe of State initiatives; and, State policy-makers must constantly adjust renewable energy goals, at times to keep up with other States and at times after being forced by pro-fossil energy groups. To illustrate this point, imagine America's interstate highway system being structured like its renewable energy market: drivers would be forced to change engines, adjust tire pressure, and replace fuel every time they crossed State lines.

State targets for reducing greenhouse gas emissions and trading carbon credits serve as another good example of vertical fragmentation. Similarly to RPS initiatives, State climate change policies display the deficit of consistency and harmony. The multitude of State greenhouse gas policies creates unnecessary complexity for investors, rendering these polices more expensive than a single, federal standard. State-by-State standards considerably raise the transactional cost for the firms attempting to conduct business in multi-State regions. In addition to the cost of "entering" an intrastate or regional carbon market (the cost of preliminary research and expenses related to filing an application and undergoing verification), firms must also incur the cost associated with the actual participation in each market. Thus, such firms are required to comply with separate inventory, monitoring, and implementation mechanisms. Program administrators must check progress against goals and provide feedback, adding to the overall costs of a program. In addition, the efficiency of State programs providing incentives to local and regional actors to support research-and-development (R&D) efforts remains highly questionable. Such incentives often duplicate the R&D efforts that the actors would have undertaken anyway, as well as financial support from the federal government. Unfortunately, because of the lack of coherent effort from the federal government paralyzed by a current political impasse, such programs, despite of all their shortcomings, are the best and the only politically feasible solution to development of climate-friendly energy in the US.

Horizontal Fragmentation

Even when one looks exclusively at the federal government, a significant degree of horizontal fragmentation exists. Decision-making processes within the more than a dozen federal agencies in charge of implementing energy policies are often stove-piped and fragmented. Channels of coordination among federal agencies such as the DOE, Department of Transportation, and Treasury remain confusing and complicated. The national research laboratories are sponsor- or donor-driven rather than mission-oriented, and energy-related research programs are poorly funded compared to other federally supported programs (Sovacool 2009a). Contrary to a historical role as an incubator of energy technologies and breakthrough energy discoveries, the DOE and its national laboratories have been subject to criticism for having ineffective management and losing their sense of mission. DOE labs suffer from overly centralized, hierarchical, and micromanaged resource allocation systems. As a result, they often lack flexibility and cost effectiveness. The following three reasons are usually offered as an explanation for the DOE's problems. First, the DOE's diverse and ambiguous mission creates an approach to energy problems that lacks coordination. Second, loss of a clear mission is further amplified by a largely dysfunctional organizational structure. Third, as a result of the problematic organizational structure, DOE research remains hampered by a weak culture of accountability.

Politicization and Manipulation

The now emergent varying views of what US energy policy should be have made it a highly contested and politicized battlefield. One recent survey of 2701 US residents conducted in the summer of 2008 concluded that familiarity with a particular type of energy generation and proximity to a power plant frequently result in public acceptance and support. This means that instead of looking at national samples, one should focus on particular local communities. Driven by concern about local environmental conditions, as well as displaying a strong correlation with preference for renewable sources of energy and opposition to fossil fuels, local communities differ greatly from one another and from a sample "All-American" community (Greenberg 2009).

As if the presence of different views were not enough, there have been notable efforts to manipulate public opinion to make energy even more politicized and contentious. Consider the following examples (Fahrenthold 2009; Lyons 2009; Mouawad 2008; Muller 1997; Radmacher 2008; Sheppard 2009; Sovacool 2008b):

- In February 1993, President Clinton proposed a broad-based energy tax to be levied on the energy content of fuels, with a higher rate for petroleum fuels than oil or natural gas. Immediately after the proposal was announced, the Clinton Administration was quick to grant concessions to different lobbies and interest groups. Under pressure from groups such as the National Association of Manufacturers and energy companies such as ExxonMobil and British Petroleum, exemptions were given related to the oil used in refineries, coke used in steelmaking, electricity used in aluminum and chlorine production, and waste-to-energy facilities, among others. The tax became so watered down that it was defeated in the Senate;
- In 2007 the Senate introduced a carbon cap-and-trade bill that would create a national market for carbon credits, but the National Association of Manufacturers called the bill tantamount to "economic disarmament." Trade groups also lobbied and convinced the US Chamber of Commerce to oppose such legislation, resulting in a government sponsored advertisement showing a man cooking breakfast over candles in a cold, darkened house, then jogging to work on empty highways, asking, "Is this really how Americans want to live?"

- When State environmental officials rejected two coal-fired power plants in Kansas because of the millions of tons of carbon dioxide they would produce in 2008, a pro-coal lobbying group, Kansans for Affordable Energy, was soon formed. The group, funded mostly from Peabody Coal (the supplier of coal to Kansas from its mines in Wyoming) and Sunflower Electric Corporation (the local utility), placed newspaper advertisements with pictures of the smiling faces of Presidents Mahmoud Ahmadinejad of Iran, Vladimir Putin of Russia, and Hugo Chavez of Venezuela, suggesting that if the coal plants were nixed, these natural gas exporting countries would benefit. Even though the ads were completely false (not one of those countries exports natural gas to Kansas), the campaign convinced the State legislature to approve the coal plants, a move that almost succeeded until it was vetoed by Governor Kathleen Sebelius;

- In Illinois, when Commonwealth Edison, a subsidiary of Exelon, was confronted with the possibility of an extension of electricity rate caps to protect consumers in 2007, the utility spent $10 million to form Consumers Organized for Reliable Electricity. The group ran a comprehensive array of television and newspaper ads taking a stance against the rate freeze on behalf of the public, but never once mentioned its ties to the utility;

- In 2008 the US coal lobby has responded to all of the negative publicity connecting coal use to climate change by creating two sleek nonprofit groups termed Americans for Balanced Energy Choices and the American Coalition for Clean Coal Electricity (ACCCE). These organizations remain dedicated to correcting misinformation about coal and reminding the public that coal keeps America's economy running strong, protects natural resources, and ensures energy security. (This claim about protecting natural resources is peculiar, since coal mining is about extracting coal from the earth, not preserving it there.) Worried that their ads were not working, the ACCCE even ran a marketing campaign during the holiday season of 2008–2009 which included singing lumps of call called The Clean Coal Carolers. Lyrics for these songs included "Frosty the Coalman," "Deck the Halls (With Clean Coal)," "Silent Night," and "Oh Christmas Tree";[2]

- During congressional debates about climate change and a pending bill attempting to create a nationwide carbon cap-and-trade system in 2009, a coal lobbying group forged at least 12 letters purporting to be from citizens' groups such as the National Association for the Advancement of Colored People or American Association of Retired Persons that were sent to congressional offices instructing them to oppose local climate change bills;

- After President Obama was elected and pledged to pass new climate legislation, the American Petroleum Institute secretly sponsored and organized a series of Energy Citizens Rallies against the legislation to provide the appearance of concerned local communities, who turned out to be hired oil lobbyists;

- When John Holdren began advocating for things like low-carbon technologies and more accurate (and thus expensive) energy prices in 2009, opponents quickly attacked him for advocating controlling population growth by putting sterilizers in drinking water and forcing women to have abortions, claims that were completely false but did succeed in evoking calls for Holdren to immediately resign;

- Only a few months after the Deep Water Horizon leaking well was sealed, the petroleum industry unleashed a campaign against the Bureau of Ocean Energy Management, Regulation and Enforcement (BOEMRE), the successor of the infamous Minerals Management Service (MMS), for unfairly slowing down the permitting process for offshore platforms. These attacks were made despite the fact that many firms continued to file permit applications hampered by the same blatant problems (copying, word by word, each other's safety plans) that, as several reports indicate, contributed to the accident.

These examples underscore that consensus on energy issues may no longer be possible in the US. The former Committee on Nuclear and Alternative Energy Systems, a panel set up by the National Academies in the 1970s to look at the energy crisis, vividly demonstrates the difficulties faced by any group trying to arrive at consensus on energy issues. "It simply can't be done," its Chair, Harvey Brooks (1980), concluded, "at least not within any group that honestly represents the spectrum of defensible views in today's academic, intellectual, and industrial community."

Interactions with Global Energy Governance

This section reflects how US energy governance has influenced, and is influenced by, energy governance pressures at the global scale. It identifies four ways US policy has been affected by global energy governance trends relating to (1) climate change, (2) trade and protectionism, (3) multi-polarity, and (4) oil dependency. It then investigates at least two ways that the US affects global energy governance through (1) idea diffusion and (2) influence on multilateral institutions.

Climate Change and Trans-Boundary Externalities

One way in which the US has been influenced by global pressure relates to climate change and the push for global, or at least regional, markets for carbon credits. When President George W. Bush decided to reverse earlier pledges that the US would ratify the Kyoto Protocol, Margot Wallstrom, the EU environmental commissioner, called his decision "very worrying" (Gelbspan 2004: 93). EU spokesperson Annika Ostergren went further when she stated that "sometimes people think that the Kyoto Protocol is only about the environment, but it's also about international relations and economic cooperation" (Gelbspan 2004: 97). As a result EU countries have agreed to set up an internal market enabling companies to trade carbon dioxide pollution permits. Since 2005, some 10,000 large industrial plants in the EU have been required to buy and sell permits to release carbon dioxide into the atmosphere, representing about 40% of the EU's total CO_2 equivalent emissions. To date, the industries covered by the scheme have included power generation, iron and steel, glass, cement, pottery, and bricks. The EU's scheme has motivated many US states to promote their own market for carbon credits, either individually (such as California) or in regional blocs (such as the Regional Greenhouse Gas Initiative in the Northeast).

Markets and Protectionism

Global trends in trade have shaped US energy policy in a variety of ways, including its stance on intellectual property, protectionist attempts to maintain economic competitiveness, and restricting foreign ownership of US energy asserts and companies. As the global energy market-place has become more connected, energy governance decisions in the US have broadened outward to include international actors, although the interaction between local/national discussions and the supranational and international sphere has been mixed. Some within the US are wary of losing sovereignty, suffering higher energy prices, and hurting economic competitiveness. These forces have exerted a more protectionist stance by creating pressure to limit technology transfer to other countries and prevent international authorities from having more sway over US decisions. Thus, recent decisions in the past 10 years to refuse to join the Kyoto Protocol, proceed with developing offshore oil and gas fields in the Gulf of Mexico and Alaska to diversify supply away from the Middle East and Africa, erect tariffs against imported Brazilian ethanol and Chinese solar panels, and set constraints on the diffusion of innovative energy technologies to the developing world.

Indeed, in early 2010 an attempt by the Chinese to partially invest in American wind farms was enough to make the "blood boil" of a few US senators. One $450 million project involving 240 turbines and 600 MW of installed capacity in Texas could create 3000 jobs in China but only 300 in the US. When they learned of this fact, four senators wrote to the Treasury Department to place a moratorium on the project until a "Buy American" clause could be mandated for all wind projects being funded, partially or wholly, with stimulus funds (Chandra 2010). Other examples concern intellectual property. Weak intellectual property right (IPR) protection has prevented some US companies from developing more advanced clean coal technologies (such as more efficient coal washing processes, advanced combustion turbines, and carbon capture and storage systems). IPR concerns connected with clean coal systems are cited as one of the most significant impediments toward diffusing such technologies to China, Indonesia, and other developing countries, especially where new technologies could be reverse engineered or copied. Research on pollution abatement technologies for sulfur dioxide and nitrogen oxide emissions in the US has been slowed by a perceived need to adapt technologies to local markets (Sovacool 2008c).

To be sure, US policy and action on energy trade and intellectual property is hardly consistent. As one example of an initiative that has both elements of protectionism and engagement simultaneously, consider the US-led Asia-Pacific Partnership on Clean Development and Climate, known as APP. APP is a multilateral partnership to "stimulate joint ventures and investment" that can lead to a "low carbon economy" (Garrison 2009: 20). Started by the George W. Bush Administration, APP has working groups that focus on clean coal, renewable energy, cement, and steel, and it works with public and private sector partners to meet the goals of improving energy security, reducing air pollution, and responding to climate change, especially in India and China. However, while in theory the APP seems to signal engagement, funding has been inconsistent and the Chinese have critiqued it for being too "soft," complaining that the partnership focuses on information sharing and capacity building, but not joint R&D projects, technology transfer, or cost sharing. China expects technology at reduced rate and cost, yet the partnership budget will not allow reduced costs and as currently funded will not cover demonstration and deployment of actual technologies. The US wants rigorous guarantees that intellectual property will be respected but the Chinese have been unable or unwilling to grant them, and both countries have refused to share some data about natural resources and geology on "national security grounds."

Diplomacy and Military Adventurism

The transition to a multipolar world, one with many strong regional powers instead of a single, dominant hegemon, is also shaping US energy diplomacy, how it engages other regimes overseas. US policy-makers have come to acknowledge more players in the international scene, and accepted that they are in fact competing with China, India, even Russia in their efforts to acquire resources abroad and form international partnerships. One excellent example concerns the Baku-Tbilisi-Ceyhan oil pipeline (BTC) traversing Azerbaijan, Georgia, and Turkey; another concerns the country's dependence on imported oil.

President Clinton and other senior staff strongly supported construction of the multi-billion dollar BTC pipeline (through bilateral aid deals and other soft incentives) for its ability to limit terrorism, promote economic development, secure access to energy resources, and enhance US leadership. US government officials openly talked about how the BTC would help limit extremism and create a bulwark against fundamentalism and terrorism by raising standards of living throughout the Caspian region. The US saw the pipeline as an important strategic asset that diversified oil supply away from Middle Eastern and Russian suppliers and helped connect emerging Caspian economies to the global market-place. It was widely believed (although later disproven) that Caspian oil reserves were the third largest in the world, after those found in Western Siberia and the Persian Gulf. Undiscovered reserves were believed to hold 200 billion barrels of oil, more than 30 times the estimated amount in the North Slope of Alaska and enough oil to meet US demand for about 30 years. The US government also strongly pushed for the BTC as a way to build a "silk road" to Central Asia where it could limit Russian geopolitical influence, restrict possible allies for Iran, and provide a long-term buffer against Chinese dominance (Sovacool and Cooper 2013).

In addition, the reliance of the US economy on imported oil has given oil-exporting countries, including the Organization of Petroleum Exporting Countries (OPEC), immense influence over US foreign and domestic policy decisions. In a policy later strengthened by the creation of the US Central Command in the early 1980s and since termed the Carter Doctrine, the administrations of Presidents Ronald Reagan, George H. Bush, Bill Clinton, and George W. Bush have each relied on the threat of military force to deter and prevent major disruption in world oil supply (Kalicki 2007; Klare 2007; Stokes 2007). Now, virtually every large deal involving major oil and gas companies spurs a spirited energy security conversation in the US. For example, the BP–Rosneft attempted stock swap in January 2011 was no exception. Representative Edward Markey, the highest ranking Democrat on the House of Representatives' Natural Resources Committee, expressed his concern over the Russian government becoming the largest shareholder of the oil giant in the following remark: "BP once stood for British Petroleum... [w]ith this deal, it now stands for Bolshoi Petroleum." Mr. Markey's concerns about Russian ownership of BP were not without reason because it serves as one of the largest fuel suppliers for the US military (Sidortsov 2011: n. 10).

The Carter Doctrine has significantly expanded since the 1970s. When Iranian forces began to attack Kuwaiti oil tankers traveling through the Persian Gulf in an attempt discourage Kuwait from supplying loans to Iraq for arms procurement at the height of the Iran–Iraq war of 1980–1988, President Reagan authorized the reflagging of Kuwaiti tankers with the US ensign to afford them naval protection. The Clinton Administration and both Bush Administrations have funneled billions of dollars into protecting the Persian Gulf and other oil-based assets. US Southern Command now promotes security cooperation activities to grow US influence and dissuade potential adversaries in oil-producing regions of South America, especially Colombia, including training, equipping, and developing security forces to protect refineries and offshore oil and gas platforms. In Central Asia, the US operates military and training programs to train and equip

Georgian and Uzbek security forces to maintain the free flow of oil. In West Africa, military aid and training has flowed to Nigeria (the third largest supplier of oil to the US) to help bolster the security of its oil infrastructure. The price of these military activities in the Persian Gulf is expected to cost the US between $29 billion and $80 billion per year (Delucchi and Murphy 2008). Even withdrawal of American troops from Iraq and Afghanistan may not provide relief to US soldiers, marines, and sailors, as well as to the taxpayers who foot war bills. As the recent heated exchange between Iran and the US over the threat of blockade of the Strait of Hormuz indicated, US military presence in the Persian Gulf will be necessary as long as oil supply channels are affected.

Norms and Institutions

The direction of influence has gone the other way – from the US outward to global actors – with regard to ideas and institutions relating to energy and environmental policy. The country was the first in the world to legislate clean air and clean water acts among other progressive environmental measures, and also the first to implement a Toxics Release Inventory to track industrial pollution rates. The US is the intellectual home for a number of remarkable energy policy tools, from tradable permits for pollutants (early programs traded leaded gasoline and water rights in the 1980s before expanding to sulfur dioxide in the early 1990s) to instruments that promote renewable energy such as system benefits charges, renewable portfolio standards, and tax credits. States within the US were also home to some of the first efforts to restructure or deregulate electricity sales by unbundling generation, transmission, and distribution and distinguishing wholesale from retail markets. These ideas have proliferated globally, with policy targets for renewable energy prevalent in more than 80 countries and specific US-invented tools, such as production tax credits and renewable portfolio standards, in 45 and 51 countries respectively. The EU Emissions Trading Scheme was designed after the US Acid Rain Program. The ideas of restructuring of the electricity industry developed in the US found a home in Europe in the form of the EU Third Energy Directive. The exploration of ideas and policy tools is certainly an area where the US exerts global influence.

A second way the US affects the global energy governance scene is through its interaction with multilateral institutions. It is a member of the International Energy Agency, providing funding, expertise, and even data from its own US Energy Information Administration. At times this influence has been criticized, as when in 2009 the International Energy Agency was accused of "deliberately underplaying shortages of oil" for fear of triggering panic buying. Anonymous employees within the organization argued that it was the US that pressured the agency to exaggerate future oil production rates, saying they would increase from 83 million barrels per day in 2008 to 105 million barrels per day by 2030, without any evidence (Macalister 2009). The US can also influence the way that organizations such as the World Bank or United Nations report or act on energy issues, because it can set the staffing rules and picks the president of the former and provides funding to the latter.

Conclusion

If there is one central lesson, energy governance in the US is currently complex, inconsistent, fragmented, politicized, and dynamic. The historic principles that guided the formative years of energy supply and use in the US – a public duty to serve, emphasis on low prices, adherence to large sophisticated technologies, responding to demand, faith in technology, trust in expertise and technocrats, and preeminence of public interest – are now in flux. No overarching theme or principle currently guides policy, and the country instead appears wedded to a portfolio approach that pushes a variety of technologies and systems, even those that trade off with each other, to appease as many diverse stakeholders and lobbyists as possible. Here, the portfolio approach makes perfect sense insofar as it accommodates different interests, more politically acceptable than one that served only limited interests.

The problem, however, is that the world's largest energy consumer is running at full steam ahead without a rudder. Currently, energy governance in the US represents more a bundle of technologies randomly juxtaposed together than any coherent energy vision. The situation illustrates quite nicely the non-technical dimensions of energy systems, that underlying social and political values can change over time and exert great influence on the course a country takes concerning energy; and focusing on governance offers a departure from commentaries exclusively about technology that so often dominate the discussion. Also, the country appears at a crossroads in terms of its role in global energy policy. Should the US choose to play a bigger role in the global energy policy, it needs to build upon its success as a policy innovator, and also learn from other participants and compensate for lost years. But while looking at governance may give us a more nuanced view of the current state of affairs, it still does little to help resolve the inconsistencies and complexities that become painstakingly apparent.

Notes

1. Driven largely by considerations of public interest and protecting rights of natural gas consumers against "exploitation at the hands of natural-gas companies," the Court subjected interstate sales of natural gas to rate regulation. Such excessive regulation split the US natural gas market into intra- and interstate markets. As a result, natural gas prices in non-producing States increased drastically (Bosselman *et al.* 2006).
2. Lyrics for some of these songs were admittedly catchy and clever: "Frosty the Coalman, he's getting cleaner every day; he's affordable and adorable and the workers keep their pay. He's abundant here in America and he helps our economy roll." Silent Night had its refrain changed from "Christ the Savior is born" to "Plenty of coal for years to come." "Oh Christmas Tree" was changed to "Technology, technology, you make the coal burn clearly."

References

Bosselman, Fred, Jim Rossi, and Jacqueline Weaver. 2006. *Energy, Economics and the Environment, Cases and Materials*. 2nd edn. New York: Foundation Press.

Brooks, Harvey. 1980. Energy: A Summary of the CONAES Report. *Bulletin of the Atomic Scientists* (February): 23.

Bupp, Irvin C., and Jean-Claude Derian. 1978. *Light Water: How the Nuclear Dream Dissolved*. New York: Basic Books.

Chandra, Nayan. 2010. Thorns Amid Green Shoots. *The Straits Times*, March 15, 2010, A18.

Clark, J. G., 1982. Federal Management of Fuel Crisis Between the World Wars. In H. D. George and H. R. Mark, eds. *Energy and Transport: Historical Perspectives on Policy Issues*. London: Sage Publications, pp. 135–147.

Clark, John G. 1990. *The Political Economy of World Energy: A Twentieth Century Perspective*. Hemel Hempstead, UK: Harvester Wheatsheaf.

Delucchi, M. A., and J. J. Murphy. 2008. US Military Expenditures to Protect the Use of Persian Gulf Oil for Motor Vehicles. *Energy Policy* 36: 2253–2264.

Fahrenthold, David A. 2009. Coal Group Reveals 6 More Forged Lobbying Letters. *Washington Post*, August 5.

Freeman, S. David. 1973. Is There an Energy Crisis? *An Overview. Annals of the American Academy of Political and Social Science* 410, 1: 1–10.

Garrison, Jean A. 2009. China's Quest for Energy Security: Political, Economic, and Security Implications. Paper Presented to the 50th Annual Meeting of the International Studies Association, New York, February 15–18.

Gelbspan, Ross. 2004. *Boiling Point: How Politicians, Big Oil and Coal, Journalists, and Activists Are Fueling the Climate Crisis – And What We Can Do to Avert Disaster*. New York: Basic Books.

Greenberg, Michael. 2009. Energy Sources, Public Policy, and Public Preferences: Analysis of US National and Site-Specific Data. *Energy Policy* 37: 3242–3249.

Gulliver, John, and D. N. Zillman. 2006. Contemporary United States Energy Regulation. In Barry Barton, Lila Barrera-Hernandez, and Alastair Lucas, eds. *Regulating Energy and Natural Resources*. Oxford: Oxford University Press, pp. 113–136.

Haas, Reinhard, Niels I. Meyer, Anne Held, *et al.* 2008. Promoting Electricity from Renewable Energy Sources – Lessons Learned from the EU, United States, and Japan. In Fereidoon P. Sioshansi, ed. *Competitive Electricity Markets: Design, Implementation and Performance*. Amsterdam: Elsevier, pp. 91–140.

Hirsh, Richard F. 1989. *Technology and Transformation in the American Electric Utility Industry*. Cambridge: Cambridge University Press.

IPCC. 2007. Summary for Policy-makers. In *Climate Change: 2007*. Washington, DC: Government Printing Office.

Josberger, Edward, William Bidlake, Rod March, and Shad O'Neel. 2009. Fifty Year Record of Glacier Change. *US Geological Survey Fact Sheet* 2009-3046.

Kalicki, J. H. 2007. Prescription for Oil Addition: The Middle East and Energy Security. *Middle East Policy* 14, 1: 76–83.

Klare, M. T. 2007. The Futile Pursuit of Energy Security by Military Force. *Brown Journal of World Affairs* 13, 2: 139–153.

Lackner, Klaus S., and Jeffrey D. Sachs. 2004. A Robust Strategy for Sustainable Energy. *Brookings Papers on Economic Activity* 2: 215–248.

Laitos, Jan G., and Thomas A. Carr. 1998. The New Dominant Use Reality on Multiple Lands. *Rocky Mountain Mineral Law Institute* 44: 1–22.

Lovins, Amory. 1976. Energy Strategy – The Road Not Taken?' *Foreign Affairs* 55: 65–96.

Lovins, Amory. 1979a. *Soft Energy Paths: Towards a Durable Peace*. New York: Harper Collins.

Lovins, Amory. 1979b. A Target Critics Can't Seem to Get in Their Sights. In Hugh Nash, ed. *The Energy Controversy: Soft Path Questions and Answers*. San Francisco, CA: Friends of the Earth.

Lutzenhiser, Loren. 1993. Social and Behavioral Aspects of Energy Use. *Annual Review of Energy and the Environment* 18: 247–289.

Lyons, Daniel. 2009. An SOS for Science: Clean Energy Should Trump Politics. *Newsweek*, October 12, 26.

Macalister, Terry. 2009. Key Oil Figures Were Distorted by US Pressure, Says Whistleblower. *The Guardian*, November 9.

Melosi, Martin V. 1985. *Coping with Abundance: Energy and Environment in Industrial America*. New York: Knopf.

Mouawad, Jad. 2008. Industries Allied to Cap Carbon Differ on the Details. *New York Times*, June 2.

Muller, Frank. 1997. Energy Taxes, the Climate Change Convention, and Economic Competitiveness. In Olav Hohmeyer, Richard L. Ottinger, and Klaus Rennings, eds. *Social Costs and Sustainability: Valuation and Implementation in the Energy and Transport Sector*. New York: Springer, pp. 465–487.

Narayanamurti, Venkatesh, Laura D. Anadon, and Ambuj D. Sagar. 2009. *Institutions for Energy Innovation: A Transformational Challenge*. Cambridge, MA: Harvard University Press.

Orr, David W. 1979. US Energy Policy and the Political Economy of Participation. *Journal of Politics* 41: 1027–1056.

Radmacher, Dan. 2008. Effort to Clean Coal's Image Won't Work. *Roanoke Times*, December 21, 7.

Rosenzweig, Cynthia, David Karoly, Marta Vicarelli, *et al.* 2008. Attributing Physical and Biological Impacts to Anthropogenic Climate Change. *Nature* 453: 353–357.

Sautter, J. A., J. Landis, and M. H. Dworkin. 2008–2009. Energy Trilemma in the Green Mountain State: An Analysis of Vermont's Energy Challenges and Policy Options. *Vermont Journal of Environmental Law* 10: 477–502.

Sheppard, Kate. 2009. Majority of Energy Citizens' Rallies Organized by Oil-Industry Lobbyists. *Grist*, August 21.

Sidortsov, Roman. 2011. Measuring Our Investment in the Carbon Status Quo: Case Study of New Oil Production Development in the Russian Arctic. LL.M thesis, Vermont Law School.

Smith, Zachary A. 2009. *The Environmental Policy Paradox*. 5th edn. Upper Saddle River, NJ: Prentice Hall.

Sovacool, Benjamin K. 2008a. The Best of Both Worlds: Environmental Federalism and the Need for Federal Action on Renewable Energy and Climate Change. *Stanford Environmental Law Journal* 27, 2: 397–476.

Sovacool, Benjamin K. 2008b. *The Dirty Energy Dilemma: What's Blocking Clean Power in the United States*. Westport, CT: Praeger.

Sovacool, Benjamin K. 2008c. Placing a Glove on the Invisible Hand: How Intellectual Property Rights May Impede Innovation in Energy Research and Development (R&D). *Albany Law Journal of Science & Technology* 18, 2: 381–440.

Sovacool, Benjamin K. 2009a. Resolving the Impasse in American Energy Policy: The Case for a Transformational R&D Strategy at the US Department of Energy. *Renewable and Sustainable Energy Reviews* 13, 2: 346–361.

Sovacool, Benjamin K. 2009b. Exploring and Contextualizing Public Opposition to Renewable Electricity in the United States. *Sustainability* 1, 3: 702–721.

Sovacool, Benjamin K. 2011. National Energy Governance in the United States. *Journal of World Energy Law and Business* 4, 2: 97–123.

Sovacool, Benjamin K., and Christopher J. Cooper. 2013. *The Governance of Energy Megaprojects: Politics, Hubris, and Energy Security*. London: Edward Elgar.

Stern, Paul C., and Elliot Aronson. 1984. *Energy Use: The Human Dimension*. New York: Freeman & Co.

Stokes, D. 2007. Blood for Oil? Global Capital, Counter-Insurgency and the Dual Logic of American Energy Security. *Review of International Studies* 33: 245–264.

Tonn, Bruce, K. C. Healy, Amy Gibson, *et al.* 2009. Power from Perspective: Potential Future United States Energy Portfolios. *Energy Policy* 37: 1432–1443.

UNDP. 1997. *Energy After Rio: Prospects and Challenges*. Geneva: United Nations Development Programme.

US EIA. 2012. United States Energy Information Administration, Country Analysis Brief, United States. http://www.eia.gov/countries/country-data.cfm?fips=US&trk=p1.

Winner, Langdon. 1982. Energy Regimes and the Ideology of Efficiency. In George H. Daniels and Mark H. Rose, eds. *Energy and Transport: Historical Perspectives on Policy Issues*. London: Sage Publications, pp. 261–277.

Yergin, Daniel. 1991. *The Prize: The Epic Quest for Oil, Money and Power*. New York: Simon & Shuster.

Chapter 27

Global Energy Policy: A View from Brazil

Suani T. Coelho and José Goldemberg[1]

Introduction

There are presently 1.3 billion people with no access to electricity worldwide and 2.7 billion people using traditional biomass from deforestation for heating and cooking, which is responsible for a significant number of deaths mainly in women and children in poor countries due to the pollutant emissions. Therefore there is an urgent need to guarantee affordable energy access in a sustainable way. According to the document "Energy for a Sustainable Future" from the UN Secretary General's Advisory Group on Energy and Climate Change (AGECC 2010), it is mandatory to guarantee energy access in poor countries using modern and clean fuels.

The two main goals established in the recommendations from the AGECC are: ensure universal access to modern forms of energy by 2030, and reduce global energy intensity by 40% by 2030. To achieve such ambitious goals it is fundamental to address new pathways, breaking with the ones used worldwide until now, mainly based on fossil fuels, and to promote the use of renewable energy in an affordable manner. According to the International Energy Agency (IEA 2011a), in 2009 it is estimated that more than US$9 billion was invested globally to supply 20 million more people with electricity access and 7 million people with advanced biomass cooking stoves. One IEA scenario (2011a) shows US$296 billion invested between 2010 and 2030, but it still leaves 1.0 billion people without electricity and 2.7 billion people without clean cooking facilities in 2030.

Transportation and energy sectors are the ones responsible for the highest greenhouse gas (GHG) emissions worldwide (22.5% and 40.7% in 2009 respectively, according to IEA 2011b). Moreover, in the case of transportation, the only renewable energy available commercially today is bioenergy (liquid biofuels). Differently from electricity production, which can be produced from several renewable energy sources (solar, wind, and small-hydro), the replacement of fossil fuels in the transportation sector through commercialized technologies is possible only with biofuels. Transportation is an integrated and essential element of our modern lifestyle, but all over the world this sector is almost exclusively dependent on petroleum-based fuels whose use results in serious burden to

The Handbook of Global Energy Policy, First Edition. Edited by Andreas Goldthau.
© 2013 John Wiley & Sons, Ltd. Published 2013 by John Wiley & Sons, Ltd.

the environment at the local, regional, and global level, particularly GHG emissions that are expected to continuously increase.

In Brazil, existing programs on energy access allowed 98.73% of the population access to electricity, being 99.7% in urban households (IBGE 2010). Also a liquefied petroleum gas (LPG) program distributed all over the country with high subsidies to make it affordable to poor people contributed to reducing significantly the use of firewood from deforestation (12% less in 2010 compared with 2006 in energy basis). Together with these programs, the biofuels programs (ethanol and biodiesel) allowed not only energy security in the country, with a significant reduction on oil imports, but also quite low GHG emissions since the Brazilian energy matrix is based mostly in renewable energy sources.

This Brazilian case has insights to offer to other developing countries (DCs) on how to tackle this challenge. Aiming to share this experience, this chapter presents a general analysis of renewables worldwide followed by the Brazilian case, and ends with a discussion on replicating the perspectives of this experience in other DCs through the introduction of adequate policies.

Renewables Worldwide

Worldwide the primary energy sources are mainly fossil fuels (34% from oil, 28% from coal and 22% from gas, corresponding to 84% from fossil fuels), renewable sources representing less than 13% of primary energy. Also it must be noticed that this share of renewable energy includes 8% from traditional biomass, obtained from deforestation in poor countries. Modern biomass (Goldemberg and Coelho 2004; Karekesi *et al.* 2006) is responsible for only 2.3% of total primary energy.

Figure 27.1 shows the contributions of the different primary energy sources in 2009, in which total consumption was 492 Exajoules, or 11.75 billion metric tons of oil equivalent (toe). For a global population of 6.79 billion inhabitants, this quantity means 1.73 toe per capita. As is also possible to observe in the same Figure, of all the energy used in the world, 85.1% is obtained from non-renewable sources (petroleum, coal, and gas), 2% from nuclear energy, and 12.9% from renewable energies (hydroelectricity, biomass, geothermal, wind, solar, and tidal).

Biomass represents the greatest contribution of renewable energies (10.2%) being consumed under different forms. More than half (60%) is consumed as traditional biomass for cooking, production of charcoal, and residential heating in rural or

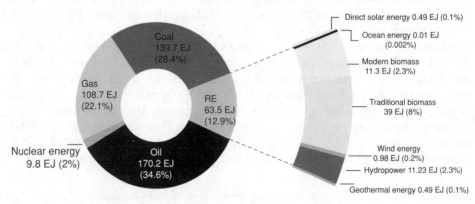

Figure 27.1   Primary energy in the world, 2009.
Source: Authors' elaboration based on IPCC SRREN (2011).

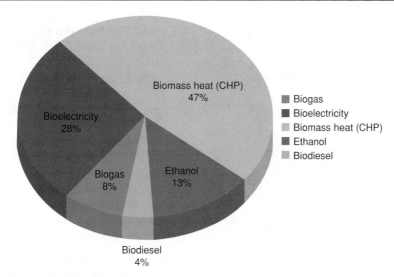

Figure 27.2   The role of biomass in the world's energy supply, 2008.
Source: Data from IPCC SRREN (2011).

semi-urban areas in developing countries. The rest is used as modern biomass in the
form of fuel (ethanol and biodiesel), biogas, bio-electricity, and cogeneration of heat and
electricity (Figure 27.2). This is the main source of renewable energy in use in the world
today. Investing in the increase of its contribution seems to be one of the policies which
would bring the greatest result in the short term.

The consumption of fossil fuels has grown at an annual rate of nearly 2% except in the
last few years. Renewable energy, on the other hand, has increased much more rapidly.
See Figure 27.3, which represents the annual average evolution of the different types of
renewable energy in the period 2004–2009.

Despite its rapid growth, this type of energy represented a modest contribution to
the global consumption in 2008. However, the weighted growth average for renewable
energies in the last five years was around 7% per year. If this situation continues the

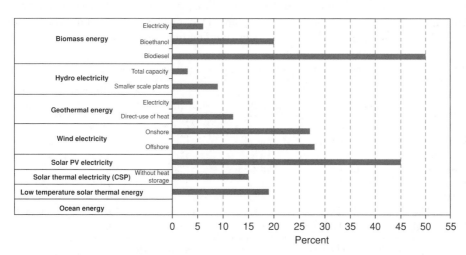

Figure 27.3   Average annual growth of renewable energy, 2004–2009.
Source: Data from IPCC SRREN (2011).

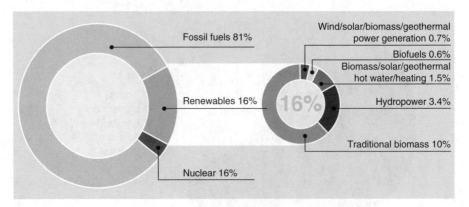

Figure 27.4   Final energy consumption in the world, 2009.

contribution of renewable energy will at least triple in the next 15 years to reach nearly 20% of global consumption by 2025. As an example, of the 300 million kilowatts added to the global electric system in 2008–2009, half came from renewable energies. The investments received by this sector (up to US$150 billion in 2009 excluding hydroelectricity) surpass those in fossil fuels for the generation of electricity.

In general, the cost of renewable energy is still greater than for fossil and nuclear fuels, but there are countries where it is already economically competitive. This tendency should increase as production increases, bringing with it a reduction of costs. The best example of this experience is the reduction of ethanol production costs in Brazil in the last 30 years. Several studies indicate that in 2050, renewable energies may represent from 20% to 50% of the global consumption of primary energy.

Considering global energy consumption in 2009 (Figure 27.4), fossil fuels are responsible for 81% against 16% from renewables, of which 10% is from traditional biomass (mainly in DCs). The situation is different when industrialized countries are compared to developing countries. Table 27.1 compares the primary energy offer in different regions with Brazil's.

This table illustrates the high consumption of biomass (up to 65% in African countries, mostly traditional biomass) in DCs, with the exception of Brazil, where "biomass" corresponds to the biofuels programs and planted wood. This high consumption of traditional biomass, together with the extremely low rates of electricity supply, are typical for

Table 27.1   Share of primary energy sources in different regions and in Brazil.

	OECD 2009	Africa 2008	Asia 2008	Latin America 2008	Brazil 2010
Oil	37.2%	21%	22%	44%	38%
Coal	19.7%	16%	51%	4.2%	5.1%
Natural gas	24.2%	13%	8%	2.3%	10.2%
Fossil fuels	**81.1%**	**50%**	**81%**	**50.5%**	**53.3%**
Hydro	2.1%	1%	2%	10%	14.2%
CRW*	5.5%	48%	15%	19.9%	27.3%**
Renewables	7.6%	49%	17%	29.9%	41.5%
Nuclear/others	**11.3%**	**1%**	**2%**	**19.6%**	**5.2%**

*CRW: Combustible renewables and waste (according to IEA definitions).
**Brazil: Modern bioenergy.
Source: IEA (2010), BEN (2011).

Table 27.2 Number and share of people without access to modern energy services in selected countries, 2008.

Countries/Regions	Without Access to Electricity		Relying on the Traditional Use of Biomass for Cooking	
	Population (million)	Share of Population	Population (million)	Share of Population
Africa	587	58%	657	65%
Nigeria	76	49%	104	67%
Ethiopia	69	83%	77	93%
DR of Congo	59	89%	62	94%
Tanzania	38	86%	41	94%
Kenya	33	84%	33	83%
Other Sub-Saharan Africa	310	68%	335	74%
North Africa	2	1%	4	3%
Developing Asia	675	19%	1921	54%
India	289	25%	836	72%
Bangladesh	96	59%	143	88%
Indonesia	82	36%	124	54%
Pakistan	64	38%	122	72%
Myanmar	44	87%	48	95%
Rest of developing Asia	102	6%	648	36%
Latin America	31	7%	85	19%
Middle East	21	11%	0	0%
DCs	1314	25%	2662	51%
World	1317	19%	2662	39%

Source: IEA (2011a).

the low values of HDI (0.361 in DR of Congo, 0.503 in Tanzania, and 0.968 in Iceland and Norway in 2006)[2] in poor countries, mainly least developing countries (LDC).

Table 27.2 shows recent results for the lack of access to electricity and use of traditional biomass for cooking in DCs compared to the world.

From the above figures we note that renewables are growing worldwide and outperform growth rates of fossil fuels. At the same time, there is a correlation between the use of renewables and a country's HDI, which expresses a strong reliance of LDCs on traditional use of biomass. In that sense, it is important to look at the composition of renewable consumption, mainly in LDCs, to see whether the bioenergy used is sustainable. Therefore it is essential to find ways to support bioenergy as well as renewables through programs that make them competitive, give them a firm place in the energy mix, even as countries develop, and make them a tool to fight poverty. Brazil's experience could collaborate in the development of such programs.

Energy in Brazil

Brazil can be considered a special case in terms of the share of renewables in its energy matrix, mainly due to the more than 74% hydro-based electric energy supply and the Brazilian programs on biofuels since 1975. At that time special policies on biofuels were introduced in the country to reduce the external dependence on imported oil (see Box 27.1). Since then several experiences have made Brazil a world leader on bioenergy, mainly regarding sugarcane ethanol production, distribution, and use in the transportation sector.

Box 27.1 Biofuel programs in Brazil.

The use of bioenergy in the Brazilian energy matrix has been a reality for a long time. Large-scale ethanol production in Brazil was initiated in 1975, and it is nowadays economically competitive with gasoline (Goldemberg *et al.* 2004; Goldemberg, 2009). Brazil is the world's second largest producer of ethanol (and the largest one using sugarcane ethanol) with 27.5 billion liters in 2010, after the US (producing ethanol from corn). In the last harvesting season there were 427 mills producing ethanol and sugar. The national average for agricultural yield in 2010 was almost 78 (metric) tonnes of sugarcane per hectare, with some regions reaching 100 tonnes per hectare. Industrial yields were in the range of 43–55 liters of ethanol per tonne of cane during the last seasons and may reach up to 82 liters of ethanol per tonne of cane in some regions (MAPA 2011).

The main advantage of sugarcane ethanol is its positive net energy balance in comparison to corn ethanol or ethanol from other crops. This energy balance is around 8.3 on average, with the best cases showing a balance of 10.2 (Macedo *et al.* 2008). Initially ethanol was used in ethanol-dedicated engines or as an octane enhancer, replacing lead and/or MTBE (methyl tertiary butyl ether). Currently, instead of ethanol-dedicated vehicles, hydrated ethanol is used in flex-fuel vehicles. More than 90% of all new cars sold in Brazil are flex-fuel, which can run on any blend of gasoline and/or ethanol, allowing drivers to make price-driven fuel choices (ANFAVEA 2010). In the domestic market, it replaces 41.5% of light duty transportation fuel in the country (Datagro 2010).

In addition to the environmental benefits, the biofuels program in Brazil has shown significant benefits in social aspects. For example, the production of ethanol in Brazil was responsible for the creation of more than one million jobs, mainly in rural areas, and the introduction of mechanical harvesting of green sugarcane has resulted in an upgrading of the technical level of the workforce.

Brazil is the world's second largest producer of biodiesel. By the end of 2010 the production was 2.3 billion liters and there were 68 plants registered with an installed capacity of 6.2 billion liters (ANP 2011). Soy is the main feedstock used for biodiesel production (counting for 80%), followed by animal fat (almost 13%) and other vegetable oils. The domestic market of biodiesel is defined by the blending mandate of 5% biodiesel (B5) in all diesel sales in the country. In 2010, the use of B5 was anticipated from the scheduled year of 2013 and there was a significant increase in biodiesel production. The large use of soy bean oil is due to its low price, a consequence of the huge national production of soy bean, this oil being a byproduct of the soy (protein) production for animal feed.[3] The increasing use of animal fat is due to the huge amount of cattle in the country (around 200 million heads) mainly to provide meat export to industrialized countries.

The Brazilian Energy Matrix

Figure 27.5 illustrates the share of different energy sources in the primary energy supply in the country. It is important to notice the participation of hydroelectricity, as well as modern biomass, corresponding to the production of biofuels for transportation and planted wood mainly for pulp/paper and pig iron sectors.

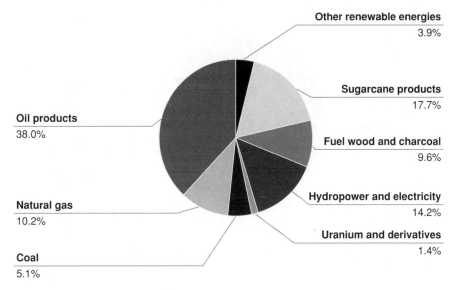

Figure 27.5 Brazilian energy supply, 2010.
Source: BEN (2011).

In the case of the electric energy offer in the country, more than 74% is from hydro, and bioenergy accounts for more than 5% (Figure 27.6).

Together with the use of sugarcane ethanol in the transportation sector in Brazil (see Box 27.1), this is the reason why GHG emissions from energy production in Brazil are quite low, being 18th highest in the world, if one excludes the emissions resulting from deforestation in Amazonia. Bagasse, the residue from sugarcane crushed to produce sugar

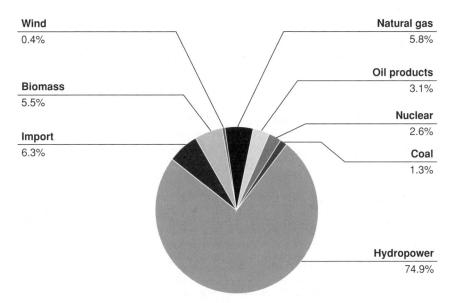

Figure 27.6 Electric energy supply in Brazil, 2010.
Source: BEN (2011).

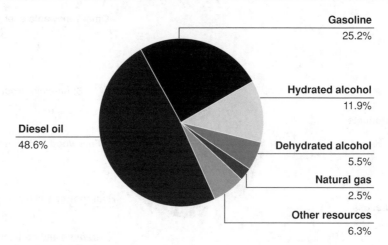

Figure 27.7 Energy consumption in transportation sector in Brazil, 2010.
Source: BEN (2011).

and alcohol, is used for coproduction of heat and power generation (cogeneration) in the mills, both for self-consumption and for the sale of electricity to the grid. The installed capacity in 2010 was almost 6000 MW and, in the 2009/2010 harvesting season, the total of electricity production from sugarcane bagasse was 20,031 GWh; 28.2% of the mills sell their surplus electricity to the grid. Over the next 10 years, in the best scenario (considering 99 bar-boilers installed in all mills, for a sugarcane production forecast at 1.04 billion tonnes), electricity production from sugarcane bagasse is expected to increase to 68,730 GWh (CONAB 2011).

When considering the transportation sector (Figure 27.7), the share of biofuels is significant, with 17.9% from sugarcane ethanol (used as pure ethanol in flexible vehicles and blended with gasoline, as presented in Box 27.1) against 25.3% gasoline. The consumption of diesel oil, however, is still high, since diesel oil is highly subsidized in the country; but (for this reason) this fuel is not allowed to be used in light vehicles, only for trucks, agricultural machines, and urban transportation. Aiming to reduce the diesel consumption in the country, the biodiesel program was introduced in 2003 and since 2010 all diesel oil in the country is blended with 5% biodiesel by volume.

In the residential sector the introduction of the LPG program allowed significant reduction in firewood consumption. Nowadays wood consumption for fuel is around 30% against 26% for LPG (in energy basis). Residential consumption of firewood has declined from 53.5% of total biomass consumption in 1970 to just 13.8% in 2002 (Figure 27.8).

Since 2002 the Brazilian government started the implementation of incentives for renewable energy. Law 10.438 of April 26, 2002 established that Eletrobras (the holding company of the Brazilian electric sector) would purchase 3300 MW through 20-year contracts equally divided among the main three renewable energy sources in Brazil (small hydro, wind, and biomass). The main requirement was that the producer should have the environmental licensing to sign the contract. The tariffs to be paid to producers were established by the Federal Government but in the case of biomass they were not considered attractive by the producers and the 1100 MW limit was not achieved (only 685 MW were commercialized).

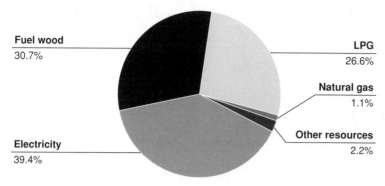

Fuel wood
30.7%

LPG
26.6%

Natural gas
1.1%

Other resources
2.2%

Electricity
39.4%

Figure 27.8 Energy consumption in Brazilian residential sector, 2010.
Source: BEN (2011).

Renewable Energies in Brazil

In contrast with the global energy matrix shown in Figure 27.1, where renewable energies represent only 12.9% of the total, in Brazil they reach 45.3% of the primary energy supplied, as indicated in Table 27.3.

In relation to electricity, the situation is even more favorable, as seen in Figure 27.3 previously. Hydroelectricity, including the fraction imported from Paraguay in Itaipu, represents more than 74% of the total in 2010 (BEN 2011). The problem is the future: although the population growth is slow in the country, the living condition of the poorer part of the population has improved, which means that more electricity will be necessary to meet their needs.

Table 27.4 shows the installed electric power installed in 2010 and the projections of the 10-year Energy Plan (PDEE-2020) and the National Energy Plan (PNE 2030): from 109 GW in 2010, power generation should grow to 171 GW in 2020 and to 236 GW in 2030. Hydroelectric energy will continue to be the basis for the expansion of electricity production in the country until 2030. However it is necessary to observe the enormous contribution of wind and biomass, in 2020, present in these projections.

Table 27.5 aggregates renewable energy and thermo energy sources and shows that the participation of renewable energy in the production of electricity by 2030 will continue to be dominant in the Brazilian energy matrix.

Table 27.3 Domestic supply of energy in Brazil, 2010.

Sources	Participation in %/(toe)
Non-renewable energy	**54.6 (147.9)**
Petroleum and derivatives	37.9 (102.8)
Natural gas	10.1 (27.6)
Mineral coal and coke	5.0 (13.7)
Uranium (U_3O_8)	1.4 (3.9)
Renewable energy	**45.3 (122.8)**
Hydro and electricity	14.1 (38.3)
Wood and vegetable coal	9.6 (26.1)
Derivatives of sugarcane	17.6 (47.8)
Other renewables	3.9 (10.6)
Total	**100 (270.8)**

Source: BEN (2011).

Table 27.4   Forecast of electric energy supply in Brazil, 2010–2030 (GW).

Sources	2010	2020	2030
Hydroelectric (with Itaipu)	82.9	115.1	148.6
Thermo-electric	17.5	28.9	42.6
Natural gas	9.2	11.7	17.5
Nuclear	2.0	3.4	7.4
Coal	1.8	3.2	4.9
Others	4.5	10.6	12.9
Alternatives	9.1	27.0	40.8
Small hydroelectric plants (PCHS)	3.8	6.4	9.0
Wind	0.8	11.5	13.5
Biomass	4.5	9.1	22.3
Total	109.6	171.1	232.0

Source: PDEE-2020 (2011) and PNE 2030 (2007).

The Energy Planning Company (EPE) recently conducted a survey of at least 20 enterprises in several regions of the country, including the Amazon region, with total power of 32 million kilowatts. It is important to note that the inventory of possible hydroelectric sites, which in the past was conducted by Eletrobras, has not been conducted since the 1990s. This has seriously hindered the possibility of participation of hydroelectricity in electricity auctions in the last few years.

There are approximately 16 million kilowatts available in the Amazon region, in addition to Belo Monte. In this region there are areas where the construction of medium-sized hydroelectric plants with 500–1000 MW is possible, which would not cause great environmental impacts. It is also important to analyze if in these locations it is possible to consider the installation of reservoirs which would regularize the flow of rivers and store water for drier periods of time. One of the great problems in the Brazilian electric sector – the main cause of the disastrous rationing in 2001 – is the fact that since 1986 hydroelectric plants constructed in the country do not have reservoirs, in order to avoid the flooding of surrounding areas. This is a situation which needs to be reassessed.

Hydroelectric plants have, in general, negative aspects with the inundation of forest areas, which affects thousands of people who live in the region, in addition to encouraging the deforestation around the plant. On the other hand, hundreds of thousands, or even millions of people who live great distances from the plant are benefited. To construct hydroelectric plants without reservoirs may help to resolve environmental problems, but it may not be very rational from an economic point of view.

This is an aspect of the problem that environmentalists have difficulty in accepting, but which they should re-examine. To flood 500 or 1000 km^2 to construct a reservoir – which is the case of Belo Monte – may seem a lot, but it is little compared to the current deforestation of the Amazon, which even if it has decreased, is still 5000 km^2 per year. In

Table 27.5   Renewable and thermo energies in the production of electricity in Brazil (Mtoe). Participation (percentage) in parentheses.

Sources	2010	2020	2030
Renewables	95.0 (84.3)	142.2 (83.1)	193.6 (82.0)
Thermals	17.1 (15.7)	28.9 (16.9)	42.6 (18.1)
Total	112.1 (100)	171.1 (100)	236.2 (100)

Source: PDEE-2020 (2011) and PNE 2030 (2007).

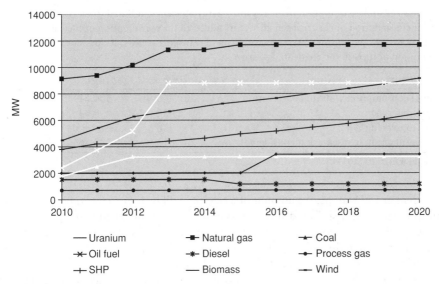

Figure 27.9 Evolution of Brazilian installed capacity excluding hydro (MW).
Source: PDEE-2020 (2011).

relation to the participation of other sources (thermo and nuclear) these were modest in the EPE plans before 2009. However, in the auctions which occurred in 2008/9, there was a significant increase in the forecast of thermo and coal source energy, since there were no hydroelectric plants able to participate in the auctions. This tendency was reversed in the EPE plans in 2010, and there was a significant increase in the participation of wind energy until 2020 (Figure 27.9).

The contribution of nuclear energy indicates an increase until 2020 due to the conclusion, expected for 2015, of the Angra III nuclear reactor. By 2030, the EPE forecasts the installation of four more nuclear reactors. However, the disaster of Fukushima is causing a reassessment of nuclear expansion plans around the world and the same should occur in Brazil. One inevitable consequence of the disaster is that nuclear energy costs should increase due to the need for additional safety measures, which will probably make this type of energy source less competitive.

In relation to the participation of biomass in the generation of electricity, the EPE forecasts for 2020 significantly underestimate this potential. Recent surveys conducted by CONAB indicate the possibility of cogeneration with sugarcane bagasse to be much higher than the numbers estimated by the EPE (see Table 27.6).

It is important to note that there is a great synergy among energy efficiency measures and renewable energies. For example, in isolated locations in the rural regions, one may produce electricity with solar panels and photovoltaic cells (PV). If these are used with

Table 27.6 Evolution of Brazilian bioelectricity supply potential (MW).

	2010	2015	2019	2025	2030	2035
São Paulo 2035 (2011)	5.130	11.646	17.322	22.382	28.614	34.464
COGEN (2010)	6.715	14.315	22.315	–	–	–
PDEE-2020 (2011)	4.496	7.353	9.163	–	–	–
CONAB (2011)	5.915	17.190	21.262	–	–	–

Table 27.7   Total consumption of energy and energy efficiency in Brazil (toe). (Corresponds to the total consumption of electricity in all sectors added to the consumption of fuels in the industrial, energy, agribusiness, services, public, and transport sectors. It does not include the consumption of fuels in the residential sector.)

Consumption	2011	2015	2020
Potential consumption without conservation	239,840	301,611	393,938
Energy conserved	2,028	9,045	22,410
Energy conserved (%)	0.8	3.0	5.7
Final consumption (conservation considered)	237,812	292,566	371,527

Source: PDEE-2020 (2011).

fluorescent or LED lamps, the amount of electricity will be four to five times smaller than with tungsten lamps, still used in some locations. Therefore, even if the production of electricity with PV is more expensive, the combination of PV with LED lamps becomes more attractive than the older lighting processes.

Considering the synergies between renewable sources and energy efficiency, it is necessary to record that energy efficiency has had a small role in energy planning in the past, despite the existence of the Energy Efficiency Law 10.295, 2001, which authorized the Executive Branch to establish maximum levels of specific energy consumption or minimum levels of energy efficiency of machines and equipment manufactured or commercialized in the country. The implementation of this law occurred only in December 2007 with the Interministerial Norm 362, which deals with energy consumption levels of refrigerators and freezers. Until then only a National Electric Energy Conservation Program (Procel) seal was put on equipment, to inform the public. No prohibition on sale of inefficient products was enforced.

On May 26, 2011 there was an expansion of the energy efficiency program with the publication of interministerial (Ministries of Development, Industry and Foreign Trade, Science and Technology, and Mines and Energy) norms which approved targets for stoves, ovens, refrigerators, freezers, and water heaters. The products which are not within the new efficiency indexes will only be able to be sold until the end of 2012. According to the EPE forecasts, which are shown in Table 27.7, the total consumption of energy forecasted for 2020 would be 5.7% lower with the measures of energy conservation considered, which represents a small reduction of 0.57% per year. In contrast, in member countries of the OECD, from 1973 to 1998 the conservation of energy was nearly 2% per year.

In the medium and long term, the measures adopted recently in these interministerial norms may have great impact in the consumption of energy, as occurred in other countries. There are, therefore, conditions for the contribution of renewable energies in the Brazilian energy matrix to remain approximately at the present level of 50%, until 2030.

Brazil's Experience in Energy Access and Poverty Alleviation

According to the Household Census of 2010 accomplished by the Federal Government (IBGE 2010), Brazil has achieved the level of 99.7% universalization of electric power access in urban areas. This means that out of over 49.2 million households in cities, only 133,000 are still with no access to electricity.

These data show that the universalization of access to electric power in cities and suburban areas is already a reality in Brazil. Nevertheless, there is still an important proportion of consumers who are not regularized, that is, who use the service by means of illegal connections.[4] The great challenge that Brazil currently faces is not to grant access to electric power, but to regularize the consumers, transforming them into clients of the electric power distributors and assuring that these new clients pay their bills regularly.

Policies for Energy Access

After decades of government ownership and operation of the electricity sector, privatization of electric companies in Brazil began in 1996 after approval of the sector's new design for an operational model and adjustment of existing legislation to permit foreign ownership of utilities. After several programs introduced in the country aiming to increase energy access, finally the "Luz para Todos" program (Light for All) was able to increase energy access substantially all over the country. Only remote villages in Amazonia still present low rates of energy access but even these are now becoming the focus of this program.

The Ministry of Energy and Mines (MME) has the overall responsibility for policy-making in the electricity sector while ANEEL, which is linked to the MME, is the Brazilian Electricity Regulatory Agency created by law in 1996. ANEEL's function is to regulate and control the generation, transmission, and distribution of power in compliance with the existing legislation and with the directives and policies dictated by the government. The National Council for Energy Policies (CNPE), is an advisory body to the MME in charge of approving supply criteria and "structural" projects while the Electricity Industry Monitoring Committee (CMSE) monitors supply continuity and security. Another important actor in the Brazilian electricity sector is Eletrobras. This company is focused on electric power generation and transmission, in which it is the leader in the Brazilian market. Eletrobras supports government strategic programs, such as the program that fosters alternative electric power sources (Proinfa), the program "Luz para Todos" (Light for All), and the National Program for Electric Power Conservation (Procel). This structure was created during the process of privatization and the agencies mentioned are the main actors in programs of energy access.[5]

Universalization of Energy Access

The "Luz para Todos" (Light for All) program was launched by the Federal Government in November 2003, aiming to eliminate the electric exclusion in the country mainly in rural areas and reach over 10 million people by 2008. The program has been coordinated by the MME, operated by Eletrobras, and executed by the utilities and rural electrification cooperatives. For meeting the initial goal, R$20 billion (around US$10 billion at November 2011) was invested. The map of electric exclusion in the country showed that the families with no energy access are mostly low-income families, in sites with lower HDI. "Luz para Todos" was extended to 2010, then again in order to complete the last one million connections. It is important to emphasize that it is mandatory for the concessionaires to provide electricity to every citizen that requests this service, always and whenever the household to be connected has its land tenure regularized or in process of regularization. And the concessionaires indeed do so. According to Table 27.8, energy access for urban households reached 99.7% in 2010.

Low Income Tariffs

Brazil has adopted performance-based regulation to ensure affordable tariffs are paid by the so-called "captive" electricity customers. A performance-based, price-capped, and multi-year tariff is used to achieve high-quality, reliable, and universal service. Tariffs in

Table 27.8   Energy access rates in urban households in Brazil in 2010.

	Households with Electricity	Households without Electricity	Electricity Access %
Brazil	49,093,032	133,097	99.7
North Region	2,993,228	19,122	99.4
Northeast Region	11,137,927	61,887	99.4
Center-West Region	3,851,820	7,655	99.8
Southeast Region	23,510,520	28,905	99.9
South Region	7,599,537	15,528	99.8

Source: IBGE (2010).

general are relatively high compared to many other countries, especially those such as Canada with a comparable proportion of hydroelectric generation in its mix. However, it should be noted that the Brazilian tariff on average is more than 30% tax, a significantly larger percentage than that imposed in most other countries (USAID 2009). The low income tariff (LIT) or "social" tariff provides large discounts to low consumers. The discount percentiles applied in energy bills are presented in Table 27.9.

Indigenous and *quilombo* people also receive the discount.[6] In these cases, the discount is 100% until the consumption equals 50 kWh/month. When consumption is higher than this, the discounts from the consumption parcel corresponding to Table 27.9 above will be applied.

During the late 1990s, the government used the RGR, a general sector fund financed by a fee on all electricity customers, for subsidizing rural electrification and tariffs for very poor consumers. More recently the CDE (Conta de Desenvolvimento Energético or Fund for Energy Development), also fee-based, replaced the RGR for subsidizing the low-income sector.[7] The resources of the CDE are also being used for urban and rural electrification purposes. According to ANEEL, R$1408 billion went to the distributors in 2006 from the CDE. Approximately 17 million customers presently receive the LIT. This represents 36% of the 50.2 million electricity customers in Brazil.[8]

This Brazilian experience on energy access mainly in rural areas shows that it is feasible for other DCs to implement similar policies. In rural areas it is fundamental that energy production can be sustainable under economic, environmental, and social aspects. That is why PV programs in other DCs often fail since there is not enough power to allow the development of economic activities (PV systems may only provide power for small activities such as light and water pump).[9]

Liquefied Petroleum Gas (LPG) Program in Brazil

Among usual ways of energy consumption for cooking, poor urban or suburban populations in Brazil rely mostly upon bottled LPG. The 13 kg bottle is available in several

Table 27.9   Discounts included in the Electric Power Social Tariff.

Range	Discount
Monthly consumption of up to 30 kWh	65%
Monthly consumption from 31 kWh to 100 kWh	40%
Monthly consumption from 101 kWh to 220 kWh	10%
Monthly consumption above 220 kWh	0%

Source: ANEEL (2011).

Table 27.10   Brazilian domestic energy consumption by source of energy.

	2010	2009	Δ 10/09
			Unit: 10^3 tep
Electricity	9.327	8.753	6.6%
Fuel wood	7.276	7.529	−3.4%
LPG	6.298	6.115	3.0%
Natural gas	255	238	7.1%
Other resources*	517	592	−12.7%
Total	**23.673**	**23.227**	**1.9%**

*Including kerosene and charcoal.
Source: BEN (2011).

specialized stores or distributed by trucks. LPG delivery infrastructure is highly developed in all regions, including rural zones. In 2004, 98.48% of the urban population owned cooking stoves and 7.61% of the rural population did not own any stove (PNAD 2004). Even so, cooking stoves are not the only means of cooking because both urban and rural poorer households keep a wood-fueled stove as a back-up in case they cannot afford LPG (Lucon *et al.* 2004). Table 27.10 shows the Brazilian domestic energy consumption evolution by source of energy. The recent reduction of firewood consumption and the growth of LPG consumption is noticeable.

In 1996 the share of Brazilian residential energy consumption from wood fuel was equivalent to the fraction using LPG, as showed in Figure 27.10. LPG consumption has been rising from the middle of the last century until 2002 when the subsidy "Gas Allowance" was removed; from this year on, its consumption decreased at same period as wood fuel consumption started to rise. From 2006, LPG consumption started to increase again since the residential use of traditional biomass has declined significantly. This inversion happened because the residential use of wood fuel for cooking in traditional wood

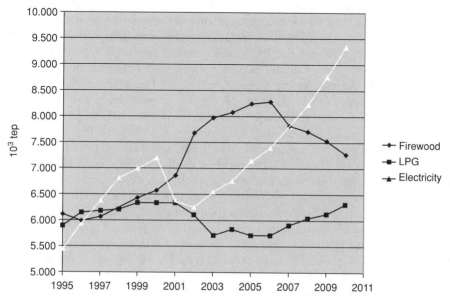

Figure 27.10   Firewood, LPG, and electricity consumption in Brazil.
Source: Authors' elaboration based on BEN (2011).

stoves has dwindled over the years. This is due to growing urbanization (currently less than 20% of the Brazilian population lives in rural areas), to the rise of the population's income, and mainly to the availability of LPG. Residential consumption of wood fuel has declined from 53.5% of total biomass consumption in 1970 to just 13.8% in 2002. Figure 27.10 shows the historic evolution of LPG versus firewood since 1995.

Proposal of Policies for Sustainable Growth in Other DCs Based on Brazilian Experience

The positive results of the Brazilian LPG program to reduce wood fuel consumption and to reduce deforestation could be replicated in other DCs. Bottles with LPG (13 kg bottles) are distributed all over the country and sold at affordable (subsidized) prices, even in Amazonian remote villages.

Related to biofuels, in general, DCs have a higher potential to produce biomass than industrialized countries due to more favorable climate conditions and lower labor costs. As a result, international trade in biofuels and/or feed stocks from developing to the developed countries is expected to increase with significant positive implications for development (UNCTAD 2009). It must be noticed that, as mentioned from the Brazilian experience, electricity from residues presents a huge opportunity with enormous benefits for the increase of energy access in rural areas in DCs. The existing "Cogen for Africa Project" funded by GEF-UNEP-AfDB[10] is an extremely important experience that is being developed in Sub-Saharan countries.[11] Through the production (and also the use) of biofuels, energy security can be improved, power access enhanced, rural development aided, and both economic development and employment accelerated. In all these regions, land availability is crucial to achieving high production levels. Higher agricultural productivity in biomass crops – sugarcane agricultural yield is less than one third of the Brazilian[12] – could allow Africa to supply about 30% of world production of biomass (FAOSTAT 2009).

Asian and African countries have seen substantial developments in their agricultural sector. Government support (adequate financial and policy instruments) allowed improved seeds and fertilizer use and these are some of the reasons behind the success, but further improvements are still needed, including an adequate infrastructure for production and distribution. Among the 39 countries already producing sugarcane in Africa, there are many countries in Southern Africa which have large potential for growing biofuel feedstock. Angola, Mozambique, Zambia, and Tanzania have low population densities and favorable soils and climate. Yet so far, commercial biofuel production in the region is limited. But this is about to change as many Southern African countries are planning to produce biofuels and have already started to grow feedstock, with the purpose of producing ethanol mainly from sugarcane and biodiesel from *Jatropha*.[13]

Another issue being discussed is the apparent conflict between small and large agrobusiness. The Brazilian experience is often seen as a large-scale monoculture of sugarcane not allowing the preservation of biodiversity and which displaces small farmers. However, in the state of Parana most sugarcane plantations are owned by small farmers organized in cooperatives and results are quite positive, allowing the replication of this model. The only economic issue related to the scale of production is in the industrial phase, where the scale of the mills is indeed an important factor to allow the economic competitiveness of biofuels.

There are several myths against biofuels (Goldemberg *et al.* 2011), most of them due to economic interest against biofuels production, but there are studies answering these questions, showing that biofuels can indeed be produced and used in a sustainable way, as well as electricity produced from biomass. There are studies (Goldemberg 2009) showing that Brazilian sugarcane plantations in most regions are implemented following strict environmental rules and adequate environmental legislation, including the preservation and recovery of riparian forest in São Paulo State, allowing the adequate preservation and recuperation of local biodiversity.

Another difficulty related to the production of biofuels in DCs and mainly in LDCs is the program "Everything But Arms" (EBA),[14] where selected countries receive huge subsidies to export sugar to industrialized countries. This system of preferences is an important contribution to the economic development of such countries, but it could be expanded to cover the production of biofuels, following certification criteria to be adequate and affordable to each country. Still in this context, it must be realized that certification is not a serious problem and can be accomplished by DCs and even LDCs if adequate capacity building and funding is provided, together with targets and timetables to allow these countries to implement these criteria step by step (UNCTAD 2008). Last but not least, UNCTAD (2008) argues that certification criteria must be used carefully, not being used as non-tariff barrier (in fact a neocolonialism).

Conclusion

Considering the above issues related to barriers to the replication of the Brazilian biofuels program in other DCs, the most significant lessons learned from the Brazilian experience include the adequate choice for biofuels crops, through the establishment of an agro-environmental-economic zoning to define the best areas for food and fuel production, ensuring food security and contributing to rural development (not only through the creation of jobs in rural areas but also through the increase of energy access from sugarcane bagasse) (Goldemberg 2009). Developing countries in Latin America, Africa, and Asia have the potential to produce the needed raw material and to produce ethanol and biodiesel, and the exploitation of such resources could be quite fast through technology transfer.[15] However, for this process to be successful two main steps are needed: adequate local incentive policies and foreign financing for the projects (and for capacity building where needed). It should be pointed out here that, when discussing the best regions for each crop, the use of *Jatropha curcas* in large-scale plantations should be carefully evaluated, because there are not yet enough varieties to insure against disease and losses, according to the Brazilian Agricultural Research Corporation/Embrapa (Sato *et al.* 2009).

Many DCs (mainly in Africa and Asia) have a small internal market but have land and climate adequate for the production of biofuels. The production of biofuels to be exported to industrialized countries could stimulate rural development, generate jobs, increase energy access, and reduce poverty. Preconditions for that include the need for capacity building to master the technologies required both in the agricultural and industrial areas. Assistance from other DCs, such as Brazil or India, which have important activities in sugarcane production (either for sugar or ethanol), could be very fruitful in this case, fostering South–South cooperation.

The introduction of a bioethanol activity in DCs should in all cases be preceded by a proper agronomic ecological zoning to identify producing areas, in order to respond to frequent criticism that biofuels production does not comply with certification criteria appropriate to local conditions. Such arguments can in reality be interpreted as non-tariff protectionism adopted by some countries in Europe and the US, to protect non-competitive agro-industrial activities in their home countries.

Notes

1. The authors acknowledge contributions from Renata Grisoli and Manuel Moreno at CENBIO (National Reference Center on Biomass), University of São Paulo.
2. Human Development Index, available at http://hdr.undp.org/en/media/HDI_2008_EN_Tables.pdf.
3. In 2009/2010 Brazil produced 68 million (metric) tonnes of soy from 23.24 million hectares, being the world's second largest soy producer (world production was 258 million tonnes). Brazil exported 28.35 million tonnes. Brazil produced in 2009/2010 6 million tonnes of soy bean oil, of which 1.26 million tonnes was for export (MAPA 2010).
4. Illegal connections: consumers who make their connections by themselves directly in the secondary network without metering.
5. The Brazilian Federal Constitution defines electric power distribution as a federal public service. In April 2002, the approval of Law 10.438/2002 attributed to ANEEL the competency to establish, for compliance by each electric power distributor, the goals to be periodically met, aiming at the universalization of electric power usage. In the use of this attribution, Resolution n. 223, from April 29, 2003, was edited, establishing the general conditions for elaboration of the Electric Power Universalization Plans aiming at serving new consumer units, and defining the responsibilities for the electric utilities.
6. A *quilombo* is a Brazilian hinterland settlement founded by people of African origin. Most of the inhabitants of *quilombos* were escaped slaves and, in some cases, a minority of marginalized Portuguese, Brazilian aboriginals, Jews, Arabs, and/or other non-black, non-slave Brazilians who experienced oppression during colonization.
7. CDE was created by Law 10.438 in 2002 as a fund aimed at fostering the energy development of the Brazilian States and the competitiveness of alternative energy projects, natural gas-fueled power stations, and Brazilian coal-fueled power stations in the locations served by the Brazilian Electric Interconnected System, and making the energy services generally available to all people throughout the Brazilian territory (the so-called universalization of the services). The CDE is regulated by the Brazilian government and administered by Eletrobras, and will exist for 25 years.
8. Of these, 14 million (or 82%) automatically receive the lowest tariff because they consume less than 80 kWh per month. In 2007 ANEEL began to tighten the eligibility procedures for the LIT by requiring that consumers be registered in government low-income programs such as Bolsa Família (BF) as proof of their low-income status. As part of this drive, ANEEL has been trying to eliminate the low-income self-declaration process for eligibility now used for electricity consumers in the 80–200 kWh per month consumption range, by requiring that instead they register in the government's Cadúnico registry and are certified as low income under Bolsa Família to receive the benefit. Using Bolsa Família and the Cadúnico is eliminating some non-poor from the LIT rolls (e.g., those with vacation homes and single occupancies). At the same time, inclusion would improve matters for very large low-income families that use more than 220 kWh per month and are now excluded from the LIT (USAID 2009).
9. Of course PVs cater to basic needs where they are lacking in rural areas and DCs. But for the follow up of the process it is necessary to provide enough power for the development of local economic activities in order to make the energy supply affordable to everyone.
10. GEF: Global Environmental Facility. UNEP: United Nations Environment Programme. AfDB: African Development Bank.

11. http://www.afrepren.org/cogen/index.htm.
12. Authors' personal field visit to West Kenya sugar mills.
13. Global Network on Energy for Sustainable Development, Bioenergy Theme, 2010. www.gnesd.org.
14. In 1968, the first UN Conference on Trade and Development (UNCTAD) recommended the creation of a "Generalized System Tariff of Preferences" under which industrialized nations would grant autonomous trade preferences to all developing countries. The European Community was the first to implement a GSP scheme (the acronym GSP sometimes refers to the system as a whole, sometimes to one of the individual schemes) in 1971. Other countries have subsequently established their own GSP schemes that differ both in their product coverage and rules of origin. In order to update its scheme on a regular basis and to adjust it to the changing environment of the multilateral trading system, the EU's GSP is implemented following a 10-year cycle. The present cycle which lasts from 2006 to 2015 was adopted in 2004. Traditionally, it has been admitted that the group of LDCs should receive more favorable treatment than other DCs. Gradually, market access for products from these countries has been fully liberalized. In February 2001, the EU adopted the so-called EBA Regulation (Everything But Arms), granting duty-free access to imports of all products from LDCs, except arms and ammunitions, without any quantitative restrictions (with the exception of bananas, sugar, and rice for a limited period). EBA was later incorporated into the GSP Regulation (EC) 2501/2001. The Regulation foresees that the special arrangements for LDCs should be maintained for an unlimited period and not be subject to the periodic renewal of the EU's scheme of generalized preferences. Available at http://ec.europa.eu/trade/wider-agenda/development/generalised-system-of-preferences/everything-but-arms/.
15. Recently, a mission of CENBIO (National Reference Center on Biomass) in Mozambique concluded that with adequate actions one can reach the sustainable production of biofuels in Africa with technologies already commercialized.

References

AGECC. 2010. *Energy for a Sustainable Future. Summary Report and Recommendations*. The Secretary General's Advisory Group on Energy and Climate Change. New York: United Nations.
ANEEL. 2011. National Regulatory Agency on Electric Energy. Available at www.aneel.gov.br.
ANFAVEA. 2010. *Official Brazilian Automotive and Autoparts Industry Guide*. Brazil Automotive Industry Yearbook.
ANP. 2011. National Petroleum Agency. Available at www.anp.gov.br.
BEN. 2011. *Brazilian Energy Balance (preliminary version)*. Brasilia: Ministry of Mining and Energy.
COGEN. 2010. Associação das Industrias de Cogeração de Energia. Private communication.
CONAB. 2011. National Company of Food Supply. Available at www.conab.gov.br.
Datagro. 2010. *Datagro Bulletin*.
FAOSTAT. 2009. *Crops. Sugarcane*. Rome: FAO.
Goldemberg, J. 2009. The Brazilian Experience with Biofuels. *Innovations Journal* 4: 91–107.
Goldemberg, J., and S. T. Coelho. 2004. Renewable Energy – Traditional Biomass vs. Modern Biomass. *Energy Policy* 32: 711–714.
Goldemberg, J., E. L. La Rovere, S. T. Coelho, *et al.* 2011. *Bioenergy Study Theme. Final Report*. Roskilde: Global Network on Energy for Sustainable Development.
Goldemberg, J., S. T. Coelho, P. M. Nastari, and O. S. Lucon. 2004. Ethanol Learning Curve – The Brazilian Experience. *Biomass and Bioenergy* 26: 301–304.
IBGE. 2010. Census 2010. Instituto Brasileiro de Geografia e Estatística. http://www.ibge.gov.br.
IEA. 2010. *Energy Statistics of Non-OECD Countries*. Paris: International Energy Agency.
IEA. 2011a. *World Energy Outlook 2011*. Paris: International Energy Agency.

IEA. 2011b. *World Energy Statistics*. Paris: International Energy Agency.

IPCC SRREN. 2011. *Renewable Energy Sources and Climate Change Mitigation*. Special report of the International Panel on Climate Change, Working Group III. Cambridge: Cambridge University Press.

Karekesi, S., K. Lata, and S. T. Coelho. 2006. Traditional Biomass Energy: Improving Its Use and Moving to Modern Energy Use. In Dirk Assmann, Ulrich Laumanns, and Dieter Uh, eds. *Renewable Energy: A Global Review of Technologies, Policies and Markets*. London: Earthscan, pp. 231–261.

Lucon, O., S. T. Coelho, and J. Goldemberg. 2004. LPG in Brazil: Lessons and Challenges. *Energy for Sustainable Development* 8, 3: 82–90.

Macedo, I. C., J. Seabra, and J. Silva. 2008. Green House Gases Emissions in the Production and Use of Ethanol from Sugarcane in Brazil: The 2005/2006 Averages and a Prediction for 2020. *Biomass and Bioenergy* 32: 582–595.

MAPA. 2010. *Anuário Estatístico da Bioenergia*. Brasília: Ministério da Agricultura, Pecuária e Abastecimento.

PDEE-2020. 2011. *Plano Decenal de Expansão de Energia 2020*. Brasilia: Ministério de Minas e Energia.

PNAD. 2004. *Pesquisa Nacional Por Amostra de Domicílios*. Instituto Brasileiro de Geografia e Estatística.

PNE 2030. 2007. *Plano Nacional de Energia 2030*. Brasilia: Ministério de Minas e Energia colaboração Empresa de Pesquisa Energética.

São Paulo 2035. 2011. *Matriz Energetica do Estado de São Paulo 2035 – Energy Matrix State of São Paulo 2035*. Available at http://www.energia.sp.gov.br/a2sitebox/arquivos/documentos/45.pdf.

Sato, M., *et al.* 2009. A Cultura do Pinhão-Manso (*Jatropha Curcas L.*): Uso para fins combustíveis e descrição agronômica. *EMBRAPA. Revista Varia Scientia* 7, 13: 47–62.

UNCTAD. 2008. *Making Certification Work for Sustainable Development: the Case of Biofuels*. Geneva and New York: United Nations.

UNCTAD. 2009. *The Biofuels Market: Current Situation and Alternative Scenarios*. Geneva and New York: United Nations.

USAID. 2009. *Transforming Electricity Consumers into Customers: Case Study of a Slum Electrification and Loss Reduction Project in São Paulo, Brazil*. Washington, DC: US Agency for International Development.

Chapter 28

Global Oil Market Developments and Their Consequences for Russia

Andrey A. Konoplyanik

Introduction

This chapter discusses the place and role of the Russian oil economy within the broader context of global oil market developments. This is why the latter has been given major analysis in the chapter. If people think about Russian energy, most of them bear in mind nowadays gas rather than oil. So the natural question arises: why only oil and not both oil and gas as the subject of analysis in this chapter? First, because it is the author's definite conviction that international oil and gas markets have been developing along common paths with some time lag, which means that it is the oil market which paves the way for the gas market's developments internationally, though, of course, key regional specificities of the latter do exist. Second, gas is a regional export good for Russia, being destined almost exclusively today and at least in the near future for Europe, through pipelines existing since Soviet times and new pipeline network, while oil is a global commodity being exported from Russia more and more by marine transportation. Moreover, even at the regional EU market, which is key for Russian exports, gas was valued in 2010, according to Eurostat, only at 9% of EU imports from Russia while oil was valued at 64% (DG Energy 2011). So oil is still a more important export good for Russia in value compared to gas. Third, the high and to a large extent artificial attention to Russian gas is mostly politically inspired, especially in Europe, after the two Russia-Ukraine gas transit and export crises in January 2006 and January 2009, and the current gas oversupply in Europe due to the economic crisis and the "silent US shale gas revolution" which resulted, inter alia, in an increased gap between lower spot gas prices and higher contract gas prices, especially of Russian gas whose contract prices have been traditionally indexed to petroleum products (Konoplyanik 2012e). Finally, the space of the chapter is limited, which means that analyses of both energies would be inevitably more superficial due to space constraints.

Russia may be an important oil producer but it effectively is and would stay as a rule/price-taker (not rule/price-maker) when it comes to global oil. This can be traced back to oil market developments and the consequences they had/have for Russia. In order to understand Russia's role in global oil, we need to understand how international markets have been developing, where they stand today, and who is responsible for that.

The Handbook of Global Energy Policy, First Edition. Edited by Andreas Goldthau.
© 2013 John Wiley & Sons, Ltd. Published 2013 by John Wiley & Sons, Ltd.

The global oil market has been transformed from a market consisting of one single segment of physical energy (where the price movement has reflected the search for supply-demand balance in physical oil) to a market consisting of a flexible combination of two segments: both physical oil and paper oil markets. The latter segment has been quickly expanding in value and has begun to dominate over the physical oil segment. It consists of mostly oil-related financial derivatives, and oil price fluctuations nowadays reflect, in this author's view, the search for supply-demand balance in oil-related financial derivatives, and not in physical oil.

This is why Russia needs to confront the challenge of the global financial, including derivatives, markets. Since the role of Russia in the global financial market is currently close to the value of a statistical discrepancy, within the current state of global oil market development, consisting of both physical and paper segments, the role of Russia is less important today than it was in the period of existence of only a physical oil market. This is why the task of diminishing dependency on the oil sector and thus on oil price fluctuations is more essential today for Russia than in the past. And of course all political speculations like Russia becoming an energy superpower should be forgotten once and for all since they indicate wrong aims and lines for action.

The country needs to embrace a different fiscal paradigm and get away from an inefficient state-dominated production coupled with unproductive consumption and an inefficient use of its oil revenues. The major challenge for Russia in this regard is how to diminish its high and increasing exploration and production costs for oil, especially bearing in mind continuous worsening of natural conditions in new Russian oil provinces. This task becomes additionally challenging with the coming development of Russian Arctic offshore oil and gas. There is no other way for this except introducing multi-dimensional revolutionary breakthroughs in Russian oil, which can only be done by bringing innovations into all aspects of the Russian oil economy (technology, corporate management, state energy policy, investment climate, etc.). It is only capital that brings technological innovations, so the improvement of the domestic investment climate in Russian energy is badly needed as the first step. This author has been arguing for multiple investment regimes for Russian subsoil use (Konoplyanik 2012d), including legal stability and differentiated oil taxation as its necessary means.

In view of the above, the structure of this chapter is as follows. It starts with the description of the general trends of oil market development which, from this author's view, can be explained by an economics-based interpretation of Hubbert's curves. The author argues that we have been living within the left-upward wing of Hubbert's curve which explains the particularities of an oil market evolving from physical to paper oil. Five major periods of this evolution of international oil markets since 1928 till nowadays are classified. Next, the chapter describes who has determined the oil price throughout these periods, from the "Seven Sisters" to OPEC to non-oil speculators. It explains why non-oil speculators began to play the key role in the paper oil market and describes the role of the US in recent damage and expected repairs to global oil futures/commodities markets. After that, the chapter analyzes economic limits of oil price fluctuations (floor and ceiling benchmarks) and, finally, draws historical conclusions for Russia and Russian energy policy.

Hubbert's Curves and Oil Market Structures

General trends of oil market development can be explained, in the author's view, by Hubbert's well-known curve. As an energy economist, I consider that the peak of

Hubbert's curve for all non-renewable energies, incuding. oil and gas, is not a fixed parameter within the time-scale but is "a moving target." It has been moving in an upward-right direction. In regard to oil, new liquid fuels have been added to the economically justified volumes of production capacities due to progress in geology (expansion of resource base of individual liquid energies due to better knowledge of the subsoil), technology (increase in technically recoverable reserves of different liquid energies and expanding possibilities for conversion into liquid fuels of non-liquid energies), and economics (decrease in costs at all the steps of investment cycle/value chain thus increasing proved recoverable reserves and/or available quantities of prospective supplies of liquid fuels). As a result, former unconventional energies become conventional ones and the cycle of added reserves (from unconventional to conventional) has been repeated constantly due to human intellect and continuous demand for energy. Hubbert's peak in liquid fuels thus is a moving upward-right target due to conversion into conventional liquids, in addition to conventional oil, of both unconventional liquid fuels (such as, historically, offshore and Arctic, heavy oil, bituminous sands, shale oil, natural gas liquids, including from shale gas, etc.) and conversion into liquids of other energies, like "gas-to-liquids" or "coal-to-liquids" technologies (see Figure 28.1).

This is why I consider that at least within the next two global investment (technological) cycles, each one equal to 15–20 years or even more, the world will not reach the Hubbert's peak in oil, in gas, or in other non-renewable energies. The first mentioned "technological cycle" is presented by available energy technologies at every stage of the energy value chain, whose commercial implementation has been financed already and whose corresponding capital expenditures (CAPEX) need to be recouped within the current economic cycle. The second "technological cycle" would be presented by already known technologies whose large-scale commercialization has not been financed yet since they are now at the stage of R&D only, and which will succeed the existing technologies after the CAPEX in the technologies of the previous investment cycle are paid back.

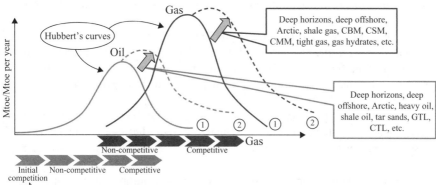

Figure 28.1 Oil and gas Hubbert's curves: upward-right supply peak movements. CBM = coalbed methane (from unmined rock), CSM = coalseam methane (from active coal mines), CMM = coalmine methane (from abandoned coal mines), GTL = gas-to-liquids, CTL = coal-to-liquids. Source: Data from Konoplyanik (2004); Dickel *et al.* (2007: 53).

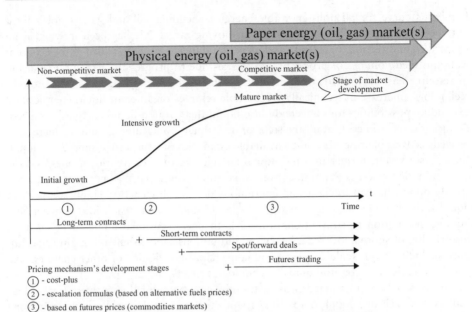

Figure 28.2 Evolution of oil and gas markets: correlations of development stages, contractual structures, and pricing mechanisms on the left (up-going) wing of Hubbert's curve.
Source: Based on Konoplyanik (2004: Fig. 28); Dickel *et al.* (2007: 60).

This is why for at least the next half century the world will continue to face energy development within the left (up-going) wing of Hubbert's oil curve.

The evolution of energy markets within the left (up-going) wing of the Hubbert's curve has its long-term tendencies and objective trends, notably (see Figure 28.2):

- Development from less competitive to more competitive energy markets,
- Development from vertical integration to term contracts and then to liquid market-places,
- Evolution of contractual structures from long-term to medium- and short-term, then to spot, then to futures trading (with U-curve development of contractual durations within the time-frame, e.g., shortening duration of the transactions at the physical market and increasing duration of the transactions at the paper market as the general trends (see Figure 28.3),
- Evolution of pricing mechanisms from "cost-plus" to "net-back replacement-value-based" and to "exchange-based" (and finally – a hypothesis – to financial derivatives-based) energy pricing.

"Competition" here means not only an increase in a number of market participants due to energy development, but a multi-faceted competition in all aspects of energy markets' functioning and development, like more competitive energy mix (no longer one single dominant fuel in the future energy balance like in the past), multiplicity/coexistence of different contractual structures and pricing mechanisms within one single economic space (on a global, regional, or a single big country level), etc. The general rule thus is: new market instruments are not implemented instead of, but in addition to incumbent ones which means that new dynamic balance needs to be reached between the new, more

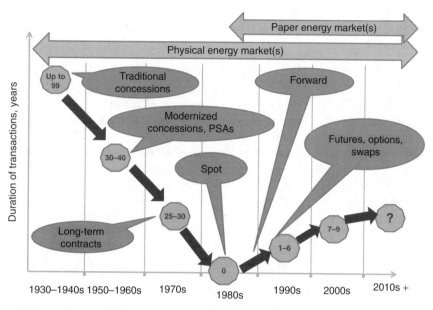

Figure 28.3 Evolution of duration of oil transactions within the time-frame.
Source: Based on Konoplyanik (2011b).

competitive combination of energies and their market instruments at each stage of energy development, at each stage of evolution of the energy markets.

Left-Upward Wing of Hubbert's Curve: Evolution from Physical to Paper Oil Markets

International oil market development through the left-upward wing of Hubbert's curve has been evolving from physical to paper oil markets (see Figure 28.2). The modern contractual structure of the global oil market and its pricing mechanisms have been developing over the past 80 years as part of the Anglo-Saxon model of open, competitive, liquid, self-regulating global markets. Within the last 20–25 years the global oil market has become an integral part of the much broader global financial market, with all key characteristics of the latter now being transferred to the world of oil deals.

Box 28.1 identifies five major periods in the evolution of international oil markets, if oil pricing mechanisms, contractual structures, and organization of market space are taken into consideration (Chevalier 1975; Dickel *et al.* 2007: 56; Konoplyanik 2000, 2004: 105, 2009–2012).

I believe that 1901 should be considered as the starting point of this process at the interregional level, when the first actually working oil concession ("D'Arcy concession") was signed in the Middle East. That was the beginning of the period of the dominance of "vertical integration" in international oil which continued until the early 1970s (see Figure 28.4). At the end of this period about 70% of international oil trade originated from concessionary agreements of the international oil companies (so-called "Seven Sisters" or "International Oil Cartel") with developing countries (mostly OPEC member states since 1960). That was the period of dominance of the major international companies in the physical oil market (a paper oil market did not exist at that time).

Box 28.1   Five periods of global oil market development and their major characteristics.

Periods	Characteristics of the Period
1928–1947 (first period)	- non-competitive physical oil market - dominance of International Oil Cartel (7 companies) - "one-base pricing" (real/virtual cost-plus) - transfer pricing/prices within vertical integration and long-term traditional concessions
1947–1969/1973 (second period)	- non-competitive physical oil market - dominance of International Oil Cartel (7 companies) - "two-base pricing" (real/virtual cost-plus in crude, net-back replacement value in petroleum products) - transfer pricing/prices within vertical integration and long-term traditional & modernized concessions & PSAs - 1969–1973 transition period from monopoly of 7 companies to monopoly of 13 states
1973–1985/6 (third period)	- non-competitive physical oil market - dominance of OPEC (cartel of 13 states) - contractual and spot pricing/prices - official selling prices (cost-plus/net-forward) within long/medium/short-term contractual structures mostly linked to spot quotations - fundamentals as key pricing factors (supply-demand balance on physical oil) - key players: participants of physical oil market - 1985–1986 transition period from net-forward to net-back crude pricing based first on net-back from petroleum products basket price at the importer's market, afterward to oil price futures quotations on key petroleum exchanges/market-places
1986–mid 2000s (approx. 2004) (fourth period)	- competitive combination of mature physical plus growing paper oil markets - commoditization of the oil market - pricing established at oil market-places mostly driven by oil hedgers - net-back from futures oil quotations - formation of the global paper oil market and its institutes based on the institutes of financial markets (instruments and institutions imported to paper oil market by financial managers from financial markets)

Periods	Characteristics of the Period
Mid 2000s+ (approx. post-2004) **(fifth period)**	- transition from physical to paper market predetermined unstable, relatively low, and volatile prices which has led to underinvestment of global oil industry, which created material preconditions for later growth of costs and prices - hedgers as key players (participants at both physical and paper oil market) - fundamentals still as key pricing factors - competitive combination of both physical and paper mature oil markets - further movement from commoditization to financialization of oil market - paper market dominates in volumes of trade - global institutions of paper oil market are formed which enable paper oil market to work in 7×24 regime - globalization, IT-technologies, broad spectrum of financial products converted crude oil into global financial asset available (accessible) to every category of professional and non-professional investors (effect of financial "vacuum sweeper") - paper oil market is an insignificant segment of global financial market - key players are non-oil speculators who have been bulling the market and have manipulated it (investment banks and their affiliated oil traders) - pricing established outside oil market-places (at non-oil financial markets) mostly by non-oil speculators - net-back from futures oil quotations & oil financial derivatives - key pricing factors are mostly financial: supply-demand balance for oil-related financial derivatives within short time-horizon

The period of dominance of the Seven Sisters in international oil was based on the Achnacarry Agreement of 1928 (Chevalier 1975; Yergin 1991) which, in my view, was a genuine managerial invention of the major oil companies, who converted their previous severe competition into long-term effective cooperation based on so-called "one-base pricing" (1928–1947) and "two-base-pricing" (1947–1969/1973) mechanisms. The ruling principles of Achnacarry effectively (from the companies' viewpoint) governed the international oil market for more than 40 years. At that period the average and marginal exploration and production (E&P) costs for oil and gas were rather low and steadily declining, since new additions to reserves (commercial discoveries) were generally provided by large and extra-large fields (economy of scale effect) located in favorable

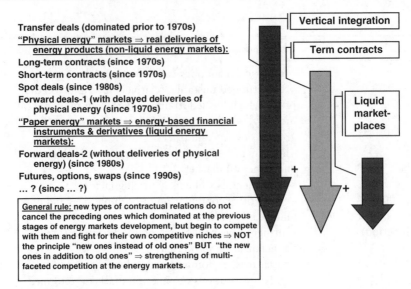

Figure 28.4 Historical evolution of contractual structure of the global oil market and its correlation with key organizational forms of market space.
Source: Based on Konoplyanik (2008a).

natural conditions and close to export marine terminals, while transportation costs to major markets (US, Western Europe, Japan) were not high and also declining due to increase in oil tankers' deadweight (another economy of scale effect). This enabled the Seven Sisters to hold prices for long at a low and stable level which stimulated an increase in oil demand while giving them (through increasing gap between falling costs and stable prices) incremental oil profits (Chevalier 1975). This was largely the "energy basis" for the Western world's post-World War II growth, especially in the "golden decade" of the 1960s.

The second genuine invention of the Seven Sisters was the introduction of the net-back replacement-value-based pricing principle in relation to the basket of petroleum products which significantly expanded demand for petroleum products and thus for oil. When Middle East oil was supplied to Western Europe it had no alternative (competitive) energies to its gasoline, diesel, jet fuel, etc. derivatives in all transportation sectors, but was facing severe competition in industry (boilers) and electricity generation from residual fuel oil (RFO) from domestic (mostly German) coal. Oil conversion rates at that time were rather low. Thus large amounts of RFO were produced which were to be utilized (marketed) in competitive segments of the market (competing with coal) in order to provide the physical possibility to produce and to market light transportation fuels (which faced no competition in their respective sectors). The genuine solution was found in implementing a replacement-value-based pricing principle in regard to RFO, linking its price by a flexible formula to levels slightly below the price of coal. Prices of light petroleum products would have been thus relatively increased so the balance price of the basket of petroleum products produced from a barrel of crude oil at the refineries of the companies of the Oil Cartel would have provided them with a necessary margin.

Subsequently, transfer deals within vertically integrated structures of the international oil companies were replaced by long-term contracts between legally independent business entities. At first, this was due to the penetration of new production companies (so-called "independents") of industrialized nations into the markets of emerging economies, access to whose subsoil was earlier blocked by the majors. Later, it was the result of nationalization of the production assets of the Vertically Integrated Oil Companies (VIOCs) in these resource-rich developing (mostly OPEC) states and the setting up of National Oil Companies (NOCs) on the basis of upstream assets of the majors; at that time NOCs did not possess their own refining capacities abroad (Chevalier 1975; Yergin 1991).

These processes coincided with the failure of the Bretton Woods System, the abolition of the gold standard and fixed dollar rate, and the start of the growth of marginal and average E&P costs in the international oil market (due to development of marginal and non-OPEC oil reserves), which eventually made it impossible to further maintain fixed prices for oil and resulted in rapid price increases.

As the range and frequency of price fluctuations increased, and the average size of the reserves additions (new commercial discoveries) were diminishing, long-term contracts were being replaced with shorter ones. A logical end to this process was the wide use of spot transactions – at first with prompt deliveries. After that (as is usually the case in the economy) the pendulum moved in the opposite direction, and the contractual mix has further evolved from spot contracts to futures transactions, which can be considered as term-deals but of a different nature to initial long-term contracts.

At this stage of physical oil market development (third period in Box 28.1) the market was governed by a monopoly of producer states united in OPEC. In the 1970s spot quotations were the drivers for OPEC official selling prices, and spot prices were driven upward by the fears of traders for lack of production capacities and/or repetition of the oil shortages, including by embargoes (like in 1973) or due to revolutions/military conflicts (like in 1979). At this time CAPEX into diversification of international oil production and related infrastructure had not yet materialized in the creation of significant non-OPEC capacities, nor was there a decline of energy (oil) intensity in the economies of the major energy-consuming and oil-importing nations or creation of adequate commercial stocks and/or other rather time-consuming and capital-intensive measures aimed at diminishing dependence on OPEC oil. International oil trade in the 1970s was still dependent (high inertia of oil industry due to its high capital intensity) on the material structure of the market organization developed earlier by the international majors, though they had lost ownership of their producing assets abroad, which became the producing assets of the NOCs.

At this stage of evolution of the oil market's contractual structure, one producer was no longer linked to one consumer "forever" (whether within one single VIOC structure, or based on long-term contractual relations between independent business entities), as used to be the case earlier. Diversification of the infrastructure of the international oil supply system allowed buyers to count on guaranteed receipt of required volumes of crude oil even without having their own production facilities and relying only on "segmented" international chains of oil trade, which were controlled by different agents and jurisdictions, and not by the unified power of the "Seven Sisters," as was the case before the early 1970s.

The first new instruments to appear were spot contracts with deferred delivery of actual goods, secured by adequate volumes of such goods in commercial stocks (forward deals-1), followed by forward transactions under which observance of this condition was not required (forward deals-2) (see Figure 28.4). This predetermined the appearance of futures and options, which do not assume the trade in actual goods (material assets), but the trade in liabilities to sell/buy them. That gave the birth to the paper oil market.

As new instruments to buy and sell oil emerged, the contractual structure of the international oil market has been constantly changing and becoming more sophisticated and more competitive. In the course of "physical" oil market development, the term of later types of contractual deals was usually shorter than for the earlier ones, evolving from concessions (up to 99 years) to long-term contracts (from an initial 15/20/30 years to just few years) to spot deals with immediate delivery (and one month for payments). The "paper" oil market developed the other way around. Terms of futures contracts grew longer, now reaching 9 years at NYMEX for the West Texas Intermediate marker (WTI) futures trade (however 80–85% of all futures mature within the first few months). Thus, the geography expanded and the set of instruments to arrange international market space grew as well: from transfer deals (within vertical integration) via term contracts to liquid market-places.

As international oil trade developed, the gap between volume of trade and volume of physical supply grew as well. On the "physical" oil market (under term contracts), the sales volume corresponded to the volume of actual supplies. Due to the continuing switch to spot transactions and abolition of the ban on arbitrage operations in long-term contracts (such as destination clauses), buyers were able to resell specific commercial batches. As a result, so-called "daisy chains" emerged which created a gap, which expanded as forward transactions developed, in the physical market between the volumes of oil traded and physical supply volumes. Trade liquidity at this stage was limited by a small number of oil tanker sizes/deadweights. Consequently, more universal trade instruments were needed, so at this point, standardized contracts started to dominate the market.

Along with evolution of the contractual structure of the oil market, the prevailing pricing system also changed. Virtual "posted" prices (the key element of transfer pricing within the concession system of "Seven Sisters" companies with corresponding host states), which were needed to optimize tax allocation of international transactions and to transfer the profit center to the mother countries of the VIOCs, and which dominated in the international oil trade until the early 1970s, were replaced in the 1970s with official selling prices (OSP) of OPEC member-states. At first OSPs were fixed, and then they appeared to be pegged to spot quotations. They were to make up a major part of the economic (price) rent in the producing states. After that, spot quotations (selling prices on the one-off deals market) became, in effect, the only and determining price benchmark. OSPs were not linked to production costs anymore (as were earlier posted prices), but rather to growing spot quotations. This aimed to compensate OPEC countries for their lost portion of economic (resource) rent that was at the previous stages of oil market development extracted by the Seven Sisters and evacuated to their centers of profits in their mother countries.

Later on, as financial managers from international financial markets came to the oil market to create and develop its paper segments, they formed a new framework of paper oil transactions similar to transactions in international financial markets. Since then, futures quotations from key petroleum exchanges (market-places) were established as price indicators for physical trade in all contractual structures, including spot, short-term, and long-term deals.

Today, pricing under all types of contractual transactions is pegged to the price levels established at the exchange – that is to quotations for marker oil grades which give prices for other grades via a differentials system. These quotations provide for the level of net-back wellhead (or net-back to delivery point) competitive prices for individual producers. This reference is utilized both in long-term contracts, which are widely used for supplies of crude from OPEC states and other producers via pipelines and by tankers, and in the spot transactions, which are usually made using maritime transportation.

Who Determines the Oil Price? From Seven Sisters to OPEC to Non-Oil Speculators

At the first and second stages of international oil market development, oil price in importing states was calculated by the companies of the International Oil Cartel as net-forward price, as if it was based on production costs. But it was in fact a fictional price since was based on virtual values. This became possible through the Achnacarry Agreement. The "one-base price" mechanism (1928–1947) determined any CIF price in any importing region worldwide as if oil was produced in the US (where the costs of production were the highest mostly due to dominance of stripper wells and high labor costs) and as if it was delivered from the Mexican Gulf to its final destination worldwide, non-dependent on its actual origin. The "two-base price" mechanism (1947–1969/73) has slightly adapted this formula (Chevalier 1975; Konoplyanik 2004: Figures 34–35) (see Table 28.1).

At the third stage of oil market development OPEC's OSP took the lead in pricing, based on growing (until early 1980s) spot quotations. The period of net-forward pricing ended in 1985 when Saudi Arabia finally lost its patience at being the swing producer within a non-disciplined OPEC. This country had been supporting declining (since mid-1981) oil prices by decreasing its actual production much below its OPEC quota,

Table 28.1 Evolution of pricing mechanisms in the international oil market.

Periods, Who Establishes the Price	*Pricing Formula for Physical Supplies*
(1) 1928–1947: International Oil Cartel (one-base pricing)	Net forward: $P_{CIF} = P_{FOB}$ (Mex. Gulf) + Freight fict. (Mex. Gulf)
(2) 1947–1969/73: International Oil Cartel (two-base pricing)	*To the West of neutral point:* Net forward: $P_{CIF} = P_{FOB}$ (Mex. Gulf) + Freight real (Mex. Gulf) *To the East of neutral point:* Net forward: $P_{CIF} = P_{FOB}$ (Mex. Gulf) + Freight real (Pers. Gulf)
(3) 1973–1986: OPEC	Net forward: $P_{CIF} = P_{FOB}$ (OPEC OSP) + Freight real (OPEC)
(4) 1986–mid-2000s: oil exchange 1 (hedgers → oil speculators)	Net back: $P_{FOB} = P_{CIF}$/exchange – Freight real $P_{CIF} = $ *Exchange quotations (oil paper market)*
(5) Mid-2000s +: oil exchange 2 (non-oil speculators)	Net back: $P_{FOB} = P_{CIF}$/exchange – Freight real $P_{CIF} = $ *Exchange quotations (non-oil non-commodities paper markets)*

P_{CIF} (net forward) – price CIF (at importer end) calculated as cost-plus; P_{FOB} (Mex. Gulf) – price FOB (at supplier end) in the Mexican Gulf area; Freight fict. (Mex. Gulf) – freight rates for fictitious oil deliveries from Mexican Gulf area to importers; Freight real (Mex. Gulf), Freight real (Pers. Gulf) – freight rates for real oil deliveries from Mexican and Persian Gulf areas; P_{FOB} (OPEC OSP) – OPEC official selling prices FOB; Freight real (OPEC) – freight rates for real oil deliveries from OPEC member-states to importers; P_{FOB} (netback) – price FOB, calculated as netback price (price CIF less transportations costs); P_{CIF} (exchange) – price CIF as exchange quotations (at consumer end); Freight real – freight rates for real oil deliveries to importers from production areas.
Source: Based on Chevalier (1975); Konoplyanik (2004, 2011c).

Table 28.2   Characteristics of spot, forward, futures, and options deals.

	Spot	Forward	Futures	Options
Trading	OTC	OTC	exchange	OTC/exchange
Derivatives	no	yes	yes	yes
Delivery	yes	(yes)	(no)	(no)

Source: Dickel *et al.* (2007: 81).

while other OPEC members were benefiting from violating their own quotas by regularly and increasingly exceeding them. At the end of 1985 Saudi Arabia in one step increased its production to its actual OPEC quota and netted-back its crude price to the price of the petroleum products basket at the consumer end of the chain (NYMEX). Having the lowest production costs, this country further protected its market share. This switch from net-forward to net-back pricing signaled the beginning of a new era in oil pricing based on quotations from the market-places, and quick development of a paper oil market (Table 28.2).

Till the mid-2000s the paper oil market was subordinate and linked to physical deliveries; hedgers dominated over speculators, and financial instruments were used for hedging price risks at the physical oil market. The oil price in the paper oil market was formed via trade in oil contracts. Since the mid-2000s the picture has changed: the paper oil market is now dominant and is de-linked from physical deliveries, speculators dominate, including from non-oil sectors of the global financial market. The oil price is formed today by financial instruments at non-oil paper markets via oil-related financial derivatives (Konoplyanik 2008–2012). How did it become possible?

Paper Oil Market: Non-Oil Speculators Begin to Play the Key Role

Since the late 1980s (starting with the fourth stage), the global oil market first developed as a commodities market, and only after that has it developed into a financial derivatives market. It provides today multiple options for trade both in physical goods aimed at actual delivery of liquid fuels as well as in its financial derivatives, aimed at earning profits from financial transactions without delivery of physical energy (see Table 28.2). The key roles are played by two groups of players having opposing interests: hedgers and speculators (*Oil Tabloid* 2010).

Hedgers are usually producers/consumers of physical goods who use financial markets to mitigate price risks. They appeared at the market with the reopening of futures trading in liquid fuels after almost a century since exchange trade was dominant at the very beginning of US oil production in the 1860s (Tarbell 1904). Futures trading in liquid fuels started at NYMEX with light fuel oil (LFO) in 1978 and with WTI in 1983. In 1988 futures trading started at IPE (London) with Brent crude which today represents the reference crude of approximately two thirds of internationally traded oil.

Oil hedgers are usually pegged to the "paper" oil market but they are not mobile (since they are linked to their physical assets) and do not tend to migrate outside the oil market (its "physical" and/or "paper" segments), except the cases when they go to financial market to raise debt (project) financing. Thus "paper" oil is much less important for them compared to "physical" oil.

Oil speculators (since the 1990s) are the players aiming to earn their profit from price fluctuations without physical deliveries/purchases. They work mostly within the paper oil market without major horizontal capital flows to other non-oil financial markets.

Non-oil speculators (since mid-2000s) are also aimed at pure monetary results but they work within the whole spectrum of global financial markets. They can enter the paper oil market from non-oil and non-commodities financial markets.

For two decades (mid-1980s to mid-2000s) oil futures markets were playgrounds for physical market players such as energy companies and major users of petroleum products (airline and maritime transport, utilities), seeking to hedge price risk in their own business (physical deliveries/purchases). Since the mid-2000s these markets started to attract a growing number of financial market traders such as banks and investment, hedge, or pension funds. They are completely foreign to the physical oil market, except for the cases when their affiliated oil-trading companies add synergy to their operations both with paper and physical oil.

The speculators' money usually consists of highly liquid financial resources (the measurement of liquidity, the so-called "churn" ratio, exceeds 2000 in oil futures trading both at NYMEX with WTI and at London's Inter-Continental Exchange (ICE: successor to IPE) with Brent crude), which are highly mobile and tend to migrate rapidly to those segments that ensure the highest returns at the moment. Thus today's speculators as a group of market players usually are not strongly linked to particular segments of financial markets, the paper oil market being just one of such segments to maximize returns in their global financial portfolios.

With regard to oil futures markets one should distinguish between regulated (petroleum exchanges) and non-regulated markets (over the counter or OTC). At regulated oil futures markets (NYMEX, ICE) contracts are standardized in terms of quality, quantity, date, and place of delivery, detailed data is available, and their operations are governed by the US Commodity Futures Trading Commission (CFTC) rules overseeing futures markets. Non-regulated OTC markets present non-standardized bilateral contracts. No CFTC rules are applied in OTC transactions. OTC markets are assumed to be much bigger than regulated oil trade markets, though they are not precisely measurable. But trades usually migrate to less regulated markets, when possible, to obtain undue advantage from the lack of restrictive regulation on traders (while the latter is a means of protection from too risky operations).

Hedgers represent a relatively stable group in terms of size and structure. Speculators are characterized by changing and unstable size and structure of players depending on changes in the oil and macroeconomic environments. Usually, in a relatively calm oil environment, the ratio of speculators to hedgers is 25–30:70–75. When the market grows their share can increase, and vice versa. According to the CFTC, the ratio between commercial (hedgers) and non-commercial (speculators) has changed from 75:15 in 2000 to 55:45 in 2007 (Konoplyanik 2011c). The dynamics of this process are "wavelike" depending on the inflow/outflow of new players from other segments of the global financial market to the paper oil market. In this case, both inflow and outflow of speculative capital can be of an explosive nature. This also explains, in my view, the nature of the 2008 price hike (see Figure 28.5).

Why do the 2000s represent a new stage in oil pricing? In my view (Konoplyanik 2012a), this can be explained by a number of consecutive developments. Underinvestment in the 1990s (when the oil price fluctuated between US \$10 and \$25/bbl) has led to cost increase since early 2000s and to decrease in spare production capacities worldwide, especially in the Middle East. Then China, India, etc. accelerated demand growth since 2003, while some major consumers (US, China) started building strategic petroleum reserves. The US Commodity Futures Modernization Act (CFMA) adopted in December 2000 triggered an inflow of a huge amount of relatively cheap money to the oil paper

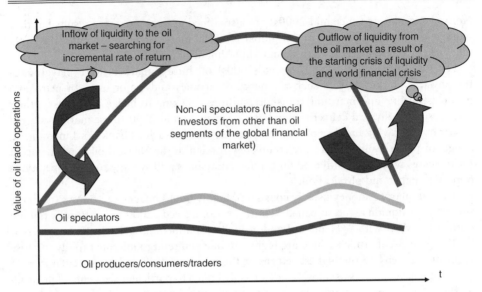

Figure 28.5   Role of non-oil speculators (global "financial investors") in forming a "price bubble" in the global oil market, 2007–2008 (principal scheme).
Source: Konoplyanik (2009b, 2010, 2011b, c, d, 2012a, b, c).

market, stemming from US pension funds and insurance companies who were prohibited earlier from investing in risky segments such as paper oil.

At the same time evolution of commodities (exchange, futures) trade has triggered a change in quality and character of the exchange. Internet and IT developments have transformed trading floors into electronic marketplaces (ICE, the former IPE, was the first to end voice floor trading). This has also led to "robotization" of electronic trading, an increase in amount of traders, and significantly eased market entry for new players.

The decrease of USD exchange rate (stimulated by increase of US oil imports leading to increase in trade and budget deficit) has led to the appearance of index oil funds, which have expanded possibilities for financial investments in oil-related instruments to hedge against fall in USD exchange rate. Paper oil markets began to be considered a safe haven against a falling USD. Globalization of financial operations has eased horizontal financial flows from/to financial (non-oil) sectors into/from the paper oil market. This ease of financial investments into the oil market (multiplicity of oil-related financial products creating a chain of derivatives on derivatives) simplified access to paper oil for the middle-income and non-professional financial investors. The "Belgian dentist" has appeared as a key private (non-institutional) financial investor on the paper oil market.

Oil-linked derivatives of index funds became a new class of financial assets aimed at compensating, inter alia, the fall in USD exchange rate. Finally, this led to switch of oil pricing from the physical market (based on supply/demand of physical oil) to the paper market (based on supply/demand of oil-related financial derivatives). This, in my view, explains why the oil-price bubble was so quickly blown up in 2007–2008, and not less abruptly blown down in 2008–2009.

US Role in Damaging and Repairing Global Oil Futures/Commodities Markets

A special word needs to be said on the US role in the turbulence of the global oil market in the 2000s. In my view (Konoplyanik 2011c), this role is two-fold. On the one hand,

I see a "damaging" role in the past: the approval of the CFMA in December 2000 left commodity transactions largely outside the reach of the CFTC and thus left companies with minimal (much lower than beforehand) regulatory obligations from too risky operations. On the other hand, I envisage a future "repairing" role of the US, alleviating its own earlier (destructive in their material results for paper oil market) actions: the Wall Street Transparency and Accountability Act (Dodd-Frank Act) enacted by US Congress (with effect from July 14, 2011) effectively replaces CFMA and makes it illegal for producers to execute trades outside forthcoming and more restrictive CFTC rules.

What were the major consequences of the adoption of the CFMA? It upgraded "excessive" speculative activity (which according to some expert estimates (Stowers 2011) could inflate oil prices up to 30%) and price manipulation in bypassing the CFTC and its regulatory reach. This led to an increase in the amount of speculators in the oil market (from 20% to 50–80%) compared to hedgers and to growth in quantity of contracts not covered by the anti-speculative limitations of the CFTC. The latter was possible due to derogation from the Commodity Exchange Act (CEA) and CFTC jurisdiction of US contracts at foreign exchanges/market-places (so-called "London loophole") and swaps (so-called "swap loophole": contracts on price differentials). CFMA stimulated increase in OTC trade in oil derivatives, outside the reach/control of the CFTC. It led to downgrading of barriers for key holders of long relatively cheap money (e.g., pensions and insurance funds) on investing into risky financial instruments (Medlock and Jaffe 2009). Finally, bidirectional growth in speculative activities (increase in "amount of speculators" multiplied by increase in "amount of available instruments for speculations") has moved speculators from "price-takers" into "price-makers" in paper oil.

What are the expected consequences of adoption of the Dodd-Frank Act? Every trade is now likely to fall under jurisdiction of the CFTC and should be evaluated to determine if it has Dodd-Frank regulatory obligations. All swap transactions must be cleared with only a few exceptions, while definition of "swaps" under Dodd-Frank is expansive: generally, "swaps" are financial products that exchange fixed for floating prices, and floating for fixed prices, but the CFTC takes a much wider view of swaps to include just about any transaction that has a price or event contingency, so under Dodd-Frank, if a producer cannot actually deliver a product, it is considered a swap and must be cleared. Any trade market players enter into may have Dodd-Frank implications. According to one of the authors of the law, (former) US Senator Chris Dodd, "we obviously needed transparency in the OTC derivatives market to restore consumer confidence and investor confidence... If we tried to pass this bill today, it wouldn't happen. It literally took the events of 2007 and 2008 to get it done. In the absence of such a crisis, the bill wouldn't have passed" (Allott 2011; Stowers 2011).

The Dodd-Frank Act can diminish speculation and decrease its inflating effect on oil prices, but the very fact of existence of the paper oil market as a tiny part of global financial markets (a "rule of thumb" comparison of volumes of global trade in physical oil, paper oil, commodities, and financial markets can be estimated as 1:3:10+:100+ (Konoplyanik 2009b, etc.), though in reality the gaps can be even bigger) and the global character of financial transactions within a computerized world still leaves the global oil market operating within a time of high price volatility. The natural question arises: but at what levels?

Economic Limits of Oil Price Fluctuations (Floor and Ceiling Benchmarks)

It is possible to identify, at least in theory, a corridor of economically justified fluctuations of oil prices. The floor level would be determined, in my view, by the upper of the two flexible parameters: (a) the long-run marginal costs (LRMC) of liquid fuels, i.e., long-run costs of currently producing fields and of those that should compensate for their natural decline and for demand growth in liquid fuels, and/or (b) break-even price of non-deficit budget of Saudi Arabia (Konoplyanik 2011a, d). The ceiling level would be determined by the lower of the two flexible parameters: (c) replacement values at the consumer end of the alternatives to liquid fuels (competitiveness of liquid fuels with other energy sources), and/or (d) purchasing power of the world economy for energy in general and oil in particular (competitiveness of energy/liquid fuels with other production factors – labor, capital, etc.).

There are a number of studies on the current production costs worldwide (e.g., Takin 2008) which calculate the current level of production costs. In my view, the bottom line for oil quotations (not the point-wise lower limit, but the average for the period determined by the payback period of an oil production project) should not be lower than the LRMC for existing and prospective reserves adequate to cover prospective demand for liquid fuels during such investment cycle period. And the spread in estimates of such costs is rather large. Moreover, such estimates are quite controversial (Konoplyanik 2009a, b, 2011a). IEA (2008) estimates LRMC for liquid fuels at US $110/bbl (calculations based on 580 major fields with cumulative reserves of 10 trillion bbl). CSM/PUCC/IIASA estimate such LRMC at $35/bbl ($ based to 2006) through 937 discovered and undiscovered oil and gas provinces with cumulative reserves of 32 trillion bbl (Aguilera *et al.* 2009). UK Energy Research Centre estimates LRMC at $90/bbl ($ based to 2000) for cumulative reserves of 19 trillion bbl (Sorrell *et al.* 2009).

The controversy over these figures is based on the fact that the general trend, existing before the end of the 1960s/early 1970, when average and marginal E&P costs were going down, has changed to its opposite since the beginning of the 1970s (Chevalier 1975; Kurenkov and Konoplyanik 1985). This means that the broader the spectrum and the higher the volume of reserves involved in LRMC assessment, the higher should be the level of LRMC, which is not the case if the three above-mentioned studies are compared with the three times difference ($35 vs. $110/bbl) between the marginal scenarios.

Moreover, not all lower-cost reserves could be involved in commercial exploitation since a number of such low-cost oil producers have been carrying out a policy of restricted access to their natural resources (though this is their sovereign right according to UNGA Resolution N 1803 as of December 14, 1962 on "Permanent Sovereignty over Natural Resources" and Art. 18 "Sovereignty over energy resources" of the Energy Charter Treaty). This means that LRMC will be even higher that just their technical assessment shows (see Figure 28.6).

But an acceptable threshold for diminution of oil price is different for the companies and the producing states: while it is LRMC for the companies, for the states it is a break-even price for their non-deficit budgets. In my view, it was the recent break in correlation between actual oil price and break-even oil price of Saudi Arabia that gave birth to the debate on the "fair oil price" and appeals about its growing level. I consider that it is the Saudi Arabian non-deficit budget break-even price that began to play the role of bottom line in international oil price fluctuations.

According to the London-based Centre for Global Energy Studies (CGES 2011; Drollas 2011), the OPEC basket price that was needed by Saudi Arabia to cover its planned expenditures has been growing steadily in line with but at levels much below the actual oil price from the early 2000s until 2009, since when the Saudi break-even price continued to grow but the actual oil price has sharply fallen below this break-even level, which was equal to US $59/bbl in 2008 (see Figure 28.7). And it was since 2009 that Saudi oil minister Al-Naimi began to repeat that a "fair" oil price (or its "optimal diapason") should be first US $60–70, then $70–80/bbl – clearly at levels higher than the non-deficit

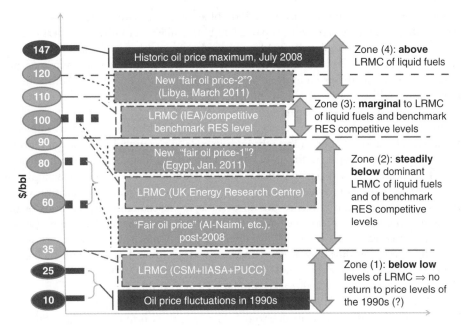

Figure 28.6   Crude oil: prices and costs, expectations and facts. LRMC = Long run marginal production costs of liquid fuels, RES = renewable energy sources.

break-even oil price and much higher than the actual oil price at that time. That has a definite influence on the market.

As Noe van Hulst, former Secretary General of the International Energy Forum (IEF), said in his presentation at the Global Commodities Forum in Geneva in January 2011, "when Saudi Arabia speaks, market listen." And the oil market not only has listened, but has been repeating the views of Al-Naimi and others at different forums, making them step by step a general view of the market participants, like at the St Petersburg Economic Forum in June 2009, when an opinion poll of senior managers of Russian and international oil companies in the presence of the President of Russia showed their almost unanimous support of the corridor US $60–70/70–80/bbl, which was in fact Al-Naimi view. In 2011 CGES has calculated the break-even price of Saudi oil first to be

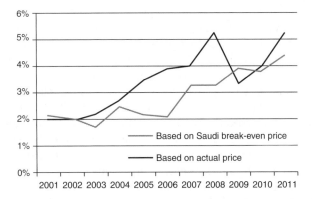

Figure 28.7   The share of crude oil in global GDP based on actual price and on break-even price for Saudi Arabian non-deficit budget.
Source: CGES (2011).

around US $83/bbl, then upgraded it to $90/bbl as result of "Arab Spring" events (and assumed that it might even be higher up to $100/bbl), while other OPEC members needed even higher prices (Konoplyanik 2011a); and since the Arab Spring these break-even prices should be inflated even further in the medium term by the intention not to permit a widening of the Arab Spring in these oil economies. But it is only Saudi Arabia, alone among other OPEC states, that can influence the oil market, and in both directions, with its production levels and spare production capacities.

Moreover, today there are only two countries in the world which can really influence the global oil market: Saudi Arabia, through its role in the global physical oil market, and the US, through its role in the global financial markets. Saudi Arabia has the highest level of oil production capacities with the lowest production costs worldwide and it is the only state that can provide a significant level of spare production capacities whose artificial fluctuations can influence the oil price level via changing expectations of the market players. And the break-even non-deficit budget price of Saudi Arabian oil (which has been presented to the world as a "fair oil price") is much higher than this state's oil production costs, thus leaving the Kingdom with a robust lever for influencing the market. I do not consider that the IEA with its commercial/strategic stocks policy has collective power at the physical market similar to Saudi Arabia's.

The US and its financial institutions has similar power on the paper oil market through a multiple number of instruments due to the US role in the world economy and financial system: since the trilateral US-UK-France agreement of 1936, oil pricing is set in USD and thus most of the oil-related financial derivatives are also priced in USD. Since the US has the monopoly on dollar emission, it is de facto controlling the global recycling of petrodollars. And this stands on the fundament of US economic power: while the country accounts for 25% of global GDP and for 30% in global financial assets, the US exchanges cover 60% of global capital stock turnover and more than 50% in global derivatives markets; the dollar accounts for two thirds of global foreign exchange reserves (Mirkin 2011). According to the Center for Energy Studies, Institute of World Economy and International Relations of the Russian Academy of Sciences, 95% of the derivatives market is controlled by four major US investment banks: JP Morgan Chase, Citibank, Bank of America, and Goldman Sachs (Zhukov 2011). Moreover, as was further argued by the Center (Zhukov 2012), the US benefits from high oil prices rather than from low oil prices, which nowadays influence negatively its economy. According to this logic, two major players at neighboring segments of the global oil market are interested in supporting relatively high, rather than low, oil prices. But below which level?

On the one hand, the upper level of oil price fluctuations does not need to exceed in the long run the evolving level of production costs of oil alternatives in the end-user segment. LRMC of oil alternatives (its replacement fuels) is considered to be equal to US $110–120/bbl (IEA 2008; Kanygin 2010). On the other hand, oil price growth cannot overcome purchasing power of the consumers, their readiness to pay for oil (energy costs to income). According to Bashmakov (2006, 2007), "sustainable variations of energy costs to GDP ratios are limited to 8–10% for the US and 9–11% for the OECD... After the upper limit is reached or exceeded (1949–1952, 1973–1985, and starting from 2005), the ratio drops, and after the lower limit is approached (1998–1999), it, on the contrary, grows." Taking into consideration evolution of structure and efficiency of energy consumption and dynamics of energy prices this means that an acceptable level of oil cost to global GDP should be within 5% limits. At the modern stages of the oil market's evolutionary changes within the period of exchange pricing, the oil cost to global GDP ratio was below 3% within the fourth period and 3–5% within the fifth period.

According to Renaissance Capital (2011), in constant 2010 USD, the average price of oil in 2008 ($98.5/bbl) corresponded to the price of oil in 1980 ($97.5/bbl), but oil burden (nominal cost of oil/nominal GDP) in 2008 (5%) was much lower than in 1980, when it exceeded 7% (the result of intensive structural changes of the world economy toward partial evacuation away from oil and due to overall increase in energy efficiency).

In the given circumstances how one can assess the speculative oil price bubbles and extremely high (in comparative terms) LRMC estimates by the IEA? Can they be considered as preparation for the period of even more costly oil than today – or for the departure away from oil? At the 20th World Petroleum Congress (December 2011, Doha, Qatar) Maria van der Hoeven, IEA Executive Director, indicated an oil price equal to $120/bbl in real terms by 2035. This figure is higher than today's LRMC estimates of non-conventional liquid fuels according to the IEA itself and LRMC of "traditional" (already commercialized) renewable energy sources ($110/bbl). This looks like an unspoken alliance of the IEA and oil speculators.

Why such an assumption? IEA forecasts de facto argue for a non-speculative character of the 2008 price peak; as such they have been placing a theoretical base under expectations of the stable high oil price. Speculators, playing bullish, have been in practice preparing the world economy for a new stage of development by bringing the world to a cross-roads: whether to continue with expensive oil or whether to deviate from the oil economy. But is the latter really possible, taking into account the huge amount of all related costs? What does a three-digit level of oil prices mean for the world, holding global oil burden within the 5–7% corridor? Today's 5% means expensive oil but a continuation of the oil era. If the price will go higher toward the 7% oil burden, that might mean the beginning of the end of the oil era. As Sheikh Yamani used to say, the Stone Age ended not because the stones ended. So I would not foresee in the long run a growth in oil price to the figures bringing the oil burden (nominal cost of oil/nominal GDP) to the critical 7% level. The interest of all market participants will be to hold oil price in the low part of zone (3) or at the edge of zones (3) and (4) in Figure 28.6.

Historical Conclusions for Russia

What consequences does the above discussion have for Russia? In the 1960s the USSR restarted large-scale oil exports to the West. In the 1970s, when oil prices increased significantly, the USSR saddled up this growth. The growing oil rent went mostly to support military parity with the West, it was not invested in R&D aimed at revolutionary technological progress outside military industries. Growing dependence on oil revenues resulted in the impossibility of overcoming the collapse in oil prices in the mid-1980s. To substitute diminished oil revenues from external markets (and lost revenues from the internal market as a result of the "dry law" campaign) and to support the "perestroika and uskoreniye" policies with simultaneous intent to support military parity with the West, the USSR began raising more and more foreign loans, resulting in deep state debt. The country became bankrupt. And finally the USSR was dissolved.

It seems that the story of the 1970s has been repeated in Russia in the 2000s to some extent. Moreover, it seems that today's role of Russia in the global oil market (combined physical plus paper market) is less significant than it was in the past, prior to the development of the paper oil market which creates additional problems (Konoplyanik 2012c).

In the past, during the second and third stages of oil market development (period of sole existence of the physical oil market), USSR oil production volumes did not play a major role in international oil trade and the state of the market. The USSR at that time was not a price-maker, but a price-taker on the international oil market. That was due to a number of objective reasons which diminished the comparative competitiveness of Soviet oil. The USSR oil provinces were unfortunately located far away from the major importing states which predetermined long-distance pipeline transportation to the markets; production costs were high due to harsh conditions in the oil-producing regions; the USSR lacked reserve production capacities, but even in case of their availability there were no possibilities within the centrally planned economy for conjunctural price-forming maneuvers with these capacities.

Today, at the fourth and fifth stages of oil market development (within the periods of dominance of paper oil markets), Russian oil production still does not play a role in defining the state of the oil market. Russia is again the price-taker and not a price-maker in global oil. In this regard it cannot be considered (and it is not necessary to try to do so), as an "energy superpower." All the factors valid earlier for the USSR (geography, costs, lack of reserve capacities) are valid today for Russia, coupled with – which is most important in the period of paper oil markets – an underdevelopment of the domestic financial system.

Russia is almost not present in the global derivatives market (including markets of oil financial derivatives). Can it play any significant role there today due to the state of development of its domestic financial market? The country is characterized by an absence of domestic oil exchanges (commodities market) due to its high level of monopolization of domestic physical oil transactions (long-term result of privatization model), an imperfection and infancy of the financial system, a long-term absence of the "quality bank" for oil, etc. International experience (including from global oil markets) shows that there is a specific sequence of actions to be taken with the aim to develop a paper market: first an effective financial system needs to be created which will be the basis for development of exchange/futures trading, and not vice versa.

When in 2009 Saudi Arabia began to promote internationally the concept of "fair oil price" (supporting the price level which was needed to provide a non-deficit budget for the Kingdom), a number of states supported this campaign, especially those where the price of non-deficit budget oil was also as high or even higher than in Saudi Arabia. Russia was among those countries. When the oil price began to grow at the beginning of the 2000s, Russia, being a price-taker and not a price-maker, found itself in the position of a free-rider on the wave of such growth. At the beginning of that decade my country carried out a rather conservative budgetary policy, holding the export non-deficit budget oil price at about US $20/bbl, within the range of price fluctuations in the previous decade of the 1990s. But later on Russia has rather quickly saddled up this growth and the level of its non-deficit budget has rocketed up, accelerating the actual oil price increases. Within four years the non-deficit Russian budget oil price increased twice and reached $40/bbl in 2007, and then in one year it made the same jump as in the previous four years, up to $60/bbl (see Figure 28.8).

This inertia was rather difficult to stop and the avalanche-like deficit-free-budget price increase has continued up until now. According to the estimates of O. Buklemishev and N. Orlova (even though they may be slightly different in details) Russia's deficit-free budget oil export price has been constantly exceeding the actual export price since 2009. And there is definitely something similar between this long-lasting trend in Russia (in the fact that the deficit-free budget oil price exceeds the factual export price) and the situation that Saudi Arabia had faced only once in 2009 when the country's deficit-free budget price became lower than the actual export price (see Figure 28.7). But here is where all the similarities between Russia and Saudi Arabia end. The main difference is

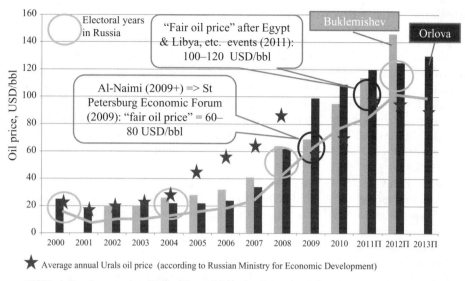

Figure 28.8 Oil price balancing the Russian budget (with and without "corruption tax") and "fair oil price."
Source: Konoplyanik (2011a). Figure created by the author based on data from presentations of O. Buklemishev and N. Orlova at the conference "20 Years after the USSR. What's Next?" (Moscow, June 9, 2011), who have kindly provided their data to the author.

that Saudi Arabia disposes of leverage to influence the world oil market (by its policy of maintaining oil production capacity reserves) while Russia doesn't (and not only because of its lack of capacity reserves).

More than that, easy life spoils people: the price growth makes them accustomed to a stable income from oil export which in turn creates inefficient national production, unproductive consumption, and inefficient usage of the resources from oil export that exceed in fact this growth. Russia has been suffering from this disease for so long that it may as well be considered chronic. And I don't think the situation has changed for the better, especially regarding the state procurement programs. The past Russian President (D. Medvedev) in desperate anger laid open to the public the figures of corruption/stealing in the country: more than 1 trillion rubles per year out of 5 trillion rubles per year in government purchases for public services, which is equivalent to a 20% "corruption tax" (Konoplyanik 2011a).

It is obvious that the deficit-free oil price correlates with the electoral cycle, i.e., in the years when the elections are held it happens to be a little higher than the trend. This is how it was in 2000, 2004, and 2008 (see Figure 28.8). But, as shown by O. Buklemishev and N. Orlova, whose figures are used in this chart, during this electoral cycle the country has been living beyond its means with the deficit-free budget price being too high. And it is expected that in 2012, which happens to be electoral as well, it will jump even further.

Supposing that the corruption tax in all the government sector does not exceed the "modest" 20% level and that it is possible to eliminate it, the deficit-free budget oil price will be 20% lower (as average in Buklemishev and Orlova's estimates). Nevertheless, this means that within the present electoral cycle the price on non-deficit Russian budget oil will still be higher compared to the level of actual Urals oil export prices (see Figure 28.8).

Russian Energy Policy: Adaptation Is Needed

It seems that the Arab Spring events happened to be extremely timely in the context of accelerated growth of the non-deficit budget petroleum price in Russia (Figures 28.6 and 28.8). But the efforts of Saudi Arabia to maintain the so-called "fair" petroleum price on a higher level than the Kingdom's non-deficit budget price are still not enough to balance Russia's budget. According to Renaissance Capital (2011), to get the share of oil in world GDP to the extremely high 7% level of the 1980s, crude prices in 2010 should have been US $152/bbl instead of $80/bbl, i.e., exceeding by only 3–4% the non-deficit budget price for Russia in the electoral year 2012 as in Buklemishev's estimate ($145–147/bbl, Figure 28.8).

The current world crude prices that correspond to the "new" deficit-free budget price of Saudi Arabian oil for 2011 ($90/bbl) are quite acceptable to the world economy as they fit perfectly into the 5% corridor of the oil costs in GDP. But the world economy won't be able to survive the level of oil prices that would ensure Russia a deficit-free budget in its electoral year.

Any country can influence the deficit-free budget oil price level by its government expenditures policy, making it increase (by raising government expenditures, building up the government share in the economy, and boosting the inefficiency of petrodollar usage by the so-called "corruption tax") or decrease (cutting government expenditures, mostly by lessening the government share in the economy and raising the efficiency of public funding, inter alia, by reducing the level of corruption).

By a well-thought investment policy the host state can influence downward the marginal costs level, stimulating investment activities of the oil companies. Capital is the carrier of innovations. That is why only direct investments (implementing the most modern and efficient technologies) can ensure control of the increase or decrease of current and marginal costs in the long term. This makes it possible for the companies to survive a decrease in oil price without critical losses, as with the help of a well-thought state tax policy (which must be flexible and adaptable in the first place), the oil companies will make profits regardless of the oil price level as long as it is maintained higher than its absolute bottom level.

A few concluding points can be mentioned in this regard. The current low efficiency in petrodollar use creates excessive demand pressure on the oil economy. It is currently combined with a non-optimal fiscal-oriented investment climate in the subsoil use, based on flat-rate mineral resource production tax and export customs duty. An optimal investment climate requires a combination of legal stability and a flexible and adaptable tax regime which has nothing to do with individual concessions given to individual projects by individual decisions of the state authorities in a "handy" manner. Moreover, oil and gas should be considered as the sixth innovative cluster of the Russian economy. But that's the subject for another story, whose major elements the author has been presenting in his other writings (Konoplyanik 2012d, etc.), available at his website at www.konoplyanik.ru.

References

Aguilera, R. F., R. G. Eggert, C. C. Lagos, and J. E. Tilton. 2009. Depletion and the Future Availability of Petroleum Resources. *The Energy Journal* 30, 1: 141–174.

Allott, Gordon. 2011. Welcome to Wall Street's Briar Patch. *Oil & Gas Financial Journal*, June: 6, 8–9.

Bashmakov, Igor. 2006. Цены на нефть: пределы роста и глубины падения [Oil prices: limits of growth and the depth of fall]. Вопросы экономики [*Economic Questions*], 3.

Bashmakov, Igor. 2007. Three Laws of Energy Transition. *Energy Policy* 35: 3583–3594.

CGES. 2011. *Arab Spring Will Impact Oil Prices in the Long Term*. London: Centre for Global Energy Studies, Monthly Oil Report, August.

Chevalier, Jean-Marie. 1975. Нефтяной кризис [Oil crisis], transl. into Russian, Москва: Мысль [Moscow: Mysl].

Dickel, Ralf, Tim Gould, Gurbuz Gunul, *et al.* 2007. *Putting a Price on Energy: International Pricing Mechanisms for Oil and Gas*. Brussels: Energy Charter Secretariat.

DG Energy. 2011. *EU-Russian relations: The Role of Gas*. Brussels: European Commission, Directorate-General for Energy (1st meeting of EU-Russia Gas Advisory Council, Vienna, 17 October 2011).

Drollas, Leo. 2011. *Saudi Arabia's Target Oil Price in 2011*. London: Centre for Global Energy Studies.

IEA. 2008. *World Energy Outlook 2008*. Paris: International Energy Agency.

Kanygin, Petr S. 2010. Экономика освоения альтернативных источников энергии (на примере ЕС) [Economic development of alternative sources of energy (case-study of the EU)]. Dissertation for Doctor of Economic Sciences, Moscow, Institute of Europe, Russian Academy of Science, 2010.

Konoplyanik, Andrey. 2000. Мировой рынок нефти: возврат эпохи низких цен? (последствия для России) [The world oil market: return of the era of low prices? (Consequences for Russia)]. Moscow: INP, Russian Academy of Science.

Konoplyanik, Andrey. 2004. Россия на формирующемся Евроазиатском энергетическом пространстве: проблемы конкурентоспособности [Russia in the emerging Eurasian energy space: problems of competitiveness]. Moscow: Nestor Academic Publishers.

Konoplyanik, Andrey. 2008a. Повышение конкурентоспособности России на мировых энергетических рынках через инструменты Энергетической Хартии [Improving the competitiveness of Russia on world energy markets through instruments of the Energy Charter Treaty]. Speech at plenary session "Global Energy Security" 8th St Petersburg International Energy Forum, April 8–10.

Konoplyanik, Andrey. 2008b. Нефтяной рынок необходимо реформировать [The oil market needs to be reformed]. *Vremia Novostey*, December 12.]

Konoplyanik, Andrey. 2009a. О причинах взлета и падения нефтяных цен [On the causes of the rise and fall of oil prices]. Нефть и газ [Oil and Gas (Ukraine)] 2: 2–4, 6–8, 10–11.

Konoplyanik, Andrey. 2009b. Кто определяет цену нефти? Ответ на этот вопрос позволяет прогнозировать будущее рынка «черного золота» [Who determines the price of oil? The answer to this question allows predicting the future market of "black gold"]. Нефть России [Oil of Russia] 3: 7–12; 4: 7–11.

Konoplyanik, Andrey. 2010. *Who Set International Oil Price? A View From Russia*. Dundee: Centre for Energy, Petroleum & Mineral Law & Policy (CEPMLP), University of Dundee. Reprinted in *Oil, Gas and Energy Law* 9, 1, January 2011.

Konoplyanik, Andrey. 2011a. В поисках «справедливости». Существует ли обоснованная цена на «черное золото» и каков может быть ее уровень? [In search of "justice." Whether there is a reasonable price of "black gold" and what may be its level?]. Нефть России [Oil of Russia] 10: 42–45; 11: 11–16.

Konoplyanik, Andrey. 2011b. Эволюция механизмов ценообразования на мировом рынке нефти: проблемы и риски движения от рынка физической к рынку бумажной энергии [Evolution of pricing mechanisms in the global oil market: challenges and risks of moving from a physical market to a paper energy market]. Speech and review presentation at plenary session 3, First Russian Petroleum Congress, Moscow, World Trade Center, March 14–16.

Konoplyanik, Andrey. 2011c. Energy Markets, Financial and Monetary Systems. Presentation at the workshop organized by Norwegian Centre for Strategic Studies and Institute of the Oil & Gas Problems, Russian Academy of Sciences, Narvik, Norway, June 24.

Konoplyanik, Andrey. 2011d. Современный мировой рынок нефти: ненефтяные спекулянты правят бал [The contemporary world oil market: non-oil speculators run the show]. Speech at the scientific-practical conference "Oil as a Special Class of Assets – Current Trends and Risks," Gazprombank and Institute of World Economy and International Relations, Moscow, December 13.

Konoplyanik, Andrey. 2012a. Эволюция международных рынков нефти и газа и инструментов защиты/стимулирования инвестиций в энергетику [Evolution of international oil and gas markets, and instruments for the protection of / incentives for investment in the energy sector]. Speech at the first scientific-educational conference "Economics of the Energy Sector as Direction of Research: Frontlines and Everyday Reality," Moscow, MSE MGU, March 23.

Konoplyanik, Andrey. 2012b. Эволюция мирового рынка нефти: закономерности развития vs. устойчивое развитие [Evolution of the global oil market: patterns of development vs. sustainable development]. Presentation at the conference "Sustainable Development in the Energy Sector," organized by NIU HSE and University of Dundee, Perm, April 17–19.

Konoplyanik, Andrey. 2012c. Эволюция контрактной структуры и механизмов ценообразования на мировом рынке нефти: кто определяет цену нефти? [Evolution of contractual structures and pricing mechanisms in the global oil market: Who determines the price of oil?] Speech at the round table "Topical Questions of Pricing Global Hydrocarbon Markets," Institute of Energy Strategy, Moscow, May 16.

Konoplyanik, Andrey. 2012d. Шестой инновационный кластер. Такую роль в российской экономике могут сыграть нефть и газ [Sixth innovation cluster. This is the role oil and gas could play in the Russian economy]. Нефть России [Oil of Russia] 4: 6–11; 5: 9–15.

Konoplyanik, Andrey. 2012e. Russian Gas in Europe: Why Adaptation Is Inevitable. *Energy Strategy Reviews* 1, 1: 42–56.

Kurenkov, Yuri, and Andrey Konoplyanik. 1985. Динамика издержек производства, цен и рентабельности в мировой нефтяной промышленности [Dynamics of production costs, prices and profitability in the global oil industry]. Мировая экономика и международные отношения [*World Economy and International Relations*] 2: 59–73.

Medlock III, Kenneth B., and Amy M. Jaffe. 2009. *Who Is in the Oil Futures Market and How Has It Changed?* Houston, TX: Rice University, James A. Baker III Institute for Public Policy.

Mirkin, Jakov M. 2011. Финансовый механизм формирования цен на нефть [Financial mechanism of forming oil prices]. Speech at seminar "Volatility of World Oil Prices – Threat to the Budgetary Process," IMEMO, Russian Academy of Sciences, June 22.

Oil Tabloid. 2010. ENI quarterly 10 (June).

Renaissance Capital. 2011. The Revolutionary Nature of Growth. Renaissance Capital, Frontier and Emerging Markets. Update, Economics and Strategy Research, June 22.

Sorrell, Steve, Jamie Speirs, Roger Bentley, *et al.* 2009. *Global Oil Depletion: An Assessment of the Evidence for a Near-Term Peak in Global Oil Production*. London: UK Energy Research Centre.

Stowers, Don. 2011. Dodd-Frank to impact producers. *Oil & Gas Financial Journal*, June: 5.

Takin, Manouchehr. 2008. *Upstream Costs and the Price of Oil*. London: Centre for Global Energy Studies.

Tarbell, Ida. 1904. *The History of the Standard Oil Company*. New York: McClure, Phillips.

Yergin, Dan. 1991. *The Prize: The Epic Quest for Oil, Money & Power*. New York: Simon & Schuster.

Zhukov, Stanislav V. 2011. Нефть как финансовый актив [Oil as a financial asset]. Speech at the scientific-practical conference "Oil as a Special Class of Assets – Current Trends and Risks," Gazprombank and Institute of World Economy and International Relations, Moscow, December 13.

Zhukov, Stanislav V. 2012. Интеграция нефтяного и финансового рынков [Integration of oil and financial markets]. Speech at the 128th session of the continuing open seminar "Economic Problems of the Energy Complex" (seminar A. N. Nekrasov), Russian Academy of Science, Moscow, March 27.

Chapter 29

Nigeria: Policy Incoherence and the Challenge of Energy Security

Ike Okonta

Introduction

Nigeria, Africa's most populous country, is also one of the world's largest oil exporters. Although oil was struck in commercial quantities in the late 1950s, Nigeria's oil industry remains in many respects a study in underdevelopment and policy chaos thereby making the longstanding quest for energy security an elusive goal. This is in spite of the fact that petroleum production accounts for 25% of the country's GDP, 95% of export earnings, and about 75% of total government revenue (EIA 2010). The Nigerian National Petroleum Corporation (NNPC), the national oil company, manages an estimated 36.2 billion barrels of proven oil reserves and also plays a role in the daily export of 2 million barrels. Even so, the NNPC is plagued by bureaucratic inefficiency, excessive interference by government officials, and a chronic inability to position the oil sector as a key driver of the country's economic development. The Niger Delta region where the bulk of the oil is produced is also embroiled in political crisis as local people mobilize to protest economic neglect and oil-induced environmental pollution.

Trapped between policy incoherence, narrowly focused vested interests, and an increasingly restive citizenry demanding more social and economic benefits, Nigeria's oil sector is rapidly becoming chaotic with the consequence that the country's energy security, and its ability to fulfill its obligations to importing countries in Europe and North America, is direly threatened. The central argument of this chapter is that the major cause of Nigeria's current energy security crisis is the persistent inability of successive governments to develop innovative and joined-up policies to bring the oil industry and other sectors of the economy into a complementary whole using the substantial oil revenues as a primary driver, thus accelerating the pace of employment generation and addressing widespread poverty. This policy incoherence, we further argue, has had profound consequences for Nigerian society, strengthening a still-ongoing regime of authoritarian and incompetent central government with an iron grip on the oil revenues, and reducing the citizenry to impoverished and disempowered bystanders.

The Handbook of Global Energy Policy, First Edition. Edited by Andreas Goldthau.
© 2013 John Wiley & Sons, Ltd. Published 2013 by John Wiley & Sons, Ltd.

"Resource Curse" theorists have drawn our attention to the tragic development trajectory of resource-rich states, particularly in Africa, afflicted by the Dutch Disease, the disruptive effects of fluctuating commodity prices, and authoritarian and ineffective political institutions propped up by oil receipts (Humphreys *et al.* 2007). While the Nigerian case provides empirical evidence for the theory, it also draws attention to its limitation as an explanatory strategy. The mere presence of abundant natural resources in a given state does not necessarily guarantee that it will inevitably be resource "cursed." Rather, the ownership structure of this resource, and the nature and resilience of the political order that governs its exploitation, are powerful factors in shaping the role of this natural resource in the country's development story. It also matters when natural resources begin to be exploited in a country, before the political institutions vital for securing accountability and transparency have taken root or after (Okonta 2006). History also plays a pivotal role in shaping these political institutions. As Eric Chaney has demonstrated in his study of ongoing political upheavals in the Arab world, the region's so-called democracy deficit is not tied to the Islamic religion but rather to the Arab world's history and institutions introduced following conquest by Arab armies over 1000 years ago (Chaney 2012). Nigeria at independence in 1960 inherited from Great Britain centralized authoritarian political institutions that had reduced local people to subjects and seized control of their natural resource endowments by force of arms. We argue in this chapter that Nigeria's "resource curse" is in a fundamental respect a colonial inheritance.

The challenge the country's political leaders and energy policy planners have been unable to meet is how to deploy creative strategies to reverse this curse. Indeed, Nigeria presents an interesting case for studying the phenomenon of countries struggling with natural resource wealth because it is an apt illustration of the persistence of the political past and its debilitating burden of privatized power, challenging the academic theorist and the policy planner to think more deeply and look more widely for the causes of certain countries' inability to transform natural resource endowments into shared prosperity.

This chapter is divided into five sections. In the first section we examine the early years of energy security policy-making in Nigeria, showing the way in which government policies in the newly independent country were mapped onto those of its authoritarian colonial predecessor, arrogating to the central government sole right to exploit the country's oil wealth and paving the ground for future trouble. Section two traces the emergence of the country's economic difficulties in the 1980s as successive military juntas proved unable to deploy oil receipts in a period of "boom" to provide a robust platform for future prosperity, leading regional and ethnic elites and organized labor to challenge the excesses of centralized power. Section three argues that the maturation of these opponents of the dominance of an all-powerful central government, epitomized by the Movement for the Survival of the Ogoni People (MOSOP) and the violent Movement for the Emancipation of the Niger Delta (MEND), drawing nourishment from deepening poverty and oil production-related environmental blight in the main oil-producing region, represents a potent threat to Nigerian policy planners' ambition to achieve energy security for the country. In section four we analyze in detail four key policy interventions that a new civilian government, elected in 1999 and desperate to head off the political and economic forces crippling the embattled oil industry, deployed and how countervailing interests and the age-old inability to bring coherence and joined-up thinking to bear on this endeavor continue to thwart the effort. Concluding, we tease out the implications of policy failure in important oil producers like Nigeria using unaccountable central authority and the vehicle of the National Oil Company to achieve energy security and shared prosperity for the citizenry.

Policy-Making and Energy Security: The Early Years

There is a striking similarity between colonial-era mining laws and those of the Nigerian state in the early years after independence. This is not a mere coincidence. The will to centralized and unaccountable power, the easier to exploit the natural resources of subjects unchallenged (the *raison d'être* of the colonial state), was replicated virtually unchanged by its indigenous successor. Just as the strictures of oppressive colonial rule were railed against by Nigerian nationalists campaigning for independence from the 1940s onward, so too did unfair and authoritarian oil decrees enacted by the post-independence state in the late 1960s come under vigorous challenge by aggrieved locals, particularly in the Niger Delta region, the country's oil belt, who felt short-changed. The subsequent struggle by the various contending parties for a dominant role in policy-making and for the lion's share of the oil prize prepared the ground for policy distortions and the emergence of corrupt and incompetent political leaders unable to deliver economic development and prosperity using this natural endowment.

The dramatic face-off between leaders of the Ijaw Youth Council (IYC), a youth-led ethnic self-determination pressure group in the Niger Delta and the Federal Government in December 1998, just when the Nigerian military junta was preparing to relinquish power to a democratically-elected government after 30 years, illustrates the dilemma of the country's oil policy planners in their effort to secure energy security.

IYC, established in the early months of 1998, is a confederation of youth organizations of the Ijaw ethnic group mainly drawn from the core delta states of Rivers, Bayelsa, and Delta (Ikelegbe 2001). Its mission is to pressure the Federal Government of Nigeria to hand over control of the substantial revenues accruing from oil production to local communities and ethnic groups in the Niger Delta region, and to also induce Royal Dutch Shell and other international oil corporations operating in the area to clean up polluted land and rivers these communities rely on for their livelihood. On December 11, 1998, leaders of IYC issued the "Kaiama Declaration" at a gathering of 5000 youths drawn from the various Ijaw clans, communities, and community-based organizations in Kaiama town, Bayelsa state. The Kaiama Declaration demanded the withdrawal of soldiers and other members of the armed forces deployed to the Niger Delta to protect oil installations and ensure the safety of the personnel of the oil companies, describing these soldiers as "forces of occupation and repression by the Nigerian state" (Ijaw Youth Council 2009). The Declaration also stated: "We, therefore, demand that all oil companies stop all exploration and exploitation activities in the Ijaw area. We are tired of gas flaring, oil spills, blowouts, and being labeled saboteurs and terrorists... We advise all oil companies' staff and contractors to withdraw from Ijaw territories by the 30th December 1998 pending the resolution of the issue of resource ownership and control in the Ijaw area of the Niger Delta."

The military government responded by declaring a state of emergency in Bayelsa state. Tanks, amphibious and fast attack craft, warships, and thousands of combat-ready soldiers poured into Yenagoa, the state capital, Kaiama, and other neighboring towns all through December 1998 and January 1999. In his annual budget speech on January 1, 1999, the Head of State, General Abdulsalaam Abubukar stated that "we cannot allow the continued reckless expression of these feelings in seizure of oil wells, rigs and platforms as well as hostage-taking in the name of expressing grievances." It is estimated that Nigerian army personnel killed 200 youths in the course of this confrontation (Human Rights Watch 1999). There was also oil production shut-down as personnel fled the area. Several of the oil companies had to declare force majeure.

It is significant that IYC leaders, in the course of their bruising battle with Nigeria's military junta, declared that while the Ijaw had resolved to remain in the Nigerian federation, they would continue to pursue the goal of self-government for the Ijaw people and a new federation of ethnic nationalities run "on the basis of equality and justice." The IYC also stated that the solution to the endemic economic and ecological crisis in the Niger Delta was the convocation of a "Sovereign National Conference" where representatives of Nigeria's estimated 350 ethnic groups would deliberate on a new political structure for the country; an arrangement that would be founded on "resource control, self-determination, and fiscal federalism." Indeed, these political demands are at the heart of the crisis in which the Nigerian oil industry is presently trapped and have impeded coherent policy-making to secure the goal of energy security since the late 1960s.

Nigeria, a former colony of Great Britain, began life as an independent country of three regions in October 1960. A complex amalgam of several ethnic, linguistic, and religious groups, the country is nevertheless dominated by three main ethnic groups: the Hausa-Fulani who are predominantly Muslim in the north, the Christian Igbo in the east, and the Yoruba who practice Islam and Christianity in the southwest. The Ijaw, the largest ethnic group in the Niger Delta, the Tiv in the central part of the country, and several other smaller ethnic groups constitute what is referred to in the local political vernacular as "ethnic minorities." At independence Nigeria operated a federal constitution and a parliamentary system, with three main political parties each controlled by the major ethnic groups dominating politics and government at the regional levels (Wolpe 1974). The Hausa-led Northern Peoples' Congress (NPC) formed the government at the federal centre. Official corruption, mounting economic problems, attempts by regional political elites to dominate the politics and resources of rival regions, and pogroms targeting Igbo public servants and artisans resident in the Northern Region, forced the Igbo to declare a new Republic of Biafra in the Eastern Region in July 1967. This triggered a bloody 30-month civil war, leading to the defeat of the secessionist Igbo in January 1970.

While control of the oilfields of the Niger Delta, at the time of the war part of Biafra, did not shape the decision of the Igbo to secede, reintegration of the oil-rich area nevertheless informed the Federal Government's war aims and strategies (Williams 1983). Cocoa, palm oil, and groundnuts were the country's major foreign exchange earners following the end of World War II in 1945. However, by the late 1960s, a decade after the Shell-BP consortium, Elf, a French oil company, and Mobil, a US firm began to produce oil from on-shore and off-shore fields in the Niger Delta, oil revenue had become the dominant revenue earner for Nigeria. Thus, policy-making was from this period onward dictated by calculations to secure the lion's share of the oil revenue for the Federal Military Government which, following a military coup and a counter-coup in 1966, had displaced the civilian democratic regime founded on the three powerful regions. The ensuing politics of revenue allocation, in the main dependent on the oilfields of the delta and the social turbulence this political struggle continue to generate in its wake, is key to understanding why energy security is, at best, in a very fragile position in Nigeria presently.

On the eve of the civil war in May 1967, the system of revenue allocation gave the central government 5% of mining rents and royalties. The central government collected but transferred 45% of mining rents and royalties to the regions on the principle of derivation. The outstanding 50% of mining rents and royalties were allocated to the regions through the Distributable Pool Account (DPA), established by the central government to even out the disparities between the regions as a result of the derivation principle (Oyovbaire 1985). The new military Head of State, General Yakubu Gowon, had created 12 new states in place of the regions in May 1967 and had also promised to appoint a commission to work out an equitable formula of revenue allocation, taking into account the desires of the various states. The imminent conclusion of the civil war to the advantage of a vastly strengthened central government led General Gowon to establish an eight-member Interim Revenue Committee headed by I. O. Dina to make recommendations on the system of fiscal federalism in the country.

The Dina committee, arguing that the fundamental lesson of the civil war was the imperative for national unity, insisted that "we have viewed all the resources of revenue of this country as the common funds of the country to be used for executing the kinds of programmes which can maintain this unity" (Federal Ministry of Information 1969). Put simply, Dina and his patrons were making a case for the centralization of fiscal resources, including oil revenue, paid in the form of rents, royalties, and other taxes, in the central government account whose economic advisors at the time had estimated that Nigeria would be earning £300 million Sterling a year in foreign exchange from oil by 1975 (*Financial Times* 1969). Indeed, they were calling for a new fiscal arrangement that would deprive the impoverished oil-bearing communities of the Niger Delta of revenue they desperately needed. Significantly, Dina's recommendations coincided with the managed shift in state earnings from oil rents and royalties to profits.

The centralization of oil revenues took several stages and was driven mainly by Gowon's permanent secretaries who had a firm grip on the policy-making apparatus of the civil service. In November 1969, in the teeth of considerable opposition from senior civil servants from the Niger Delta unhappy that the derivation principle had been jettisoned, thus considerably reducing their region's take of the oil revenue, the central government published the Petroleum Decree (Decree no. 51 of 1969) stating that "the entire ownership and control of all petroleum in, under or upon any lands to which this section applies" be vested in the state. The decree also applied to land under the territorial waters of Nigeria including its continental shelf (*Petroleum Intelligence Weekly* 1969).

Shortly after the Biafra war ended in January 1970, the victorious central government followed up the Petroleum Decree with yet another, Decree no. 13 of 1970, radically altering the structure of revenue allocation. Decree 13, among other changes, divided one half of the DPA equally among the states, while the other half was shared on the basis of relative population. These changes had two important effects. The amount of revenue accruing to the DPA, and the financial strength of the central government *vis-à-vis* the states, particularly oil-producing states like Rivers, Midwest, and Southeastern, was increased. But there was a deeper consequence. As the central government tightened its grip on power and the instruments with which to allocate profitable opportunities in the burgeoning oil economy, the link between wealth and the labor needed to generate it snapped. Access to important government officials became the all-important ticket to riches. A commercial triangle developed, comprising representatives of Western multinationals trying to sell all sorts of manufactured goods, some of dubious value, to an oil-rich but underdeveloped country; local middlemen; and powerful state officials whose role was to assist both multinationals and local businessmen alike (Turner 1978: 167). Bribery, conspicuous consumption, and a new commercial capitalist ethos that encouraged the importation of foreign goods to the detriment of local manufacture became firmly established (Williams 1976).

The first efforts at serious policy-making in the oil industry necessarily had to contend with this inclement arena where the culture of rent-seeking and easy profits had taken the place of entrepreneurship and productive capitalism anchored on utilization of the country's natural resources, including oil. The 1914 Colonial Mineral Ordinance was the first oil-related law in colonial Nigeria and ensured that only British subjects and British companies were granted mining leases and licenses. In 1937 this ordinance granted Shell D'Arcy (precursor of today's Royal Dutch Shell) exclusive exploration and prospecting rights covering all of Nigeria (Khan 1994: 16). However, in 1955, following pressure from the US government, Mobil was granted prospecting rights. Gulf, Agip, and Safrap were also granted oil prospecting licenses following the end of colonial rule in 1960. General Gowon's 1969 decree annulled the 1937 colonial ordinance, only to replicate its centralizing spirit by placing ownership of this vital natural resource in the state (Frynas *et al.* 2000). It also specified the form, rights, powers, and restrictions on the various types of oil production licenses. Further, the decree stipulated that the oil exploration license (OEL), Oil Prospecting License (OPL), and Oil Mining Lease (OML) could only be granted to Nigerian citizens or companies registered in Nigeria.

Gowon's military government took Nigeria into OPEC as a full member in 1971. In its Declaratory Statement of Petroleum Policy in Member Countries during the 16th OPEC conference in June 1968 (also known as Resolution XVI 90) the organization had advised member countries to acquire "participation in and control over all aspects of oil operations." It was in a bid to implement this resolution that Abdulazeez Atta, Gowon's Permanent Secretary of the Ministry of Finance and later Secretary of the Federal Military Government, led efforts to establish two institutions: the Petroleum Advisory Board, an open forum to promote informed policy-making, and the Nigerian National Oil Corporation (NNOC) to implement oil policy in a coordinated and efficient way (Turner 1978: 179). Nigeria had emerged as one of the largest producers of sweet (almost sulfur-free) crude oil at this time. Further, the country's geographical location in the Atlantic Basin, closer (in comparison to Middle East producers) to two of the world's largest markets for crude and refined petroleum products – Western Europe and the US – conferred price premiums on its oil. The goals that Atta spelt out for the country's oil policy institutions were to increase and diversify oil production, increase indigenous ownership and control of the growing industry, and integrate it into the government's economic development plans by encouraging the emergence of important downstream sectors like local refining, a petrochemical industry, and production and distribution of liquefied petroleum gas (LPG). These three goals, Atta felt, would not only ensure energy security but also increase value-added in the sector.

The end of the civil war did not diminish the vicious struggle for economic perks and political influence between regional and ethnic elites in Gowon's military government. Well-educated bureaucrats had risen to positions of power and influence following the demise of the elected First Republic in 1966. Where previously these civil servants took orders directly from elected politicians who served in a cabinet of ministers under the Prime Minister, they were now able to work directly with the head of the military junta and to wield considerable influence over him, sometimes relegating to the background the politicians that Gowon had appointed to serve as commissioners (ministers) to super-intend the various government ministries and departments (Osaghae 1998: 77–79). In the vicious two-cornered struggle for power between the senior bureaucrats and the commissioners and between rival bureaucrats themselves for access to Gowon, coherent and forward-looking oil policy was the victim. In particular, there was open rivalry between administrators in the Ministry of Mines and Power and the Ministry of Finance over control of policy-making in the oil industry. While the Ministry of Finance under Atta favored interventionist policies and the government's direct participation in the industry, the Ministry of Mines and Power preferred the policy of regulation and non-intervention. The former won the battle following Nigeria's entry into OPEC.

Even so, the new NNOC, established in 1971 as the primary instrument with which to secure the government's participation in the oil industry, was not given a free hand to perform this function. First, it was designated a parastatal under the oil ministry (Ministry of Mines and Power). Second, it could not spend money to service its operations beyond a limit approved by the Federal Executive Council nor did it have exclusive powers in all stages of oil operations including production and marketing. Third, the chairman of its board was the Permanent Secretary (top civil servant) of the oil ministry (Turner 1978: 180). As the corporation evolved into a semi-autonomous organ with a comple-ment of well-trained technocrats knowledgeable in the various segments of this highly complex industry, tension developed between the latter and general administrators in the ministry.

While the technocrats advocated a fundamental revision of oil policy to reflect the new desire of the state to control the industry and use it as a major instrument to develop critical sectors of the economy, bureaucrats in the ministry, having over the years developed a close relationship with officials of the international oil corporations, whose interests they now sought to safeguard by pushing policies favorable to their continued dominance in the Nigerian oil industry, stoutly opposed the technocrats under them. Suspicions of corruption on the part of the bureaucrats were widespread, and this played a major role in undermining morale among technocrats. Excessive bureaucratic intervention in these early years also prepared the ground for the emergence of a state oil corporation marked by incompetence, low productivity, and inability to grow the NNOC into a major player in the upstream and downstream sectors of the Nigerian oil industry like its foreign counterparts.

Even as Nigeria underwent political upheaval from the end of the civil war in 1970 to 1985 as military juntas replaced each other after bloody coups and an elected Second Republic was sacked by yet another military strongman after only four years in power, oil policy remained surprisingly stable. Nationalization of the industry was only partial, and took place mainly in the 1970s. The central government had acquired a 35% equity in the French-owned Safrap shortly after the end of the war, because the oil company was perceived as having supported Biafra. In the ensuing years the government established joint ventures with the oil majors, increasing to 60% its stake in these ventures (NNPC 1981). Royal Dutch Shell, which accounted for half of total yearly production, was the operator of Shell Producing Development Company (SPDC), a joint venture between the company, the Nigerian government, Agip, and Elf (formerly Safrap). Five majors – Shell, Mobil, Chevron, Agip, and Elf, dominated and continue to dominate the industry. The only major policy initiative undertaken by the government that was inimical to the interests of the oil majors was the 1979 nationalization of BP's stake in the Shell-BP partnership because BP had violated the embargo on trading with apartheid South Africa. The Nigerian government, mobilizing global efforts toward achieving majority black rule in both South Africa and Zimbabwe, wanted to influence British foreign policy in favor of this goal. Successive governments were anxious to ensure the well-being of the industry and the oil majors as state reliance on oil revenues grew.

However, this stability in policy-making did not translate into initiatives to grow effective state participation in the oil industry, increase value-added, and provide the enabling environment for a secure energy future. As the central government tightened its grip on oil revenues it sought to coopt ethnic and regional elites likely to resist this policy by offering them lucrative contracts and sundry perks in the oil industry. The NNOC, transformed into the Nigerian National Petroleum Corporation (NNPC) by military decree in 1977 following the merger of the commercial and regulatory arms of the industry, became increasingly inefficient and its transactions opaque and shielded from public scrutiny. *The Punch*, a leading newspaper, reported in September 1979 that nearly 3 billion dollars was missing from the corporation's account with the Midland Bank in London (*The Punch* 1979). Whistleblowers also charged that the IOCs were routinely engaged in overlifting between 1975 and 1978 in collusion with senior officials of the corporation (Khan 1994: 52). Inter-elite squabbles also delayed project planning and execution in domestic refining, export of refined products, petrochemicals, and liquefied natural gas, initiatives that could have provided additional revenue and diversified the economy. Electricity generation that could have been boosted by increased production of gas oil remained comatose.

With natural gas reserves estimated at between 3.0 and 3.4 trillion cubic meters (tcm), the country ranks eighth in the world and fifth in OPEC (*Africa Oil and Gas Report* 2012). However, poor planning and lack of foresight has led the oil companies to flare the gas associated with oil in their areas of production. Moreover, poor infrastructural support systems, unfeasible gas utilization schemes, and prohibitive costs of extracting, processing, and separating associated gas have constituted formidable stumbling blocks in the development of this important downstream sector. Indeed, it was only in 1999, after the NNPC had been demoted to a junior partner in the country's liquefied natural gas company, that the IOCs were able to begin to produce and export the product. Successive governments were also unable to tackle the Dutch Disease symptoms that had occurred during the "oil boom" years as government capital expenditure, mostly on white elephant projects, ballooned even as the main tradable sectors – agriculture and manufacturing – collapsed.

Intimations of a Future Storm

A major consequence of authoritarian rule in underdeveloped but resource-rich countries is that citizens are prevented from participating in the process of policy-making to secure their interests, leading unaccountable leaders to pursue self-serving goals that further pauperize the country even in the midst of abundance. The resulting incoherence in policy-making, rooted in this regime of unaccountable power, tends to impact social and economic life negatively, triggering social unrest which in turn goes on to erode the pillars of political stability that energy security requires to thrive and prosper.

Nigeria slid into economic recession in 1982 following reduced demand for its oil as Western importing countries battled with problems at home. This also coincided with production increase in North Sea oil. Official corruption and capital flight, which had peaked during the government of Shehu Shagari, elected in 1979 before it was replaced by yet another military junta in December 1983, further reduced room for policy innovation to revive the economy. Experts have noted that the key problem with the country's oil industry has been the persistent failure by government officials to develop an efficient framework and goal-driven implementation of a cohesive sectoral plan to diversify revenues away from crude oil exports. Now the citizenry began to pay the price for these sundry policy failures. Social and economic indicators plunged. As living conditions worsened and unemployment grew, protests led by organized labor and other segments of civil society proliferated (Jega 2000: 11–18). Regional opposition to the central government's dominance of politics and oil, the main source of external revenue, also grew sharply. It was this latter development that was to provide the platform for the near crippling of the oil industry, threatening energy security.

General Ibrahim Babangida who mounted a palace coup and took over in August 1985 inherited a comatose economy and an increasingly restive citizenry. Crude oil export revenues had fallen from a peak of $24.9 billion in 1980 to $9.9 billion in 1983. While they recovered slightly in 1985, the bottom fell off crude exports in 1986, earning the country only $5.7 billion, about one-fifth the level of oil export receipts in 1980 (Khan 1994: 52). Babangida pursued two main policies during this period in an attempt to increase oil production and also enhance the country's energy security. Between 1986 and 1991 he signed Memorandums of Understanding with the IOCs offering better incentives for exploration and production. These changes in notional profit margins and production cost allowances, in addition to new production sharing contracts to supplement the original joint ventures, encouraged Shell, Elf, Mobil, Exxon, BP, and Statoil to sign new production deals with the government. Shell established a new wholly-owned company, SNEPCO, to explore and produce oil off-shore. The production sharing scheme, unlike the joint venture, permitted oil firms to retain 80% of an oilfield's production, after taxes had been paid and costs recovered, for a period of 10 years (Ofoh 1992). Still, the absence of transparency and accountability in public expenditure meant that corruption continued unchecked, squandering the revenue that accrued from this new initiative.

Following discussions with the IMF and World Bank, Babangida also adopted a "Structural Adjustment Programme" designed to introduce market mechanisms into the economy. The naira was sharply devalued; subsidies on health, education, and other social sectors were reduced; and privatization of publicly-owned enterprises was begun. Still, the Structural Adjustment Programme failed to diversify the economy and reverse the increasing dependency on oil revenue. By the early 1990s, over 90% of foreign exchange receipts, 97% of total export receipts, 70% of budgetary receipts, and 25% of GDP were still accounted for by oil (Forrest 1995: 30–42). The share of agriculture in GDP instead of increasing as envisaged by SAP, decreased from 41% in 1986 to 37% in 1991. The share of industry increased slightly from 29% in 1986 to 38% in 1991 but subsequently slumped to 7% where it has since remained. To worsen matters, the external debt increased from $9 billion in 1980 to $30 billion in 1992, and the country fell $6 billion behind on debt-servicing payments (Okogu 1992). Its debts to its joint-venture partners also began to mount, making it increasingly difficult for the oil majors to embark on new exploration and production projects. As the central government proved unable to manage the country's mounting economic difficulties, regional, ethnic, and religious tensions began to grow, particularly in the oil-rich delta where the Ijaw, Ogoni, and other ethnic minority groups had for four decades borne the brunt of oil production but received very little of the proceeds.

The Years of Turmoil, 1990–2006

The grid of unaccountable power whose emergence and broad contours we traced in the preceding sections of this paper was driven primarily by and nourished on oil. Democratization is therefore always a difficult and hazardous process in authoritarian resource-rich states like Nigeria, as military incumbents adroitly maneuver the levers of powers using the oil revenues at their disposal to stall the democratization process, or failing that, transform the emerging "democratic" government into a pliant vehicle for their continuing power projects. Even as the grievances of those excluded from the oil bonanza exploded into violent disruption of oil production from the 1990s, the military junta was still able to successfully transfer power to civilian authoritarians that were its mirror image, ensuring that the regime of incompetence and its menu of policy incoherence and social chaos continued unabated.

A major feature of oil production in the Niger Delta is environmental degradation. Since the first barrels of crude were shipped from the region by Shell in 1958, farmlands and fish-bearing rivers and streams have been subjected to relentless pollution (Okonta and Douglas 2001). The 1969 Petroleum Decree and the Land Use Decree of 1972 (later revised in 1978 as the Land Use Act) had effectively removed ownership of oil-bearing land from local communities in the region and placed it under the control of the central government. This legislation made it difficult for community leaders to pressure the IOCs and make a case for remediation and monetary compensation for losses sustained in the course of oil exploration and production. Government policy on environmental protection was not guided by the gruesome realities on the ground in the Niger Delta, nor did legislation seek to effectively punish transgressing oil companies.

Driven from the cities by the deepening economic recession in the late 1980s, thousands of unemployed urban workers returned to their rural villages in the Niger Delta to find the rivers emptied of fish and other aquatic life and the farms drenched with spilled oil and unable to sustain crops. The region began to heave with resentment, directed at the oil companies. Local people began to attack oil workers and to disrupt oil production in their vicinities. By 1988, 344 production hold-ups had been recorded, out of which 211 resulted in extensive damage to oil pipelines. Shell's production activities in the Niger Delta were held up no less than 22 times between January and August 1991 alone (*African Guardian* 1992). The company's production fell between 1990 and 1993 because of these disruptions. Its Eastern division lost 1269 man-days in 1993. The Western division suffered similar losses. Of 2645 spill incidents in 1992, 1837 were due to community protest (*African Guardian* 1994).

In August 1990, Ken Saro-Wiwa, a writer and minority rights activist, launched the Movement for the Survival of the Ogoni People (MOSOP). The Ogoni, comprising an estimated 500,000 people, are one of the smaller ethnic groups in the delta, and their land has been the site of major Shell installations since 1958 when the giant Bomu oilfield was struck. Saro-Wiwa and other MOSOP officials spelt out the grievances of the people against Shell and the Nigerian government in the Ogoni Bill of Rights, a document presented to the international community, and demanded that a fair portion of the oil revenue taken from the Ogoni fields be given back to the community to enable the people meet their social and economic needs (Okonta 2008: 179–180). The Bill of Rights also made a case for the ending of the company's pollution of the Ogoni ecology and also asked for "political control of Ogoni affairs by the Ogoni people." When neither the company nor the central government paid attention to their demands, MOSOP officials mobilized a large swathe of the Ogoni in a peaceful march in January 1993, and succeeded in expelling Shell workers from the Ogoni oilfields.

The military junta, worried that the Ogoni example would be replicated in similarly aggrieved communities in other parts of the delta leading to crippling of oil production, sought to meet the challenge in two ways: deploying a military expeditionary force to the Ogoni community, and establishing the Oil Minerals Producing Areas Development Commission (OMPADEC). The decree bringing OMPADEC into existence increased the 1.5% of the oil revenue allocated to the oil-producing communities to 3% and transferred this fund to the Commission to administer on their behalf (Federal Ministry of Information 1992). Although OMPADEC was conceived as a development agency with the additional responsibility of monitoring and managing ecological problems associated with the production and exploration activities of the oil companies, the military junta also saw it as a security agency with which to manage the growing unrest in the region. Consequently, a senior intelligence official was redeployed to Port Harcourt, chief city of the Niger Delta, in July 1993 to superintendend OMPADEC. Since the Commission was under no supervisory authority other than the head of the military junta, inefficiency and financial mismanagement quickly set in.

A World Bank team that studied OMPADEC's activities in the Niger Delta in 1995 concluded that it would be difficult for the Commission to fulfill its mandate as a development agency because it did not have the requisite personnel to enable it to meet its ecological and sustainable development mandate; there was an absence of long-term planning; and there was no integrated approach to development planning, which should have involved the local communities and other government agencies in the area (Greenpeace Nederland 1996: 33). OMPADEC officials also regularly collaborated with Shell officials to secretly channel funds to the military task force the junta had sent to Ogoni to suppress MOSOP (*The Observer* 1995.) The military task force proceeded on a campaign of mass murder, rape, and terror in Ogoniland. Following the killing of four Ogoni community leaders by a mob charging that the four had plotted with the junta for Shell workers to return to the Ogoni oilfields, two Ogoni artisans were bribed by OMPADEC to claim that Saro-Wiwa had instigated the killings. Saro-Wiwa and eight other MOSOP officials were hanged by a kangaroo military tribunal set up by the military junta in November 1995.

But Ken Saro-Wiwa's murder only broadened the scope of the uprising of the oil-producing minority ethnic groups, leading them to more fervently demand adequate compensation for environmental degradation and other hazards of exploration and production activities, a greater share of the oil revenue which they charged was going to the majority ethnic groups, and greater political autonomy within the Nigerian federation (Osaghae 1998: 245). The Movement for the Survival of Ijaw Ethnic Nationality in the Niger Delta (MOSIEND) and the Movement for Reparation to Ogbia (MORETO) subsequently emerged in the Ijaw territory, drawing on MOSOP's organizational structure and mobilization strategies. Clashes between these youth-led groups, the oil companies, and Nigerian soldiers sent to protect oil installations and workers became more frequent, leading to disruption in oil production. By the time the Ijaw Youth Council emerged on the scene in late 1998 demanding that the oil companies leave the region it had become clear that the central government's policy of using military might to suppress legitimate protest in the Niger Delta and ensure the integrity of the oilfields had become an embarrassing failure.

General elections in early 1999 marked the end of military rule and paved the way for new policies designed to placate enraged delta communities and secure the oilfields. The constitution that ushered in Nigeria's Fourth Republic, while written by the departing generals without input from the citizenry, nevertheless allocated an unprecedented 13% of total oil revenue directly to the oil-producing states of the Niger Delta (International Crisis Group 2006). The Niger Delta Development Commission (NDDC) was also established to take the place of the failed OMPADEC and coordinate community development efforts in the restive region where poverty, despair, and rage against a corrupt and incompetent central government had deepened. The law establishing NDDC also required the oil companies to contribute to a common fund the new Commission would utilize in financing development projects in the local communities. Former US President Jimmy Carter and other officials of the Atlanta-based Carter Center had monitored the elections in the Niger Delta and had reported widespread rigging and other malpractices in favor of the Peoples Democratic Party (PDP) and Olusegun Obasanjo, a former military head of state whose candidature for President the outgoing junta was backing (Kew 1999). Local human rights groups also reported intimidation of voters and outright stuffing of ballot boxes in other parts of the country. Obasanjo was sworn in as President in May 1999 amid widespread protests. Governors and legislative assembly members, the bulk of them known supporters of the corrupt military junta that had just ceded power, also took office in the core delta states of Rivers, Bayelsa, and Delta. Thus, the new "democratic" dispensation began life without the legitimate consent of ordinary citizens in the troubled delta region.

The consequence of this democratic deficit in the oil region was that "elected" officials, instead of mobilizing personnel and the new 13% derivation to address the pressing needs of the impoverished communities, began to use these public funds for their personal use. There was neither transparency nor accountability. Also left unaddressed was the environmental blight which resulted from oil production. Local PDP politicians in the region had armed unemployed youth and elements drawn from the self-determination groups that had proliferated following the demise of Saro-Wiwa and used them to rig elections in 1999 and again in 2003. The IOCs, looking to secure their installations, also recruited these groups to "protect" their oil wells and flow stations. Politicians and oil executives alike facilitated a thriving local arms industry, unwittingly paving the way for the proliferation of armed youth gangs in the region who soon resorted to kidnapping of oil workers for ransom, terrorizing of local people, and oil bunkering (illegal siphoning of oil in commercial quantity from pipelines for sale on the underground oil market). A consultant's report for Shell in 2003 stated that an estimated 275,000–685,000 barrels of oil was being stolen by thieves in the delta every day (SPDC 2003). Some of the money obtained from bunkering was recycled to purchase more arms. The report also stated that "Shell has become an integral part of the Niger Delta conflict" and that violence in the region was now on par with conflict in Columbia and Chechnya. Two armed groups, Niger Delta Vigilante led by Tom Ateke, and Niger Delta Peoples Volunteer Force (NDPVF), led by Asari Dokubo, a former president of IYC and both with links to corrupt politicians in Rivers state, emerged during this period ostensibly to fight for "self-determination" and "resource control" for the Ijaw.

In 2005 world oil prices rose above $50 a barrel for the first time when Dokubo and fellow militant fighters in NDPVF threatened to take on the soldiers deployed to the region and cripple the oil industry. Militant attacks subsequently took half a million barrels of Nigerian oil out of world markets (Shaxson 2007). The Movement for the Emancipation of the Niger Delta (MEND), a coalition of armed militant organizations drawn mainly from Ijaw clans in the western delta, came into existence in late 2005, following the arrest and detention of Asari Dokubo on treason charges by the central government. In January 2006 MEND militants attacked naval officers protecting a Shell complex near Warri, the country's second most important oil city. A Shell pipeline in the same vicinity was also blown up (Ukiwo 2007: 587) Oil production stoppages became routine as skirmishes between militants and Nigerian soldiers in the delta creeks intensified. Militants bombed an important Shell terminal in the early months of 2006 and also kidnapped nine expatriate staff of Wilbros, a US servicing company. MEND demanded the release of Dokubo and the former governor of Bayelsa state who it claimed had been detained because of their championing of the cause of Ijaw control of the oil on their land. Said Gbomo Jomo, MEND's spokesman said in an interview in February 2006: "All pipelines, flow stations, and crude loading platforms will be targeted for destruction. We are not communists, just a bunch of extremely bitter men" (*Wall Street Journal* 2006).

The Graveyard of Incoherent Policies

Illegitimate power sustained by purloined oil wealth is inherently unstable, always too preoccupied with fighting off or placating challengers to articulate forward-looking and joined up policy and secure the public good. A common thread running through the various efforts of Nigeria's PDP government to transform the NNPC and the oil industry into efficient vehicles for prosperity and energy security since it muscled its way to power in 1999 is their democratic deficit, ensuring that these policies neither find firm purchase in the citizenry nor speak meaningfully to their pressing problems (Zalik 2011: 184– 199). The challenges of a comatose NNPC and a restive oil region and the failed policies to speak to them deployed by this unaccountable government demonstrate the hazard authoritarian and corrupt-ridden states face when they utilize the vehicle of the NOC, anchored on dubious and fiercely contested governmental power, to drive the process of energy security. In this section, we examine four of these policies, related to the extent that they attempt to address the "resource curse" discussed in the introduction of this chapter with the same authoritarian tools and centralizing impulse that triggered the country's energy security crisis in the first place.

As has already been argued, at the heart of Nigeria's current energy security crisis is the persistent inability of successive governments, military and civilian, to craft policies to integrate the oil industry into the other sectors of the economy using the substantial oil revenues as a primary instrument, and in so doing accelerate the pace of job creation and address deepening poverty. These policy failures are, in turn, born of the crisis of illegitimate power in the country, driven by the desire of competing elites to seize the oil prize. In 1970, just before the advent of the oil boom, only 19 million Nigerians lived below the poverty line. When the civilian government of Olusegun Obasanjo took over in 1999, some three decades and $400 billion in oil earnings later, 90 million Nigerians were living below that line (Shaxson 2007: 4). The government-controlled Economic and Financial Crimes Commission estimated that $300 billion of this wealth was spirited out of the country by corrupt Nigerians during the same period (Ribadu 2009).

Aware of the deepening poverty and anger in the land, the Obasanjo government established the Oil and Gas Reform Implementation Committee in 2000, a broad consultative process designed to restructure the oil industry along market lines, make it more transparent, reposition it as a vital driver of economic regeneration, and in so doing ensure energy security. Four key policies constitute the pillars on which the new oil regime was to pivot. These include the Niger Delta Development Commission (NDDC), the Nigerian Extractive Industries Transparency Initiative (NEITI), Local Content/Petroleum Industry Bill, and Amnesty Programme/Gulf of Guinea Energy Security Strategy. Nigerian citizens, and local people resident in the Niger Delta in particular, were neither consulted when these policies were being developed nor involved in their implementation process. We examine these policies to the extent that they have attempted to achieve their goals.

Niger Delta Development Commission

A bill, signed into law by the National Assembly in 2000, created the NDDC to replace the moribund OMPADEC as the primary platform for development initiatives in the increasingly restive oil region. Four years later the Commission produced a 15-year master plan for the region, estimated to cost $2.9 billion (International Crisis Group 2006). Local communities in the core delta states (Rivers, Bayelsa, and Delta) criticized the central government for extending the plan to the periphery states of Abia, Akwa Ibom,

Cross Rivers, Edo, Imo, and Ondo. MOSOP leaders also charged that decision-making was taken out of the hands of local communities and put in the central government's, and that this would make NDDC unaccountable to stakeholders in the delta region.

Although the 2000 law gives NDDC 3% of oil company budgets, 15% of member states' statutory federal allocations, and 50% of their ecological fund allocations, Commission officials persistently complain that funds actually received are not adequate for projects in the region. Between 2001 and 2004 the central government gave NDDC an average of $64 million per annum, 77% of the budgeted sum. In 2006 federal legislators, under pressure from delta communities, increased the Commission's annual spend to $185 million (International Crisis Group 2006). Even so, NDDC officials have regularly been accused of fraud. Federal lawmakers have also been accused of accepting bribes from Commission officials to cover up a missing $68.5 million (*Daily Champion* 2004). Local journalists and NGOs also say they are routinely denied access to information about NDDC finances and how they are allocated to projects.

With the emergence of MEND in early 2006 and subsequent attacks on oil infrastructure which drastically reduced exports, an embattled President Obasanjo announced a $1.6 billion plan to build new roads in the region, provide electricity, and boost recruitment of civil servants from the area (*The Guardian* 2006). He also appointed the retired secret service chief who had supervised the failed OMPADEC as chairman of a new Consolidated Council on Socio-Economic Development of the Coastal States of the Niger Delta, an apparent sop to the charge by core delta states that NDDC was too "omnibus."

NDDC has, however, failed to address the fundamental challenge of poverty and social unrest in the region. A MEND spokesman explained in response to the central government's failed development initiatives in the region: "There will be no rest for the Nigerian government and collaborating oil companies until the stolen oil is returned to its rightful owners with compensation for all the years of theft and slavery" (International Crisis Group 2006).

Nigeria Extractive Industries Transparency Initiative

This initiative, like NDDC, is sorely handicapped to the extent that it is being driven by a central government that does not put a premium on the accountability and transparency it preaches. Concerned that corruption and lack of transparency in public affairs was impeding economic development in the poorer countries, James Wolfensohn, President of the World Bank, declared in 1996 that the Bank would mobilize global efforts to tackle this blight (World Bank 1996.) This marked the beginning of the international transparency campaign. In June 2002, the Hungarian financier and philanthropist George Soros joined Global Witness, a London-based campaigning NGO, to set up the Publish What You Pay coalition to "ask for oil companies to be forced by Western laws and regulations to disclose payments to all governments where they operate" (Shaxson 2007: 215). The Extractive Industries Transparency Initiative, a similar effort, was established in the same year following the World Summit on Sustainable Development in Johannesburg. Olusegun Obasanjo, President of Nigeria, was persuaded to join the initiative and NEITI was inaugurated in February 2004. The enabling Act, passed by the National Assembly in May 2007, mandates NEITI to promote due process and transparency in extractive revenues paid to and received by the central government as well as ensure transparency and accountability in the application of these revenues. The objective is to ensure that these revenues are "managed transparently to promote development, reduce poverty, conflict, ignorance, deprivation and disease" (NEITI 2011).

Shortly after its inauguration in 2004 NEITI commissioned a financial, physical, and process audit of the Nigerian petroleum industry for the period 1999–2004, the first comprehensive audit of the industry since oil production began in 1958. It followed up with audits of the oil and gas sector for the period 2005–2008. NEITI officials communicated the findings of these audits to various stakeholders, including government officials, journalists, civil society organizations, and legislators. These findings only confirmed what critics of the industry had been saying for years: NNPC officials were mostly incompetent and involved in corrupt practices; modalities for tax assessment of IOCs in the joint ventures were neither clear nor transparent; and the quantity of oil being produced per year could not be reliably verified (NEITI 2009).

While NEITI has been partially successful in bringing to light financial transactions in the country's premier industry – information hitherto denied the citizenry – the initiative has itself come under criticism from civil society activists who charge that the audit reports are not promptly published. NEITI officials themselves have also been accused of diverting public funds for personal use. In July 2010 the international secretariat of EITI threatened that it would revoke Nigeria's membership "if adequate consideration is not given to effectiveness and efficiency of the initiative" (*The Guardian* 2010). An ad hoc investigative committee of the National Assembly, inaugurated in January 2012 in the wake of public protests over the removal of fuel subsidies by the government, revealed that NEITI has not been successful in curbing corruption and administrative laxity in the national oil corporation nor encouraged government officials to channel the revenue from oil into addressing the development needs of the country (*The Vanguard* 2012).

Local Content/Petroleum Industry Bill

The age-old struggle for power and dominance of the policy process between unaccountable elites during the years of military rule did not abate even as a "democratically-elected" government assumed the reins in 1999. A government that derives its mandate from flawed elections is always prey to factional in-fighting as cynical political actors move to grab their share of the pie, trampling over the public good in the process. In this case, the NNPC is the chief vehicle for securing this pie, and any policy designed to transform it into an efficient and impartial player is bound to kiss the dust if it is not powered by the peoples' will legitimately obtained through the process of free and fair elections.

One of the chief complaints of civil servants in the oil ministry when the country began to take tentative steps toward growing indigenous participation in the industry in the 1970s was that the IOCs owned all the assets and that the bulk of money spent on procurement was done outside Nigeria. Local entrepreneurs, investors, and artisans were denied a stake in the oil industry and there was little or no integration with the other sectors of the economy. The picture had not changed much when the country returned again to civilian rule in 1999. Emulating Norway which sought to build up in-country technological capability in the industry as a route to broader economic growth and industrial development, the Nigerian government in 2005 set goals of achieving 45% local content in the oil and gas industry by 2006 and 70% by 2010 (Thurber *et al.* 2010). It was the government's hope that this move would generate industrial and technological spin-offs in the restive delta region where unemployment, particularly among youths, was high. The goal was regional stability and energy security.

The NNPC, the primary driver of the local content initiative, has been moderately successful in growing local participation in the industry. While by 2011 the percentage of local content was well below the 2010 target of 70%, some observers have estimated local content to have reached 35% in 2009, as compared with about 5–8% when the initiative started in 2005 (*The Guardian* 2011). However, as a 2010 study of the NNPC noted, "one of the most nettlesome ongoing challenges is how to separate legitimate Nigerian value addition from cases that involve no more than the establishment of a Nigerian middleman (Thurber *et al.* 2010). Significantly, the middleman culture, nurtured into existence as the central government seized control of the burgeoning oil revenues in the early 1970s, is alive and well and the NNPC itself remains the epicenter of struggle between competing elites for lucrative middleman roles in the oil industry.

Conceived alongside the local content initiative, the Petroleum Industry Bill (PIB) was introduced in the National Assembly in 2009 in an attempt to restructure the NNPC and tackle the perennial problems of the oil sector. The NNPC is something of an anomaly. While formally it is a vertically integrated oil company, it is neither a genuine commercial company nor a meaningful oil producer. NNPC does not control the revenue it generates and is thus unable to devise an independent business strategy. The IOCs, its joint venture partners, are left to perform nearly all the important and difficult functions of exploring and producing oil. Even so, the NNPC's portfolio of activities are so wide-ranging and incoherent that the central government is unable to impose control over it and also use it as an effective policy-making platform.

The PIB proposes to transform the NNPC into a limited liability company, which would control and reinvest its own revenue, working with the joint venture partners in a new commercial arrangement rid of bureaucratic red tape. Two new independent bodies would be created to regulate the industry and also set policy, resolving the problem wherein NNPC functioned as both operator and regulator (*Thisday* 2011.) Local communities in the delta region would be allocated shares in the new NNPC Ltd. The new model is to be based on Statoil, Norway's commercially successful state-owned oil company. The government has not been able to get the bill enacted. Ethnic bickering in the National Assembly, stonewalling by entrenched NNPC officials defending the status quo, and lobbying by IOC executives concerned that the new legislation would translate into higher taxes and reduced profits have combined to stall the legislative process.

Even if the bill is eventually passed, it will not prove an antidote to the vicious struggle between competing regional and ethnic elites for revenues generated by the national oil company. It is instructive that the PIB does not address the fundamental question of who will control these revenues – the central government or officials of the new NNPC Ltd.

Amnesty Programme/Gulf of Guinea Energy Security Strategy

The illegitimacy and abuse of public office that began to haunt Nigeria's democratization project from inception in 1999 ensured that policy initiatives targeted at the growing crisis in the Niger Delta did not find traction. This was at a time when the strategic interests of importing countries, particularly the US and Western Europe, required that political stability reign in key oil-exporting countries like Nigeria. In 2005 the US imported more oil from the Gulf of Guinea, Africa's oil belt of which Nigeria is a pivotal part, than it did from the Middle East (Shaxson 2007: 2). Indeed, by 2008, Exxon Mobil was producing more oil in Angola than it did in the US. It was this consideration that informed the launching in October 2004 of the Task Force on Gulf of Guinea Security by the Center for Strategic and International Studies, the influential US think tank. In its report, published in 2005, the Task Force recommended that "the United States should make security and governance in the Gulf of Guinea an explicit priority" (CSIS 2005: vii). The US Africa Command (AFRICOM) was established three years later specifically to police the volatile but oil-rich region.

US officials recognize that a key cause of local resentment and political instability in Africa's oil belt is the lack of transparency and accountability in the disbursement of the oil riches. Since the turn of the millennium they have been quietly pressing Nigerian government officials to open up their books and firm up the institutions of democratic governance as these steps would guarantee energy security for both Nigeria and importing countries in the long term. The US government mediated between Asari Dokubo and the central government in 2004 when the bloody face-off between militants and Nigerian troops cut oil production and sent prices soaring. By March 2009, following MEND's bombing campaigns, exports had fallen to 1.6 million barrels per day, from 2.6 million in the early months of 2006 (*The Guardian* 2009). In October 2009 a chastened Nigerian government launched an Amnesty Programme for "repentant militants" who would surrender their arms and renounce violence in return for rehabilitation, job training, and monthly allowances (Nwajiaku-Dahou 2010).

The Amnesty Programme does not address the root causes of the conflict in the Niger Delta: corrupt and unaccountable government officials at both central and local levels who embezzle funds allocated for the development of the oil region; widespread poverty and youth unemployment; worsening oil-induced environmental degradation, despoiling farmlands and water sources; and deep resentment by local people and the region's elites alike that authoritarian government policies since the end of the civil war in 1970 has taken away oil that is rightfully theirs without their consent. The peace secured by the government-brokered Amnesty Programme is only temporary, to the extent that this policy does not address the fundamental issues.

Conclusion

Nigeria's quest for energy security and a meaningful role in the oil industry since the 1970s is bound up with efforts in other sectors of social and economic life to secure prosperity and political order for the citizenry. This journey has been marked by persistent policy failure, not least because Nigerian politics and the institutions and processes that make for impartial rule-making have been degraded over the years by ethnic and regional elites competing for their share of the oil "cake." The resulting policy incoherence has had damaging consequences for social and economic life in Nigeria, leading embittered citizens, aggregated in ethnic, regional, and civic groups to increasingly challenge the dominance of the central government in the oil industry.

Nigeria displays the classic symptoms of the "resource curse" – Dutch Disease, the boom and bust inherent in commodity pricing, and a centralized and authoritarian state manned by incompetent and self-serving officials. Departing in 1960, colonial Britain handed over to local successors a regime of rapine and unaccountable power. Unfair colonial-era laws that dispossessed local people of their oil-rich land were replicated by authoritarian and corrupt incumbents in post-independence Nigeria, preparing the ground for a vicious free-for-all contest for the oil riches between powerful elites. In this struggle, coherent and joined-up policy-making to secure energy security and the public interest are the inevitable casualties.

The key site for this self-destructive competition is the NNPC, unable to transform into a competitive and efficient national oil company, generating more revenue and driving wider economic development in an impoverished African country where 90% of the population now live below the poverty line. The Niger Delta, the country's oil region which is at the sharp end of the stick, is now a cauldron of resentment and protest, putting at risk continued oil production. Nigeria therefore presents a challenge for global energy policy, putting into question the efficacy of the NOC as a reliable vehicle for stable oil production and nation-wide prosperity in countries where the fundamental questions of natural resource ownership and appropriate institutions to govern this valuable resource, are still very much hotly contested. These are matters policy-makers and academics alike should be paying close attention to in the years ahead.

Nigeria may still be Africa's leading oil producer and a reliable ally of the US and other oil-importing countries, but the obstacles in the way of its ambition to ramp up production as Middle East producers succumb to political conflict are gargantuan and are even likely to increase in the coming years, given the country's dysfunctional politics and feckless policy elites.

References

Africa Oil and Gas Report. 2012. How NNPC Cripples Nigeria's Gas to Power Aspirations. Lagos, April 4.

Chaney, Eric. 2012. Democratic Change in the Arab World, Past and Present. Paper presented at the Spring 2012 Conference of the Brookings Institution, Washington, DC.

CSIS. 2005. *A Strategic US Approach to Governance and Security in the Gulf of Guinea*. Washington, DC: Center for Strategic and International Studies.

Daily Champion. 2004. Mbah Aja on NDDC. Lagos, August 3.

EIA. 2010. Nigeria. US Energy Information Administration Country Analysis Briefs, www.eia.gov.

Federal Ministry of Information. 1969. *Nigeria, Report of the Interim Revenue Allocation Committee*. Lagos: Federal Ministry of Information.

Federal Ministry of Information. 1992. *OMPADEC: New Dawn for the Oil-Producing Communities*. Abuja: Federal Ministry of Information.

Financial Times. 1969. Oil: The Boom Now, Bonanza to Come. London, August 4, Special Supplement on Nigeria.

Forrest, Tom. 1995. *Political and Economic Development in Nigeria*. Boulder, CO: Westview Press.

Frynas, Jedrzej George, Matthias P. Beck, and Kamel Mellahi. 2000. Maintaining Corporate Dominance after Decolonization: the "First Mover Advantage" of Shell-BP in Nigeria. *Review of African Political Economy* 85: 407–425.

Greenpeace Nederland. 1996. *The Niger Delta: A Disrupted Ecology. The Role of Shell and Other Oil Companies*. Amsterdam: Greenpeace.

Human Rights Watch. 1999. *Nigeria: Crackdown in the Niger Delta*. New York: Human Rights Watch.

Humphreys, Macartan, Jeffrey Sachs, and Joseph Stiglitz. 2007. *Escaping the Resource Curse*. New York: Columbia University Press.

Ijaw Youth Council. 2009. *The Kaiama Declaration*. Port Harcourt: Isis Press.

Ikelegbe, Augustine. 2001. The Perverse Manifestation of Civil Society: Evidence from Nigeria. *Journal of Modern African Studies* 39, 1: 1–24.

International Crisis Group. 2006. *The Swamps of Insurgency: Nigeria's Delta Unrest*. Africa Report No. 115. Brussels: International Crisis Group.

Jega, Attahiru, ed. 2000. *Identity Transformation and Identity Politics Under Structural Adjustment in Nigeria*. Uppsala: Nordiska Afrikainstitutet.

Kew, Darren. 1999. "Democracy, Dem Go Craze, O": Monitoring the 1999 Nigerian Elections. *Issue* 27, 1: 29–33.

Khan, Sarah Ahmad. 1994. *Nigeria: The Political Economy of Oil*. Oxford: Oxford University Press.

NEITI. 2009. *NEITI: Its Structure and Activities*. Abuja: Nigeria Extractive Industries Transparency Initiative.

NEITI. 2011. *NEITI Audit Report, 2005*. Abuja: Nigeria Extractive Industries Transparency Initiative.

NNPC. 1981. *The Nigerian Oil Industry: Facts and Data*. Lagos: Nigerian National Petroleum Corporation.

Nwajiaku-Dahou, Kathryn. 2010. *The Politics of Amnesty in the Niger Delta: Challenges Ahead*. Paris: Institut Français des Relations Internationales.

Ofoh, E. P. 1992. *Trends in Production Sharing Contracts in Nigeria*. London: Society of Petroleum Engineers.

Okogu, B. E. 1992. *Africa and Economic Structural Adjustment: Case Studies of Ghana, Nigeria, and Zambia*. Pamphlet Series 29. Vienna: OPEC Fund for International Development.

Okonta, Ike. 2006. *Behind the Mask: Explaining the Emergence of the MEND Militia in Nigeria's Oil-Producing Niger Delta*. Berkeley, CA: University of California, Working Papers on the Economies of Violence.

Okonta, Ike. 2008. *When Citizens Revolt: Nigerian Elites, Big Oil, and the Ogoni Struggle for Self-Determination*. Trenton, NJ: Africa World Press.

Okonta, Ike, and Oronto Douglas. 2001. *Where Vultures Feast: Shell, Human Rights and Oil in the Niger Delta*. San Francisco, CA: Sierra Club Books.

Osaghae, Eghosa. 1998. *Crippled Giant: Nigeria Since Independence*. London: Hurst & Co.

Oyovbaire, Egite. 1985. *Federalism in Nigeria*. London: Macmillan.

Petroleum Intelligence Weekly. 1969. Sudden New Oil Law Issued by Nigeria to Oil Firms. December 22.

Ribadu, Nuhu. 2009. Nigeria and the Challenge of Corruption. Paper presented at the Centre for the Study of African Economies, University of Oxford.

Shaxson, Nicholas. 2007. *Poisoned Wells: The Dirty Politics of African Oil*. London: Palgrave Macmillan.

SPDC. 2003. *Peace and Security in the Niger Delta: Conflict Expert Group Baseline Report*. Lagos: Shell Producing Development Company.

The African Guardian. 1992. Oil-Producing Communities Disrupt Production. Lagos, August 17.

The African Guardian. 1994. Chaos, Disruptions in the Oil Industry. Lagos, July 25.

The Guardian. 2006. The Return of Horsfall. Lagos, November 4.

The Guardian. 2009. Militants Cut Oil Production. Lagos, May 12.

The Guardian. 2010. Nigeria May Lose EITI Status. Lagos, July 14.

The Guardian. 2011. NNPC and Local Content: Hope at Last? Lagos, December 12.

The Observer. 1995. Shell Letters Shows Nigerian Army Links. London, December 17.

The Punch. 1979. Scam Uncovered in the Oil Industry. Lagos, September 3.

The Vanguard. 2012. NNPC and the Cult of Corruption. Lagos, January 20.

Thisday. 2011. Oil and Gas: PIB – Will It Ever Be Enacted? Lagos, November 18.

Thurber, Mark C., Ifenyinwa M. Emelife, and Patrick R. P. Heller. 2010. *NNPC and Nigeria's Patronage System*. Palo Alto, CA: Stanford University, Freeman Spogli Institute for International Studies.

Turner, Terisa. 1978. Commercial Capitalism and the 1975 Coup. In Keith Panter-Brick, ed. *Soldiers and Oil: The Political Transformation of Nigeria*. London: Frank Cass, pp. 166–200.

Ukiwo, Ukoha. 2007. From "Pirates" to "Militants": A Historical Perspective on Anti-State and Anti-Oil Company Mobilization Among the Ijaw of Warri, Western Niger Delta. *African Affairs* 106/425: 587–610.

Wall Street Journal. 2006. As Oil Supplies are Stretched, Rebels, Terrorists, Get New Clout. April 10.

Williams, Gavin. 1976. Nigeria: A Political Economy. In Gavin Williams, ed. *Nigeria: Economy and Society*. London: Rex Collins.

Williams, Gavin. 1983. *The Origins of the Nigerian Civil War*. Milton Keynes: Open University Press.

World Bank. 1996. Ten Things You Did Not Know About the World Bank and Anti-Corruption. Washington, DC: World Bank.

Wolpe, Harold. 1974. *Urban Politics in Nigeria: A Study of Port Harcourt*. Berkeley, CA: University of California Press.

Zalik, Anna. 2011. Labeling Oil, Contesting Governance: Legal Oil, the GMoU, and Profiteering in the Niger Delta. In Cyril Obi and Siri Aas Rustad, eds. *Oil and Insurgency in the Niger Delta: Managing the Complex Politics of Petro-Violence*. Uppsala and London: Nordic Africa Institute and Zed Books.

Conclusion: Global Energy Policy: Findings and New Research Agendas

Andreas Goldthau

This book has offered a global policy perspective on energy. So far, a predominant focus on nation states, academic stovepipes, and policy silos characterized debates on global energy. While some existing works approached energy challenges from a foreign policy perspective, others conceptualized them in governance frameworks, or adopted a public economics angle. Yet, as it was argued in the introduction, accommodating new consumer heavyweights in global energy markets and lifting some 1.3 billion people out of energy poverty while at the same time safeguarding crucial climate targets requires a more holistic approach to global energy. In fact, supply security, the traditional focus of (national) energy policies, has clearly been complemented by a sustainability agenda; more recently, it has been charged even further by a development agenda. In short, and as aptly put by an author in this book, "events over the last two decades have changed the focus of global energy policy such that current and future strategies must be aimed at delivering secure, affordable, environmentally sustainable, and socially equitable access to energy services" (Bradshaw, p. 50).

This book has approached energy as a cross-cutting theme, simultaneously touching upon aspects of markets and finance, poverty and development, climate and sustainability, as well as traditional security policy. While states can be all but disregarded in addressing these transnational challenges, additional energy actors matter – private companies, international organizations or global policy networks. At the same time, transnational energy challenges need to be simultaneously addressed on global, regional, and local levels. In light of this, analyses in this book had to leave the prevalent and narrow focus prescribed by scholarly prisms and instead adopt a multilevel perspective, opened up to a multitude of actors. In this, this book faced a double challenge: providing for a scholarly rigorous and empirically rich assessment of crucial issues arising in global energy whilst at the same time going the "extra mile" and extracting policy implications. This task became even more challenging as the book at the same time aspired to map a new policy field.

The Handbook of Global Energy Policy, First Edition. Edited by Andreas Goldthau.
© 2013 John Wiley & Sons, Ltd. Published 2013 by John Wiley & Sons, Ltd.

In order to account for the multi-faceted nature of global energy challenges, analyses were organized around four deeply intertwined key dimensions: markets, security, sustainability, and development. This approach allowed for an interdisciplinary perspective on each dimension, taking discussions beyond prevailing scholarly prisms. The book also gathered insights from a highly diverse group of scholars and practitioners. This allowed an interdisciplinary view on global energy policy as it contributed to bridging theory and policy practice. Further, bringing together authors from all five continents, and by prominently featuring insights from non-OECD countries, the book centrally acknowledged the ongoing shift in global energy towards emerging economies. Finally, seven case studies complemented "dimensional" analyses: China, India, the EU, the US, Brazil, Russia, and Nigeria. The case selection reflected key trends in global energy: a current and significant shift away from incumbent energy heavyweights – the OECD countries – to newly emerging Asian consumers; the emergence of a low carbon paradigm in energy production, consumption, and use; and the rising importance of the social equality aspect in energy policy agendas. At the same time, case studies also represented key challenges facing global energy policy: resource governance; fossil fuel policy lock-in; incoherent policy design; or trade-offs between competing energy policy goals. In this, a top-down approach along dimensions was flanked by a bottom-up approach, providing for rich empirical insights from a national perspective.

The two-pronged approach adopted in this book has proven successful. Taken together, analyses provide for a comprehensive and detailed assessment of global energy challenges and the call on policy stemming from them. First, it has become clear that the systemic competition between state- and market-based models is gaining momentum again. This directly reflects above-mentioned shifts and their repercussions on the institutions, rules, and regimes structuring interaction of energy players. Recent trends in upstream petroleum agreements, for instance, reveal a change in fiscal regimes governing hydrocarbon extraction. This is due to higher price volatility characterizing crude markets but is also a function of general high price environments in which National Resource Companies tend to regain strength. Adding to this, a growing consensus on the inability of energy markets to properly address negative externalities, notably greenhouse gas emissions, brings a stronger dose of top-down regulation into the game. While not fully unfolding at a global level, regional approaches (such as European attempts to decarbonize its electricity sector) or domestic ones (such as China's policies aimed at enhancing energy efficiency) clearly rely on the regulatory state toolbox, if not on outright state intervention. Development related goals, finally, tend to come with additional support for state involvement in the energy sector. India's experience regarding energy subsidies is a case in point. That said, as analyses have revealed, the importance of market players in driving technological innovation or in providing for crucial funding for energy investment cannot be discarded. While certainly influenced by policies and the surrounding regulatory environment, strategic choices taken by private energy actors remain key for channeling capital into energy projects, renewable or fossil; and while certainly incentivized by government policies, it is the risk taking attitude and innovative potential of energy entrepreneurs that keeps on triggering important technological progress. A clear trend toward more state intervention in global energy therefore seems to be flanked by an undaunted importance of private market actors.

Second, the international energy governance architecture has seemingly proven able to adapt to changing patterns in global energy. While national security challenges may remain by and large defined by energy as a goal, means, and end of national policy, international oil organizations (OPEC and IEA) have been complemented by a Gas

Exporting Countries Forum (GECF), a reaction to changing international gas market structures. Bridging the producer/consumer divide by at the same time embracing emerging powers, new fora such as the G20 have begun to complement existing ones (G8). In addition, significant interlinkages between the global climate governance architecture and international institutions addressing energy security concerns have emerged, notably between UNFCCC, IEA, and G8. Moreover, global policy networks and public–private partnerships have lately complemented traditional multilateral formats. This reflects an increasing multipolarity characterizing global energy but at the same time also shows clear limits to dealing with international energy security issues in the absence of a global Leviathan. As several authors note, these new fora tend to be highly heterogeneous, hence characterized by strongly diverging interests, in addition to lacking clear mandates to authoritatively address crucial issues such as energy poverty or climate change. As a consequence, the ability of existing or emerging institutions to comprehensively tackle or govern global energy challenges remains limited, and so does their capacity to create win-win outcomes in an increasingly complex global energy landscape. On top of this, the effectiveness of some institutions tends to be challenged by technological progress. The international non-proliferation regime, for instance, grapples with lowered technological thresholds to acquisition of nuclear technology. Still, as some authors note, it remains important to be clear about the nature of the security challenge. Fears of "resource scrambles," for instance, may trigger ill-informed policies, both on a national and an international level.

Third, the increasing complexity of global energy challenges reveals an unmatched need for innovative policy tools employed on a global level. This comprises mechanisms to spread global best practice in low carbon policies, incentive schemes fostering the global deployment and dissemination of clean energy technologies, or tailored institutional models to introduce carbon pricing to established energy markets. Analyses in this book revealed a striking variety of possible combinations of market-based mechanisms coupled with economic and regulatory instruments requiring strong state leadership. While the important role of national policies is acknowledged, effective supra-national regimes are stressed as being of utmost importance for facilitating smart energy choices, both of socio-economic entities as of individuals. This need for innovative global policy tools is most evident at the intersection of the sustainability and development challenges. As the case of Brazil reveals, policies aimed at low carbon energy supply may have positive externalities on the development of rural communities and on fostering energy access. Energy efficiency may provide countries with great opportunity for pursuing a sustainable (i.e., low carbon) path to development, and for mobilizing the full potential of available resources. At the same time, policies geared at fostering renewables and biofuels may come with negative side-effects for food security if not well designed and managed. In this context, and again, the potential for global best practice can hardly be overstated. Emerging economies such as Brazil may have an important role to play when it comes to sharing the lessons learnt on managing the fuel/food challenge. Rising powers such as India may profit but also provide insights into what to avoid in order to not fall into the fuel subsidies trap. Oil-rich nations may be able to learn from successful cases how to avoid being "cursed" with resource endowments. And established economies such as the EU may provide insights on the effectiveness of policies related to energy efficiency and how to decarbonize the power system – a daunting regulatory and financial challenge for industrialized nations and emerging ones alike. For this to succeed, global mechanisms need to flank domestic ones – through facilitation (such as global energy partnerships), funding (e.g., by multilateral development organizations), or by exerting soft coercion (EITI).

In all, highlighting markets, security, development, and sustainability as the key dimensions of global energy policy, the book successfully mapped a new policy field; accounted for the multi-faceted nature of global energy policy challenges; broadened discussions on global energy policy beyond the prevalent "oil supply" dimension; and acknowledged regionally different needs and visions on global energy policy. Still, this book is only a start. Various avenues of further research on global energy policy emerge from it, and call for scholarly investigation. First, energy has apparently started to become mainstreamed into various other policy fields, some of which are reflected in the book – fighting poverty or combating climate change. Yet, these may not fully reflect the entire spectrum of policy fields relevant in this regard. Selected policy areas such as equity and justice may provide for interesting starting points for further research and inform discussions on global energy policy beyond the four dimensions investigated in the book. Second, a growing literature links energy policy to paradigms. In essence, paradigms provide guidance on how to make sense of an observed phenomenon. In this, a given policy problem may be addressed by various policy tools, depending on the paradigm adopted. Given the retreat of the West in global energy affairs, the predominant paradigm so far – the liberal market model – is about to give way to new ones. It may therefore be worth investigating the role paradigms play in informing global energy policies in an increasingly multipolar world. Third, and finally, policy initiatives aimed at increasing energy efficiency, reducing carbon emissions, or providing energy access abound on all levels. While providing for rich empirical material to study multilevel policy linkages, global energy also allows for scholarly investigation in almost laboratory-like conditions to study the cause, effect, and outcome of various policies – a rare opportunity for social science research.

It is hoped that this book will trigger further investigations along these lines and beyond, and that the fine selection of contributions gathered in the present volume helps firmly establish global energy policy as a field of academic inquiry.

Index

Figures are indexed as, for example, 268f.

The Handbook of Global Energy Policy, First Edition. Edited by Andreas Goldthau.
© 2013 John Wiley & Sons, Ltd. Published 2013 by John Wiley & Sons, Ltd.